# Lecture Notes in Computer Science 10256

Commenced Publication in 1973
Founding and Former Series Editors:
Gerhard Goos, Juris Hartmanis, and Jan van Leeuwen

More information about this series at http://www.springer.com/series/7412

Valentin E. Brimkov · Reneta P. Barneva (Eds.)

# Combinatorial Image Analysis

18th International Workshop, IWCIA 2017
Plovdiv, Bulgaria, June 19–21, 2017
Proceedings

Springer

*Editors*
Valentin E. Brimkov
SUNY Buffalo State
Buffalo, NY
USA

Reneta P. Barneva
State University of New York at Fredonia
Fredonia, NY
USA

and

Institute of Mathematics and Informatics
Bulgarian Academy of Sciences
Sofia
Bulgaria

ISSN 0302-9743          ISSN 1611-3349 (electronic)
Lecture Notes in Computer Science
ISBN 978-3-319-59107-0      ISBN 978-3-319-59108-7 (eBook)
DOI 10.1007/978-3-319-59108-7

Library of Congress Control Number: 2017940841

LNCS Sublibrary: SL6 – Image Processing, Computer Vision, Pattern Recognition, and Graphics

Printed on acid-free paper

This Springer imprint is published by Springer Nature
The registered company is Springer International Publishing AG
The registered company address is: Gewerbestrasse 11, 6330 Cham, Switzerland

# Preface

It is a great pleasure to welcome you to the proceedings of the 18th International Workshop on Combinatorial Image Analysis (IWCIA 2017) held in Plovdiv, Bulgaria, June 19–21, 2017.

Image analysis is a scientific discipline providing theoretical foundations and methods for solving real-life problems that appear in various, often societally sensitive, areas of human practice, such as medicine, robotics, defense, and security. Since typically the input data to be processed are discrete, the "discrete" or "combinatorial" approach to image analysis appears to be a natural one and therefore its applicability is expanding. The fact is that combinatorial image analysis often provides various advantages in terms of efficiency and accuracy over the more traditional approaches based on continuous models and requiring numeric computation.

For over 25 years, the IWCIA workshop series has been providing a forum for researchers throughout the world to present cutting-edge results in combinatorial image analysis, to discuss recent advances and new challenges in this research field, and to promote interaction with researchers from other countries. IWCIA had successful prior meetings in Paris (France) 1991, Ube (Japan) 1992, Washington DC (USA) 1994, Lyon (France) 1995, Hiroshima (Japan) 1997, Madras (India) 1999, Caen (France) 2000, Philadelphia, PA (USA) 2001, Palermo (Italy) 2003, Auckland (New Zealand) 2004, Berlin (Germany) 2006, Buffalo, NY (USA) 2008, Playa del Carmen (Mexico) 2009, Madrid (Spain) 2011, Austin, TX (USA) 2012, Brno (Czech Republic) 2014, and Kolkata (India) 2015. The workshop in Plovdiv retained and enriched the international spirit of these workshops. The IWCIA 2017 Program Committee was very international; its members are renowned experts coming from 17 different countries from Asia, Australia and Oceania, Europe, North and South America. Submissions came from 19 different countries from Africa, Asia, Europe, and North America.

Each submitted paper was sent to three reviewers. EasyChair provided a convenient platform for smoothly carrying out the review process, which was quite rigorous, conducted in a double-blind review mode. The most important selection criterion for acceptance or rejection of a paper was the overall score received. Other criteria included: relevance to the workshop topics, correctness, originality, mathematical depth, clarity, and presentation quality. We believe that as a result, only high-quality papers were accepted for presentation at IWCIA 2017 and for publication in the present volume.

The program of the workshop included presentations of contributed papers and keynote talks by five distinguished scientists. Alfred (Freddy) Bruckstein (Technion, IIT, Israel) surveyed some models of stochastic multi-agent interactions, involving simple ant-like a(ge)nts moving in grid or general planar graph environments, leading to interesting results concerning the average number of visits to various sites and to connections between Euclidean and discrete geometry. Edwin Hancock (University of York, UK) presented the edge-based Laplacian and quantum graphs and their use for developing more sophisticated heat diffusion and wave propagation. He also discussed possible applications to shape modeling and recognition. Marc van Kreveld (Utrecht University, The Netherlands) addressed questions related to geometric representations. He showed that a simple polygon can always be represented on the grid with constant Hausdorff distance and sometimes with constant Frechet distance, and discussed relations to certain mathematical games, such as the Japanese picture puzzles. Christian Ronse (Université de Strasbourg, France) investigated partial order relations on partial partitions of a set. He discussed their usefulness to guiding image analysis operations, such as filtering, reduction, or segmentation. Günter Rote (Freie Universität Berlin, Germany) first reviewed the existing congruence testing algorithms in two and three dimensions. Then he presented new algorithmic techniques and geometric insights that lead to fast algorithms in four dimensions.

The contributed papers are grouped into two parts. The first part includes 17 papers devoted to theoretical foundations of combinatorial image analysis, in particular studies on discrete geometry and topology, tilings and patterns, array grammars and languages, graphical models, and other technical tools for image analysis. The second part includes ten papers presenting application-driven research on topics such as image segmentation, classification, reconstruction, and compression, texture analysis, and bioimaging. We believe that all presented works were of high quality and the workshop participants benefited from the scientific program. We hope that many of these papers are of interest to a broader audience, including researchers in scientific areas such as pattern analysis and recognition, computer vision, shape modeling, and computer graphics.

A poster session provided some authors with the opportunity to present their ongoing research projects and original works in progress. The texts of these works are not included in this volume.

Many individuals and organizations contributed to the success of IWCIA 2017. First of all, the chairs are indebted to IWCIA's Steering Committee for endorsing the candidacy of Plovdiv for the 18th edition of the workshop. We wish to thank everybody who submitted their work to IWCIA 2017. Thanks to their contributions, we succeeded in having a technical program of high scientific quality. We are indebted to all participants and especially to the contributors of this volume. Our most sincere thanks go to the IWCIA 2017 Program Committee whose cooperation in carrying out high-quality reviews was essential in establishing a strong scientific program. We express our sincere gratitude to the keynote speakers, Alfred Bruckstein, Edwin Hancock, Marc van Kreveld, Christian Ronse, and Günter Rote, for their remarkable talks and overall contribution to the workshop program.

The success of the workshop would not be possible without the hard work of the local Organizing Committee. Special thanks go to the co-chair of the Organizing Committee, Georgi Vragov (Bulgarian Academy of Sciences) for the considerable amount of time and effort he devoted to the workshop organization, and to the other committee members, Veselin Igrachev (Rakursy, Plovdiv), Marian Iliev (Union of Bulgarian Scientists), Ivan Koychev (University of Sofia St. Kliment Ohridski), Ilia Kozhukharov (AMDFA, Plovdiv), Simeon Marlokov (Milara Int., Plovdiv), and Georgi Totkov (University of Plovdiv Paisii Hilendarski), for their valuable work. We remember with gratitude the assistance provided by the three students of Vessela

Statkova from the Academy of Music, Dance and Fine Arts in Plovdiv and by the students from the Plovdiv Mathematical High School Acad. K. Popov, who helped make this conference an enjoyable and fruitful event. We appreciate the support of Plovdiv Municipality and the personal involvement of Mr. Stefan Stoyanov, Deputy Mayor of the City of Plovdiv. We also acknowledge with gratitude the help we received from the Association for Development of the Information Society and its chair, Ivan Koychev. Thanks go to SUNY Fredonia and SUNY Buffalo State for their support. Finally, we wish to thank Springer, Computer Science Editorial, and especially Alfred Hofmann and Anna Kramer, for their efficient and kind cooperation in the timely production of this book.

June 2017                                                                              Valentin E. Brimkov
                                                                                        Reneta P. Barneva

# Organization

IWCIA 2017 was held in Plovdiv, Bulgaria, June 19–21, 2017.

## General Chairs

Valentin E. Brimkov      SUNY Buffalo State, USA
Reneta P. Barneva      SUNY Fredonia, USA

## Steering Committee

Bhargab B. Bhattacharya    Indian Statistical Institute, Kolkata, India
Valentin E. Brimkov    SUNY Buffalo State, USA
Gabor T. Herman    CUNY Graduate Center, USA
Kostadin Koroutchev    Universidad Autonoma de Madrid, Spain
Josef Šlapal    Technical University of Brno, Czech Republic

## Keynote Speakers

Alfred M. Bruckstein    Technion, IIT, Israel
Edwin R. Hancock    University of York, UK
Marc van Kreveld    Utrecht University, The Netherlands
Christian Ronse    Université de Strasbourg, France
Günter Rote    Freie Universität Berlin, Germany

## Program Committee

Eric Andres    Université de Poitiers, France
Tetsuo Asano    JAIST, Japan
Péter Balázs    University of Szeged, Hungary
Jacky Baltes    University of Manitoba, Canada
George Bebis    University of Nevada at Reno, USA
Bhargab B. Bhattacharya    Indian Statistical Institute, Kolkata, India
Partha Bhowmick    IIT Kharagpur, India
Alfred M. Bruckstein    Technion, IIT, Israel
Jean-Marc Chassery    Université de Grenoble, France
Li Chen    University of the District of Columbia, USA
David Coeurjolly    Université de Lyon, France
Mousumi Dutt    St. Thomas College of Engineering and Technology, India
Fabien Feschet    Université d'Auvergne, France

# Organizing Committee

| | |
|---|---|
| Reneta Barneva (Co-chair) | SUNY Fredonia, USA |
| Ivan Koychev (Co-chair) | University of Sofia St. Kliment Ohridski, Bulgaria |
| Georgi Vragov (Co-chair) | Bulgarian Academy of Sciences, Bulgaria |
| Veselin Igrachev | Rakursy, Bulgaria |
| Marian Iliev | Union of Bulgarian Scientists, Bulgaria |
| Ilia Kozhukharov | AMDFA, Plovdiv, Bulgaria |
| Simeon A. Marlokov | Milara Int., Bulgaria |
| Georgi Totkov | University of Plovdiv Paisii Hilendarski, Bulgaria |

# Additional Reviewers

Khaled Abuhmaidan
Piyush Bhunre
Ranita Biswas
Federico Gelsomini
Yangwei Liu
Hamid Mir-Mohammad-Sadeghi
Oliver Müller
Sanjoy Pratihar
Mohammad Reza Saadat

# Partner Institutions

Association for Development of the Information Society, Bulgaria
Plovdiv Municipality, Bulgaria
SUNY Buffalo State, Buffalo, NY, USA
SUNY Fredonia, Fredonia, NY, USA

# Contents

**Theoretical Foundations: Discrete Geometry and Topology, Tilings and Patterns, Grammars, Models, and Other Technical Tools for Image Analysis**

Simplifier Points in 2D Binary Images.............................. 3
  Kálmán Palágyi

Trajectories and Traces on Non-traditional Regular Tessellations
of the Plane..................................... 16
  Benedek Nagy and Arif Akkeleş

On Sets of Line Segments Featuring a Cactus Structure................ 30
  Boris Brimkov

Construction of Thinnest Digital Ellipsoid Using Inverse Projection
and Recursive Integer Intervals.................................. 40
  Papia Mahato and Partha Bhowmick

On the Chamfer Polygons on the Triangular Grid.................... 53
  Hamid Mir-Mohammad-Sadeghi and Benedek Nagy

Verification of Hypotheses Generated by Case-Based Reasoning
Object Matching.............................................. 66
  Petra Perner

Template-Based Pattern Matching in Two-Dimensional Arrays............ 79
  Yo-Sub Han and Daniel Průša

Construction of Persistent Voronoi Diagram on 3D Digital Plane.......... 93
  Ranita Biswas and Partha Bhowmick

Extension of a One-Dimensional Convexity Measure to Two Dimensions ... 105
  Sara Brunetti, Péter Balázs, and Péter Bodnár

Algorithms for Stable Matching and Clustering in a Grid............... 117
  David Eppstein, Michael T. Goodrich, and Nil Mamano

A Relational Generalization of the Khalimsky Topology ............... 132
  Josef Šlapal

Toward Parallel Computation of Dense Homotopy Skeletons
for nD Digital Objects......................................... 142
  Pedro Real, Fernando Diaz-del-Rio, and Darian Onchis

Polynomial Time Algorithm for Inferring Subclasses of Parallel Internal
Column Contextual Array Languages . . . . . . . . . . . . . . . . . . . . . . . . .    156
   *Abhisek Midya, D.G. Thomas, Alok Kumar Pani, Saleem Malik,*
   *and Shaleen Bhatnagar*

Parallel Contextual Array Insertion Deletion P System. . . . . . . . . . . . . . .    170
   *S. James Immanuel, D.G. Thomas, Robinson Thamburaj,*
   *and Atulya K. Nagar*

A 3D Curve Skeletonization Method . . . . . . . . . . . . . . . . . . . . . . . . . .    184
   *Nilanjana Karmakar, Sharmistha Mondal, and Arindam Biswas*

Inscribing Convex Polygons in Star-Shaped Objects . . . . . . . . . . . . . . . . .    198
   *Nikolay M. Sirakov and Nona Nikolaeva Sirakova*

On Characterization and Decomposition of Isothetic Distance Functions
for 2-Manifolds. . . . . . . . . . . . . . . . . . . . . . . . . . . . . . . . . . . . . . .    212
   *Piyush K. Bhunre, Partha Bhowmick, and Jayanta Mukhopadhyay*

**Theory and Applications: Image Segmentation, Classification,
Reconstruction, Compression, Texture Analysis, and Bioimaging**

Topological Data Analysis for Self-organization of Biological Tissues . . . . . .    229
   *M.J. Jimenez, M. Rucco, P. Vicente-Munuera, P. Gómez-Gálvez,*
   *and L.M. Escudero*

Distance Between Vector-Valued Representations of Objects in Images
with Application in Object Detection and Classification . . . . . . . . . . . . . . .    243
   *Nataša Sladoje and Joakim Lindblad*

A Statistical-Topological Feature Combination for Recognition of Isolated
Hand Gestures from Kinect Based Depth Images . . . . . . . . . . . . . . . . . . .    256
   *Soumi Paul, Hayat Nasser, Mita Nasipuri, Phuc Ngo, Subhadip Basu,*
   *and Isabelle Debled-Rennesson*

Image Segmentation via Weighted Carving Decompositions . . . . . . . . . . . .    268
   *Derek Mikesell and Illya V. Hicks*

An Image Texture Analysis Method for Minority Language Identification . . .    280
   *Darko Brodić, Alessia Amelio, and Zoran N. Milivojević*

JPEG Quantization Table Optimization by Guided Fireworks Algorithm . . . .    294
   *Eva Tuba, Milan Tuba, Dana Simian, and Raka Jovanovic*

Shape Matching for Rigid Objects by Aligning Sequences Based
on Boundary Change Points . . . . . . . . . . . . . . . . . . . . . . . . . . . . . . . .    308
   *Abdullah N. Arslan and Nikolay M. Sirakov*

Gradient and Graph Cuts Based Method for Multi-level Discrete
Tomography . . . . . . . . . . . . . . . . . . . . . . . . . . . . . . . . . . . . . . . . . .    322
    *Tibor Lukić and Marina Marčeta*

Reconstruction of Nearly Convex Colored Images . . . . . . . . . . . . . . . . . . .    334
    *Fethi Jarray and Ghassen Tlig*

A Greedy Algorithm for Reconstructing Binary Matrices with Adjacent 1s . . .    347
    *Fethi Jarray and Ghassen Tlig*

**Author Index** . . . . . . . . . . . . . . . . . . . . . . . . . . . . . . . . . . . . . . . . . .    357

# Theoretical Foundations: Discrete Geometry and Topology, Tilings and Patterns, Grammars, Models, and Other Technical Tools for Image Analysis

# Simplifier Points in 2D Binary Images

Kálmán Palágyi(✉)

Department of Image Processing and Computer Graphics,
University of Szeged, Szeged, Hungary
palagyi@inf.u-szeged.hu

**Abstract.** The concept of a simple point is well known in digital topology: a black point in a binary picture is called a simple point if its deletion preserves topology. This paper introduces the notion of a simplifier point: a black point in a binary picture is simplifier if it is simple, and its deletion turns a non-simple border point into simple. We show that simplifier points are line end points for both $(8,4)$ and $(4,8)$ pictures on the square grid. Our result makes efficient implementation of endpoint-based topology-preserving 2D thinning algorithms possible.

**Keywords:** Discrete geometry · Digital topology · Topology preservation · Thinning algorithms

## 1 Introduction

A *digital binary picture* assigns a color of black or white to each point of the considered digital space [5,8]. A *reduction* [2] transforms a binary picture only by changing some black points to white ones, which is referred to as *deletion*. *Parallel reductions* can delete a set of black points simultaneously, while a *sequential reduction* traverses the black points of a picture, and considers the actually visited point for possible deletion at a time [9].

A 2D reduction is *topology-preserving* if each object in the input picture contains exactly one object in the output picture, and each white component in the output picture contains exactly one white component in the input picture [5]. A black point is called *simple point* for a set of black points if its deletion is a topology-preserving reduction [4,5].

*Thinning* [2,7,12] is a frequently used method for making an approximation to the *skeleton* in a topology–preserving way [5]: the border points of a binary object that satisfy certain topological and geometric constraints are deleted in iteration steps. The entire process is then repeated until only the 'skeleton' is left. The greater part of existing 2D thinning algorithms preserve *line end points* (i.e., black points that are adjacent to exactly one black point). Note that Bertrand and Couprie proposed an alternative approach by accumulating curve interior points that are called *isthmus*es [1].

In this paper we introduce the concept of a *simplifier point*: a black point in a binary picture is simplifier if it is simple, and its deletion turns a non-simple

© Springer International Publishing AG 2017
V.E. Brimkov and R.P. Barneva (Eds.): IWCIA 2017, LNCS 10256, pp. 3–15, 2017.
DOI: 10.1007/978-3-319-59108-7_1

border point into simple. We show that all simplifier points are line end points for the considered two kinds of binary pictures on the 2D square grid. This result makes efficient implementation of endpoint-based topology-preserving 2D thinning algorithms possible.

The rest of this paper is organized as follows. Section 2 briefly reviews the relevant notions and results. Then in Sect. 3 we prove that simplifier points are line end points. An efficient scheme for endpoint-based 2D thinning algorithms is proposed in Sect. 4.

## 2   Basic Notions and Results

In this paper, we use the fundamental concepts of digital topology as reviewed by Kong and Rosenfeld [4,5]. Note that there are other approaches that are based on cellular/cubical complexes [6], but we insist on the 'historical paradigm'.

Let us denote by $S$ the square grid (that is dual to $\mathbb{Z}^2$, i.e., the set of points in the 2D plane with integer coordinates). The elements of the considered grid (i.e., regular squares) are called *points*. Two points are 4-*adjacent* if they share an edge, and they are 8-*adjacent* if they share an edge or a vertex. Note that both adjacency relations are reflexive and symmetric. Let us denote by $N_j(p)$ the set of points being $j$-adjacent to a point $p$, and let $N_j^*(p) = N_j(p)\backslash\{p\}$ $(j = 4, 8)$, see Fig. 1.

**Fig. 1.** The considered adjacency relations on the square grid. Set $N_4(p)$ contains point $p$ and the four points marked "•" (left), and set $N_8(p)$ is formed by $p$ and the eight points marked "★" (right).

A sequence of distinct points $\langle p_0, p_1, \ldots, p_m \rangle$ is called a $j$-*path* from $p_0$ to $p_m$ in a non-empty set of points $X \subseteq S$ if each point of the sequence is in $X$ and $p_i$ is $j$-adjacent to $p_{i-1}$ for each $i = 1, 2, \ldots, m$. Two points are said to be $j$-*connected* in a set $X$ if there is a $j$-path in $X$ between them. A set of points $X$ is $j$-*connected* in the set of points $Y \supseteq X$ if any two points in $X$ are $j$-connected in $Y$. A $j$-*component* of a set of points $X$ is a maximal (with respect to inclusion) $j$-connected subset of $X$.

Let $(k, \bar{k})$ be an ordered pair of adjacency relations $((k, \bar{k}) = (8, 4), (4, 8))$. A $(k, \bar{k})$ *binary digital picture* on grid $S$ is a quadruple $(S, k, \bar{k}, B)$ [5], where $B \subseteq S$ denotes the set of *black points*, and each point in $S\backslash B$ is said to be a *white point*. A *black component* or *object* is a $k$-component of $B$, while a *white component* is a $\bar{k}$-component of $S\backslash B$.

A black point $p$ is an *interior point* if all points in $N_{\bar{k}}^*(p)$ are black. A black point is said to be a *border point* if it is not an interior point (i.e., it is $\bar{k}$-adjacent to at least one white point). A black point $p$ is a *line end point* if $N_k^*(p)$ contains exactly one black point.

A black point is said to be *simple* for a set of black points (or in a picture) if its deletion is a topology-preserving reduction [4,5]. Kardos and Palágyi gave easily visualized characterizations of simple points in $(8,4)$ and $(4,8)$ pictures by sets of matching templates [3]. The base matching templates depicted in Figs. 2 and 3 are the rephrased versions of the templates presented in [3]. Notations: each black template position matches a black point; each white element matches a white point; each position depicted in gray matches any point (i.e., either a white point or a black point). Note that all the rotated and reflected versions of the base matching templates also match simple points. For the sake of brevity if a point is matched by a rotated/reflected version of a base matching template, we say that the given point is matched by that base template.

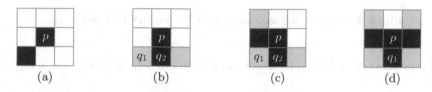

(a)          (b)          (c)          (d)

**Fig. 2.** Base matching templates for characterizing simple points in $(8,4)$-pictures. (Note that notions $p$, $q_1$, and $q_2$ help us to prove Theorem 1.)

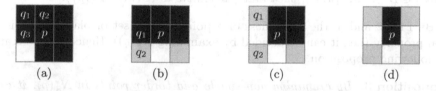

(a)          (b)          (c)          (d)

**Fig. 3.** Base matching templates for characterizing simple points in $(4,8)$-pictures. (Note that notions $p$, $q_1$, $q_2$, and $q_3$ help us to prove Theorem 2.)

We can state that interior points are not simple, some border points may only be simple, and line end points are simple.

## 3 Simplifier Points

In this section we introduce the notion of a simplifier point and it is shown that simplifier points are line end points for both $(8,4)$ and $(4,8)$ pictures.

First, let us establish two useful properties of simple points.

**Proposition 1.** *Let $p$, $q \in N_8^*(p) \backslash N_4(p)$, and $r = N_4^*(p) \cap N_4^*(q)$ be three black points in picture $(\mathcal{S}, 8, 4, B)$ (see Fig. 4a). Then $q$ is simple in picture $(\mathcal{S}, 8, 4, B)$ if and only if $q$ is simple in picture $(\mathcal{S}, 8, 4, B \backslash \{p\})$ (i.e., the simpleness of $q$ does not depend on the color of $p$).*

**Proposition 2.** *Let $p$ and $q \in N_8^*(p) \backslash N_4(p)$ be two black points, and $r = N_4^*(p) \cap N_4^*(q)$ be a white point in picture $(\mathcal{S}, 4, 8, B)$ (see Fig. 4b). Then $q$ is simple in picture $(\mathcal{S}, 4, 8, B)$ if and only if $q$ is simple in picture $(\mathcal{S}, 4, 8, B \backslash \{p\})$ (i.e., the simpleness of $q$ does not depend on the color of $p$).*

(a)                          (b)

**Fig. 4.** Configurations associated with Proposition 1(a) and 2(b).

Propositions 1 and 2 can be readily seen with the help of Figs. 2 and 3, respectively.

Let us now define the notion of a simplifier point.

**Definition 1.** *A point $p \in B$ in picture $(\mathcal{S}, k, \bar{k}, B)$ $((k, \bar{k}) = (8, 4), (4, 8))$ is a simplifier point if $p$ is simple in $(\mathcal{S}, k, \bar{k}, B)$, and there is a non-simple and border point $q \in B$ in $(\mathcal{S}, k, \bar{k}, B)$, such that $q$ is simple in $(\mathcal{S}, k, \bar{k}, B \backslash \{p\})$.*

By Figs. 2 and 3, the simpleness of a point $q$ for a set of black points is a local property (i.e., it can be decided by examining $N_8^*(q)$). Hence we can state the following proposition.

**Proposition 3.** *By examining non-simple and border points in $N_8^*(p)$, it can be decided whether a point $p$ is simplifier or not.*

The following two theorems are to characterize simplifier points in $(8, 4)$ and $(4, 8)$ pictures.

**Theorem 1.** *If a point in a $(8, 4)$-picture is a simplifier point, then it is a line end point.*

*Proof.* Let $(\mathcal{S}, 8, 4, B)$ be a picture and $p$ be a simple point for $B$. Assume that there is a point $q \in B$, that is a border point in $(\mathcal{S}, 8, 4, B)$, it is not simple in $(\mathcal{S}, 8, 4, B)$, but it is simple in $(\mathcal{S}, 8, 4, B \backslash \{p\})$. (In other words, it is assumed that $p$ is a simplifier point.) By Proposition 3, we can suppose that $q \in N_8^*(p)$.

Since simple points in $(8, 4)$-pictures are characterized by the matching templates depicted in Fig. 2, the following cases are to be investigated:

- If $p$ is matched by the template in Fig. 2a, then $p$ is a line end point.
- If $p$ is matched by the template in Fig. 2b, then consider the two template positions marked $q_1$ and $q_2$.
  - If $q = q_1$, then the simpleness of $q$ does not depend on the color of $p$ by Proposition 1.
  - Let $q = q_2$. Since $q$ is simple in $(S, 8, 4, B \backslash \{p\})$, $q$ is matched by a template in Fig. 2.
    * If $q$ is matched by the template in Fig. 2a, then $p$ is a line end point as it is depicted in Fig. 5a.
    * If $q$ is matched by the template in Fig. 2b, then $p$ is a line end point as it is depicted in Fig. 5b, or $q$ is matched by the template in Fig. 2c in picture $(S, 8, 4, B)$ (in which $p$ is a black point) as it is shown in Fig. 5b' (i.e., $q$ is simple for $B$). In the latter case we arrived at a contradiction.
    * If $q$ is matched by the template in Fig. 2c, then $q$ is matched by the template in Fig. 2d in picture $(S, 8, 4, B)$, see Fig. 5c (i.e., $q$ is simple for $B$). Thus we arrived at a contradiction.
    * If $q$ is matched by the template in Fig. 2d, then $q$ is an interior point in picture $(S, 8, 4, B)$, see Fig. 5d (i.e., $q$ is not a border point). Since $q$ is a border point in $(S, 8, 4, B)$, we arrived at a contradiction.
- If $p$ is matched by the template in Fig. 2c, then consider the two template positions marked $q_1$ and $q_2$.
  - If $q = q_1$, then the simpleness of $q$ does not depend on the color of $p$ by Proposition 1.
  - Let $q = q_2$. Since $q$ is simple in $(S, 8, 4, B \backslash \{p\})$, $q$ is matched by a template in Fig. 2.
    * If $q$ is matched by the template in Fig. 2a, then $q$ is matched by the template in Fig. 2b in picture $(S, 8, 4, B)$, see Fig. 6a (i.e., $q$ is simple for $B$). Thus we arrived at a contradiction.
    * If $q$ is matched by the template in Fig. 2b, then $q$ is matched by the template in Fig. 2c in picture $(S, 8, 4, B)$, see Fig. 6b (i.e., $q$ is simple for $B$). Thus we arrived at a contradiction.
    * If $q$ is matched by the template in Fig. 2c, then $q$ is matched by the template in Fig. 2d in picture $(S, 8, 4, B)$, see Fig. 6c (i.e., $q$ is simple for $B$). Thus we arrived at a contradiction.
    * If $q$ is matched by the template in Fig. 2d, then $q$ is an interior point in picture $(S, 8, 4, B)$, see Fig. 6d. Since $q$ is a border point in $(S, 8, 4, B)$, we arrived at a contradiction.
- If $p$ is matched by the template in Fig. 2d, then consider the template position marked $q_1$. Let $q = q_1$. Since $q$ is simple in $(S, 8, 4, B \backslash \{p\})$, $q$ is matched by a template in Fig. 2.
  - It is easy to check that $q$ is not matched by the template in Fig. 2a, and it is not matched by the template in Fig. 2b, see Fig. 7ab.
  - If $q$ is matched by the template in Fig. 2c, then $q$ is matched by the template in Fig. 2d in picture $(S, 8, 4, B)$, see Fig. 7c (i.e., $q$ is simple for $B$). Thus we arrived at a contradiction.

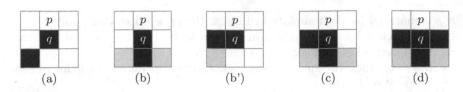

Fig. 5. Configurations associated with Theorem 1 when (the originally black) simple point $p$ is matched by the template in Fig. 2b and $q = q_2$.

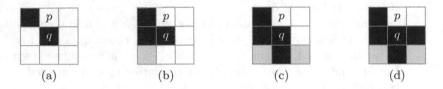

Fig. 6. Configurations associated with Theorem 1 when (the originally black) simple point $p$ is matched by the template in Fig. 2c and $q = q_2$.

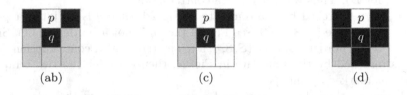

Fig. 7. Configurations associated with Theorem 1 when (the originally black) simple point $p$ is matched by the template in Fig. 2d and $q = q_1$.

- If $q$ is matched by the template in Fig. 2d, then $q$ is an interior point in picture $(\mathcal{S}, 8, 4, B)$, see Fig. 7d. Since $q$ is a border point in $(\mathcal{S}, 8, 4, B)$, we arrived at a contradiction.                                                                    □

**Theorem 2.** *If a point in a $(4, 8)$-picture is a simplifier point, then it is a line end point.*

*Proof.* Let $(\mathcal{S}, 4, 8, B)$ be a picture and $p$ be a simple point in that picture. Assume that there is a point $q \in B$, that is a border point in $(\mathcal{S}, 4, 8, B)$, it is not simple in $(\mathcal{S}, 4, 8, B)$, but it is simple in $(\mathcal{S}, 4, 8, B \setminus \{p\})$. (In other words, it is assumed that $p$ is a simplifier point.) By Proposition 3, we can suppose that $q \in N_8^*(p)$.

Since simple points in $(4, 8)$-pictures are characterized by the matching templates depicted in Fig. 3, the following cases are to be investigated:

- If $p$ is matched by the template in Fig. 3a, then consider the three template positions marked $q_1$, $q_2$, and $q_3$.
  - Let $q = q_1$. Since $q$ is simple in $(\mathcal{S}, 4, 8, B \setminus \{p\})$, $q$ is matched by a template in Fig. 3.

* If $q$ is matched by the template in Fig. 3a, then $q$ is an interior point in picture $(\mathcal{S}, 4, 8, B)$, see Fig. 8a. Since $q$ is a border point in $(\mathcal{S}, 4, 8, B)$, we arrived at a contradiction.
* It is easy to check that $q$ is not matched by the templates in Figs. 3b, 3c, and 3d, see Fig. 8bcd.

- Let $q = q_2$. Since $q$ is simple in $(\mathcal{S}, 4, 8, B \backslash \{p\})$, $q$ is matched by a template in Fig. 3.
  * It is easy to check that $q$ is not matched by the templates in Figs. 3a, 3c, and 3d, see Fig. 9acd.
  * If $q$ is matched by the template in Fig. 3b, then $q$ is an interior point in picture $(\mathcal{S}, 4, 8, B)$, see Fig. 9b. Since $q$ is a border point in $(\mathcal{S}, 4, 8, B)$, we arrived at a contradiction.
- Let $q = q_3$. Since $q$ is simple in $(\mathcal{S}, 4, 8, B \backslash \{p\})$, $q$ is matched by a template in Fig. 3.
  * It is easy to check that $q$ is not matched by the templates in Figs. 3a and 3b, see Fig. 10ab.
  * If $q$ is matched by the template in Fig. 3c, then $q$ is matched by the template in Fig. 3b in picture $(\mathcal{S}, 4, 8, B)$, see Fig. 10c (i.e., $q$ is simple for $B$). Thus we arrived at a contradiction.
  * If $q$ is matched by the template in Fig. 3d, then $q$ is matched by the template in Fig. 3c in picture $(\mathcal{S}, 4, 8, B)$, see Fig. 10d (i.e., $q$ is simple for $B$). Thus we arrived at a contradiction.
- If $p$ is matched by the template in Fig. 3b, then consider the two template positions marked $q_1$ and $q_2$.
  - Let $q = q_1$. Since $q$ is simple in $(\mathcal{S}, 4, 8, B \backslash \{p\})$, $q$ is matched by a template in Fig. 3.
    * It is easy to check that $q$ is not matched by the template in Fig. 3a, see Fig. 11a.
    * If $q$ is matched by the template in Fig. 3b, then $q$ is matched by the template in Fig. 3a in picture $(\mathcal{S}, 4, 8, B)$, see Fig. 11b (i.e., $q$ is simple for $B$). Thus we arrived at a contradiction.
    * If $q$ is matched by the template in Fig. 3c, then $q$ is matched by the template in Fig. 3b in picture $(\mathcal{S}, 4, 8, B)$, see Fig. 11c (i.e., $q$ is simple for $B$). Thus we arrived at a contradiction.
    * If $q$ is matched by the template in Fig. 3d, then $q$ is matched by the template in Fig. 3c in picture $(\mathcal{S}, 4, 8, B)$, see Fig. 11d (i.e., $q$ is simple for $B$). Thus we arrived at a contradiction.
  - If $q = q_2$, then the simpleness of $q$ does not depend on the color of $p$ by Proposition 2.
- If $p$ is matched by the template in Fig. 3c, then consider the two template positions marked $q_1$ and $q_2$. In both cases the simpleness of $q$ does not depend on the color of $p$ by Proposition 2.
- If $p$ is matched by the template in Fig. 3d, then $p$ is a line end point.

$\square$

(a)    (bcd)

**Fig. 8.** Configurations associated with Theorem 2 when (the originally black) simple point $p$ is matched by the template in Fig. 3a and $q = q_1$.

(acd)    (b)

**Fig. 9.** Configurations associated with Theorem 2 when (the originally black) simple point $p$ is matched by the template in Fig. 3a and $q = q_2$.

(ab)    (c)    (d)

**Fig. 10.** Configurations associated with Theorem 2 when (the originally black) simple point $p$ is matched by the template in Fig. 3a and $q = q_3$.

(a)    (b)    (c)    (d)

**Fig. 11.** Configurations associated with Theorem 2 when (the originally black) simple point $p$ is matched by the template in Fig. 3b and $q = q_1$.

Theorems 1 and 2 state that if a point is simplifier, then it is a line end point. The following proposition formulates that the converse of our theorems does not hold:

**Proposition 4.** *Let $p \in B$ be a line end point in picture $(\mathcal{S}, k, \bar{k}, B)$ $((k, \bar{k}) = (8, 4), (4, 8))$, and let $N_k^*(p) \cap B = \{q\}$ be a non-simple and border point in that picture. Then $q$ may be non-simple in $(\mathcal{S}, k, \bar{k}, B \backslash \{p\})$ (i.e., $p$ may not be a simplifier).*

Figure 12 is to illustrate Proposition 4.

A sequential reduction is topology-preserving if its *deletion rule* deletes only simple points [4,5]. Endpoint-based sequential 2D thinning algorithms are

**Fig. 12.** Configurations associated with Proposition 4. Point $q$ is a non-simple and border point, and $p$ is a line end point in $(8,4)$ pictures (left) and $(4,8)$ pictures (right). We can state that $q$ remains non-simple after the deletion of $p$. Hence line end point $p$ is not a simplifier.

composed of topology-preserving sequential reductions that do not delete line end points. The following proposition is an easy consequence of Theorems 1 and 2:

**Proposition 5.** *The produced 'skeleton' of an endpoint-based sequential 2D thinning algorithm (working on $(8,4)$- or $(4,8)$-pictures) contains all non-simple border points in the original input picture (and in the intermediate pictures of the iterative thinning process).*

Since that algorithm may delete only simple points and preserves line end points, by Theorems 1 and 2, all non-simple border points remain non-simple border points. Hence those points are in the produced 'skeleton'.

## 4 Efficient Implementation of Endpoint-Based 2D Thinning Algorithms

Proposition 5 provides us an efficient method to implement endpoint-based 2D thinning algorithms. The proposed method is sketched in Algorithm 1.

The input of Algorithm 1 is array $A$ which stores the $(k, \bar{k})$-picture to be thinned. In input array $A$, the value "1" corresponds to black points and the value "0" is assigned to white ones. Both input and the output pictures are stored in the same array (i.e., array $A$ will contain the produced 'skeleton'), so the proposed method is memory saving.

In order to speed up the process Algorithm 1 uses the list *border_list* that stores the border points to be checked in the actual picture of the iterative thinning process. In order to avoid storing more than one copy of a border point in *border_list*, and checking again and again points in the final 'skeleton', array $A$ represents a four-color picture:

- a value of "0" corresponds to white points,
- a value of "1" is assigned to interior points,
- a value of "2" corresponds to border points to be checked (i.e., elements of the current *border_list*), and
- a value of "3" is assigned to the detected line end points and non-simple border points (that are elements of the produced 'skeleton' by Proposition 5, hence their re-checking is not needed).

---

**Algorithm 1.**    Efficient Implementation of Endpoint-Based 2D Thinning

---

**Input**: array $A$ storing the $(k, \bar{k})$-picture to be thinned
**Output**: array $A$ containing the picture with the produced 'skeleton'
`// collect border points by a single scan of array` $A$
$border\_list \leftarrow\ <$ empty list $>$
**foreach** element $p$ in array $A$ **do**
$\quad$ **if** $p$ is a border point **then**
$\qquad$ $border\_list \leftarrow border\_list\ +\ <p>$
$\qquad$ $A[p] \leftarrow 2$

`// thinning process`
**repeat**
$\quad$ `// one iteration / one sequential reduction`
$\quad$ $number\_of\_deleted\_points \leftarrow 0$
$\quad$ **foreach** point $p$ in $border\_list$ **do**
$\qquad$ **if** $p$ is a line end point or a non-simple point **then**
$\qquad\quad$ `// a point in the 'skeleton' is found`
$\qquad\quad$ $A[p] \leftarrow 3$
$\qquad\quad$ $border\_list \leftarrow border\_list\ -\ <p>$
$\qquad$ **else if** $T(p) =$ **true then**
$\qquad\quad$ `// deletion`
$\qquad\quad$ $A[p] \leftarrow 0$
$\qquad\quad$ $border\_list \leftarrow border\_list\ -\ <p>$
$\qquad\quad$ $number\_of\_deleted\_points \leftarrow number\_of\_deleted\_points\ +1$
$\qquad\quad$ `// updating the list`
$\qquad\quad$ **foreach** point $q$ being $\bar{k}$-adjacent to $p$ **do**
$\qquad\qquad$ **if** $A[q] = 1$ **then**
$\qquad\qquad\quad$ $A[q] \leftarrow 2$
$\qquad\qquad\quad$ $border\_list \leftarrow border\_list\ +\ <q>$
**until** $number\_of\_deleted\_points = 0$;

---

Note that Palágyi et al. proposed a similar method for implementing 3D fully parallel thinning algorithms [10]. It uses two lists to speed up the process: one for storing the border points in the current picture, the other list is to collect all deletable points in the actual phase of the process.

First, the input picture is scanned and all the border points are inserted into the list *border_list*. We should mention here that it is the only time consuming scan of the entire array $A$. Since only a small part of points in a usual picture belong to the objects, the thinning procedure is much faster if we just deal with the set of border points in the actual picture.

Then the iterative thinning process itself is performed. The kernel of the **repeat** cycle corresponds to one iteration (i.e., one sequential reduction), and variable *number_of_deleted_points* is to store the number of deleted points within the actual iteration. If a point $p$ is deleted, then *border_list* is updated since all interior points that are $\bar{k}$-adjacent to $p$ become border points. The algorithm

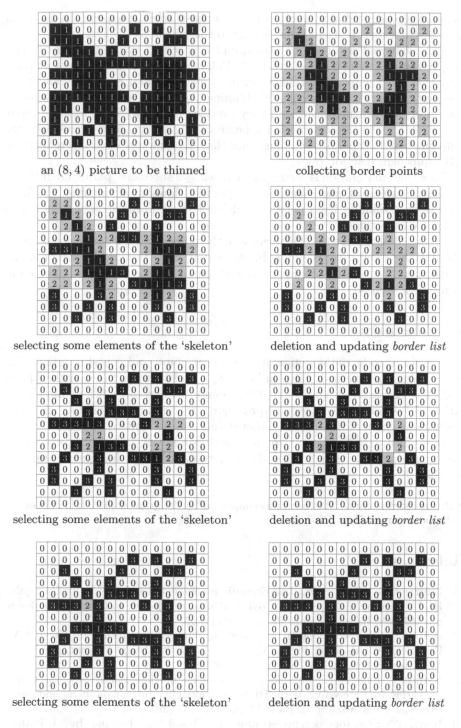

an (8, 4) picture to be thinned

collecting border points

selecting some elements of the 'skeleton'

deletion and updating *border list*

selecting some elements of the 'skeleton'

deletion and updating *border list*

selecting some elements of the 'skeleton'

deletion and updating *border list*

**Fig. 13.** Illustration of the proposed implementation scheme (see Algorithm 1).

terminates when stability is reached (i.e., $number\_of\_deleted\_points = 0$). Then all points having a nonzero value belong to the produced 'skeleton'.

The sequential reduction in Algorithm 1 is specified by deletion rule $T$ (i.e., $T(p) = $ **true** if $p$ is to be deleted). In [9] Palágyi has studied *equivalent deletion rules* that yield pairs of equivalent parallel and sequential reductions and pairs of equivalent parallel and sequential thinning algorithms as well. (Two thinning algorithms are called equivalent if they produce the same result for each input picture [11].) The implementation scheme of Algorithm 1 can be adapted for endpoint-based 2D parallel thinning algorithms that comprise parallel reductions with equivalent deletion rules.

Note that an iteration is decomposed into $k \geq 2$ successive parallel reductions in *subiteration-based* and *subfield-based* parallel thinning algorithms [2]. Those algorithms terminate if no points are deleted in an entire iteration (i.e., in a cycle of $k$ reductions).

Figure 13 illustrates the effectiveness of the proposed implementation scheme, where $T(p) = $ **true** if $p$ is simple in the actual $(8, 4)$ picture. Elements in the actual *border_list* are depicted in grey, and row-by-row ordering was assumed in each deletion phase.

## 5    Conclusions

This paper introduces the notion of a simplifier point, and it is shown that simplifier points are line end points in both $(8, 4)$ and $(4, 8)$ pictures on the square grid. The characterization of simplifier points involves an efficient implementation of endpoint-based topology-preserving 2D thinning algorithms in which multiple checking of non-simple border points can be omitted.

In a future work we are to deal with simplifier points in pictures on the remaining two regular 2D grids (i.e., triangular and hexagonal sampling schemes).

**Acknowledgements.** This work was supported by the grant OTKA K112998 of the National Scientific Research Fund.

## References

1. Bertrand, G., Couprie, M.: Transformations topologiques discrètes. In: Coeurjolly, D., Montanvert, A., Chassery, J. (eds.): Géométrie discrète et images numériques, pp. 187–209. Hermès Science Publications (2007)
2. Hall, R.W.: Parallel connectivity-preserving thinning algorithms. In: Kong, T.Y., Rosenfeld, A. (eds.) Topological Algorithms for Digital Image Processing, pp. 145–179. Elsevier Science B.V, Amsterdam (1996)
3. Kardos, P., Palágyi, K.: On topology preservation in triangular, square, and hexagonal grids. In: Proceedings of the 8th International Symposium on Image and Signal Processing and Analysis, ISPA 2013, pp. 782–787 (2013)
4. Kong, T.Y.: On topology preservation in 2-D and 3-D thinning. Int. J. Pattern Recognit. Artif. Intell. **9**, 813–844 (1995)

5. Kong, T.Y., Rosenfeld, A.: Digital topology: introduction and survey. Comput. Vis. Graph. Image Process. **48**, 357–393 (1989)
6. Kovalevsky, V.A.: Geometry of Locally Finite Spaces. Publishing House, Berlin (2008)
7. Lam, L., Lee, S.-W., Suen, S.-W.: Thinning methodologies – a comprehensive survey. IEEE Trans. Pattern Anal. Mach. Intell. **14**, 869–885 (1992)
8. Marchand-Maillet, S., Sharaiha, Y.M.: Binary Digital Image Processing - A Discrete Approach. Academic Press, New York (2000)
9. Palágyi, K.: Equivalent sequential and parallel reductions in arbitrary binary pictures. Int. J. Pattern Recognit. Artif. Intell. **28**, 1460009-1–1460009-16 (2014)
10. Palágyi, K., Németh, G., Kardos, P.: Topology preserving parallel 3D thinning algorithms. In: Brimkov, V.E., Barneva, R.P. (eds.) Digital Geometry Algorithms, vol. 2, pp. 165–188. Springer, Dordrecht (2012)
11. Palágyi, K., Németh, G., Kardos, P.: Topology-preserving equivalent parallel and sequential 4-subiteration 2D thinning algorithms. In: Proceedings of the 9th International Symposium on Image and Signal Processing and Analysis, ISPA 2015, pp. 306–311 (2015)
12. Suen, C.Y., Wang, P.S.P. (eds.): Thinning Methodologies for Pattern Recognition. Series in Machine Perception and Artificial Intelligence. World Scientific, Singapore (1994)

# Trajectories and Traces on Non-traditional Regular Tessellations of the Plane

Benedek Nagy[(✉)] and Arif Akkeleş

Department of Mathematics, Faculty of Arts and Sciences,
Eastern Mediterranean University, Mersin-10, Famagusta, North Cyprus, Turkey
nbenedek.inf@gmail.com

**Abstract.** In this paper, shortest paths on two regular tessellations, on the hexagonal and on the triangular grids, are investigated. The shortest paths (built by steps to neighbor pixels) between any two points (cells, pixels) are described as traces and generalized traces on these grids, respectively. In the hexagonal grid, there is only one type of usual neighborhood and at most two directions of the steps are used in any shortest paths, and thus, the number of linearizations of these traces is easily computed by a binomial coefficient based on the coordinate differences of the pixels. Opposite to this, in the triangular grid the neighborhood is inhomogeneous (there are three types of neighborhood), moreover this grid is not a lattice, therefore, the possible shortest paths form more complicated sets, a kind of generalized traces. The linearizations of these sets are described by associative rewriting systems, and, as a main combinatorial result, the number of the shortest paths are computed between two triangles, where two cells are considered adjacent if they share at least one vertex.

**Keywords:** Combinatorics · Traces · Trajectories · Non-traditional grids · Triangular grid · Generalized traces · Shortest paths · Number of shortest paths · Enumerative combinatorics

## 1 Introduction

In 1977, analyzing basic networks, Mazurkiewicz introduced the concept of partial commutations. Two independent parallel events commute, i.e., their executing order can be arbitrary in a sequential simulation. By using the concept of commutations, the work of the concurrent systems can be described by traces. In these systems some (pairs of) elementary processes (i.e., atomic actions; they are represented by the letters of the alphabet) may depend on each other, and some of them can be pairwise independent. The order of two consecutive independent letters can be arbitrary, in this way the traces are a kind of generalizations of words. Traces and trace languages play important roles in describing parallel events and processes. An automata theoretic approach on rational trace languages can be found in [18,21], and in [19,20] for context-free trace languages. Actually, linearizations of trace languages, that are sets of words representing

© Springer International Publishing AG 2017
V.E. Brimkov and R.P. Barneva (Eds.): IWCIA 2017, LNCS 10256, pp. 16–29, 2017.
DOI: 10.1007/978-3-319-59108-7_2

traces of the trace languages, are accepted by various type of automata. A special two-dimensional representation of some traces is given by trajectories on the square grid. These trajectories were also used to describe syntactic constraints for shuffling two parallel events described by words in [9]. Trajectories on other regular grids could also play a somewhat similar important role. A kind of generalization of traces is presented in [7], where apart from the usual permutation rules, some other rewriting rules are also allowed. We show that the sets of shortest paths in the triangular grid can be seen as generalized traces.

Considering the set of (types and directions of the) possible steps as the alphabet, a path, a sequence of steps, is actually a word. This gives the link between formal languages, trace theory and digital geometry.

Path counting, an interesting and important technique in digital geometry and in digital image processing, was invented in [23]. The number of shortest paths on the square grid was computed by an algorithm in [1], while recursive formulae were given in [2]. The square grid is counted as the traditionally used tessellation of the plane. There are other two regular tessellations, namely, the hexagonal and the triangular grids. They are usually called non traditional or unconventional grids, since they are less used in practice. On the other side they have symmetries and more interesting combinatorial structures than the square grid has. In several cases they proved to be more efficient in applications as well. Recently, in [4], counting the shortest paths based on the two closest types of neighborhood on the triangular grid was also considered. In this paper, these two non-traditional but regular tessellations of the two-dimensional plane are used. They are duals of each other in graph theoretic sense. The main achievements of this work are the following ones: The shortest paths between any two pixels of these grids are described as (generalized) traces and the number of the shortest paths are computed by enumerative combinatorial techniques. As we will see, the case of the hexagonal grid is very simple, it is, actually, shown only for the analogy. The shortest paths between two hexagons form a trace, in which the order of the two types of steps can be arbitrary. The triangular grid is not a lattice, therefore, as we will see, the shortest paths based on the third widely used neighborhood (that is each pixel having 12 neighbors) form more complicated sets. We also present formulae to compute the number of shortest paths, in this way complementing the results of [4].

We note here that various digital, i.e., path based distances are investigated for the triangular grid, e.g., distances by neighborhood sequences [10,11,14] and weighted distances [16]. In this paper, we use one of the most natural digital distance functions which is a special case of the previously mentioned distance functions. Nevertheless, it already has very interesting theoretical, combinatorial properties, as we will see later on.

We assume that the readers are familiar with traces and rewriting systems, otherwise they are referred to, e.g., [3,6,24]. As usual, in this paper, the traces are also represented by sets of words.

## 2   Traces and Trajectories on the Hexagonal Grid

The hexagonal grid can be elegantly described by three coordinates such that every hexagon has a unique triplet with 0-sum [5], see Fig. 1: formally, the (set of pixels of the) hexagonal grid can be described as $\{p(p(1), p(2), p(3)) \mid p(1), p(2),$ $p(3) \in \mathbb{Z}, p(1) + p(2) + p(3) = 0\}$. Consequently, by stepping from a pixel to a *neighbor* one, two of the coordinate values are changing by $\pm 1$. This formulation coincides the well known and widely used concept of neighbor pixels. (On the hexagonal grid two pixels are neighbors if they share a side of a hexagon, see, e.g., the pixels $(-2, 0, 2)$ and $(-2, 1, 1)$ on Fig. 1. Actually, the same neighborhood concept for the hexagons is resulted if it is required to share at least one point on their border.) A *path* connects two pixels by a sequence of neighbor pixels. The (digital) *distance* of two pixels is the length, that is the number of steps, of a/the shortest path between them. It is easy to prove (see, e.g., [8,11]) that the distance, i.e., the number of steps in a/the shortest path, of any two pixels can be computed as

$$d(p(p(1), p(2), p(3)), q(q(1), q(2), q(3))) = \max_{i \in \{1,2,3\}} \{|p(i) - q(i)|\}.$$

A *(hexagonal stepping) lane* is a set of pixels with a fixed coordinate value, e.g., $y = -1$ for the top hexagons on Fig. 1. One can very easily generate a shortest path connecting the two given pixels $p$ and $q$: if they share a coordinate value, i.e., $p(i) = q(i)$ (for any $i \in \{1, 2, 3\}$), then keeping that coordinate value fixed, the pixels can be connected through a lane [11,17], i.e., by the pixels of $\{r(r(1), r(2), r(3)) \mid r(i) = p(i), \min\{p(j), q(j)\} \le r(j) \le \max\{p(j), q(j)\}, j \ne i\}$. If there is no shared coordinate, then let $i \in \{1, 2, 3\}$ be a value such that $|p(i) - q(i)| \ge |p(j) - q(j)|$ for every $j \in \{1, 2, 3\}$, then a shortest path is the concatenation of the paths connecting, e.g., $p$ to $r$ and $r$ to $q$ with $r(j) = p(j), r(k) = q(k), r(i) = -p(j) - q(k)$ where $i, j, k$ are pairwise different elements of $\{1, 2, 3\}$. In the next section we show how the number of shortest paths can be computed.

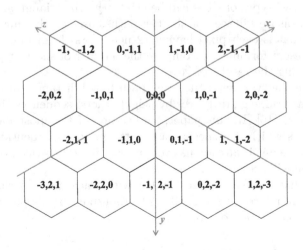

**Fig. 1.** A part of the hexagonal grid with a symmetric coordinate frame.

## 2.1    Number of Shortest Paths on the Hexagonal Grid

Since the hexagonal grid is a lattice, one can step from any pixel to any of the six directions. Thus the order of the steps in a shortest path is not important, the reached pixel depends on only their respective numbers. Using the alphabet $\Sigma = \{\rightarrow, \searrow, \swarrow, \leftarrow, \nwarrow, \nearrow\}$ for the steps in the six directions, each letter is independent of each other, therefore one can freely move (permute) them in a path. One can observe that shortest path contains at most two distinct letters and their numbers are determined by the coordinate differences of the pixels. Therefore, we can state the following result that can be proven by elementary combinatorics.

**Theorem 1.** *The number of shortest paths between $p$ and $q$ is given by the binomial coefficient*

$$\binom{\max_i\{|p(i) - q(i)|\}}{\min_i\{|p(i) - q(i)|\}}.$$

Actually, an equivalence set of shortest paths is the commutative closure of any singleton language of a shortest path, i.e., these traces are based on the maximal independency relations (commutations): instead of the words, their Parikh-vectors [22], the multiset of their letters can be used to describe shortest paths as traces on the hexagonal grid.

As the main contribution of the paper a similar question is answered: it is shown how the number of shortest paths can be counted on the triangular grid (based on neighborhood relation of 12 neighbors).

# 3    Preliminaries: Description of the Triangular Grid

The triangular grid, preserving the symmetry of the grid, can also be described by three integer coordinates [10,11,25]. There are two types of pixels (by orientation): the *even* pixels have zero-sum triplets, while the *odd* pixels have one-sum triplets. The neighborhood relations are formally defined: Let $p(p(1), p(2), p(3))$ and $q(q(1), q(2), q(3))$ be two pixels, they are $m$-neighbors ($m \in \{1, 2, 3\}$) if

- $|p(i) - q(i)| \leq 1$, for $i \in \{1, 2, 3\}$, and
- $|p(1) - q(1)| + |p(2) - q(2)| + |p(3) - q(3)| = m$.

Two pixels are *neighbors*, if they are $m$-neighbors for some $m \in \{1, 2, 3\}$. Various neighborhoods and the coordinate system used are shown in Fig. 2. The set of pixels sharing a fixed coordinate is called a *lane*, e.g., $y = -2$ for the topmost pixels of Fig. 2. *Paths*, their lengths and *distances* of pixels are also defined analogously to the hexagonal case.

A step to a 2-neighbor does not modify the parity, while step to a 1-neighbor or a 3-neighbor pixel changes (inverts) the parity. The basic motions, the possible steps form our alphabet: Let $\Sigma = \{\uparrow_1, \downarrow_1, \nwarrow_1, \searrow_1, \nearrow_1, \swarrow_1, \leftarrow_2, \rightarrow_2, \searrow_2, \nwarrow_2, \nearrow_2, \swarrow_2, \uparrow_3, \downarrow_3, \searrow_3, \nearrow_3, \nwarrow_3, \swarrow_3\}$. The arrows show the directions, while the indices indicate the used neighborhood of the given step. The steps, the letters of the alphabet correspond to grid-vectors:

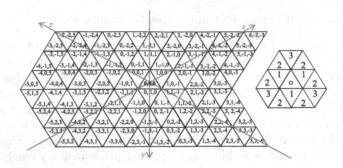

**Fig. 2.** A part of the triangular grid with coordinate values (left) and various neighborhood of an even pixel (right).

$$\uparrow_1 = (0, -1, 0), \quad \downarrow_1 = (0, 1, 0), \quad \searrow_1 = (0, 0, 1),$$
$$\searrow_1 = (0, 0, -1), \quad \nearrow_1 = (1, 0, 0), \quad \swarrow_1 = (-1, 0, 0),$$
$$\leftarrow_2 = (-1, 0, 1), \quad \rightarrow_2 = (1, 0, -1), \quad \searrow_2 = (0, -1, 1),$$
$$\searrow_2 = (0, 1, -1), \quad \nearrow_2 = (1, -1, 0), \quad \swarrow_2 = (-1, 1, 0),$$
$$\uparrow_3 = (1, -1, 1), \quad \downarrow_3 = (-1, 1, -1), \quad \searrow_3 = (-1, -1, 1),$$
$$\searrow_3 = (1, 1, -1), \quad \nearrow_3 = (1, -1, -1), \quad \swarrow_3 = (-1, 1, 1).$$

For any two pixels $p(p(1), p(2), p(2))$ and $q(q(1), q(2), q(3))$, their respective positions and their shortest paths are isometrically transformed, and, therefore, their distance is kept by using the following transformations of the grid (see [15] for details).

- If $p$ is an odd pixel, then by a mirroring both $p$ and $q$ to the center of the edge shared by the pixels $(0, 0, 0)$ and $(0, 1, 0)$ one obtains $p'(-(p(1) - 1), -p(3), -p(2))$ and $q'(-(q(1) - 1), -q(3), -q(2))$ (By this transformation the parities of the pixels are also changed.)
- If $p$ is an even pixel, then by a translation one can obtain $p'(0, 0, 0)$ and $q'(q(1) - p(1), q(2) - p(2), q(3) - p(3))$.
- Now, let $p$ be the origin and let $q$ be a pixel such that $i \in \{1, 3\}$ is the direction for which $|q(i)| \geq |q(j)|$ for any $j \in \{1, 2, 3\}$. Then the mirroring to the axis corresponding to the direction $k \in \{1, 3\}$ such that $k \neq i$ transforms $q$ to $q'$ such that $q'(2) = q(j), q'(j) = q(2), q'(k) = q(k)$. (The image of the origin is itself.)

Based on the previous transformations, w.l.o.g., we can assume that $p$ is the origin $(0, 0, 0)$, i.e., further in this paper, we measure the distance, the shortest paths from the origin to any pixel $q$ of the grid with the following property: the second coordinate of $q$ has the largest absolute value among its coordinates, i.e., $|q(2)| \geq |q(1)|$ and $|q(2)| \geq |q(3)|$. Pixels $q$ with this property form the *analyzed part* of the grid.

In the next subsection we give a shortest path from the origin to any pixel $q$ with the above property based on a greedy algorithm.

## 3.1    A Shortest Path

We refer to [10,11] for the detailed description of a more general algorithm (using various digital distances based on neighborhood sequences) that provides a shortest path; it can be applied for our case using any type of the defined neighborhoods in each step: in terms of neighborhood sequences, it can be written as a sequence with only 3's meaning that every type of neighbors is allowed in each step. In this section we provide a shortest path from the origin to any pixel of the analysed part.

**Proposition 1.** *Let $q(q(1), q(2), q(3))$ be an even pixel such that $|q(2)| \geq |q(1)|$ and $|q(2)| \geq |q(3)|$.*

- *If $q(2) > 0$, then a shortest path from $(0,0,0)$ to $q$ is obtained as the concatenation of $-q(1)$ $\nearrow_2$ steps and $-q(3)$ $\searrow_2$ steps: $\nearrow_2^{|q(1)|} \searrow_2^{|q(3)|}$.*
- *If $q(2) < 0$, then a shortest path from $(0,0,0)$ to $q$ is obtained as the concatenation of $q(3)$ $\searrow_2$ steps and $q(1)$ $\nearrow_2$ steps: $\searrow_2^{|q(3)|} \nearrow_2^{|q(1)|}$.*

*Proof.* Let $q(q(1), q(2), q(3))$ be an even pixel such that $|q(2)| \geq |q(1)|$ and $|q(2)| \geq |q(3)|$.

We show a formal proof for the case $q(2) > 0$; a similar proof suffices for the other case.

Thus $q$ has the coordinate values, $(q(1), q(2), q(3)$ with $q(1) \leq 0$ and $q(3) \leq 0$ and $q(2) = |q(1)| + |q(3)|$.

Then, a shortest path from $(0,0,0)$ to $q$ cannot have a length less than $q(2)$ since in every step a coordinate value is changed by at most 1.

Let us consider the path consisting of $-q(1)$ $\nearrow_2$ steps and then $-q(3)$ $\searrow_2$ steps. The path $\nearrow_2^{|q(1)|} \searrow_2^{|q(3)|}$ is a valid path, since steps to 2-neighbors are allowed at any pixels (and thus, also in even pixels). Using the coordinate representations of the steps: from $(0,0,0)$, it goes through on $(-1,1,0), (-2,2,0), \dots,$ $(q(1), |q(1)|, 0)$ and then from $(q(1), |q(1)|, 0)$ it goes through $(q(1), |q(1)|+1, -1)$, $(q(1), |q(1)| + 2, -2), \dots, (q(1), |q(1)| + |q(3)|, q(3)) = q$. It actually uses $|q(1)|$ steps in a lane and then $|q(3)|$ steps on another lane, together $q(2)$ steps. The proof of the case is done.                                   □

The next proposition can be proven with a similar technique.

**Proposition 2.** *Let $q(q(1), q(2), q(3))$ be an odd pixel such that $|q(2)| \geq |q(1)|$ and $|q(2)| \geq |q(3)|$.*

- *If $q(2) > 0$, then a shortest path from $(0,0,0)$ to $q$ is obtained as the concatenation of 1 $\downarrow_1$ step, $-q(1)$ $\nearrow_2$ steps and $-q(3)$ $\searrow_2$ steps: $\downarrow_1 \nearrow_2^{|q(1)|} \searrow_2^{|q(3)|}$.*
- *If $q(2) < 0$, then a shortest path from $(0,0,0)$ to $q$ is obtained as the concatenation of 1 $\uparrow_3$ step, $q(3) - 1$ $\searrow_2$ steps and $q(1) - 1$ $\nearrow_2$ steps: $\uparrow_3 \searrow_2^{|q(3)|-1} \nearrow_2^{|q(1)|-1}$.*

Observe that in the obtained paths the largest coordinate difference (the second coordinate in our case) is decreased in every step. Consequently, the (digital) distance of any two pixels can be computed as:

**Lemma 1**

$$d(p(p(1),p(2),p(3)),q(q(1),q(2),q(3))) = \max_{i\in\{1,2,3\}}\{|p(i)-q(i)|\}.$$

We note here, that this result can be seen as a special case of one of the main results of [12–14], but here we have used a much simpler formulation. (Instead of allowing neighborhood sequences to generate distance functions, in our case, in each step we could step to any neighbors of the actual pixel. This allows us to derive the simple formula established in the previous lemma.)

In this paper we concentrate on the shortest paths. Some of them can be generated by a greedy algorithm (related to the algorithm of [11]), since some steps of the algorithm may contain a non-deterministic choice. However, there could be some of them that cannot be generated by the greedy algorithm.

A shortest path in the triangular grid may contain various steps. Notice that some of the vectors (representing elements of $\Sigma$) can be used to any pixels (zero-sum vectors). Some vectors can be applied only for even pixels (vectors with sum 1) and some of them can be applied only for odd pixels (vectors with sum $-1$). Thus, for instance the sequence of steps $\uparrow_1\uparrow_3$ can be applied for odd pixels only, while the sequence $\uparrow_3\uparrow_1$ works only starting from an even pixel. The order of the steps becomes important, because the triangular grid is not a lattice, and thus, some of the grid-vectors do not translate the grid to itself.

## 4    Generalized Traces Describing Shortest Paths

In this section we present an associative calculus that provides all the shortest paths equivalent to the one the process starts with. First, we recall the definition: $\mathcal{C} = (\Sigma, P)$ is an associative calculus, where $\Sigma$ is a finite alphabet and $P$ is a finite set of productions (rewriting rules). Each rewriting rule is an element of $\Sigma^* \times \Sigma^*$. A rule is usually written in the form $u \leftrightharpoons v$, where $u, v \in \Sigma^*$. Let $w \in \Sigma^*$ be given, we say that $w'$ is obtained from $w$ applying the rewriting rule $u \leftrightharpoons v$, if there exist $w_1, w_2 \in \Sigma^*$ such that either $w = w_1 u w_2$ and $w' = w_1 v w_2$, or $w = w_1 v w_2$ and $w' = w_2 u w_2$. Actually, $w$ can also be obtained from $w'$ by the same production, thus we may use the notation $w \leftrightarrow w'$. By the reflexive and transitive closure of $\leftrightarrow$, the relation $\leftrightarrow^*$ is defined.

Observe that the calculus $\mathcal{C}$ defines an equivalence relation on $\Sigma^*$. The equivalence class represented by $w$ is denoted by $\mathcal{C}(w) = \{w' \mid w \leftrightarrow^* w'\}$.

Now, let us see how such a calculus can be applied to describe shortest paths. In any shortest path the largest coordinate difference of the two endpoints (that is the second coordinate value in our case) must decrease in each step.

Observe that the cardinality of $\Sigma$ is 18, but every pixel has only 12 neighbors. The triangular grid is not a lattice, the steps $\Sigma_\triangle = \{\downarrow_1, \seararrow_1, \nearrow_1, \leftarrow_2, \rightarrow_2, \searrow_2, \searrow_2, \nearrow_2, \nearrow_2, \uparrow_3, \searrow_3, \nearrow_3\}$ can be used at even, and the steps $\Sigma_\triangledown = \{\uparrow_1, \searrow_1, \nearrow_1, \leftarrow_2, \rightarrow_2, \searrow_2, \searrow_2, \nearrow_2, \nearrow_2, \downarrow_3, \searrow_3, \nearrow_3\}$ can be used at odd pixels. As one may observe, only the steps to 2-neighbor pixels can be applied for every pixel, the possible directions of steps to 1- and 3-neighbors depend on the parity of the actual pixel. The independence relation contains pairs of letters such that at least one of the letters indicates a step to a 2-neighbor pixel. In terms of traces, this fact can be concluded in the following way:

**Lemma 2.** *In any paths, any letter* $a \in \Sigma_\triangle \cap \Sigma_\triangledown = \{\leftarrow_2, \rightarrow_2, \nwarrow_2, \searrow_2, \nearrow_2, \swarrow_2\}$ *commutes with any letter* $b \in \Sigma$ *(a ≠ b).*

Moreover, in the triangular grid there are composite steps that can be broken to two steps in various ways. That is related to the serializations in generalized traces [7]. In the next lemma all of them are listed that are needed in shortest paths in the analyzed part of the grid, i.e., the second coordinate is modified by two (during these steps). For the other parts of the grid the description is analogous.

**Lemma 3.** *The following equivalences hold:*

- $\nearrow_2 \nearrow_2$ *is equivalent to* $\uparrow_3 \nearrow_3$ *for even pixels;*
- $\nearrow_2 \nearrow_2$ *is equivalent to* $\nearrow_3 \uparrow_3$ *for odd pixels;*
- $\nwarrow_2 \nwarrow_2$ *is equivalent to* $\uparrow_3 \nwarrow_3$ *for even pixels;*
- $\nwarrow_2 \nwarrow_2$ *is equivalent to* $\nwarrow_3 \uparrow_3$ *for odd pixels;*
- $\nearrow_2 \nwarrow_2$ *is equivalent to* $\uparrow_3 \uparrow_1$ *for even pixels;*
- $\nearrow_2 \nwarrow_2$ *is equivalent to* $\uparrow_1 \uparrow_3$ *for odd pixels;*
- $\nwarrow_2 \nwarrow_2$ *is equivalent to* $\searrow_3 \downarrow_3$ *for even pixels;*
- $\searrow_2 \searrow_2$ *is equivalent to* $\downarrow_3 \searrow_3$ *for odd pixels;*
- $\swarrow_2 \swarrow_2$ *is equivalent to* $\swarrow_3 \downarrow_3$ *for even pixels;*
- $\swarrow_2 \swarrow_2$ *is equivalent to* $\downarrow_3 \swarrow_3$ *for odd pixels;*
- $\searrow_2 \swarrow_2$ *is equivalent to* $\downarrow_1 \downarrow_3$ *for even pixels;*
- $\searrow_2 \swarrow_2$ *is equivalent to* $\downarrow_3 \downarrow_1$ *for odd pixels;*
- $\downarrow_1 \nwarrow_2$ *is equivalent to* $\searrow_3 \swarrow_2$ *for even pixels;*
- $\downarrow_1 \nearrow_2$ *is equivalent to* $\swarrow_3 \searrow_2$ *for even pixels.*

*Proof.* We show the formal proof of the first equivalences. The others can be proven in a similar manner.

$$\nearrow_2 \nearrow_2 \text{ is equivalent to } \uparrow_3 \nearrow_3 \text{ for even pixels.}$$

On the left side there are two consecutive steps to 2-neighbors, they are defined for all pixels. On the right side the first step is $\uparrow_3$ that is available only at even pixels, thus the statement has meaning only for even pixels.

Now, let us see the coordinate representations of these steps:
on the left side: $2(1, -1, 0) = (2, -2, 0)$, while
on the right side: $(1, -1, 1) + (1, -1, -1) = (2, -2, 0)$. The equivalence is established. □

It can be proven, e.g., by a combinatorial way, that there are no more equivalences of two consecutive steps (in our shortest paths) needed. The equivalences that are not listed in the previous lemmas, e.g., $\nwarrow_2 \nearrow_2$ is equivalent to $\uparrow_1 \uparrow_3$ for odd pixels, can be obtained by using some of the listed equivalences, e.g.,

$\nwarrow_2\nearrow_2$ is equivalent to $\nearrow_2\nwarrow_2$ (by Lemma 2) and that is equivalent to $\uparrow_1\uparrow_3$ for odd pixels (by Lemma 3). Of course, there are longer sequences of steps that are equivalent to each other, e.g., $\downarrow_1\downarrow_3\nearrow_3\nearrow_2$ is equivalent to $\nearrow_2\nearrow_2\nearrow_2\downarrow_1$, but their equivalence is based on the listed equivalences by two consecutive steps. It can be proven that the equivalence of every (consecutive) sequence of steps (in our shortest paths) to another sequence is based on the listed equivalences.

Based on the previous lemmas we are ready to present a rewriting system (especially, an associative calculus) that can obtain all the shortest paths that are equivalent to an initial one. Since the parity of the pixels plays an important role, in a path we keep track of them by allowing it to write this information after any step to a 2-neighbor pixel, e.g., the shortest path $\nearrow_2\uparrow_3\nearrow_2\nearrow_2\nwarrow_2$ to the pixel $(4,-5,2)$ can also be written in the following forms: $\nearrow_2$ (e) $\uparrow_3\nearrow_2\nearrow_2$ (o) $\nwarrow_2$ or $\nearrow_2$ (e) $\uparrow_3\nearrow_2$ (o) $\nearrow_2$ (o) $\nwarrow_2$ (o), etc. The latter form when all the steps to 2-neighbors are extended with this information is called *fully informed description* of the path. For steps to 1-neighbor or 3-neighbor pixels we do not need additional information, since not the same steps are allowed for even and for odd pixels, i.e., $\Sigma_\triangle \cap \Sigma_\triangledown$ does not contain any steps to a 1- or a 3-neighbor pixel. In the following theorem this extended form is used (however, one may get any correct forms by deleting any/all these information). The theorem is a consequence of the previous results, especially of Lemmas 2 and 3.

**Theorem 2.** *Let $\mathcal{C}(\Sigma, P)$ be an associative calculus with rewriting rules*
$P = \{a(x)b(x) \leftrightharpoons b(x)a(x) \mid a, b \in \{\nearrow_2, \nwarrow_2, \swarrow_2, \searrow_2\}, \ x \in \{e, o\}, \ a \neq b\} \cup$
$\{a(x)b \leftrightharpoons ba(y) \mid a \in \{\nearrow_2, \nwarrow_2, \swarrow_2, \searrow_2\}, b \in \{\uparrow_1, \downarrow_1, \uparrow_3, \downarrow_3, \nearrow_3, \nwarrow_3, \swarrow_3, \searrow_3\},$
$x, y \in \{e, o\}, \ x \neq y\} \cup \{ \quad \nwarrow_3\swarrow_2 \ (o) \leftrightharpoons \downarrow_1\nwarrow_2 \ (o), \quad \nearrow_3\nwarrow_2 \ (o) \leftrightharpoons \downarrow_1\swarrow_2 \ (o),$
$\nearrow_2$ (e) $\nearrow_2$ (e) $\leftrightharpoons \uparrow_3\nearrow_3, \quad \nearrow_2$ (o) $\nearrow_2$ (o) $\leftrightharpoons \nearrow_3\uparrow_3, \quad \nwarrow_2$ (e) $\nwarrow_2$ (e) $\leftrightharpoons \uparrow_3\nwarrow_3,$
$\nwarrow_2$ (o) $\nwarrow_2$ (o) $\leftrightharpoons \nwarrow_3\uparrow_3, \quad \nearrow_2$ (e) $\nwarrow_2$ (e) $\leftrightharpoons \uparrow_3\uparrow_1, \quad \nearrow_2$ (o) $\nwarrow_2$ (o) $\leftrightharpoons \uparrow_1\uparrow_3,$
$\swarrow_2$ (e) $\swarrow_2$ (e) $\leftrightharpoons \searrow_3\downarrow_3, \quad \swarrow_2$ (o) $\swarrow_2$ (o) $\leftrightharpoons \downarrow_3\searrow_3, \quad \searrow_2$ (e) $\searrow_2$ (e) $\leftrightharpoons \searrow_3\downarrow_3,$
$\swarrow_2$ (o) $\swarrow_2$ (o) $\leftrightharpoons \downarrow_3\searrow_3, \quad \searrow_2$ (e) $\searrow_2$ (e) $\leftrightharpoons \downarrow_1\downarrow_3, \quad \searrow_2$ (o) $\searrow_2$ (o) $\leftrightharpoons \downarrow_3\downarrow_1\}.$
*Let $\mathcal{C}(w)$ denote the set of all words that can be obtained from a given fully informed description of a word $w \in \Sigma^*$ by applying any (finite number) of the rewriting rules of $P$.*

*Let $w \in \Sigma^*$ be a fully informed description of a shortest path from $(0,0,0)$ to a pixel $q$ $(|q(2)| \geq |q(1)|, |q(2)| \geq |q(3)|)$. Then, applying $\mathcal{C}$ to $w$, the set $\mathcal{C}(w)$ contains exactly those strings that describe shortest paths from $(0,0,0)$ to $q$ (by a fully informed description).*

Actually, the system can be understood as a generalized trace [7] as we detail below. The system has several permutative rules, the rules by which a step to a 2-neighbor can be interchanged in the path with the previous or the next step (if it is not the same). However, the system has another types of productions, e.g., $\nwarrow_2$ (e) $\nwarrow_2$ (e) $\leftrightharpoons \uparrow_3\nwarrow_3$ or $\nwarrow_3\swarrow_2$ (o) $\leftrightharpoons \downarrow_1\nwarrow_2$ (o). To have our system in a more similar fashion as the description in [7], we can introduce new letters abbreviating these "double steps". In the mentioned examples they could be: $\nwarrow_{2+2}$ (e) and $\nwarrow_{1+2}$. These new letters show the unbroken "double step" referring the motion with vectors $(0,-2,2)$, and $(0,2,-1)$ in our case from even pixels (it is indicated in the first case, since this vector works for both even and odd pixels, while in the second case it is obviously working only for even pixels). By breaking the original productions to two parts using these intermediate "double steps", we got the productions $\searrow_{2+2}(e) \leftrightharpoons \searrow_2 (e) \nwarrow_2 (e)$, $\searrow_{2+2}(e) \leftrightharpoons \uparrow_3\searrow_3$ and $\searrow_{1+2}\leftrightharpoons\searrow_3\swarrow_2 (o)$, $\searrow_{1+2}\leftrightharpoons\downarrow_1\searrow_2 (o)$, respectively. In this form the system is a generalized trace, but only shortest paths without any "double steps" can be counted as real shortest paths.

## 5 The Number of Shortest Paths

In this section, by complementing the results of [4], we count the number of shortest paths by a combinatorial approach. As we have already seen, there are two cases by the sign of $q(2)$, and also by the parity of pixel $q$. In each case we gave a shortest path in Propositions 1 and 2, and now, enumerations are provided to compute the number of shortest paths.

Let us start with the case when $q$ is even.

### 5.1 The Case of Even Paths

In this case, the parity of $p$ and $q$ are the same, i.e., both of them are even. There is a shortest path containing only steps to 2-neighbor pixels (see Proposition 1). However, there can be some other shortest paths in which some of the steps to 2-neighbors are replaced based on some of the productions shown in Theorem 2.

First, let us analyze the case $q(2) < 0$. By Proposition 1 we have a shortest path $\searrow_2^{|q(3)|} \nearrow_2^{|q(1)|}$. From this path, by the calculus $\mathcal{C}$ given in Theorem 2, one can obtain any shortest path $\Pi$. Observe that in the calculus, apart from the permutative rules (that changes only the order of two consecutive steps) there are three types of real rewriting rules that can be applied. Based on them, let us introduce the following notations (for a shortest path $\Pi$).

**Notation 1.** *In case $q(2) < 0$, the letters $\gamma, \delta$ and $\varepsilon$ are defined as follows.*

- *Let $\gamma$ be the number of application of rewriting rules $\nearrow_2 (e) \searrow_2 (e) \to \uparrow_3\uparrow_1$ and $\nearrow_2 (o) \searrow_2 (o) \to \uparrow_1\uparrow_3$ (minus the number of their applications in reverse directions).*
- *Let $\delta$ be the number of applications of $\searrow_2 (e) \searrow_2 (e) \to \uparrow_3\searrow_3$ and $\searrow_2 (o) \searrow_2 (o) \to \searrow_3\uparrow_3$ (minus the number of their applications in reverse directions).*
- *Let $\varepsilon$ be the number of applications of $\nearrow_2 (e) \nearrow_2 (e) \to \uparrow_3\nearrow_3$ and $\nearrow_2 (o) \nearrow_2 (o) \to \nearrow_3\uparrow_3$ (minus the number of their applications in reverse directions).*

Then $\Pi$ contains the following numbers of the following types of steps:

$\searrow_2$ steps: $|q(3)| - \gamma - 2\delta$;     $\nearrow_2$ steps: $|q(1)| - \gamma - 2\varepsilon$;     $\uparrow_1$ steps: $\gamma$;

$\uparrow_3$ steps: $\gamma + \delta + \varepsilon$;     $\searrow_3$ steps: $\delta$;     $\nearrow_3$ steps: $\varepsilon$.

Moreover, the order of some steps, the ones that changes the parity, takes matter: these steps are alternating in the following way: starting by a step $\uparrow_3$ (if any), then one step from the set $\{\uparrow_1, \searrow_3, \nearrow_3\}$, then again a step $\uparrow_3$ (if any), etc.

The steps $\searrow_2$ and $\nearrow_2$ can be anywhere in any order. Consequently, to compute the number of shortest paths, we have

$$\frac{(\gamma+\delta+\varepsilon)!}{\gamma!\,\delta!\,\varepsilon!} \times \binom{|q(1)|+\gamma+2\delta}{|q(1)|-\gamma-2\varepsilon} \times \binom{|q(2)|}{|q(3)|-\gamma-2\delta}$$

different ones for a fixed value of $\gamma,\delta,\varepsilon$, $(\gamma,\delta,\varepsilon \geq 0,\ \gamma+2\delta \leq |q(3)|,\ \gamma+2\varepsilon \leq |q(1)|)$. This can be seen as follows: let the $\gamma+\delta+\varepsilon$ many $\uparrow_3$ steps are given. The first term $\frac{(\gamma+\delta+\varepsilon)!}{\gamma!\,\delta!\,\varepsilon!}$ refers for the possibility to arrange the appropriate number of $\uparrow_1, \searrow_3, \nearrow_3$ steps to have an alternating order of parity changing steps, as it is requested. The second term $\binom{|q(1)|+\gamma+2\delta}{|q(1)|-\gamma-2\varepsilon}$ gives the number of possibilities to place the $\nearrow_2$ steps into the path. Finally, the third term, $\binom{|q(2)|}{|q(3)|-\gamma-2\delta}$ gives the number of ways the $\searrow_2$ steps can be put into the path.

Observe that for these pixels $|q(2)| = q(1) + q(3)$. Finally, using the possible values of $\gamma, \delta$ and $\varepsilon$, we obtain the following theorem.

**Theorem 3.** *Let* $q(q(1), q(2), q(3))$ *be a pixel of the triangular grid such that* $q(1) + q(2) + q(3) = 0$ *and* $q(2) < 0$. *Then, the number of shortest paths from the origin* $(0,0,0)$ *to* $q$ *is given by*

$$\sum_{\gamma=0}^{\min\{|q(1)|,|q(3)|\}} \sum_{\delta=0}^{\lfloor\frac{|q(3)|-\gamma}{2}\rfloor} \sum_{\varepsilon=0}^{\lfloor\frac{|q(1)|-\gamma}{2}\rfloor} \frac{(\gamma+\delta+\varepsilon)!\binom{|q(2)|}{|q(3)|-\gamma-2\delta}\binom{|q(1)|+\gamma+2\delta}{|q(1)|-\gamma-2\varepsilon}}{\gamma!\,\delta!\,\varepsilon!}.$$

The case when $q(2) > 0$ is analogous (with steps to downward directions).

**Notation 2.** *In case* $q(2) > 0$, *let the* $\gamma, \delta$ *and* $\varepsilon$ *be defined as follows.*

- *Let* $\gamma$ *be the number of application of* $\searrow_2$ *(e)* $\nearrow_2$ *(e)* $\rightarrow\downarrow_1\downarrow_3$ *and* $\searrow_2$ *(o)* $\nearrow_2$ *(o)* $\rightarrow\downarrow_3\downarrow_1$ *(minus the number of their applications in reverse directions).*
- *Let* $\delta$ *be the number of applications of* $\nearrow_2$ *(e)* $\nearrow_2$ *(e)* $\rightarrow\nearrow_3\downarrow_3$ *and* $\nearrow_2$ *(o)* $\nearrow_2$ *(o)* $\rightarrow\downarrow_3\nearrow_3$ *(minus the number of their applications in reverse directions).*
- *Let* $\varepsilon$ *be the number of applications of* $\searrow_2$ *(e)* $\searrow_2$ *(e)* $\rightarrow\searrow_3\downarrow_3$ *and* $\searrow_2$ *(o)* $\searrow_2$ *(o)* $\rightarrow\downarrow_3\searrow_3$ *(minus the number of their applications in reverse directions).*

The only other difference (based on the possible rewriting) is that in this case the "alternation" of steps from the set $\{\downarrow_1, \nearrow_3, \searrow_3\}$ and $\downarrow_3$ starts with a step from the first set and finishes by a $\downarrow_3$ step. The number of possible paths is given by the following formula; and it can be proven by the application of multiplication and addition rules, in a similar manner as in the previous case.

**Theorem 4.** *Let* $q(q(1), q(2), q(3))$ *be a pixel of the triangular grid such that* $q(1) + q(2) + q(3) = 0$ *and* $q(2) > 0$. *Then, the number of shortest paths from the origin* $(0,0,0)$ *to* $q$ *is computed as*

$$\sum_{\gamma=0}^{\min\{|q(1)|,|q(3)|\}} \sum_{\delta=0}^{\lfloor\frac{|q(1)|-\gamma}{2}\rfloor} \sum_{\varepsilon=0}^{\lfloor\frac{|q(3)|-\gamma}{2}\rfloor} \frac{(\gamma+\delta+\varepsilon)!\binom{|q(2)|}{|q(1)|-\gamma-2\delta}\binom{|q(3)|+\gamma+2\delta}{|q(3)|-\gamma-2\varepsilon}}{\gamma!\,\delta!\,\varepsilon!}.$$

## 5.2   The Case of Odd Paths

This case is also divided into two parts based on the sign of $q(2)$. Let us start with $q(2) < 0$. By Proposition 2 a shortest path to $q$ is given as $\uparrow_3 \nwarrow_2^{|q(3)-1|} \nearrow_2^{|q(1)-1|}$. In these shortest paths, the three types of rewriting of Notation 1 can be applied that change the types of the steps (not only their order). Consequently, let us define and use $\gamma$, $\delta$ and $\varepsilon$ similarly as in the case of even paths for $q(2) < 0$, see Notation 1. Then the numbers of various steps that a shortest path $\Pi$ contain (with a fixed value of $\gamma, \delta$ and $\varepsilon$) are as follows:

$\nwarrow_2$ steps: $|q(3) - 1| - \gamma - 2\delta$;   $\nearrow_2$ steps: $|q(1) - 1| - \gamma - 2\varepsilon$;   $\uparrow_1$ steps: $\gamma$;
$\uparrow_3$ steps: $1 + \gamma + \delta + \varepsilon$;   $\nwarrow_3$ steps: $\delta$;   $\nearrow_3$ steps: $\varepsilon$.

Here the alternating sequence of steps $\uparrow_3$ and steps from the set $\{\uparrow_1, \nwarrow_3, \nearrow_3\}$ starts and ends with an $\uparrow_3$ step. Consequently, the final formula for the number of shortest paths in this case is computed:

**Theorem 5.** *Let $q(q(1), q(2), q(3))$ be a pixel of the triangular grid such that $q(1) + q(2) + q(3) = 1$ and $q(2) < 0$. Then, the number of shortest paths from the origin $(0, 0, 0)$ to $q$ is*

$$\sum_{\gamma=0}^{\min\{q(1)-1, q(3)-1\}} \sum_{\delta=0}^{\lfloor \frac{q(3)-1-\gamma}{2} \rfloor} \sum_{\varepsilon=0}^{\lfloor \frac{q(1)-1-\gamma}{2} \rfloor} \frac{(\gamma + \delta + \varepsilon)! \binom{|q(2)|}{q(1)-1-\gamma-2\varepsilon} \binom{q(3)+\gamma+2\varepsilon}{q(3)-1-\gamma-2\delta}}{\gamma! \, \delta! \, \varepsilon!}.$$

Note that in this case $q(1), q(3) \geq 1$ and $|q(2)| = q(1) + q(3) - 1$.

Now, let us consider the last case: $q$ is odd, $q(2) > 0$ and $q(2) = |q(1)| + |q(3)| + 1$. This case is the most complex, since the parity of the pixels must be changed during the path, and in this direction any of the elements of the set $\{\downarrow_1, \nearrow_3, \nwarrow_3\}$ can be applied for such reason. Therefore, actually, it is easier to break the set of the shortest paths $\mathcal{C}(w)$ – where $w$ could be, e.g., from Proposition 2, $w = \downarrow_1 (\nearrow_2 (o))^{|q(1)|}(\nwarrow_2 (o))^{|q(3)|}$ – to three disjoint sets. In this way, we can compute the number of shortest paths when the first parity changing step is $\downarrow_1$, is $\nearrow_3$ and is $\nwarrow_3$, separately. We may use the calculus $\mathcal{C}$ keeping the first parity changing step (but maybe moving it by permutative steps interchanging its place with some steps to 2-neighbors) starting from the following words, respectively: $\downarrow_1 (\nearrow_2 (o))^{|q(1)|}(\nwarrow_2 (o))^{|q(3)|}$, $\nearrow_3 (\nearrow_2 (o))^{|q(1)|-1}(\nwarrow_2 (o))^{|q(3)|+1}$ and $\nwarrow_3 (\nearrow_2 (o))^{|q(1)|+1}(\nwarrow_2 (o))^{|q(3)|-1}$.

Defining $\gamma$, $\delta$ and $\varepsilon$ in the same way, as they were in the case of even paths with $q(2) > 0$ (Notation 2), one can obtain the final result for this, most complex case.

**Theorem 6.** *Let $q(q(1), q(2), q(3))$ be a pixel of the triangular grid such that $q(1) + q(2) + q(3) = 1$ and $q(2) > 0$. Then, the number of shortest paths from the origin $(0, 0, 0)$ to $q$ is given by*

$$\min\{|q(1)|-1,|q(3)|-1\} \sum_{\gamma=0} \left\lfloor\frac{|q(3)|-1-\gamma}{2}\right\rfloor \sum_{\delta=0} \left\lfloor\frac{|q(1)|-1-\gamma}{2}\right\rfloor \sum_{\varepsilon=0} \frac{(\gamma+\delta+\varepsilon)!\binom{q(2)-2}{|q(1)|-1-\gamma-2\varepsilon}\binom{|q(3)|+\gamma+2\varepsilon}{|q(3)|-1-\gamma-2\delta}}{\gamma!\;\delta!\;\varepsilon!}+$$

$$+\min\{|q(1)|,|q(3)|-2\}\sum_{\gamma=0}\left\lfloor\frac{|q(3)|-2-\gamma}{2}\right\rfloor\sum_{\delta=0}\left\lfloor\frac{|q(1)|-\gamma}{2}\right\rfloor\sum_{\varepsilon=0}\frac{(\gamma+\delta+\varepsilon)!\binom{q(2)-2}{|q(1)|-\gamma-2\varepsilon}\binom{|q(3)|-1+\gamma+2\varepsilon}{|q(3)|-2-\gamma-2\delta}}{\gamma!\;\delta!\;\varepsilon!}+$$

$$+\min\{|q(1)|-2,|q(3)|\}\sum_{\gamma=0}\left\lfloor\frac{|q(3)|-\gamma}{2}\right\rfloor\sum_{\delta=0}\left\lfloor\frac{|q(1)|-2-\gamma}{2}\right\rfloor\sum_{\varepsilon=0}\frac{(\gamma+\delta+\varepsilon)!\binom{q(2)-2}{|q(1)|-2-\gamma-2\varepsilon}\binom{|q(3)|+1+\gamma+2\varepsilon}{|q(3)|-\gamma-2\delta}}{\gamma!\;\delta!\;\varepsilon!}.$$

## 6   Concluding Remarks

First, we summarize our main results in Fig. 3 showing the number of shortest paths (red color) for the indicated pixels. As we have seen instead of the binomial coefficients that are obtained on the hexagonal lattice, more complex formulae can be used to compute these values. These numbers can be considered, as a kind of generalizations of the binomial coefficients using the triangular grid, and actually, they are the cardinalities of the sets representing trajectories of the shortest paths on the triangular grid. As we have seen they are closely connected to generalized traces.

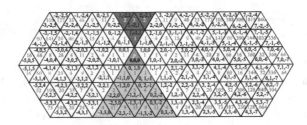

**Fig. 3.** A part of the triangular grid with the number of shortest paths from the origin. (The region for which the formulae are directly provided is shown with yellow background, for the other pixels the values can be obtained by rotating the grid.) (Color figure online)

We note here that a related result, namely, the number of shortest paths is computed by using only 1-neighbors (closest neighbors) and at most 2-neighbors for the triangular grid in [4], in this way, we have completed the task started there.

## References

1. Das, P.P.: An algorithm for computing the number of minimal paths in digital images. Pattern Recognit. Lett. **9**, 107–116 (1988)
2. Das, P.P.: Computing minimal paths in digital geometry. Pattern Recognit. Lett. **10**, 595–603 (1991)
3. Diekert, V., Rozenberg, G. (eds.): The Book of Traces. World Scientific, Singapore (1995)

4. Dutt, M., Biswas, A., Nagy, B.: Number of shortest paths in triangular grid for 1- and 2-neighborhoods. In: Barneva, R.P., Bhattacharya, B.B., Brimkov, V.E. (eds.) IWCIA 2015. LNCS, vol. 9448, pp. 115–124. Springer, Cham (2015). doi:10.1007/978-3-319-26145-4_9
5. Her, I.: Geometric transformations on the hexagonal grid. IEEE Trans. Image Process. 4(9), 1213–1222 (1995)
6. Herendi, T., Nagy, B.: Parallel Approach of Algorithms. Typotex, Budapest (2014)
7. Janicki, R., Kleijn, J., Koutny, M., Mikulski, L.: Generalising traces, CS-TR-1436, Technical report, Newcastle University (2014). Partly presented at LATA 2015
8. Luczak, E., Rosenfeld, A.: Distance on a hexagonal grid. Trans. Comput. **C–25**(5), 532–533 (1976)
9. Mateescu, A., Rozenberg, G., Salomaa, A.: Shuffle on trajectories: syntactic constraints. Theor. Comput. Sci. **197**, 1–56 (1998)
10. Nagy, B.: Finding shortest path with neighborhood sequences in triangular grids. In: Proceedings of the 2nd ISPA, pp. 55–60 (2001)
11. Nagy, B.: Shortest path in triangular grids with neighbourhood sequences. J. Comput. Inf. Technol. **11**, 111–122 (2003)
12. Nagy, B.: Calculating distance with neighborhood sequences in the hexagonal grid. In: Klette, R., Žunić, J. (eds.) IWCIA 2004. LNCS, vol. 3322, pp. 98–109. Springer, Heidelberg (2004). doi:10.1007/978-3-540-30503-3_8
13. Nagy, B.: Digital geometry of various grids based on neighbourhood structures. In: KEPAF 2007, 6th Conference of Hungarian Association for Image Processing and Pattern Recognition, Debrecen, pp. 46–53 (2007)
14. Nagy, B.: Distances with neighbourhood sequences in cubic and triangular grids. Pattern Recognit. Lett. **28**, 99–109 (2007)
15. Nagy, B.: Isometric transformations of the dual of the hexagonal lattice. In: Proceedings of the 6th ISPA, pp. 432–437 (2009)
16. Nagy, B.: Weighted distances on a triangular grid. In: Barneva, R.P., Brimkov, V.E., Šlapal, J. (eds.) IWCIA 2014. LNCS, vol. 8466, pp. 37–50. Springer, Cham (2014). doi:10.1007/978-3-319-07148-0_5
17. Nagy, B.: Cellular topology and topological coordinate systems on the hexagonal and on the triangular grids. Ann. Math. Artif. Intell. **75**, 117–134 (2015)
18. Nagy, B., Otto, F.: CD-systems of stateless deterministic R(1)-automata accept all rational trace languages. In: Dediu, A.-H., Fernau, H., Martín-Vide, C. (eds.) LATA 2010. LNCS, vol. 6031, pp. 463–474. Springer, Heidelberg (2010). doi:10.1007/978-3-642-13089-2_39
19. Nagy, B., Otto, F.: An automata-theoretical characterization of context-free trace languages. In: Černá, I., Gyimóthy, T., Hromkovič, J., Jefferey, K., Králović, R., Vukolić, M., Wolf, S. (eds.) SOFSEM 2011. LNCS, vol. 6543, pp. 406–417. Springer, Heidelberg (2011). doi:10.1007/978-3-642-18381-2_34
20. Nagy, B., Otto, F.: CD-systems of stateless deterministic R(1)-automata governed by an external pushdown store. RAIRO-ITA **45**, 413–448 (2011)
21. Nagy, B., Otto, F.: On CD-systems of stateless deterministic R-automata with window size one. J. Comput. Syst. Sci. **78**, 780–806 (2012)
22. Parikh, R.: On context-free languages. J. ACM **13**(4), 570–581 (1966)
23. Rosenfeld, A., Pfaltz, J.L.: Distance functions on digital pictures. Pattern Recognit. **1**, 33–61 (1968)
24. Rozenberg, G., Salomaa, A. (eds.): Handbook of Formal Languages, vol. 1–3. Springer, Heidelberg (1997)
25. Stojmenovic, I.: Honeycomb networks: topological properties and communication algorithms. IEEE Trans. Parallel Distrib. Syst. **8**, 1036–1042 (1997)

# On Sets of Line Segments
# Featuring a Cactus Structure

Boris Brimkov[✉]

Computational and Applied Mathematics, Rice University, Houston, TX 77005, USA
boris.brimkov@rice.edu

**Abstract.** In this paper we derive sharp upper and lower bounds on
the number of intersections and closed regions that can occur in a set of
line segments whose underlying planar graph is a cactus graph. These
bounds can be used to evaluate the complexity of certain algorithms
for problems defined on sets of segments in terms of the cardinality of
the segment sets. In particular, we give an application in the problem
of finding a path between two points in a set of segments which travels
through a minimum number of segments.

**Keywords:** Set of segments · Cactus graph · Segment intersections ·
Segment cycle

## 1 Introduction

Sets of straight line segments with special structures and properties appear
in various applications of geometric modeling, such as scientific visualization,
computer-aided design, and medical image processing. Such segment sets with a
special structure are investigated in the present paper.

Let $M$ be a set of line segments in the plane, and $\overline{M}$ be the union of all
points of segments in $M$; exclude from consideration the degenerate case of
intersecting collinear segments which can be merged into a single segment. Let
$P$ be the set of all intersection points of segments in $\overline{M}$ and $C$ be the set of closed
bounded regions into which the segments in $\overline{M}$ partition the plane. Let $|M| = m$,
$|P| = p$, and $|C| = c$, where $|\cdot|$ denotes set cardinality. The following are well-
known relations between these quantities which hold for arbitrary segment sets:
$p \leq \frac{m(m-1)}{2}$, $c \leq \frac{(m-1)(m-2)}{2}$. Both bounds are sharp, i.e., there are classes
of segment sets for which the bounds hold with equality. For special classes of
segment sets, better bounds can be derived. For example, if a set of segments
features a "tree" structure, then $p \leq m - 1$ and $c = 0$.

These kinds of bounds can be used to analyze the time and space complexity of algorithms for finding the intersections and bounded regions occurring
in a set of segments in terms of $m$, $p$, and $c$; these are fundamental tasks in
computational geometry and have been widely studied (cf. [1,3,13,20,21]). For
example, Balaban's algorithm for finding segment intersections [1] runs in optimal $O(m \log m + p)$ time and $O(m)$ space. For special classes of graphs, an

© Springer International Publishing AG 2017
V.E. Brimkov and R.P. Barneva (Eds.): IWCIA 2017, LNCS 10256, pp. 30–39, 2017.
DOI: 10.1007/978-3-319-59108-7_3

estimation of $p$ in terms of $m$ can yield a better bound on the time and space complexity of existing algorithms. For instance, for the aforementioned class of segment sets which feature a tree structure, the time complexity of Balaban's algorithm simplifies to $O(m \log m)$, while for general segment sets the algorithm may require $\Omega(m^2)$ arithmetic operations.

In the present paper, we obtain sharp bounds on $p$ and $c$ for the class of segment sets which feature a cactus structure, i.e., a structure where the boundaries of any two closed regions share at most one point. We show that for such segment sets, $p \leq 2(m - k_1) - 3k_2$ and $c \leq (m - k_1) - 2k_2$, where $k_1$ and $k_2$ are the numbers of connected components of $\overline{M}$ consisting, respectively, of a single segment and multiple segments. Both bounds are sharp, as they are attained for certain classes of segment sets. We also illustrate how these bounds can be used to estimate and compare the running times of certain algorithms for computation of intersections in a set of segments and for finding a path between two points in a set of segments which travels through a minimum number of segments. See [5, 7–11, 14, 17] and the bibliographies therein for other applications of computing $p$ and $c$, as well as for techniques and results on other problems defined on segment sets and on graphs constructed through segment sets.

## 2    Preliminaries

Let $M$ be a set of $m$ segments in the plane, and $\overline{M}$ be the union of all points of segments in $M$. Let $P(M)$ and $J(M)$ be the set of all intersections and the set of all end-points of segments from $M$, respectively (note that $P \cap J$ may be non-empty); when there is no scope for confusion, dependence on $M$ will be omitted. Let $G_M = (V, E)$ be a plane graph whose vertex set is $P \cup J$ and where vertices $u$ and $v$ are adjacent whenever there is a segment $s \in M$ which contains $u$ and $v$, such that there is no $w \in V \cap s$ that is between $u$ and $v$.

By a *cycle* of $M$ we will mean any closed simple polygonal curve in $\overline{M}$. By a *cycle segment set* of $M$ we will mean the set of segments in $M$ that contribute to a cycle of $M$ by more than a single point. There is a one-to-one correspondence between the cycles of $M$, the cycle segment sets of $M$, and the bounded faces of $G_M$ (in the planar embedding induced by $\overline{M}$). We will call a connected component of $\overline{M}$ *trivial* if it consists of a single segment, and *nontrivial* if it contains two or more segments. Let $k_1$ denote the number of trivial components of $\overline{M}$, and $k_2$ denote the number of nontrivial components of $\overline{M}$. Given a segment $s \in M$, $\overline{M} \backslash s$ denotes the union of all points of segments in $M \backslash \{s\}$.

A *cut vertex* of a graph $G$ is a vertex whose deletion increases the number of connected components of $G$. A *biconnected component* or *block* of $G$ is a maximal subgraph of $G$ which has no cut vertices. An isomorphism between graphs $G_1$ and $G_2$ will be denoted by $G_1 \simeq G_2$. Given a vertex $v$ of $G$, $G - v$ will denote $G$ with $v$ removed, along with all edges incident to $v$. A vertex of $G$ is a *leaf* if it has a single neighbor in $G$.

A graph $G$ is called a *cactus graph* (or simply a *cactus*) if any two cycles of $G$ have at most one vertex in common. Every edge of a cactus graph belongs to at most one cycle, and the biconnected components of a cactus graph are either cycles or single edges; see Fig. 1, left, for an example of a cactus graph. Properties of cactus graphs have been studied with some applications in mind; for example, cactus graphs arise in the theory of condensation in statistical mechanics, and in the design of telecommunication systems, material handling networks, and local area networks. For more applications, properties, and problems solved on cactus graphs, see [2,6,15,16,18,19] and the bibliographies therein.

We will say that a set of segments $M$ is a *segment cactus* if the graph $G_M$ is a cactus; see Fig. 1, right, for an example. By definition, two cycles of a segment cactus can have at most one vertex in common, i.e., they cannot share a portion of a segment different from a point. Thus, there is a one-to-one correspondence between the cycles of $M$, the cycle segment sets of $M$, and the cycles of $G_M$ (as well as the faces of $G_M$ in the embedding induced by $\overline{M}$). If a segment cactus has no cycles, it is a *segment forest*; if the set of segments is also connected, then it is a *segment tree*. Thus $T$ is a segment tree if and only if the corresponding graph $G_T$ is a tree.

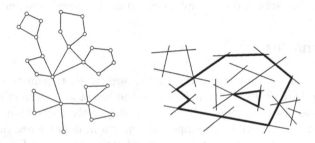

**Fig. 1.** *Left:* An example of a cactus graph. *Right:* An example of a segment cactus with two of its cycles marked by thick lines.

When there is no scope for confusion, some of the definitions introduced in this section may be applied to $M$ and $\overline{M}$ interchangeably, e.g., we may refer to a cycle segment set of $\overline{M}$ or a connected component of $M$.

## 3    Main Results

We begin this section with some preliminary observations about segment sets, and then present several structural results on segment cacti. It is well-known (cf. [22]) that every planar graph has an embedding where its edges are mapped to straight line segments. Let $G$ be an arbitrary planar graph, and $\widehat{G}$ be a straight-line embedding of $G$. If the edges incident to some degree 2 vertex $v$ of $G$ are drawn in $\widehat{G}$ as collinear segments, then $v$ can be slightly shifted so that the segments incident to it are no longer collinear. This implies the following observation.

**Observation 1.** *For any planar graph $G$, there exists a segment set $M$ such that $G_M \simeq G$.*

By definition, each of the $k_1$ trivial components of a segment set $M$ consist of a single segment, and each of the $k_2$ nontrivial components of $M$ consist of at least two segments; thus, we have the following observation.

**Observation 2.** *For any segment set $M$, $m \geq k_1 + 2k_2$.*

Moreover, since the segments in each nontrivial component are connected, each nontrivial component must have at least one intersection point.

**Observation 3.** *For any segment set $M$, $p \geq k_2$.*

If $M$ is a segment forest with trivial components $t_1, \ldots, t_{k_1}$ and nontrivial components $T_1 \ldots, T_{k_2}$, then there can be at most $|T_i| - 1$ intersections in each nontrivial component $T_i$, which occurs when no three segments have a common intersection point. This implies the following bound on the number of intersections in a segment forest.

**Observation 4.** *If $M$ is a segment forest, then $p \leq m - k_1 - k_2$.*

Note that if $M$ is a connected segment cactus different from a segment tree, the graph $G_M = (V, E)$ satisfies the inequality $|V| \leq |E|$. However, it is not necessarily the case that $|P| \leq |M|$: while a graph edge is incident to exactly two vertices, a segment from $M$ can contain arbitrarily many intersections with other segments.

   The next result concerns segments whose removal does not affect the connectivity of an arbitrary segment set; a consequence of this result will be used in the sequel.

**Proposition 1.** *For any nontrivial connected segment set $M$, there are at least two segments $s_a$ and $s_b$ in $M$ such that $\overline{M} \backslash s_a$ and $\overline{M} \backslash s_b$ are connected.*

*Proof.* Let $H$ be a graph which has a vertex for each segment in $M$, and where two vertices are adjacent whenever the corresponding segments intersect in $\overline{M}$.

   Let $s_x$ and $s_y$ be any two vertices of $H$, and $x$ and $y$ be non-intersection points respectively belonging to the segments $s_x$ and $s_y$ in $\overline{M}$. Since $\overline{M}$ is connected, there is a path $x, p_1, \ldots, p_k, y$ between $x$ and $y$, where $p_1, \ldots, p_k$ are parts of segments (or entire segments) of $M$. In particular, let $p_t \subseteq s_{i_t}$ for $1 \leq t \leq k$ (where $s_{i_1} = s_x$ and $s_{i_k} = s_y$). By construction of $H$, for $1 \leq t \leq k - 1$, $s_{i_t}$ is adjacent to $s_{i_{t+1}}$ in $H$. Thus, the path $x, p_1, \ldots, p_k, y$ in $\overline{M}$ corresponds to a path $s_x, s_{i_1}, \ldots, s_{i_k}, s_y$ in $H$, so $H$ is connected.

   Since any connected graph with at least two vertices has at least two non-cut vertices, $H$ has two non-cut vertices $s_a$ and $s_b$. We claim that $\overline{M} \backslash s_a$ and $\overline{M} \backslash s_b$ are connected. To see why, let $x$ and $y$ be any two points in $\overline{M} \backslash s_a$. If $x$ and $y$ belong to the same segment, clearly there is a path between them. Otherwise, let $s_x$ and $s_y$ respectively be segments containing $x$ and $y$. Since $s_a$ is a non-cut vertex of $H$, $H - s_a$ is connected. Let $s_x, s_{i_1}, \ldots, s_{i_k}, s_y$ be a simple path between $s_x$ and $s_y$ in $H - s_a$. By construction of $H$, segments $s_x$ and $s_{i_1}$ intersect in $\overline{M}$;

thus, there is a path between $x$ and every point in $s_{i_1}$. Similarly, segments $s_{i_1}$ and $s_{i_2}$ intersect in $\overline{M}$, so there is also a path between $x$ and every point in $s_{i_2}$. Continuing in this fashion, we see that there is a path between $x$ and $y$ in $\overline{M}\backslash s_a$, so $\overline{M}\backslash s_a$ is connected; similarly, $\overline{M}\backslash s_b$ is connected.                    □

**Corollary 1.** *Any nontrivial segment tree $M$ contains at least two segments $s_a$ and $s_b$ such that $\overline{M}\backslash s_a$ and $\overline{M}\backslash s_b$ are connected, and such that $s_a$ and $s_b$ each contain a single intersection point.*

*Proof.* By Proposition 1, there are two segments $s_a$ and $s_b$ such that $\overline{M}\backslash s_a$ and $\overline{M}\backslash s_b$ are connected; we claim that each of these segments contains a single intersection point. Indeed, since $M$ is a segment tree and is therefore connected, $s_a$ and $s_b$ must each contain at least one intersection point. Suppose for contradiction that $s_a$ contains two (or more) intersection points $x$ and $y$. Since $M$ is a segment tree, there is only one path, namely along $s_a$, between the segments which intersect $s_a$ at $x$ and $y$. Then, there will be no path between these segments in $\overline{M}\backslash s_a$, a contradiction.                    □

Let $M$ be a set of segments and $s$ be a segment of $M$ with endpoints $\ell$ and $r$. Let $\ell'$ be the first intersection point in $s$ encountered when moving along $s$ in a straight line from $\ell$ to $r$ in $\overline{M}$, and $r'$ be the last intersection point encountered. We will say that *trimming* $s$ is the operation of replacing $s$ by a segment $s'$ with endpoints $\ell'$ and $r'$; if $s$ has fewer than two intersection points, then trimming $s$ means deleting $s$. We will say that *trimming* $M$ means repeatedly trimming the segments in $M$ until further trimming yields no difference. Note that it may be possible to trim a segment, then trim another segment, and then trim the first segment again. See Fig. 2 for an illustration of trimming.

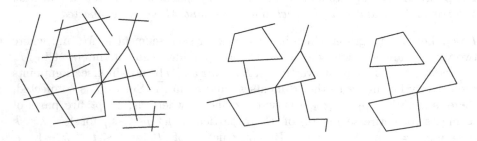

**Fig. 2.** *Left:* Set of segments $M$. *Middle:* Trimming every segment of $M$ once. *Right:* Trimming $M$.

**Proposition 2.** *A segment cactus $M$ with $c \geq 1$ cycles contains at least two segments $s_1$ and $s_2$, such that for $i \in \{1, 2\}$,*

*(A) $s_i$ belongs to a single cycle segment set $S_i$,*
*(B) the connected components of $\overline{M}\backslash s_i$ which do not contain segments of $S_i$ are segment trees.*

*Proof.* If $c = 1$, every segment in the single cycle segment set of $\overline{M}$ satisfies properties $(A)$ and $(B)$; thus, assume henceforth that $c \geq 2$.

Let $Q = \{s_1, \ldots, s_q\}$ be a maximal set of segments of $M$ such that for $1 \leq i \leq q$, $s_i$ does not belong to any cycle segment set of $\overline{M}$, and $s_i$ is a segment whose deletion does not disconnect $\overline{M} \backslash \{s_1, \ldots, s_{i-1}\}$. Let $M' = M \backslash Q$. By construction, $\overline{M}$ and $\overline{M}'$ have the same cycle segment sets; moreover, the connected components of $\overline{M} \backslash M'$ (i.e. of $\overline{Q}$) are segment trees. Hence, for any segment $s \in M'$, the connected components of $\overline{M} \backslash s$ which do not contain segments of $M'$ are segment trees. Let $M''$ be the set of segments obtained by trimming $M'$ (in fact, $M''$ is identical to the set of segments obtained by trimming $M$). Note that $\overline{M}$, $\overline{M}'$, and $\overline{M}''$ have the same cycles.

$G_{M''}$ has no leaves, since a leaf of $G_{M''}$ would have to be an endpoint of a segment in $\overline{M}''$, and all endpoints of segments in $\overline{M}''$ are also intersection points. Thus, all outer blocks of $G_{M''}$ (i.e., biconnected components with a single cut vertex) are cycles. Since $c \geq 2$ and since $\overline{M}$ and $\overline{M}''$ have the same cycles, it follows that $G_{M''}$ has at least two cycles; thus, $G_{M''}$ has at least two outer blocks which are cycles, say $C_1$ and $C_2$. Let $S_1$ and $S_2$ be the cycle segment sets in $\overline{M}$ corresponding to $C_1$ and $C_2$, respectively. For $i \in \{1, 2\}$, exactly two edges of $C_i$ in $G_{M''}$ are incident to the cut vertex $v_i$ of $C_i$; thus, in $\overline{M}$, $v_i$ corresponds to an intersection point of at most two segments of $S_i$. Since $S_i$ contains at least three segments, there is a segment $s_i \in S_i$ which does not contain $v_i$ as an intersection point in $\overline{M}$. Then, since $C_i$ is an outer cycle, $s_i$ does not belong to any other cycle segment set of $M$, i.e., $s_i$ satisfies property $(A)$. Furthermore, the connected components of $\overline{M} \backslash s_i$ which do not contain segments of $S_i$ also do not contain segments of $M'$; however, as shown above, the connected components of $\overline{M} \backslash s_i$ which do not contain segments of $M'$ are segment trees. Thus, $s_i$ satisfies property $(B)$. $\square$

Combining the previous results, we will now derive sharp bounds on the number of intersections and cycles in a segment cactus.

**Theorem 1.** *If $M$ is a segment cactus, then:*

$$k_2 \leq p \leq 2(m - k_1) - 3k_2 \tag{1}$$

$$0 \leq c \leq (m - k_1) - 2k_2. \tag{2}$$

*Proof.* The lower bound in (1) follows from Observation 3, and the lower bound in (2) follows from the fact that a segment forest is also a segment cactus.

If $\overline{M}$ is a segment forest, then $p \leq 2p - k_2 \leq 2(m - k_1 - k_2) - k_2 = 2(m - k_1) - 3k_2$, where the first inequality follows from Observation 3 and the second inequality follows from Observation 4; this establishes the upper bound in (1). Likewise, if $\overline{M}$ is a segment forest, then the upper bound in (2) follows from Observation 2 and the fact that $c = 0$. Thus, it remains to be shown that the upper bounds in (1) and (2) hold for the case when the segment cactus is not a segment forest, i.e., when $c \geq 1$, and hence $m \geq 3$. We will proceed by induction on $m$. Both inequalities clearly hold for $m = 3$. Assume the inequalities hold for some $m \geq 3$ and let $M$ be a segment cactus with $m + 1$ segments.

By Proposition 2, $M$ contains a segment $s_1$ which belongs to a single cycle segment set $S_1$, such that the connected components of $\overline{M}\backslash s_1$ which do not contain segments of $S_1$ are segment trees. If $\overline{M}\backslash s_1$ does not have any connected components which do not contain segments of $S_1$, let $s_* = s_1$. Note that in this case, deleting $s_*$ from $\overline{M}$ decreases the number of intersection points by at most two, and the number of cycles by one. If $\overline{M}\backslash s_1$ has at least one connected component $T$ which does not contain segments of $S_1$, $T$ is a segment tree which can only intersect $s$ in a single point, since otherwise $s$ would be part of at least two cycles. If $T$ consists of a single segment, let $s_*$ be that segment. If $T$ contains at least two segments, then by Corollary 1, $T$ contains two segments $s_a$ and $s_b$, each having a single intersection point, such that removing either one of them from $T$ does not disconnect $T$. If neither $s_a$ nor $s_b$ intersect $s$, let $s_* = s_1$. If exactly one of $s_a$ and $s_b$ intersects $s$, let $s_*$ be the segment among $s_a$ and $s_b$ which does not intersect $s$. If both $s_a$ and $s_b$ intersect $s$, then $s$, $s_a$, and $s_b$ must all intersect in the same point; in this case, let $s_* = s_1$. In each of these cases, deleting $s_*$ from $\overline{M}$ decreases the number of intersection points by at most one, and does not affect the number of cycles.

Thus, the segment cactus $\overline{M}\backslash s_*$ has $m$ segments, $p - i$ intersection points for some $i \in \{0, 1, 2\}$, and $c - j$ cycles for some $j \in \{0, 1\}$. By the induction hypothesis, $p - i \leq 2(m - k_1) - 3k_2$. Then, for the segment cactus $M$ with $m + 1$ segments and $p$ intersections, we obtain $p \leq 2(m - k_1) - 3k_2 + i \leq 2(m - k_1) - 3k_2 + 2 = 2(m + 1 - k_1) - 3k_2$. Similarly, by the induction hypothesis, $c - i \leq (m - k_1) - 2k_2$. Then, for the segment cactus $M$ with $m + 1$ segments and $c$ cycles we obtain $c \leq (m - k_1) - 2k_2 + i \leq (m - k_1) - 2k_2 + 1 = (m + 1 - k_1) - 2k_2$. This concludes the inductive step and establishes the inequalities.                    □

**Observation 5.** *The upper and lower bounds in (1) and (2) hold with equality for classes of segment cacti like the ones in Fig. 3.*

**Fig. 3.** *Left:* A class of segment sets for which the lower bounds in (1) and (2) hold with equality. *Right:* A class of segment sets for which the upper bounds in (1) and (2) hold with equality.

## 4   Applications

### 4.1   Finding Intersections

The inequalities derived in Theorem 1 can be used to evaluate and compare the running times of certain algorithms when these are applied to segment sets with a cactus structure. Consider the algorithms of Bentley-Ottmann [3], Chazelle [13], and Balaban [1] which compute all intersections in a given set of segments. The time complexities of these algorithms are respectively $O((m + p) \log m)$, $O(p + \frac{m \log^2 m}{\log \log m})$, and $O(p + m \log m)$, the last one being optimal for general segment sets. The worst case performance of these algorithms is achieved for sets of segments with $\Omega(m^2)$ intersections, and is respectively $\Omega(m^2 \log m)$ for Bentley-Ottmann's algorithm, and $\Omega(m^2)$ for Chazelle's and Balaban's algorithms. Thus, regarding worst case time complexity, Chazelle's and Balaban's algorithms are superior to Bentley-Ottmann's algorithm. However, if a segment set has a cactus structure, Bentley-Ottmann's and Balaban's algorithms run in $O(m \log m)$ time and are superior to Chazelle's algorithm, which runs in $O(\frac{m \log^2 m}{\log \log m})$ time.

Chen and Chan [12] modified Bentley-Ottmann's algorithm to an $O((m + p) \log m)$-time in-place algorithm, i.e., an algorithm which uses $O(1)$ cells of memory in addition to the input array, and whose output is printed in write-only space; likewise, Vahrenhold [4,23] presented an in-place modification of Balaban's algorithm with an $O(m \log^2 m + p)$ time complexity. As in the original versions of the two algorithms discussed above, in terms of worst-case time complexity, the in-place version of Balaban's algorithm is superior to the in-place version of Bentley-Ottmann's algorithm, as they require $\Omega(m^2)$ and $\Omega(m^2 \log m)$ time, respectively. However, on a segment set with a cactus structure, the latter runs in $O(m \log m)$ time and is superior to the former which runs in $O(m \log^2 m)$ time.

## 4.2 Constrained Shortest Path

Let $M$ be an arbitrary connected segment set, and $x$ and $y$ be two points in $\overline{M}$. Consider the problem of finding a path between $x$ and $y$ such that the path consists of a minimum number of segments (or parts of segments) of $M$. This problem could model a scenario where segments represent different lines of public transportation, where transfer times between lines is high compared to travel time along a line; thus, one would be interested to find a route which requires the fewest transfers.

To find such a path, let $H$ be the graph defined in Proposition 1. Given a simple path $s_1, \ldots, s_k$ in $H$, for $1 \leq i \leq k - 1$, let $x_i$ be the intersection point between the segments in $\overline{M}$ corresponding to $s_i$ and $s_{i+1}$. For $2 \leq i \leq k - 1$, let $p_i \subseteq s_i$ be the segment with endpoints $x_{i-1}$ and $x_i$; for some points $x_0 \in s_1$ and $x_k \in s_k$, let $p_1 \subseteq s_1$ and $p_k \subseteq s_k$ respectively be the segments with endpoints $x_0, x_1$, and $x_{k-1}, x_k$. Then, $p_1, \ldots, p_k$ is a path in $\overline{M}$ corresponding to the path $s_1, \ldots, s_k$ in $H$.

Now, if a path between points $x$ and $y$ in $\overline{M}$ passes through the smallest number of segments, it must also pass through the smallest number of intersections; thus, such a path in $\overline{M}$ corresponds to a path with the smallest number of edges between $s_x$ and $s_y$ in $H$, where $s_x$ and $s_y$ are respectively segments containing $x$ and $y$. If one or both of $x$ and $y$ are intersection points in $\overline{M}$, then $H$ can be modified by adding new nodes $s'_x$ and $s'_y$ which are respectively adjacent to all of the nodes corresponding to segments which intersect at $x$ and $y$ in $\overline{M}$; then, the shortest path between $s'_x$ and $s'_y$ would correspond to the path in $\overline{M}$ with the minimum number of segments.

Since $H$ has $m$ vertices and $p$ edges, the path with the smallest number of edges between $s_x$ and $s_y$ in $H$ can be found by breadth first search in $O(m + p)$ time. For a general segment set, the worst case run time of this procedure could be $\Omega(m^2)$; however, if the segment set has a cactus structure, by Theorem 1, the run time would be $O(m)$.

## 5    Concluding Remarks

In this paper, we derived bounds on the number of intersections and closed regions that can occur in segment cacti. These bounds can be used to evaluate the complexity of certain algorithms for problems defined on sets of segments, and, in some cases, to conclude that a generally sub-optimal algorithm outperforms a generally optimal algorithm when applied to a segment cactus. It would be interesting to derive similar upper and lower bounds on $p$ and $c$ for other special classes of segment sets, for example those corresponding to maximal outerplanar or maximal planar graphs.

**Acknowledgements.** We thank the three anonymous reviewers for their valuable comments. This material is based upon work supported by the National Science Foundation under Grant No. 1450681.

## References

1. Balaban, I.J.: An optimal algorithm for finding segment intersections. In: Proceedings of 11-th Annual ACM Symposium on Computational Geometry, pp. 211–219 (1995)
2. Ben-Moshe, B., Dvir, A., Segal, M., Tamir, A.: Centdian computation in cactus graphs. J. Graph Algorithms Appl. **16**(2), 199–224 (2012)
3. Bentley, J.L., Ottmann, T.A.: Algorithms for reporting and counting geometric intersections. IEEE Trans. Comput. **28**, 643–647 (1979)
4. Bose, P., Maheshwari, A., Morin, P., Morrison, J., Smid, M., Vahrenhold, J.: Space-efficient geometric divide-and-conquer algorithms. Comput. Geometry **37**(3), 209–227 (2007)
5. Brévilliers, M., Chevallier, N., Schmitt, D.: Triangulations of line segment sets in the plane. In: Arvind, V., Prasad, S. (eds.) FSTTCS 2007. LNCS, vol. 4855, pp. 388–399. Springer, Heidelberg (2007). doi:10.1007/978-3-540-77050-3_32
6. Brimkov, B., Hicks, I.V.: Memory efficient algorithms for cactus graphs and block graphs. Discrete Appl. Math. **216**, 393–407 (2017)
7. Brimkov, V.E.: Approximability issues of guarding a set of segments. Int. J. Comput. Math. **90**(8), 1653–1667 (2013)
8. Brimkov, V.E., Leach, A., Mastroianni, M., Wu, J.: Guarding a set of line segments in the plane. Theoret. Comput. Sci. **412**(15), 1313–1324 (2011)
9. Brimkov, V.E., Leach, A., Wu, J., Mastroianni, M.: Approximation algorithms for a geometric set cover problem. Discrete Appl. Math. **160**, 1039–1052 (2012)

10. de Castro, N., Cobos, F.J., Dana, J.C., Márquez, A., Noy, M.: Triangle-free planar graphs and segment intersection graphs. J. Graph Algorithms Appl. **6**(1), 7–26 (2002)
11. Chan, T.M., Chen, E.Y.: Optimal in-place and cache-oblivious algorithms for 3-D convex hulls and 2-D segment intersection. Comput. Geometry **43**(8), 636–646 (2010)
12. Chen, E.Y., Chan, T.M.: A space-efficient algorithm for line segment intersection. In: Proceedings of the 15th Canadian Conference on Computational Geometry, pp. 68–71 (2003)
13. Chazelle, B.M.: Reporting and counting arbitrary planar intersections. Report CS-83-16, Department of Computer Science, Brown University, Providence, RI, USA (1983)
14. Francis, M.C., Kratochvíl, J., Vyskočil, T.: Segment representation of a subclass of co-planar graphs. Discrete Math. **312**(10), 1815–1818 (2012)
15. Harary, F., Uhlenbeck, G.: On the number of Husimi trees I. Proc. Natl. Acad. Sci. **39**, 315–322 (1953)
16. Husimi, K.: Note on Mayers' theory of cluster integrals. J. Chem. Phys. **18**, 682–684 (1950)
17. Kára, J., Kratochvíl, J.: Fixed parameter tractability of independent set in segment intersection graphs. In: Bodlaender, H.L., Langston, M.A. (eds.) IWPEC 2006. LNCS, vol. 4169, pp. 166–174. Springer, Heidelberg (2006). doi:10.1007/11847250_15
18. Kariv, O., Hakimi, S.L.: An algorithmic approach to network location problems, part 1: the p-center. SIAM J. Appl. Math **37**, 513–537 (1979)
19. Koontz, W.L.G.: Economic evaluation of loop feeder relief alternatives. Bell Syst. Tech. J. **59**, 277–281 (1980)
20. Preparata, F., Shamos, M.I.: Computational Geometry: An Introduction. Springer, New York (1985)
21. Tiernan, J.C.: An efficient search algorithm to find the elementary circuits of a graph. Commun. ACM **13**, 722–726 (1970)
22. Wagner, K.: Bemerkungen zum Vierfarbenproblem. Jahresbericht der Deutschen Mathematiker-Vereinigung **46**, 26–32 (1936)
23. Vahrenhold, J.: Line-segment intersection made in-place. Comput. Geometry **38**, 213–230 (2007)

# Construction of Thinnest Digital Ellipsoid Using Inverse Projection and Recursive Integer Intervals

Papia Mahato and Partha Bhowmick$^{(\boxtimes)}$

Department of Computer Science and Engineering,
Indian Institute of Technology, Kharagpur, India
papiamahatostar@gmail.com, bhowmick@gmail.com

**Abstract.** In this paper, we investigate the problem of characterization and construction of digital ellipsoid to its thinnest (2-minimal) topological model. We show how this ellipsoid model admits certain characterization based on isothetic-distance and functional-plane properties. Based on this novel characterization, we derive certain recurrences on the integer intervals that contain the values of a specific integer expression corresponding to the integer points comprising the digital ellipsoid. This, in turn, helps in designing an efficient algorithm for its construction in the integer space. The algorithm, in principle, is based on inverse projection of digital elliptical discs and the functional-plane relation of voxels comprising the digital ellipsoid.

**Keywords:** Digital ellipsoid · Digital geometry · Integer intervals · Integer algorithm

## 1 Introduction

Ellipsoid, also known as spheroid, is an important primitive in 3D geometry. However, unlike other 3D primitives like plane and sphere that have been studied in digital geometry in great detail, ellipsoid has not been studied up to its merit till date. Although some work related to lattice point distribution on real ellipsoid have been reported in [7,14], they do not closely relate to digital-geometric models of ellipsoid. In this paper, we present a study on an interesting characterization of the topologically thinnest model (2-minimal) of digital ellipsoid, which eventually leads to designing an efficient algorithm for its construction.

We consider an ellipsoid with integer specification. Further, for brevity, we take its canonical form, which means its center is $(0,0,0)$ and its axes are simply the coordinate axes. Hence, its equation is

$$\frac{x^2}{a^2} + \frac{y^2}{b^2} + \frac{z^2}{c^2} = 1, \tag{1}$$

where $a$, $b$, and $c$ are integers representing the respective lengths of its semi-principal axes along $x$-, $y$-, and $z$-directions. Without loss of generality, we

© Springer International Publishing AG 2017
V.E. Brimkov and R.P. Barneva (Eds.): IWCIA 2017, LNCS 10256, pp. 40–52, 2017.
DOI: 10.1007/978-3-319-59108-7_4

assume that $a \geqslant b \geqslant c$. For an early reckoning, let us mention here that our objective is to construct a digital ellipsoid as a topologically well-formed set of voxels of minimum cardinality such that each voxel in this set lies as much close as possible to the corresponding real ellipsoid. Figure 1 shows an example of digital ellipsoid produced by our algorithm.

## 1.1   Metrics and Topology

We fix here some basic definitions and metrics that are used in the sequel. Let $\mathbb{R}^3$ be the 3-dimensional euclidean space and $\mathbb{Z}^3$ the 3-dimensional integer space. A *voxel* or *3-cell* is perceived as a unit cube centered at a point in $\mathbb{Z}^3$ and is thus also uniquely identified by its center. As shown in Fig. 1, two distinct voxels are said to be 0-*adjacent* if they share a vertex (0-*cell*), 1-*adjacent* if they share an edge (1-*cell*), and 2-*adjacent* if they share a face (2-*cell*). According to this, for $k = 1, 2$, two voxels are also $(k-1)$-adjacent whenever they are $k$-adjacent. So, in Fig. 1, the 1-adjacent voxels are 0-adjacent too, and hence the 2-adjacent voxels are both 1- and 0-adjacent.

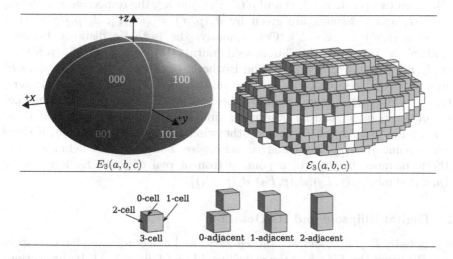

**Fig. 1.** Top: A real ellipsoid with its octants (left) and the corresponding digital ellipsoid for $a = 9, b = 7, c = 4$ (right). Bottom: Different adjacency relations.

For $k = 0, 1, 2$, a $k$-*path* means a sequence of voxels where every two consecutive voxels are $k$-adjacent. A voxel set $S$ is $k$-*connected* if every two voxels of $S$ are connected by a $k$-path. Let $S'$ be a subset of voxel set $S$. If $S \setminus S'$ is not $k$-connected, then the set $S'$ is said to be $k$-*separating* in $S$. A voxel $p$ of $S'$ is a *simple voxel* if $S' \setminus p$ is also $k$-separating in $S$. The set $S'$ is $k$-*minimal* if it is $k$-separating in $S$ and does not contain any simple voxel. In particular, $S'$ is 2-*minimal* if it is 2-*separating* in $S$ and does not contain any simple voxel. In the context of our work, if we consider $S'$ as the digital ellipsoid and $S$ as $\mathbb{Z}^3$, then $S'$

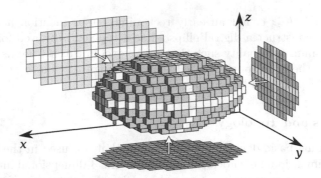

**Fig. 2.** The principle of our algorithm: construction of $\mathcal{E}_3(9,7,4)$ from the inverse projection of digital elliptical discs, $\mathcal{E}_{xy}^*(9,7)$, $\mathcal{E}_{yz}^*(7,4)$, and $\mathcal{E}_{xz}^*(9,4)$.

is 2-minimal in $\mathbb{Z}^3$ (Definition 1). Removal of any voxel from a 2-minimal surface produces a *tunnel* in the surface, thereby destroying the 2-separating property of the digital surface [9].

Between two points $p(i,j,k)$ and $p'(i',j',k')$ in $\mathbb{R}^3$, the respective *x-distance*, *y-distance*, and *z-distance* are given by $d_x(p,p') = |i - i'|$, $d_y(p,p') = |j - j'|$, and $d_z(p,p') = |k - k'|$. Consequently, the isothetic distance between $p$ and $p'$ is taken as the Minkowski norm [13], given by $d_\infty(p,p') = \max\{d_x(p,p'), d_y(p,p'), d_z(p,p')\}$. The isothetic distance of the point $p(i,j,k)$ from a surface $\Gamma$ is given by $d_\perp(p,\Gamma) = \min\{d_x(p,\Gamma), d_y(p,\Gamma), d_z(p,\Gamma)\}$; here, $d_x(p,\Gamma) = d_x(p,q)$ if there exists a (the nearest, if there is more than one) point $q(x,j,k)$ on $\Gamma$, and $\infty$ otherwise; similarly, $d_y(p,\Gamma) = d_y(p,q)$ if there exists a point $q(i,y,k)$ on $\Gamma$, and $\infty$ otherwise; and $d_z(p,\Gamma) = d_z(p,q)$ if there exists a point $q(i,j,z)$ on $\Gamma$, and $\infty$ otherwise. Following this definition, the isothetic distance between of a point $p$ from a real ellipsoid $E_3$ is given by $d_\perp(p,E_3) = \min\{d_x(p,E_3), d_y(p,E_3), d_z(p,E_3)\}$.

## 1.2  Digital Ellipsoid and Its Octants

We denote by $F_{xy}$, $F_{xz}$, and $F_{yz}$ the $xy$-, $xz$-, and $yz$-coordinate planes, respectively. We denote by $E_3(a,b,c)$ the real ellipsoid that follows Eq. 1. Its projections on $F_{xy}$, $F_{xz}$, and $F_{yz}$ are 2D ellipses, which are denoted by $E_{xy}(a,b)$, $E_{xz}(a,c)$, and $E_{yz}(b,c)$, respectively. The respective digital ellipses of these 2D ellipses are denoted by $\mathcal{E}_{xy}(a,b)$, $\mathcal{E}_{xz}(a,c)$, $\mathcal{E}_{yz}(b,c)$.

We denote by $\mathcal{E}_{xy}^{(t)}(a,b)$ the arc of the digital ellipse $\mathcal{E}_{xy}(a,b)$ lying in the $t$-th quadrant, where $1 \leqslant t \leqslant 4$. For $t = 1$ in particular, $\mathcal{E}_{xy}^{(t)}(a,b)$ contains all integer points of $\mathcal{E}_{xy}(a,b)$ with $x, y \geqslant 0$. As explained in [16,19], a digital ellipse in canonical form is 4-symmetric by constitution, as it comprises four symmetric digital arcs lying in four quadrants. Hence, each digital elliptical disc (i.e., a digital elliptical disc given by the union of the digital ellipse and its interior integer points), namely $\mathcal{E}_{xy}^*$, $\mathcal{E}_{xz}^*$, or $\mathcal{E}_{yz}^*$, is also 4-symmetric on its containing plane. An example is shown in Fig. 2.

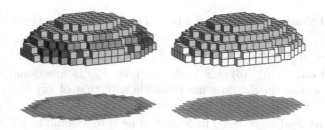

**Fig. 3.** Hemi-ellipsoid $\mathcal{H}_3(9,7,4)$. Left: yellow = voxels with one axis-parallel distance $(d_x, d_y,$ or $d_z)$ from $E_3 \leqslant \frac{1}{2}$, saffron = with two axis-parallel distances $\leqslant \frac{1}{2}$, red = with $\max\{d_x, d_y, d_z\} \leqslant \frac{1}{2}$. Right: yellow = $d_x \leqslant \frac{1}{2}$, white = $d_x > \frac{1}{2}$. (Color figure online)

We use the notation $E_3$ or $\mathcal{E}_3$ instead of $E_3(a, b, c)$ or $\mathcal{E}_3(a, b, c)$, for notational simplicity. Similar simplicity is also followed for other notations whenever it is clear from the context.

A digital ellipsoid means a set of voxels or integer points obtained by discretization/voxelization of a real ellipsoid. Based on the topological framework of discretization, a digital ellipsoid can be modeled as *naive*, *standard*, or *supercover*—the three usual models found for other geometric primitives like plane or sphere. Out of these three, naive model is the thinnest and it is the model we consider in our work. Consequently, for our work, we define a digital ellipsoid as follows.

**Definition 1.** *A digital ellipsoid $\mathcal{E}_3$ is a 2-minimal set of voxels such that $\max\limits_{p \in \mathcal{E}_3} d_\perp(p, E_3)$ is minimized.*

With respect to the three coordinate planes, a real or a digital ellipsoid can be divided into $2^3 = 8$ symmetric octants. We denote the $t$-th octant of $\mathcal{E}_3$ by $\mathcal{E}_3^{(t)}$, where $1 \leqslant t \leqslant 8$, and represent it by a 3-bit number of value $t-1$, as shown in Fig. 1. Owing to the 8-symmetry, we characterize only the first octant $\mathcal{E}_3^{(1)} (0 \leqslant x \leqslant a, 0 \leqslant y \leqslant b, 0 \leqslant z \leqslant c)$ of a digital ellipsoid, discretize it based on this characterization, and then take the reflection of the resultant set about the coordinate planes in order to construct the full ellipsoid.

## 2    Inverse Projection

The projection of a digital ellipsoid on each of the coordinate planes is a digital ellipse. Further, as we show in this section, each coordinate plane acts as a functional plane for a subset of the voxel set comprising the digital ellipsoid.

### 2.1    Functional Plane

A coordinate plane is said to be *functional to a voxel set $S$* if every two voxels in $S$ have distinct projections (pixels) on that plane [5]. We extend this to define the functional plane(s) of each voxel $p$ in $S$. For this, we denote by $\mathbb{A}_x^{(2)}(p)$ (resp., $\mathbb{A}_y^{(2)}(p)$ and $\mathbb{A}_z^{(2)}(p)$) the pair of 2-adjacent voxels of $p$ along $x$-axis (resp., along $y$- and $z$-axes). A coordinate plane, say $F_{xy}$, is said to be *functional to a voxel* $p \in S$ if and only if $\mathbb{A}_z^{(2)}(p) \cap S = \emptyset$. That is, $F_{xy}$ is functional to $p$ if and only if its projection on $F_{xy}$ does not coincide with that of any of its 2-adjacent voxels from $S$. Clearly, each of the coordinate planes is functional to $p$ if and only if it has no 2-adjacent voxel in $S$, and none is functional to $p$ if $\mathbb{A}_x^{(2)}(p) \cap S$, $\mathbb{A}_y^{(2)}(p) \cap S$, and $\mathbb{A}_z^{(2)}(p) \cap S$ are all nonempty. In Fig. 3, we have shown a hemi-ellipsoid/hemispheroid given by $\mathcal{H}_3 = \bigcup_{t=1}^{4} \mathcal{E}_3^{(t)}$. We have the following theorem on the functional-plane property of digital ellipsoid.

**Theorem 1 (Voxel functional plane).** *For each voxel of a digital ellipsoid, there exists at least one functional plane.*

*Proof.* Let $p$ be a voxel in $\mathcal{E}_3$. Assume that none of the coordinate planes is functional to $p$. So, $p$ has at least one 2-adjacent voxel in $\mathcal{E}_3$ from each of $\mathbb{A}_x^{(2)}, \mathbb{A}_y^{(2)}, \mathbb{A}_z^{(2)}$. Let these voxels be $p_x, p_y, p_z$, respectively. One such configuration (out of eight possible) is shown in the inset figure. Clearly, if $p$ is removed from $\mathcal{E}_3$, then $\mathcal{E}_3$ still remains 2-separating, as the set $\{p_x, p_y, p_z\}$ does not permit any 2-path to pass through. This means $\mathcal{E}_3$ is not 2-minimal, which contradicts Definition 1, whence the proof. □

We use Theorem 1 for construction of a digital ellipsoid using an inverse projection of the digital elliptical discs from their corresponding functional planes, that is, from $\mathbb{Z}^2$ to $\mathbb{Z}^3$. For construction of the digital elliptical discs, we use a standard algorithm, e.g., [12,16,19]. The mapping from digital ellipsoid to a elliptical disc on its functional plane is surjective in nature, wherefore an integer point of the elliptical disc does not necessarily map to a unique

voxel of the digital ellipsoid during inverse projection. Herein comes the challenge of identifying the right voxels while constructing the digital ellipsoid by inverse projection from three functional planes. For this, we define the voxel set $\mathcal{E}_3^+ = \{p \in \mathbb{Z}^3 : d_\perp(p, E_3) \leqslant \frac{1}{2}\}$. We first show that $\mathcal{E}_3^+$ is 2-separating and thereby contains $\mathcal{E}_3$ as its subset. Subsequently, on removing the simple voxels from $\mathcal{E}_3^+$, we get $\mathcal{E}_3$. We first have the following lemma for this.

**Lemma 1.** *If $p_{in}$ and $p_{ex}$ are 2-adjacent to each other with $p_{in}$ in the interior and $p_{ex}$ in the exterior or on the surface of $E_3$, then either $d_\perp(p_{in}, E_3) \leqslant \frac{1}{2}$ or $d_\perp(p_{ex}, E_3) \leqslant \frac{1}{2}$.*

*Proof.* As $p_{in}$ and $p_{ex}$ are 2-adjacent to each other, we have $d_z(p_{in}, p_{ex}) = 1$. Let, w.l.o.g., the respective coordinates of $p_{in}$ and $p_{ex}$ be $(i, j, k)$ and $(i, j, k+1)$. Hence, if $d_z(p_{in}, E_3) \geqslant \frac{1}{2}$, then $d_z(p_{ex}, E_3) \leqslant \frac{1}{2}$. Since $d_\perp(p_{ex}, E_3) \leqslant d_z(p_{ex}, E_3)$, the result follows. □

**Theorem 2 (2-separating).** *The voxel set $\mathcal{E}_3^+$ is 2-separating and hence tunnel-free.*

*Proof.* We prove by contradiction. Let us assume that $\mathcal{E}_3^+$ is not tunnel-free. Then there exists a 2-path in $\mathbb{Z}^3 \setminus \mathcal{E}_3^+$ connecting the interior and the exterior of $\mathcal{E}_3^+$. Let us consider one such 2-path $p_1 \rightsquigarrow p_n$, where $p_1$ lies inside of $E_3$ and $p_n$ on the surface or outside of $E_3$. The path $p_1 \rightsquigarrow p_n$ can be partitioned into two sub-paths: $p_1 \rightsquigarrow p_{\text{in}}$ lying inside $E_3$ and $p_{\text{ex}} \rightsquigarrow p_n$ outside $E_3$. Clearly, $p_{\text{in}}$ and $p_{\text{ex}}$ are 2-adjacent to each other. Hence, by Lemma 1, either $p_{\text{in}} \geqslant \frac{1}{2}$ or $p_{\text{ex}} \geqslant \frac{1}{2}$, which implies either $p_{\text{in}}$ or $p_{\text{ex}}$ belongs to $\mathcal{E}_3^+$—a contradiction. □

To determine the necessary and sufficient condition of deciding whether a voxel is simple in $\mathcal{E}_3^+$, we need the following theorem.

**Theorem 3 (Simpleness).** *A voxel $p$ in $\mathcal{E}_3^+$ is 'simple' if and only if*

$$\mathcal{A}_{\mathcal{E}_3^+}^{(2)}(p) := \left| \mathbf{A}_x^{(2)}(p) \cap \mathcal{E}_3^+ \right| \cdot \left| \mathbf{A}_y^{(2)}(p) \cap \mathcal{E}_3^+ \right| \cdot \left| \mathbf{A}_z^{(2)}(p) \cap \mathcal{E}_3^+ \right| \geqslant 1. \tag{2}$$

*Proof.* For the forward proof, let $p$ satisfy Eq. 2. Then by this equation, each of the sets $\mathbf{A}_x^{(2)}(p)$, $\mathbf{A}_y^{(2)}(p)$, and $\mathbf{A}_z^{(2)}(p)$ contains at least one voxel from $\mathcal{E}_3^+ \setminus \{p\}$, or equivalently, there is at least one 2-adjacent voxel of $p$ along each of the three principal (i.e., $x, y, z$) directions. Hence, removal of $p$ from $\mathcal{E}_3^+$ does not give rise to a 2-path, since the 2-adjacent voxels of $p$ along either of these directions around $p$ does not permit any 2-path to cross the surface of $E_3$. Thus, $p$ is a simple voxel.

Conversely, if $p$ is a simple voxel, then it must have at least one 2-adjacent voxel in each of the three sets, $\mathbf{A}_x^{(2)}(p) \cap \mathcal{E}_3^+, \mathbf{A}_y^{(2)}(p) \cap \mathcal{E}_3^+, \mathbf{A}_z^{(2)}(p) \cap \mathcal{E}_3^+$, because otherwise the union of these sets would contain a voxel $q$ such that $d_\perp(q, E_3) \leqslant \frac{1}{2}$, which violates the definition of $\mathcal{E}_3^+$. This sets Eq. 2 in place. □

**Theorem 4 (Digital ellipsoid).** *The voxel set comprising a digital ellipsoid is given by*

$$\mathcal{E}_3 = \left\{ p : \left( d_\perp(p, E_3) \leqslant \frac{1}{2} \right) \wedge \left( \mathbf{A}_{\mathcal{E}_3^+}^{(2)}(p) = 0 \right) \right\}. \tag{3}$$

*Proof.* Follows from the definition of $\mathcal{E}_3^+$, its 2-separating property (Theorems 2 and 3). □

## 2.2 Isothetic Distance

The relation between the isothetic distance of a voxel of a digital ellipsoid and its functional plane aids in framing the integer intervals that are required during construction of the digital ellipsoid. We first put here the following lemma.

**Lemma 2.** *The axis-parallel distances of each voxel $p(i, j, k) \in \mathcal{E}_3(a, b, c)$ from $E_3(a, b, c)$ are given as follows.*

$$d_x(p, E_3) = \left| |i| - \frac{a}{bc} \sqrt{b^2 c^2 - c^2 j^2 - b^2 k^2} \right| \qquad \text{if } F_{yz} \text{ is functional.} \tag{4a}$$

$$d_y(p, E_3) = \left| |j| - \frac{b}{ac} \sqrt{a^2 c^2 - c^2 i^2 - a^2 k^2} \right| \qquad \text{if } F_{xz} \text{ is functional.} \tag{4b}$$

$$d_z(p, E_3) = \left| |k| - \frac{c}{ab} \sqrt{a^2 b^2 - b^2 i^2 - a^2 j^2} \right| \qquad \text{if } F_{xy} \text{ is functional.} \tag{4c}$$

*Proof.* The above equations can easily be derived from the definition of distance metrics given in Sect. 1.1, using elementary algebraic steps.    □

Using Lemma 2, we get the isothetic distance for each voxel $p$ in the digital ellipsoid $\mathcal{E}_3$ from the corresponding real ellipsoid $E_3$, which is given by $d_{\perp}(p, E_3) = \min\{d_x(p, E_3), d_y(p, E_3), d_z(p, E_3)\}$. However, for efficient computation, we do not directly use this distance metric in the algorithm for construction of $\mathcal{E}_3$. Instead, we use the inverse projection from the digital elliptical discs on the three coordinate planes as follows. The *inverse projection* of a pixel in a digital elliptical disc is a voxel set that satisfies Eq. 3. Union of all these voxels obtained by inverse projection yields the required solution. The rationale lies in the following theorem.

**Theorem 5 (Inverse projection).** *If a voxel $p \in \mathcal{E}_3^{(1)}$ has more than one functional plane, then the inverse projections of its corresponding pixels from these functional planes map to a unique and same voxel in $\mathcal{E}_3^{(1)}$, which is $p$.*

*Proof.* Let, w.l.o.g., both $F_{xy}$ and $F_{xz}$ be functional to the voxel $p \in \mathcal{E}_3^{(1)}$. Let the respective projections of $p$ on these two functional planes be $p'$ and $p''$. Since inverse projection maps a pixel to one or more voxels satisfying Eq. 3, the inverse $(p')^{-1}$ of $p'$ would belong to $\mathcal{E}_3$ and thus would be non-simple. Hence, as per Eq. 4c, the $z$-coordinate of $(p')^{-1}$ can be positive or negative, which implies it is a unique voxel in $\mathcal{E}_3^{(1)}$. A similar argument holds for $p''$ too, whence $(p')^{-1} = (p'')^{-1} = p$.    □

## 3    Integer Intervals

We derive here the recurrences on integer intervals that are used in the algorithm for digital ellipsoid construction discussed in Sect. 4. As the ellipsoid is 8-symmetric, we discuss here the result for $\mathcal{E}_3^{(1)}$.

**Lemma 3.** *$F_{xy}$ is functional to a voxel $p(i, j, k) \in \mathcal{E}_3^{(1)}(a, b, c)$ if and only if $p$ is non-simple and*

$$4a^2b^2c^2 - (2abk + ab)^2 \leqslant 4c^2(b^2i^2 + a^2j^2) < 4a^2b^2c^2 - (2abk - ab)^2. \quad (5)$$

*Proof.* By Theorem 4, the two conditions "$p \in \mathcal{E}_3^{(1)}$" and "$F_{xy}$ is functional" equivalently imply that $p$ is non-simple and $d_z(p, E_3) \leqslant \frac{1}{2}$. By Lemma 2 and Eq. 4c, $d_z(p, E_3) = |k - \frac{c}{ab}\sqrt{a^2b^2 - b^2i^2 - a^2j^2}|$. So, equivalently,

$$-\frac{1}{2} \leqslant k - \frac{c}{ab}\sqrt{a^2b^2 - b^2i^2 - a^2j^2} < \frac{1}{2}$$

$$\iff k - \frac{1}{2} < \frac{c}{ab}\sqrt{a^2b^2 - b^2i^2 - a^2j^2} \leqslant k + \frac{1}{2}$$

$$\iff ab(2k - 1) < 2c\sqrt{a^2b^2 - b^2i^2 - a^2j^2} \leqslant ab(2k + 1)$$

$$\iff 4a^2b^2c^2 - (2abk + ab)^2 \leqslant 4c^2(b^2i^2 + a^2j^2) < 4a^2b^2c^2 - (2abk - ab)^2,$$

as $a, b, c, i, j, k$ are all integers.                                                          □

In line with the above theorem, the following two corollaries are symmetrically provable.

**Corollary 1.** $F_{yz}$ *is functional to a voxel* $p(i, j, k) \in \mathcal{E}_3^{(1)}(a, b, c)$ *if and only if $p$ is non-simple and*
$$4a^2b^2c^2 - (2bci + bc)^2 \leqslant 4a^2(c^2j^2 + b^2k^2) < 4a^2b^2c^2 - (2bci - bc)^2.$$

**Corollary 2.** $F_{xz}$ *is functional to a voxel* $p(i, j, k) \in \mathcal{E}_3^{(1)}(a, b, c)$ *if and only if $p$ is non-simple and*
$$4a^2b^2c^2 - (2acj + ac)^2 \leqslant 4b^2(c^2i^2 + a^2k^2) < 4a^2b^2c^2 - (2acj - ac)^2.$$

We refine Lemma 3 to deduce the recursive intervals, as stated next.

**Theorem 6.** $F_{xy}$ *is functional to a voxel* $p(i, j, k) \in \mathcal{E}_3(a, b, c)$ *if and only if $p$ is non-simple and* $4c^2(b^2i^2 + a^2j^2)$ *lies in the interval* $I_n = [u_n, v_n := u_n + l_n)$, *where* $k = c - n$, $n \geqslant 0$, *and* $u_n$ *and* $l_n$ *are given as follows.*

$$u_n = \begin{cases} 4a^2b^2c^2 - (2abc + ab)^2 & \text{if } n = 0 \\ u_{n-1} + l_{n-1} & \text{otherwise} \end{cases}$$

$$l_n = \begin{cases} 8a^2b^2c & \text{if } n = 0 \\ l_{n-1} - 8a^2b^2 & \text{otherwise} \end{cases}$$

$$(6)$$

*Proof.* We get $u_0$ and $l_0$ corresponding to $n = 0$ by substituting $k = c$ in Eq. 5. To get the recurrence of $l_n$ for $n > 0$, observe that $l_n = 4a^2b^2c^2 - (2ab(c - n) - ab)^2 - 4a^2b^2c^2 + (2ab(c - n) + ab)^2 = 8a^2b^2(c - n)$, as per Eq. 5. Hence, $l_{n-1} - l_n = 8a^2b^2(c - n + 1) - 8a^2b^2(c - n) = 8a^2b^2$. To get the recurrence of $u_n$, we substitute $k = c - n$ in Eq. 5 to get $v_{n-1} = 4a^2b^2c^2 - (2ab(c - n + 1) - ab)^2$, and substitute $k = c - n$ to get $u_n = 4a^2b^2c^2 - (2ab(c - n) + ab)^2 = v_{n-1}$. Thus, $u_n = v_{n-1} = u_{n-1} + l_{n-1}$.                                                          □

For other two functional planes, we have the following corollaries.

**Corollary 3.** $F_{yz}$ *is functional to a voxel* $p(i, j, k) \in \mathcal{E}_3(a, b, c)$ *if and only if $p$ is non-simple and* $4a^2(c^2j^2 + b^2k^2)$ *lies in the interval* $I_n = [u_n, v_n := u_n + l_n)$, *where* $i = a - n$, $n \geqslant 0$, *and* $u_n$ *and* $l_n$ *are given as follows.*

$$u_n = \begin{cases} 4a^2b^2c^2 - (2bca + bc)^2 & \text{if } n = 0 \\ u_{n-1} + l_{n-1} & \text{otherwise} \end{cases}$$

$$l_n = \begin{cases} 8b^2c^2a & \text{if } n = 0 \\ l_{n-1} - 8b^2c^2 & \text{otherwise} \end{cases}$$

$$(7)$$

**Corollary 4.** $F_{xz}$ *is functional to a voxel* $p(i, j, k) \in \mathcal{E}_3(a, b, c)$ *if and only if* $p$ *is non-simple and* $4b^2(c^2 i^2 + a^2 k^2)$ *lies in the interval* $I_n = [u_n, v_n := u_n + l_n)$, *where* $j = b - n$, $n \geqslant 0$, *and* $u_n$ *and* $l_n$ *are given as follows.*

$$
u_n = \begin{cases} 4a^2 b^2 c^2 - (2acb + ac)^2 & \text{if } n = 0 \\ u_{n-1} + l_{n-1} & \text{otherwise} \end{cases}
$$
$$
l_n = \begin{cases} 8a^2 c^2 b & \text{if } n = 0 \\ l_{n-1} - 8a^2 c^2 & \text{otherwise} \end{cases}
\tag{8}
$$

## 4    Algorithm for Digital Ellipsoid

As mentioned in Sect. 1, we consider the canonical form whereby the ellipsoid is 8-symmetric. The center can be an integer point, since it simply means a translation on the voxel set of $\mathcal{E}_3$ centered at $o := (0, 0, 0)$. For simplicity, however, we show here in Algorithm 1 the steps with center at $o$.

In Line 1 of Algorithm 1, we use three 2D arrays for $\mathcal{E}_{xy}^{*1}$, $\mathcal{E}_{yz}^{*1}$, and $\mathcal{E}_{xz}^{*1}$. Their respective sizes are $(a + 1) \times (b + 1)$, $(b + 1) \times (c + 1)$, and $(a + 1) \times (c + 1)$. They contain the pixel sets of 1st quadrants of the corresponding digital elliptical discs on $F_{xy}$, $F_{yz}$, and $F_{xz}$. These pixel sets are generated by the ellipse-drawing algorithm mentioned earlier.

In Lines 2–4, the procedure GenerateVoxels maps the pixel sets $\mathcal{E}_{xy}^{*1}$, $\mathcal{E}_{yz}^{*1}$, $\mathcal{E}_{xz}^{*1}$ to (partial) voxel sets of the 1st octant of the digital ellipsoid. Theorem 6, Corollaries 3 and 4 are used here.

A demonstration of the algorithm is shown in Fig. 4 for construction of $\mathcal{E}_3(9, 7, 4)$. The results produced by the procedure GenerateVoxels (Lines 2–4) are shown step by step. In Line 5 of Algorithm 1, the full voxel set is generated by symmetry.

Procedure GenerateVoxels first initializes the necessary parameters (Line 1–Line 3) to generate voxels from $\mathcal{E}_s^{*1}$ (here 's' signifies the coordinate plane). Procedure InitializeParameters is called for this initialization. In Line 4 of GenerateVoxels, the first voxel for $\mathcal{E}_3^{(1)}$ is added, based on the value of the octant $t$. In the outer **while** loop (Line 5), $i$ is incremented at unit step along a particular axis of $\mathcal{E}_s^{*1}$. In the **repeat-until** loop (Line 7), $j$ is incremented at

---

**Algorithm 1.** DIGITAL ELLIPSOID (int $a, b, c$)

1  Construct $\mathcal{E}_{xy}^{*1}, \mathcal{E}_{yz}^{*1}, \mathcal{E}_{xz}^{*1}$ on $F_{xy}, F_{yz}, F_{xz}$

2  $\mathcal{E}_3^{(1)} \leftarrow$ GenerateVoxels$(a, b, c, \mathcal{E}_{xy}^{*1}, 0)$

3  $\mathcal{E}_3^{(1)} \leftarrow \mathcal{E}_3^{(1)} \cup$ GenerateVoxels$(b, c, a, \mathcal{E}_{yz}^{*1}, 1)$

4  $\mathcal{E}_3^{(1)} \leftarrow \mathcal{E}_3^{(1)} \cup$ GenerateVoxels$(a, c, b, \mathcal{E}_{xz}^{*1}, 2)$

5  $\mathcal{E}_3 \leftarrow \{(i, j, k) : (|i|, |j|, |k|) \in \mathcal{E}_3^{(1)}\}$

6  **return** $\mathcal{E}_3$

---

---

**Procedure** GenerateVoxels$(a, b, c, \mathcal{E}_s^{*1}, t)$

---

1  int  $i \leftarrow 0, j \leftarrow 1, k, u, v, l, n \leftarrow 0, m, k_0, n_i \leftarrow 0, n_j \leftarrow 1, r_i, r_j, r_l$
2  InitializeParameters$(i, j, k, u, v, l, n, m, r_i, r_j, r_l, a, b, c)$
3  $m_0 \leftarrow 0, l_0 \leftarrow l, k_0 \leftarrow k$
4  **if**  $t = 0$ **then** $\mathcal{E}_3^{(1)} \leftarrow \{(0, 0, k)\}$ **else if** $t = 1$ **then** $\mathcal{E}_3^{(1)} \leftarrow \{(k, 0, 0)\}$ **else**
    $\mathcal{E}_3^{(1)} \leftarrow \{(0, k, 0)\}$
5  **while** $i \leqslant a$ **do**
6  $\quad$ **while** $(u \leqslant m) \wedge (m \leqslant v) \wedge (k \geqslant 0)$ **do**
7  $\quad\quad$ **repeat**
8  $\quad\quad\quad$ **if** $\mathcal{E}_s^{*1}[i][j] = 1$ **then**
9  $\quad\quad\quad\quad$ **if** $j = 0$ **then**
10 $\quad\quad\quad\quad\quad$ $k_0 \leftarrow k$
11 $\quad\quad\quad\quad$ **if** NotSimple$(i, j, k, t)$ **then**
12 $\quad\quad\quad\quad\quad$ **switch** $t$ **do**
13 $\quad\quad\quad\quad\quad\quad$ **case** 0
14 $\quad\quad\quad\quad\quad\quad\quad$ $\mathcal{E}_3^{(1)} \leftarrow \mathcal{E}_3^{(1)} \cup (i, j, k)$
15 $\quad\quad\quad\quad\quad\quad$ **case** 1
16 $\quad\quad\quad\quad\quad\quad\quad$ $\mathcal{E}_3^{(1)} \leftarrow \mathcal{E}_3^{(1)} \cup (k, i, j)$
17 $\quad\quad\quad\quad\quad\quad$ **case** 2
18 $\quad\quad\quad\quad\quad\quad\quad$ $\mathcal{E}_3^{(1)} \leftarrow \mathcal{E}_3^{(1)} \cup (i, k, j)$

19 $\quad\quad\quad$ $m \leftarrow m + r_j(2n_j + 1), n_j \leftarrow n_j + 1, j \leftarrow j + 1$
20 $\quad\quad$ **until** $m \geqslant v$
21 $\quad\quad$ $k \leftarrow k - 1, n \leftarrow n + 1$
22 $\quad\quad$ $u \leftarrow u + l, l \leftarrow l - r_l, v \leftarrow u + l$
23 $\quad$ $i \leftarrow i + 1, j \leftarrow 0, n_j \leftarrow 0, m_0 \leftarrow m_0 + r_i(2n_i + 1), m \leftarrow m_0, n_i \leftarrow n_i + 1, k \leftarrow k_0$
24 $\quad$ UpdateParameters$(u, v, k, n, a, b, c)$
25 $\quad$ **if** $(u > m) \vee (m \geqslant v)$ **then**
26 $\quad\quad$ $k \leftarrow k - 1$
27 $\quad\quad$ UpdateParameters$(u, v, k, n, a, b, c)$
28 $\quad$ $l \leftarrow l_0 - n r_l$
29 **return** $\mathcal{E}_3^{(1)}$

---

**Procedure** InitializeParameters$(i, j, k, u, v, l, n, m, r_i, r_j, r_l, a, b, c)$

---

1  $k \leftarrow c, u \leftarrow 4a^2 b^2 c^2 - 2(abk + ab)^2$
2  $v \leftarrow 4a^2 b^2 c^2 - 2(abk - ab)^2, m \leftarrow 4c^2(b^2 i^2 + a^2 j^2)$
3  $l \leftarrow 8a^2 b^2 c, r_l \leftarrow 8a^2 b^2$
4  $r_i \leftarrow 4c^2 b^2, r_j \leftarrow 4c^2 a^2$

---

unit step along another axis of $\mathcal{E}_s^{*1}$. And in the inner **while** loop (Line 6), $k$ is used to compute the value of the third coordinate—as the inverse projection—of the current pixel $(i, j)$ of $\mathcal{E}_s^{*1}$. The working mechanism of these loops is based on

---

**Procedure** UpdateParameters$(u, v, k, n, a, b, c)$

---

1  $u \leftarrow 4a^2b^2c^2 - 2(abk + ab)^2, v \leftarrow 4a^2b^2c^2 - 2(abk - ab)^2$
2  $n \leftarrow c - k$

---

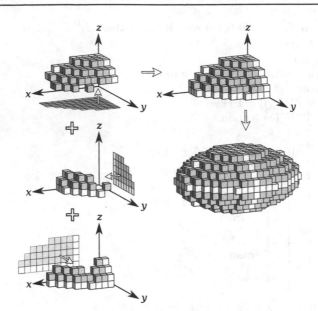

**Fig. 4.** A demonstration of our algorithm.

Theorem 6, Corollaries 3, and 4. In the inner **while** loop, the condition whether $m$ lies in the interval $[u, v)$ is verified. Inside the **repeat-until** loop, $m$ is updated with increasing value of $j$ until $m \geqslant v$. Here $\mathcal{E}_s^{*1}[i][j] = 1$ (Line 8) means the current pixel $(i, j)$ belongs to $\mathcal{E}_s^{*1}$. In Lines 13–17, the procedure NotSimple verifies whether the current voxel is simple or not; if not, then based on the value of $t$, the requisite voxel $(i, j, k)$ or $(k, i, j)$ or $(i, k, j)$ is added to $\mathcal{E}_3^{(1)}$. In Line 19, $m$ is updated as $j$ value is increased.

As initially $j = 0$, for $n_j + 1$ increments of $j$, $m$ will be increased by $r_j(2n_j + 1)$. Similarly, for $n_i + 1$ increments of $i$, $m$ will be increased by $r_i(2n_i + 1)$. After the **repeat–until** loop, $k$ is decreased by 1, and the upper bound, lower bound, and interval length of the interval are updated (Line 21–22). Outside of the inner **while** loop (Line 6), $i$ is increased by unity, and $m$ is updated accordingly. Other necessary parameters are updated in Line 23–28. UpdateParameters is called to update the necessary parameters.

## 5   Concluding Notes

We have proposed here a proven algorithm for construction of the thinnest/2-minimal model of digital ellipsoid. This is the first algorithm in the literature of digital geometry for construction of the thinnest model of ellipsoid in the integer space and we would like to make further analysis of the algorithm for deriving some tight bounds on the number of operations used in it.

As we have shown in this paper how topological analysis of digital ellipsoid results in interesting characterization, which may also be explored further for a deeper understanding of its geometric and topological properties. Although several work have been reported in recent time related to voxelization of implicit surfaces under different topological conditions, e.g., [11,17], construction of the thinnest model of digital ellipsoid in the 2-minimal topology remained an open problem, which is addressed by us in this paper. Apart from the 2-minimal model, there are other models like 'standard' and 'graceful', which are also used for voxelization of 3D primitives like line, plane, and sphere [1–6,8,10,15,18]. Designing efficient algorithms for these models of ellipsoid in the voxel space also deem to be useful and can be pursued in continuation and enhancement of the work proposed in this paper.

# References

1. Andres, E., Jacob, M.: The discrete analytical hyperspheres. IEEE Trans. Vis. Comput. Graph. **3**(1), 75–86 (1997)
2. Bera, S., Bhowmick, P., Bhattacharya, B.B.: On the characterization of absentee-voxels in a spherical surface and volume of revolution in $\mathbb{Z}^3$. J. Math. Imaging Vis. **56**(3), 535–553 (2016)
3. Biswas, R., Bhowmick, P.: From prima quadraginta octant to lattice sphere through primitive integer operations. Theoret. Comput. Sci. **624**, 56–72 (2016)
4. Biswas, R., Bhowmick, P., Brimkov, V.E.: On the polyhedra of graceful spheres and circular geodesics. Discrete Appl. Math. **216**, 362–375 (2017)
5. Brimkov, V.E., Barneva, R.P.: Graceful planes and lines. Theoret. Comput. Sci. **283**(1), 151–170 (2002)
6. Brimkov, V.E., Coeurjolly, D., Klette, R.: Digital planarity–a review. Discrete Appl. Math. **155**(4), 468–495 (2007)
7. Chamizo, F., Cristóbal, E., Ubis, A.: Lattice points in rational ellipsoids. J. Math. Anal. Appl. **350**(1), 283–289 (2009)
8. Chamizo, F., Cristóbal, E., Ubis, A.: Visible lattice points in the sphere. J. Number Theor. **126**(2), 200–211 (2007)
9. Cohen-Or, D., Kaufman, A.: Fundamentals of surface voxelization. Graph. Models Image Process. **57**(6), 453–461 (1995)
10. Fiorio, C., Toutant, J.-L.: Arithmetic discrete hyperspheres and separatingness. In: Kuba, A., Nyúl, L.G., Palágyi, K. (eds.) DGCI 2006. LNCS, vol. 4245, pp. 425–436. Springer, Heidelberg (2006). doi:10.1007/11907350_36
11. Gérard, Y., Provot, L., Feschet, F.: Introduction to digital level layers. In: Debled-Rennesson, I., Domenjoud, E., Kerautret, B., Even, P. (eds.) DGCI 2011. LNCS, vol. 6607, pp. 83–94. Springer, Heidelberg (2011). doi:10.1007/978-3-642-19867-0_7
12. Haiwen, F., Lianqiang, N.: A hybrid generating algorithm for fast ellipses drawing. In: International Conference on Computer Science and Information Processing (CSIP), pp. 1022–1025. IEEE (2012)
13. Klette, R., Rosenfeld, A.: Digital Geometry: Geometric Methods for Digital Picture Analysis. Morgan Kaufmann, San Francisco (2004)

14. Kühleitner, M.: On lattice points in rational ellipsoids: An omega estimate for the error term. Abhandlungen aus dem Mathematischen Seminar der Universität Hamburg **70**(1), 105–111 (2000)
15. Magyar, A.: On the distribution of lattice points on spheres and level surfaces of polynomials. J. Number Theor. **122**(1), 69–83 (2007)
16. Mahato, P., Bhowmick, P.: Construction of digital ellipse by recursive integer intervals. In: Normand, N., Guédon, J., Autrusseau, F. (eds.) DGCI 2016. LNCS, vol. 9647, pp. 295–308. Springer, Cham (2016). doi:10.1007/978-3-319-32360-2_23
17. Toutant, J.-L., Andres, E., Largeteau-Skapin, G., Zrour, R.: Implicit digital surfaces in arbitrary dimensions. In: Barcucci, E., Frosini, A., Rinaldi, S. (eds.) DGCI 2014. LNCS, vol. 8668, pp. 332–343. Springer, Cham (2014). doi:10.1007/978-3-319-09955-2_28
18. Toutant, J.L., Andres, E., Roussillon, T.: Digital circles, spheres and hyperspheres: From morphological models to analytical characterizations and topological properties. Discrete Appl. Math. **161**(16–17), 2662–2677 (2013)
19. Yao, C., Rokne, J.G.: Run-length slice algorithms for the scan-conversion of ellipses. Comput. Graph. **22**(4), 463–477 (1998)

# On the Chamfer Polygons
# on the Triangular Grid

Hamid Mir-Mohammad-Sadeghi$^{(\boxtimes)}$ and Benedek Nagy$^{(\boxtimes)}$

Department of Mathematics, Faculty of Arts and Sciences,
Eastern Mediterranean University, Mersin-10, Famagusta, North Cyprus, Turkey
ha.sadeghi@gmail.com, nbenedek.inf@gmail.com

**Abstract.** Weighted (or with other name, chamfer) distances on the
triangular grid was introduced recently based on the three well-known
neighborhoods. By having various values of the three used weights, the
approximation of the Euclidean disks are shown, based on the isoperimetric ratio. Our results are also compared to similar results on the
square grid. It is shown that the triangular grid, with three weights,
overperforms the quality of the approximation on the square grid by
both two and three weights (i.e., by the traditional $3 \times 3$ and the $5 \times 5$-
neighborhoods, respectively) in terms of maximal and average relative
errors.

**Keywords:** Digital distances · Chamfer distances · Digital disks ·
Approximation of the Euclidean distance · Non-traditional grids · Chamfer polygons

## 1 Introduction

Distance functions and metrics play important roles in several fields including
theoretical ones, e.g., mathematics and geometry, and also, in applications in
engineering and various disciplines related to computer science. The most usual
metric is the Euclidean distance and that is the base of Euclidean geometry.
However, in image processing and computer graphics discrete space (based on a
grid/tessellation) is preferred, and fast computation is needed. These discrete or
digital spaces have some inherently different properties from the Euclidean space.
In the Euclidean space there are infinitely many distinct points between any two
distinct points; opposite to this, there are neighbor points (pixels) in digital
spaces. The points of a discrete grid having Euclidean distance $r$ from a given
point of the grid (e.g., the Origin) do not form a circle (in the usual sense), but
usually they form a small finite set that is not connected in any sense. Therefore,
digital distances are of high importance; they are used in various applications
instead of the Euclidean distance [11]. Digital disks, in this paper, are based
on digital distance functions. We may mention here, for the completeness that
digital versions of disks can also be obtained by digitizing Euclidean circles/disks
[11,18] they are the "digitized" circles and disks and they are not the topic of
this paper.

© Springer International Publishing AG 2017
V.E. Brimkov and R.P. Barneva (Eds.): IWCIA 2017, LNCS 10256, pp. 53–65, 2017.
DOI: 10.1007/978-3-319-59108-7_5

There are three regular tessellations of the plane: the square, the hexagonal and the triangular grids. The points (pixels) of these regular grids are usually addressed by integer coordinate values. In the square grid, two independent coordinates are used. The pixels of the hexagonal grid can be addressed with two integers [12], or with a more elegant solution, with three coordinate values whose sum is zero reflecting the symmetry of the grid [10,13]. Similarly, in the triangular grid three coordinate values can effectively be used which are not linearly independent [15,16,29]; and in this way, the three types of neighborhood ([7], see also Fig. 1), are easily captured in a mathematical way.

Digital distances are path-based and they are defined by connecting pixels/points by paths through neighbor pixels/points. The cityblock and the chessboard distances [27], the first two digital distances, are based on the number of steps connecting the points where 4-neighbor or 8-neighbor pixels are considered in each step on the square grid, respectively. Since they are very rough approximations of the Euclidean distance, the theory of digital distances are developed in various ways. As already recommended in, as a kind of alternating use of the two neighborhoods, the neighborhood sequences allow that the steps may vary in a path [5,19]. In this way, a family of octagonal distances is obtained, with octagons as digital disks [6,9]. Weighted or chamfer distances were also introduced to have a good approximation to the Euclidean distance, and at the same time, to have low computational cost, for e.g., distance transforms [1,2,28,30]. It is well known that the approximation of the Euclidean disk/distance becomes better and better when larger and larger neighborhood is allowed, i.e., a larger number of weighted steps are used (see, e.g., [3]). With the traditional two neighborhoods, with two weights (one for the cityblock, and other for the diagonal movements) the obtained disks are octagons. Instead of this, $3 \times 3$ neighborhood, $5 \times 5$ neighborhood is introduced and used giving a third weight on knight movements. In this way, the digital disks become hexadecagons. Further, by $7 \times 7$ neighborhood and 5 weights, 32-gons are obtained, etc. We refer to [3] where optimal weights are computed for various sizes of neighborhood. We just mention here, that the weighted distances and the neighborhood sequences could also be mixed, e.g., the weight sequences were introduced in [26] to have an errorless estimation on a perimeter of a square (with enough large number of weights, but with only two types of neighborhood, i.e., using $3 \times 3$ neighborhood only).

Another way to obtain better digital distances, e.g., the lower their rotational dependency, is based on non-traditional grids. Both the hexagonal and the triangular grids have better symmetric properties than the square grid has: they have more symmetry axes and rotations with smaller angles already transform the grid into itself. The theory of distances based on neighborhood sequences on the triangular grid is also well developed. Digital circles/disks and their types are analyzed in [17]; while the approximation of the Euclidean circles/distance is done in [25] using the dual grid notation. The weighted distances have also been investigated, recently, on the triangular grid [21].

Some of the goodness measures of digital distances used in various applications give values how good are the approximations of the Euclidean distance by them [4]. It can be done by measuring the compactness ratio of the polygons of the digital disks obtained by digital distances. It is known that the (Euclidean) circles/disks are the most compact objects in the plane, the ratio of the perimeter square over the area, is $4\pi \approx 12.566$ for them, the smallest among all objects'. By measuring this value for the digital disks, the approximation of the Euclidean distance is measured. The highest compactness of the circles can also be used to define another type of digital disks: the most compact grid objects try to inherit this characteristic property of the Euclidean circles/disks; they are characterized in [22,23,31] on various grids. Digital disks (spheres) are analyzed in [14] in $n$D rectangular grids based on weighted distances. Other frequently used measure is the maximal absolute error and its normalized version [2,3] comparing the chamfer polygon to the Euclidean disk. In this paper, the computation of chamfer polygons (digital disks), and some notes and comparisons on the approximation of the Euclidean distance are provided.

## 2   The Triangular Grid

In this section, we briefly recall the description of the triangular grid and the definition of weighted distances.

The triangular grid is a regular tessellation of the plane with same size equilateral triangles. Actually, it is not a lattice, since there are grid vectors that do not transform the grid to itself. This is due to the fact that there are two types of orientations of the triangles. The grid is described by three coordinate axes $x, y$, and $z$ (see Fig. 1, right). In this paper we refer for the triangle pixels, as points, and usually, we will use their center, i.e., the dual, hexagonal (also knows as honeycomb) grid notation. Each point of grid is described by a unique coordinate triplet using only integer values. However, the three values are not independent: the sum of coordinate values can be 0 (even point, shape $\triangle$) or 1 (odd point, $\triangledown$). The vector through the mid-point of the edge to the opposite corner point is parallel/antiparallel with one of the axes (see also Fig. 1, right). Further we refer to the set of points of the triangular grid by $\mathbb{Z}_*^3$ ($\mathbb{Z}_*^3 = \{(x, y, z) | x + y + z \in \{0, 1\}\}$).

There are three different types of widely used neighborhoods on the triangular grid. (In the rectangular grid there are only two types of basic neighborhood.) Two distinct points (triangles) are 1-neighbors iff they have a side in

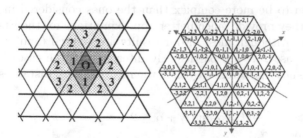

**Fig. 1.** Type of neighbors (left); coordinate system for the triangular grid (right).

common; they are strict 2-neighbors if there is a triangle which is 1-neighbor of both of them; further, the two points are strict 3-neighbors iff they share a corner, but they are not 1- and 2-neighbors. Two pixels are 2-neighbor if they are strict 2-neighbors or 1-neighbors, and two pixels are 3-neighbors if they are strict 3-neighbors or 2-neighbors. Formally: Let $p = (p(1), p(2), p(3))$ and $q = (q(1), q(2), q(3))$ be two distinct points of $\mathbb{Z}_*^3$, they are $m$-neighbors $(m = 1, 2, 3)$, if the following two conditions hold:

(1) $|p(i) - q(i)| \leq 1$ for every $i \in \{1, 2, 3\}$, and

(2) $\sum_{i=1}^{3} |p(i) - q(i)| \leq m$.

Equality in the second equation defines the strict $m$-neighborhood relation. In the following we describe some notations which are needed later on. By $\alpha$-movement, i.e., by movement with weight $\alpha$, we denote a step from a point to one of its 1-neighbors; by $\beta$-movement (i.e., by weight $\beta$) a step to a strict 2-neighbor, and similarly by $\gamma$-movement a step to a strict 3-neighbor point. We say that two lanes are parallel if the same coordinate is fixed, e.g., the lane $\{p(p(1), 0, p(3))\}$ is parallel to $\{p(p(1), 3, p(3))\}$; both of them are perpendicular to axis $y$.

Let $p = (p(1), p(2), p(3))$ and $q = (q(1), q(2), q(3))$ be two 3-neighbor points, then the number of their coordinate differences gives the order of their strict neighborhood (as we have defined above). Weighted distances are path based distances, thus we need paths connecting the points. A path from $p$ to $q$ can be defined by a finite point sequence $p = p_0, ..., p_n = q$ in which the points $p_{i-1}$ and $p_i$ are 3-neighbors for every $1 \leq i \leq n$. Thus this path can also be seen as sequence of $n$ steps, such that in each step we move to a 3-neighbor point of the previous one. The weight of the path is equal to $\alpha n_1 + \beta n_2 + \gamma n_3$, where $n_i$ is the number of steps to strict $i$-neighbors in the path $(n = n_1 + n_2 + n_3)$. There are several different paths from $p$ to $q$ with various weights. The weighted distance $d(p, q; \alpha, \beta, \gamma)$ of $p$ and $q$ with weights $\alpha, \beta, \gamma$ is the sum of weights of a/the minimal weighted path between $p$ and $q$. In this paper, the natural condition of the weights, that is $0 < \alpha \leq \beta \leq \gamma$, used. There are various cases regarding the relative ratio of the weights (see [21]).

In this paper, somewhat complementing the results of [24], those cases are considered which allow to use 1-steps and also 3-steps in the same paths. These cases are proven to be more complex than the ones considered in [24] and, on the other side, they provide much better approximations for the Euclidean disks as we will show. Although most of our results are general (and we prove them for the general case), we are particularly interested in the cases for which the digital disks are not characterized yet: in the next sections, distances with weight conditions:

- $2\alpha > \beta$, $3\alpha > \gamma$, $\alpha + \beta > \gamma$ and $\gamma + \alpha \leq 2\beta$
- $2\alpha \leq \beta$ and $3\alpha > \gamma$

are considered.

# 3   Preliminaries: Technical Notions and Notations

Now we recall our central concept: the chamfer balls (chamfer polygons) or digital disks are defined as

$$D(o, r; \alpha, \beta, \gamma) = \{p \mid d(o, p; \alpha, \beta, \gamma) \le r\}.$$

Obviously they depend not only on the radius $r$, but also on the used weights $\alpha, \beta, \gamma$. The centre of the disk is the point $o$. Since we work in the dual representation of the triangular grid, the elements of $\mathbb{Z}_*^3$ will refer for the center points of the triangle pixels.

**Proposition 1.** *Let the length of the sides of each equilateral triangle on a grid be one unit. Let the coordinate axes of the triangular grid go through on the origin (as in Fig. 4). Let $\bar{p} = (x, y)$ be the Cartesian coordinate pair of the point in the middle of the triangle addressed by $p = (p(1), p(2), p(3))$. Then the Cartesian coordinates of $\bar{p}$ can be obtained in the following way:*

- *if $p$ is a even point ($p(1) + p(2) + p(3) = 0$), then $x = \frac{p(1) - p(3)}{2}$ and $y = -\frac{\sqrt{3}}{2} p(2)$.*
- *if $p$ is a odd point ($p(1) + p(2) + p(3) = 1$), then $x = \frac{p(1) - p(3)}{2}$ and $y = -\frac{\sqrt{3}}{2} p(2) + \frac{\sqrt{3}}{6}$.*

*Proof.* It is a simple geometrical calculation.                                        □

In this way, we can easily define the convex hull of any (finite) set $X$ of grid points: let $\bar{X} \subset \mathbb{R}^2$ be the convex polygon with smallest area such that each point of $X$ is included in $\bar{X}$. Formally:

$$\bar{X} = \left\{ \sum_{j=1}^{n} \lambda_j \bar{p}_j \,\middle|\, \sum_{j=1}^{n} \lambda_j = 1, \lambda_j \ge 0 \text{ and } p_j \in X \right\}.$$

Further, the digital set $X \subset \mathbb{Z}_*^3$ is (digitally) *H-convex* if $X = \bar{X} \cap \mathbb{Z}_*^3$ [8,11]. In this paper we are interested in convex hulls of digital disks: $\bar{D}$.

Let $l_{p_1 p_2}$ be a straight line segment between $\bar{p}_1$ and $\bar{p}_2$. Let $m_{p_1 p_2}$ denote the slope of that line segment connecting $\bar{p}_1$ and $\bar{p}_2$. Further, let $S_{\triangle poq} = \{p' \mid \bar{p}' \text{ is inside or on the border of triangle } \triangle \bar{p} \bar{o} \bar{q} \text{ and } p' \ne p, p' \ne q\}$.

Let $L_{x, -y}$ denote the half-lane $\{p = (p(1), p(2), 0) \in \mathbb{Z}_*^3 \mid p(2) \le 0\}$, that is actually a half lane perpendicular to axis $z$ (between the positive part of axis $x$ and the negative part of axis $y$) starting from $o$. Let $L_x$ denote the half diamond chain $\{p = (p(1), -\lfloor \frac{p(1)}{2} \rfloor, -\lfloor \frac{p(1)}{2} \rfloor) \in \mathbb{Z}_*^3 \mid p(1) \ge 0\}$ which is a diamond chain lying on the nonnegative part of axis $x$, see also Fig. 2(a).

Further, we define other subsets of the triangular grid:
$S_{L_x, L_{x, -y}} = \{p = (p(1), p(2), p(3)) \mid p(1) \ge 0, p(2) \le p(3) \le 0\}$ that is the set of points between $L_x$ and $L_{x, -y}$ including the points on the borders, see Fig. 2(b).

Similarly, $S_{L_{x,-y},L_{-y}} = \{p = (p(1), p(2), p(3)) \mid p(2) \leq 0, p(1) \geq p(3) \geq 0\}$ as it is shown in Fig. 2(c). Let $F = S_{L_x,L_{x,-y}} \cup S_{L_{x,-y},L_{-y}}$ denote one sixth of the triangular grid and let $W(r; \alpha, \beta, \gamma) = D(o, r; \alpha, \beta, \gamma) \cap F$ the part of the disc inside this sixth. Now, let

$S_\alpha = \{p = (p(1), p(2), p(3)) \mid r - \alpha < d(o, p; \alpha, \beta, \gamma) \leq r, p \in F\}$.

Let $S_{1\alpha} = \{p \in W(r; \alpha, \beta, \gamma) \mid p$ is 1-neighbor of $q$ and $q \in S_\alpha\} \backslash S_\alpha$.

Let $S_{2\alpha} = \{p \in W(r; \alpha, \beta, \gamma) \mid p$ is 1-neighbor of $q, q \in S_{1\alpha}\} \backslash (S_\alpha \cup S_{1\alpha})$.

Further, this notation is continued for

$S_{n\alpha} = \left\{p \in W(r; \alpha, \beta, \gamma) \mid p \text{ is 1-neighbor of } q, q \in S_{n-1\alpha}\right\} \backslash \left(S_{n-1\alpha} \cup S_{n-2\alpha}\right).$

Let $S_{\alpha_1} = \{p \notin D(0, r; \alpha, \beta, \gamma) \mid p$ is 1-neighbor of $q, p \in F$ and $q \in S_\alpha\}$, see Fig. 2 (d). Let $V$ be the set of corner points $p$ of the polygon (i.e., $\bar{p}$ of the convex hull $\bar{D}$) in $F$. Since the symmetry of the grid, the digital disks and also their convex hulls are symmetric. Therefore, in the following lemmas, we consider only points in $F$ which gives a sixth of the grid such that each digital disk (and its convex hull) has six similar, rotated parts. See, e.g., [20] for rotations and other isometric transformations of the grid.

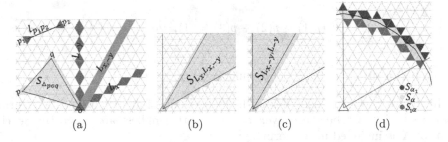

(a)         (b)         (c)         (d)

**Fig. 2.** Various notations: (a) $S_{\triangle_{poq}}$, $l_{p_1p_2}$, $L_{-y}$, $Lx, -y$ and $L_x$ (b) $S_{L_x, Lx, -y}$ (c) $S_{L_{x,-y}, L_{-y}}$ (d) $S_{1\alpha}$, $S_\alpha$ and $S_{\alpha_1}$

Let us assume that the sidelength of the triangle pixels be 1 unit (in the Euclidean plane $\mathbb{R}^2$). By using the dual grid notation, i.e., the center of each triangle pixel instead of the pixel itself, a movement to a

– 1-neighbor means a step by length $\frac{\sqrt{3}}{3} \approx 0.57735$,
– to a strict 2-neighbor means a step by length exactly 1, and
– to a strict 3-neighbor means a step by length $\frac{2\sqrt{3}}{3} \approx 1.1547$.

One may observe that in the triangular grid, the step to a strict 3-neighbor has length twice than the length of a step to a 1-neighbor. In some of our examples the sidelength of the triangles, i.e., the length of the strict 2-steps may be chosen not to be exactly 1 (unit).

# 4  Digital Disks and Their Corners

In this section digital disks are described as convex hulls of the point sets reached with paths having weights at most a given value, called, radius. This way, the usage of dual representation of the triangular grid, i.e., the vertices of the hexagonal grid, allows us to compute standard geometric measures, such as the usual perimeter and area of these objects. Thus let us consider the centers of the pixels instead of them (as we have shown already). The weights $\alpha$, $\beta$ and $\gamma$ can have various relations defining various cases. In different cases different paths become optimal, i.e., with minimum weight, and thus, the formula for the distance depends on the considered case. In some cases, only hexagons, enneagons and dodecagons can be obtained [21,24]. Note that in the cases considered in [24], digital disks with at most twelve corners are obtained. The formulae for computing the weighted distance may depend not only on the weights, but on the type (parity) of the points.

Let $\bar{D}(o, r; \alpha, \beta, \gamma)$ denote the convex hull of $D(o, r; \alpha, \beta, \gamma)$ in $\mathbb{R}^2$. Since various formulae are used for computing the distance $d$, convex hulls can be obtained in various shapes. The relative values of the weights define the various cases. In the following part we describe some of the shapes of the possible objects $\bar{D}(o, r; \alpha, \beta, \gamma)$. One of the characterization of these digital disks goes by measuring their side lengths $l$, perimeters $P$, areas $A$, and thus, their isoperimetric ratios $\kappa$. The isoperimetric ratio is defined as $\kappa = \frac{P^2}{A}$ which can be used to compare and approximate the Euclidean circle.

Let $d_p^E$ be the Euclidean distance of $\bar{o} = (0, 0)$ and $\bar{p} = (x, y)$, then, obviously, $d_p^E = \sqrt{x^2 + y^2}$.

**Proposition 2.** *Let* $p \in S_{1}\alpha$ *and* $q \in S_\alpha$, *then, obviously,* $d(o, p; \alpha, \beta, \gamma) < d(o, q; \alpha, \beta, \gamma)$. *If* $p$ *and* $q$ *are 1-neighbors, then* $d_p^E < d_q^E$.

*Proof.* Suppose to the contrary that $d_p^E > d_q^E$, then either

- $d(o, p; \alpha, \beta, \gamma) = d(o, q; \alpha, \beta, \gamma) + \alpha > r$ or
- $d(o, p; \alpha, \beta, \gamma) = d(o, q; \alpha, \beta, \gamma) - \alpha + \beta > r - \alpha$ or
- $d(o, p; \alpha, \beta, \gamma) = d(o, q; \alpha, \beta, \gamma) - \beta + \gamma > r - \alpha$.

In each of these cases $d(o, p; \alpha, \beta, \gamma) > r$ or $d(o, p; \alpha, \beta, \gamma) > r - \alpha$, hence $p \in S_\alpha$ or $p \notin D(o, r; \alpha, \beta, \gamma)$ which made contradiction with the assumption.  □

**Lemma 1.** *If* $p \in F$ *is a corner of* $\bar{D}(o, r; \alpha, \beta, \gamma)$ *(i.e.,* $p \in V$*), then* $p \in S_\alpha$.

*Proof.* We show that any point $p \notin S_\alpha$ is not in $V$. Suppose that $p \in S_{1}\alpha$. Let $q$ be a 1-neighbor of $p$ such that $q \in S_\alpha$. From Proposition 2, it is clear that $d_p^E < d_q^E$, therefore $p$ cannot be a corner. With the same process it can be shown that the other points of $S_{2}\alpha \cup S_{2}\alpha \cup \cdots \cup S_{n}\alpha$ ($S_{n}\alpha$ is the origin) have distance less than $r$, but they cannot be a corner. Now suppose $p \in S_{\alpha_1}$ and $q \in S_\alpha$. Since $p \notin D(o, r; \alpha, \beta, \gamma)$, therefore $d(o, p; \alpha, \beta, \gamma) > r$ and it cannot be a corner. Hence each corner point (in $F$) belongs to $S_\alpha$.  □

**Lemma 2.** *For any point $p \in D(o, r; \alpha, \beta, \gamma)$, $p \notin V$ if and only if there exist two distinct points $p_1, p_2 \in D(o, r; \alpha, \beta, \gamma)$, $(p_1 \neq p, p_2 \neq p)$ such that $p \in S_{\triangle_{p_1 o p_2}}$.*

*Proof.* Suppose $p \in S_{\triangle_{p_1 o p_2}}$ $(p \neq p_1, p \neq p_2)$. It is clear that in this case $p \notin V$, i.e., it is not a corner.

Now suppose $p \in D(o, r; \alpha, \beta, \gamma)$, $p \notin V$. Considering the geometric shape of $D(o, r; \alpha, \beta, \gamma) = S_{\triangle_{p_1^c o p_2^c}} \cup \cdots \cup S_{\triangle_{p_{n-1}^c o p_n^c}}$, with $p_i^c \in V$ (for $i = 1, \ldots, n$ where $n$ is the number of corners). Hence, $\exists i$ such that $p \in S_{\triangle_{p_i^c o p_{i+1}^c}}$. □

**Theorem 1.** $\bar{D}(o, r; \alpha, \beta, \gamma)$ *is H-convex with every possible parameters $\alpha, \beta$ and $\gamma$.*

*Proof.* Suppose that for a given $\alpha, \beta, \gamma$ and $r$ the convex hull $\bar{D}$ is not H-convex. Then, by definition, $D \neq \bar{D} \cap \mathbb{Z}_*^3$ and $\exists p, q \in D(o, r; \alpha, \beta, \gamma)$ and $p_1, p_2, p_3 \in V$ such that $l_{pq} \cap l_{p_1 p_2} \neq \emptyset$ and $l_{pq} \cap l_{p_2 p_3} \neq \emptyset$ (see Fig. 3). Since $\exists p' \in S_{\triangle_{p_1 o p_3}}$ such that $p' \notin D(o, r; \alpha, \beta, \gamma)$ and that makes contradiction with Lemma 2, hence $\bar{D}$ is H-convex. □

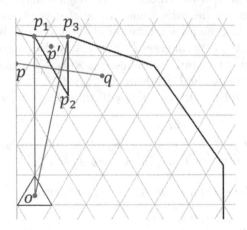

**Fig. 3.** Example for $p' \in S_{\triangle_{p_1 o p_3}}$ but $p' \notin D(o, r; \alpha, \beta, \gamma)$. (Note that in this example the used shape is not a disk.)

**Lemma 3.** *Suppose $p_1, p_2, p_3 \in S_\alpha$ and $m_{op_1} < m_{op_2} < m_{op_3}$. If $m_{p_2 p_3} \leq m_{p_1 p_2} < 0$, then $p_2 \notin V$.*

*Proof.* If $m_{p_2 p_3} = m_{p_1 p_2}$ then $p_2$ is on the line between $p_1$ and $p_3$, therefore $p_2$ is not a corner. Suppose, now, $m_{p_2 p_3} < m_{p_1 p_2}$. Since $p_1, p_2, p_3 \in S_\alpha$, then $p_1, p_2, p_3 \in D(o, r; \alpha, \beta, \gamma)$ and $p_2 \in S_{\triangle_{p_1 o p_2}}$ (consider Fig. 4), therefore by Lemma 2, $p_2 \notin V$. □

*Remark 1.* If $p_1, p_2, p_3 \in S_\alpha$ and $m_{p_2 p_3} < m_{p_1 p_2}$, then from geometry it is clear that $m_{p_2 p_3} < m_{p_1 p_3} < m_{p_1 p_2}$ (see Fig. 4).

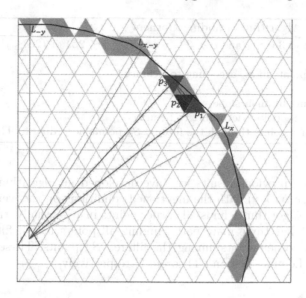

**Fig. 4.** Example for $p_2 \in S_{\triangle_{p_1 o p_3}}$, hence it cannot be a corner.

**Theorem 2.** *Let $p \in V$ and $q \in S_\alpha, m_{op} < m_{oq}$. If $\forall p' \in S_\alpha, m_{pq} < m_{pp'}$ and $m_{op} < m_{op'}$, then $q \in V$ and $\nexists p^c \in V$ such that $m_{op} < m_{op^c} < m_{oq}$.*

*Proof.* Suppose to the contrary that $\exists p^c \in V$ such that $m_{op} < m_{op^c} < m_{oq}$, then on the basis of Lemma 3, $m_{pp^c} < m_{p^c q}$ and by Remark 1, $m_{pp^c} < m_{pq}$ (see Fig. 5(a)) which makes contradiction with the assumption. Now suppose to the contrary that $q$ is not a corner and by Lemma 2 there exists $p_1^c \in V$ such that $q \in S_{\triangle_{p o p_1^c}}$, $m_{op} < m_{op_1^c}$ and $l_{pp_1^c}$ is a side of $D(o, r; \alpha, \beta, \gamma)$. Since $m_{pq}$ is minimum slope, $m_{pq} < m_{pp_1^c}$, this makes a contradiction with $q \in S_{\triangle_{p o p_1^c}}$ (consider Fig. 5(b)). Hence $q \in V$. □

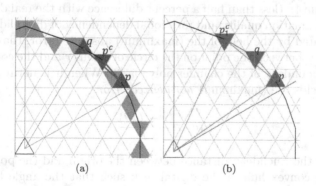

(a)                                    (b)

**Fig. 5.** Comparing slopes (a) $m_{pp^c} < m_{pq}$ (b) $m_{pq} < m_{qp_1^c}$, hence $q \notin S_{\triangle_{p o p_1^c}}$.

---

**Algorithm 1.** Algorithm of finding Corners

1: **if** $(2\alpha > \beta$ and $3\alpha > \gamma$ and $\alpha + \beta > \gamma$ and $\alpha + \gamma \leq 2\beta)$ or $(2\alpha \leq \beta$ and $3\alpha > \gamma)$
    **then**
2:     Find all points in $S_\alpha$ according to Lemma 1
3: **end if**
4: Sort the points by their slope
5: Filtering the $S_\alpha$ points according to Lemma 3
6: Find all corner points through the filtered $S_\alpha$ points according to Theorem 2

---

Based on the previous results we have developed an algorithm to find the corners of $\bar{D}(o, r; \alpha, \beta, \gamma)$ with the mentioned conditions on the weights for any nonnegative $r$: Algorithm 1 finds all corner points in $F$, actually, the set $V$.

Algorithm 1 has three main parts, finding all points in $S_\alpha$, filtering them, and finding all corner points among the filtered $S_\alpha$ points. These three parts work based on Lemmas 1, 3 and Theorem 2, respectively.

## 5   Approximation of Euclidean Disks

Digital disks in the triangular grid of various weights and radius are considered, they can have many different shapes with various number of corners. In this section we show some interesting ones. It is known that, in the square grid, by considering three weights ($5 \times 5$ neighborhood with 24 local neighbors) the digital disks have 16 corners [3] (in special cases the polygon may be degenerated, e.g., if all weights equal to each other, specially, a square is obtained). By five weights, i.e., $7 \times 7$ neighborhood, (maybe degenerated) 32-gons are obtained. Comparing them to the digital disks on the triangular grid, we present the convex hull of the digital disks $D(o, 723; 8, 15, 18)$ having a large number of corners, actually it has 63 corners.

One of the "goodness measures" of the digital disks is their isoperimetric ratio. The digital disk $D(o, 892; 29, 56, 68)$ has 42 corners and its isoperimetric ratio is 12.628 (see Fig. 6), which measure is just 1.0049 times larger than the optimal Euclidean value $4\pi$ (less than half a percent difference with the real disk).

The approximation quality and rotational dependency of digital distances are usually measured by the help of the (maximum) absolute error, that is, the distance difference between the point on the circle of radius $r$ and the corresponding point on the boundary of the chamfer polygon [3]. When it is normalized by the radius $r$, it yields the (maximum) *normalized error*:

$$E(\theta) \equiv \frac{r - L(\theta)}{r},$$

where $L(\theta)$ is the Euclidean distance between the origin and the point $p$ on the border of the convex hull of the digital disk such that the angle between the line connecting the origin to point $p$ and axis $x$ is $\theta$. Normalized error is usually measured in percentages. For $5 \times 5$ neighborhood, i.e., with 3 weights, in [3] the

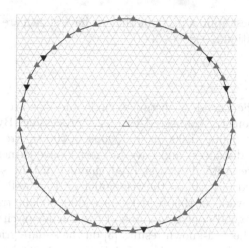

**Fig. 6.** The digital disk $D(o, 892; 29, 56, 68)$ with $\kappa \approx 12.628$.

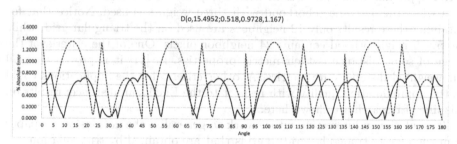

**Fig. 7.** The rotational dependency of the distances on the square and on the triangular grids with three weights: the relative error in the function of the angle $\theta \in [0, \pi]$. The solid graph (our results in the triangular grid) and dotted graph (best result on the square grid, [3]).

optimal weights are computed to obtain minimal normalized error. For these weights disks with average error 0.699% and with maximal relative error 1.356% are obtained, this latter occurs at 13.5°. The value of this relative error as the function of the degree $\theta$ can be seen in Fig. 7 with dotted line.

In the triangular grid, with three weights, considering the usual 12 neighbors of a pixel, the digital disk $D(o, 892; 29, 56, 68)$ results the average error 0.464% and the maximum relative error 0.792% that occurs at 46.1°. See also Fig. 7, where the value of the error is shown for various direction comparing it also to the previously mentioned best known case for the square grid.

Another digital disk $D(o, 1116; 29, 56, 68)$ has 48 corners and its isoperimetric ratio is $\kappa \approx 12.636$. It has the average error 0.577% and the maximum relative error is 1.119% that occurs at 2.91°. Observe that the maximal error in this example occurs in a different angle than at our previous example.

Notice that the best approximations that we have shown are obtained approximately at the condition $\alpha + \gamma = \sqrt{3} \cdot \beta$.

## 6   Conclusions

Approximation of the Euclidean distance, circle and disk are frequent topics of papers in digital geometry connected to image processing. By using neighborhood sequences, the digital disks are octagons in the square grid [5,6] (hence the name octagonal distances) and have at most twelve corners in the triangular grid [17]. The literature about weighted/chamfer distances is also rich. This concept has appeared recently on the triangular grid, as well. In this paper, we have continued the work on this field by providing weight triplets and digital disks that give very good approximations of the Euclidean circle/disk. To compare our results to the well-known results [3] on the square grid, we have shown that by the traditional three neighbor relations, on the triangular grid, much better approximations can be done than by optimal weights on the square grid, even when the $5 \times 5$ neighborhood with three weights is used. Notice that for obtaining these results only 12 neighbors are used on the triangular grid, while the $5 \times 5$ neighborhood contains 24 neighbor pixels. One of the criteria of the usage of some distance is their metric properties. In [21] it is proven that all our weighted distances are metrics. Another criteria for the applications could be the approximation quality with respect to the Euclidean distance; results on this line of research are shown here.

It would be an interesting future work to generalise the notions of the paper from the regular triangular grid to grids that are obtained by triangulation.

## References

1. Borgefors, G.: Distance transformations in arbitrary dimensions. Comput. Vis. Gr. Image Process. **27**(3), 321–345 (1984)
2. Borgefors, G.: Distance transformations in digital images. Comput. Vis. Graph. Image Process. **34**(3), 344–371 (1986)
3. Butt, M.A., Maragos, P.: Optimum design of chamfer distance transforms. IEEE Trans. Image Process. **7**(10), 1477–1484 (1998)
4. Celebi, M.E., Celiker, F., Kingravi, H.A.: On Euclidean norm approximations. Pattern Recogn. **44**(2), 278–283 (2011)
5. Das, P.P., Chakrabarti, P.P., Chatterji, B.N.: Generalised distances in digital geometry. Inform. Sci. **42**, 51–67 (1987)
6. Das, P.P., Chatterji, B.N.: Octagonal distances for digital pictures. Inform. Sci. **50**, 123–150 (1990)
7. Deutsch, E.S.: Thinning algorithms on rectangular, hexagonal and triangular arrays. Comm. ACM **15**, 827–837 (1972)
8. Eckhardt, U.: Digital lines and digital convexity. In: Bertrand, G., Imiya, A., Klette, R. (eds.) Digital and Image Geometry. LNCS, vol. 2243, pp. 209–228. Springer, Heidelberg (2001). doi:10.1007/3-540-45576-0_13
9. Farkas, J., Baják, Sz., Nagy, B.: Notes on approximating the Euclidean circle in square grids. Pure Math. Appl. - PU.M.A. **17**, 309–322 (2006)

10. Her, I.: Geometric transformations on the hexagonal grid. IEEE Trans. Image Proc. **4**, 1213–1221 (1995)
11. Klette, R., Rosenfeld, A.: Digital geometry. Geometric methods for digital picture analysis. Morgan Kaufmann Publishers, Elsevier Science B.V. (2004)
12. Luczak, E., Rosenfeld, A.: Distance on a hexagonal grid. Trans. Comput. **C–25**(5), 532–533 (1976)
13. Middleton, L., Sivaswamy, J.: Hexagonal Image Processing: A Practical Approach. Springer, London (2005)
14. Mukherjee, J.: Hyperspheres of weighted distances in arbitrary dimension. Pattern Recogn. Lett. **34**, 117–123 (2013)
15. Nagy, B.: Metrics based on neighbourhood sequences in triangular grids. Pure Math. Appl. - PU.M.A. **13**, 259–274 (2002)
16. Nagy, B.: Shortest path in triangular grids with neighbourhood sequences. J. Comput. Inf. Technol. **11**, 111–122 (2003)
17. Nagy, B.: Characterization of digital circles in triangular grid. Pattern Recogn. Lett. **25**, 1231–1242 (2004)
18. Nagy, B.: An algorithm to find the number of the digitizations of discs with a fixed radius. Electron. Notes Discrete Math. **20**, 607–622 (2005)
19. Nagy, B.: Distance with generalized neighbourhood sequences in $n$D and $\infty$D. Disc. Appl. Math. **156**, 2344–2351 (2008)
20. Nagy, B.: Isometric transformations of the dual of the hexagonal lattice. In: ISPA 2009, Salzburg, Austria, pp. 432–437 (2009)
21. Nagy, B.: Weighted distances on a triangular grid. In: Barneva, R.P., Brimkov, V.E., Šlapal, J. (eds.) IWCIA 2014. LNCS, vol. 8466, pp. 37–50. Springer, Cham (2014). doi:10.1007/978-3-319-07148-0_5
22. Nagy, B., Barczi, K.: Isoperimetrically optimal polygons in the triangular grid. In: Aggarwal, J.K., Barneva, R.P., Brimkov, V.E., Koroutchev, K.N., Korutcheva, E.R. (eds.) IWCIA 2011. LNCS, vol. 6636, pp. 194–207. Springer, Heidelberg (2011). doi:10.1007/978-3-642-21073-0_19
23. Nagy, B., Barczi, K.: Isoperimetrically optimal polygons in the triangular grid with Jordan-type neighbourhood on the boundary. Int. J. Comput. Math. **90**, 1629–1652 (2013)
24. Nagy, B., Mir-Mohammad-Sadeghi, H.: Digital disks by weighted distances in the triangular grid. In: Normand, N., Guédon, J., Autrusseau, F. (eds.) DGCI 2016. LNCS, vol. 9647, pp. 385–397. Springer, Cham (2016). doi:10.1007/978-3-319-32360-2_30
25. Nagy, B., Strand, R.: Approximating Euclidean circles by neighbourhood sequences in a hexagonal grid. Theoret. Comput. Sci. **412**, 1364–1377 (2011)
26. Nagy, B., Strand, R., Normand, N.: A weight sequence distance function. In: Hendriks, C.L.L., Borgefors, G., Strand, R. (eds.) ISMM 2013. LNCS, vol. 7883, pp. 292–301. Springer, Heidelberg (2013). doi:10.1007/978-3-642-38294-9_25
27. Rosenfeld, A., Pfaltz, J.L.: Distance functions on digital pictures. Pattern Recogn. **1**, 33–61 (1968)
28. Sintorn, I.-M., Borgefors, G.: Weighted distance transforms in rectangular grids. ICIAP **2001**, 322–326 (2001)
29. Stojmenovic, I.: Honeycomb networks: topological properties and communication algorithms. IEEE Trans. Parallel Distrib. Syst. **8**, 1036–1042 (1997)
30. Svensson, S., Borgefors, G.: Distance transforms in 3D using four different weights. Pattern Recogn. Lett. **23**, 1407–1418 (2002)
31. Vainsencher, D., Bruckstein, A.M.: On isoperimetrically optimal polyforms. Theoret. Comput. Sci. **406**, 146–159 (2008)

# Verification of Hypotheses Generated by Case-Based Reasoning Object Matching

Petra Perner[✉]

Institute of Computer Vision and Applied Computer Sciences,
IBaI, PSF 301114, 04251 Leipzig, Germany
pperner@ibai-institut.de
http://www.ibai-institut.de

**Abstract.** Case-based reasoning object-matching consists of the methods at choice when the objects can be identified by case models. The result of the matching process is a number of hypotheses for the true shape of the objects. These hypotheses have to be verified in a hypothesis-verification process. In this paper we review what has been done so far and present our hypothesis-verification rules. The rules are evaluated and the results are discussed and presented in images. We consider two different hypothesis-verification rules, one is based on set-theory and the other one is based on statistical measures. Finally, we describe the results achieved so far and give an outlook about further work.

**Keywords:** Case-based reasoning object-matching · Hypothesis-test verification · Set theory · Statistical measures

## 1 Introduction

Case-based object-matching [1] is the methods at choice when the objects can be identified by case models. These case models can be learnt from the raw data by case mining [2]. For the case-matching procedure, we need a proper similarity measure that depends of the case model description. In our case, the case models are object contours such as round, ellipse-like, or more fuzzy-like geometric figures. The chosen similarity measure in this work is the cosine-similarity measure [6]. The properties of this similarity measure have been described in detail in [6]. The case matcher takes the case models and matches them against the objects in the image. In case the similarity measure is high the found contour will be marked in the image. Often the matcher does not bring out only one contour for an object, instead of the matcher fires several times at slightly different spatial positions in the image for the same object. These multiple matches have to be evaluated after the matching in a hypothesis verification procedure. The aim of this hypothesis-verification procedure is to obtain only the considered object.

We describe in this paper what kind of hypothesis verification methods we have developed and tested on our image database. The state-of-the-art of hypothesis verification methods is described in Sect. 2. In Sect. 3 we describe the hypothesis

© Springer International Publishing AG 2017
V.E. Brimkov and R.P. Barneva (Eds.): IWCIA 2017, LNCS 10256, pp. 66–78, 2017.
DOI: 10.1007/978-3-319-59108-7_6

generation and the problems concerned with it. In Sect. 4 we describe kinds of hypothesis-verification based on Set Theory. Hypothesis-verification rules based on statistics and results are given in Sect. 5. Finally we summarize our work in Sect. 6 and we give an outlook to further work on improvements of the matching results.

## 2  State-of-the Art

The aim of the hypothesis verification is to decide whether a match can be accepted as correct or not. Therefore hypothesis verification is closely related to the object recognition process. In literature we can find different approaches for this process. Grimson and Huttenlocher [1] as well as Jurie [2] and Kartatzis et al. [3] refer to features in the form of points or line segments. The common target of all these researchers is to find the best pose for the detected data features. However, all papers follow different strategies: Grimson and Huttenlocher [1] focus on the question how random matches can be prevented. They developed a formal means for finding the fraction of model features that have to be evaluated in order to ensure that the match occurs only with a given probability at random. The derivation of this fraction is done in three steps whereas the type of feature, the type of transformation from model to image and a bound on the positional and orientational error are known. First for every pairing of a model feature to data feature the set of transformations is determined. This set defines a particular volume in the transformation space. In the next step the probability of a common point of intersection between $l$ and more volumes is calculated. This probability corresponds to match of at least $l$ pairings of model and image features. Last a second probability that describes that $l$ or more volumes will intersect at random, is used to specify a threshold for the fraction of model features that have to be evaluated at least in order to ensure that the probability of a random match is lower than a given value.

The aim of the research of Jurie [2] is to find the pose of the model features that best matches the data features. The pose hypotheses are generated by correspondences between the model and the data features. Early researches propose to evaluate only some correspondences in order to find an initial pose hypothesis $P$ that is refined by iteratively enlarging the number of correspondences. Jurie [2] describes that this way of hypothesis generation and verification is not optimal. Therefore the paper suggests the opposite approach: A pose space is generated from different model-data-pairings. A "box" of the pose space is computed including the initial position $P$ that is large enough to compensate the data errors. Assuming that the distribution of model-data-correspondences is Gaussian, the maximal probability of the object to be matched is determined. Then the box can be refined. The process repeats until the "box" only contains one pose.

A simpler method of model-based pose estimation and verification is described in Shahrokni et al. [7]. They deal with the automatic detection of polyhedral objects. Hypotheses are generated by the knowledge-based connection of corners and line segments. The model and the transformed hypotheses are evaluated with the method of the least squares. The best hypothesis minimizes the sum of squared differences between the model and the transformed hypothesis.

Katartzis et al. [3] discuss the automatic recognition of rooftops, which are characterized by lines and their connections. After detecting the line segments in the image, they are grouped in a hierarchical graph. The highest level of the hierarchy contains closed contours. Every node of the graph is assigned a value that on the one hand assigns the saliency of the hypothesis and on the other hand represents the likelihood of the presence of a 3D-structure, which depends on domain-specific knowledge. Based on the hierarchical graph is defined a Markov Random Field (MRF). By maximization of the a posteriori probability of the MRF for the concrete graph a consistent configuration of the data is found.

In general the verification process for object hypotheses based on line segments is a widely discussed field.

An approach that totally differs from the discussed ones is given in Leibe et al. [4]. The heart of the described object recognition system is a database with different appearances of parts of the object that should be recognized. Additionally an "Implicit Shape Model" is learnt in order to combine the parts to a correct object. If multiple objects are located in the image then some hypotheses may overlap each other so that a verification step is required. The method follows the principle of Minimal Description Length (MDL) that is borrowed from the information theory. The description length of a hypothesis depends on its area and the probability that the pixels inside the hypothesis are no object pixels. The description length of two overlapping hypotheses is generated in the same way. From the resulting values is concluded whether two overlapping hypotheses refer to two objects or only to one.

## 3   Hypothesis Generation and Problems

### 3.1   Hypothesis Generation

In this Section, we want to give you an overview about the model-based object recognition method that we use to generate our hypotheses. The method is extensively discussed in Perner and Buehring [5].

A model-based object recognition method uses templates that generalize the original objects and matches these templates against the objects in the image. During the match a score is calculated that describes the goodness of the fit between the object and the template.

We determine the similarity measure based on the cross correlation by using the direction vectors of an image. This requires the calculation of the dot product $\tilde{l}_k$ between each direction vector of the model $\vec{m}_k = (v_k, w_k)^T$, $k = 1, \ldots, n$, and the corresponding image vector $\vec{i}_k = (d_k, e_k)^T$:

$$\tilde{l}_k = \langle \vec{m}_k, \vec{i}_k \rangle = \vec{m}_k \cdot \vec{i}_k = (v_k \cdot d_k + w_k \cdot e_k), \quad k = 1, \ldots, n \tag{1}$$

Note that the dot product $\tilde{l}_k$ (see Eq. 1) takes also into account the length of the vectors $\vec{m}_k$ and $\vec{i}_k$. That means that $\tilde{l}_k$ is influenced by the intensity of the contrast in the image and the model. In order to remove this influence, the direction vectors are normalized to unit length by dividing them through their gradient:

$$l_k = \left\langle \frac{\vec{m}_k}{\|\vec{m}_k\|}, \frac{\vec{i}_k}{\|\vec{i}_k\|} \right\rangle = \frac{\vec{m}_k \cdot \vec{i}_k}{\|\vec{m}_k\| \cdot \|\vec{i}_k\|} = \frac{v_k \cdot d_k + w_k \cdot e_k}{\sqrt{v_k^2 + w_k^2} \cdot \sqrt{d_{k_i}^2 + e_k^2}}, \quad k = 1, \ldots, n \quad (2)$$

The score $l_k$ (see Eq. 2) takes into account only the directions of the model and the image vector, i.e. it is invariant against changes of the contrast intensity. We can get the angle between the direction vectors by determining the value of $l_k$. Therefore we can conclude that the value of $l_k$ ranges from $-1$ to $1$. The vectors $\vec{m}_k$ and $\vec{i}_k$ have the same direction if $l_k = 1$, the vectors are orthogonal if $l_k = 0$ and both vectors have opposite direction if $l_k = -1$. In the rest of the paper we say that the value of $l_k$ is the *local similarity score* of the two direction vectors $\vec{m}_k$ and $\vec{i}_k$.

Usually, we are mainly interested in the similarity score between the complete model and the image. We want to define this *global similarity scores*$_1$ between the model and the image as the mean of all local similarity scores:

$$s_1 = \frac{1}{n} \sum_{k=1}^{n} l_k \quad (3)$$

Just like the local similarity score $l_k$ the global score $s_1$ is invariant against illumination changes and it ranges from $-1$ to $1$. In case of $s_1 = 1$ and $s_1 = -1$ the model and the image object are identical. If $s_1 = 1$ then all vectors in the model and the corresponding image vectors have exactly the same direction. If $s_1 = -1$ then all the vectors have exactly opposite directions, that is only the contrast between the model and the image is changed.

In general we have to subdivide between global and local contrast changes. If the contrast between the model and the image is globally inversed then all the model and image vectors have opposite directions. If the contrast is locally inversed then only some model and image vectors have opposite direction. With some little modifications the similarity measure $s_1$ becomes invariant to global contrast changes (see Eq. 4) and local contrast changes (see Eq. 5), respectively.

$$s_2 = \left| \frac{1}{n} \sum_{k=1}^{n} l_k \right| \quad (4)$$

$$s_3 = \frac{1}{n} \sum_{k=1}^{n} |l_k| \quad (5)$$

In contrast to range of $s_1$, the values of $s_2$ and $s_3$ are non-negative.

The aim is to store only one model for objects with similar shapes of different scale and rotation. Therefore a transformed model must be compared to the image at a particular location. The value of arccos $s_2$ indicates the mean angle between the model and the image vectors.

## 3.2   Kinds of Hypothesis Verification Based on Set-Theory

The matching process determines each possible match between the image pixels and the model. In the following we consider the found object as a hypothesis. To each hypothesis is assigned a matching score based on the similarity measure $s_3$. The score from the observation of Fig. 1 we can see that the models often match the same object, i.e. we have a superimposition of models. All the hypotheses in this image have scores greater than 0.8. Now we need to find a rule which allows us to remove determines the similarity between the model and the image pixels. It can range from 0 to 1 whereas the value of 1 says identity and the value of 0 dissimilarity. By defining a threshold for the score we can exclude hypotheses. This is the simplest hypothesis verification process. If the threshold is set to 0.8 then 734 hypotheses remain. They are shown in Fig. 1 false hypotheses. The hypotheses in this particular case overlap, touch or are inside of each other. From that we can develop special relationships of the hypotheses.

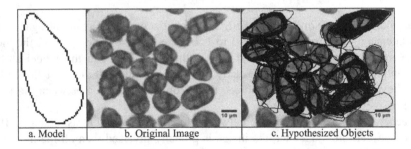

**Fig. 1.** Contour Model, original image and hypothesized objects.

The definition of the relationships is based on two hypothesized objects A and B. S (A) is the set of all image pixels that are inside the contour of the object A including also the image pixels of the contour. Equally, S(B) is the set of all image pixels inside the contour of object B including all image pixels of the contour. We want to distinguish between three relationships that are described in Table 1.

Determining the number of common pixels of every pair of hypotheses we conclude to their relation. In Sect. 4 we analyse the generated hypotheses using the defined relationships.

**Table 1.** Relationships between two Hypotheses.

| Relation | Figure | Description |
|---|---|---|
| a. Inside | | We say that the set $S(B)$ is *inside* the set $S(A)$ if all elements that are included in the set $S(B)$ are also included in the set $S(A)$, i.e. $S(B) \subset S(A)$ and there is $$S(A) \cap S(B) = B$$ and $S(A) \cup S(B) = S(A)$. |
| b. Overlapping | | We say that the sets $S(A)$ and $S(B)$ *overlap* each other if they have some equal elements, i.e. $$S(A) \cap S(B) \neq \emptyset, \ S(A) \cap S(B) \neq S(A)$$ and $S(A) \cap S(B) \neq S(B)$. |
| c. Almost Inside | | We say that the hypothesis $B$ is *almost inside* the hypothesis $A$ if almost all elements of $S(B)$ are also elements of $S(A)$, i.e. $$S(A) \cap S(B) \neq \emptyset, \ S(A) \cap S(B) \neq S(A)$$ and $S(A) \cap S(B) \neq S(B)$ and $|S(A) \cap S(B)| \approx |S(B)|$ and $|S(A) \cup S(B)| \approx |S(A)|$. This relation is a special case of the relation "overlapping". |
| d. Touching | | We say that hypotheses $A$ and $B$ are *touching* if their contours $C(A)$ and $C(B)$ have some equal elements, whereas the equal elements are neighboured. Touching Hypotheses are a special case of overlapping hypotheses. But it is also possible that two sets are touched and one set is inside the other set. |

# 4   Results

This Section focuses on the reduction of initial hypotheses.

## 4.1   The Relationship "Inside"

In this Section we want to investigate in the relationship "inside" (Table 1a). Given a sorted list[1] of hypotheses, we first extract the hypothesis pairs that fulfil the relationship "inside" (see Fig. 2b). From the 734 hypothesized matches we can create 138 hypotheses pairs that fulfil the relationship inside, i.e. one of both hypotheses is totally overlaid by the

---

[1] The matching process lists all matched objects sorted by the scale whereas objects with the same scale are sorted by the rotation.

other. Note that these pairs are only based on 136 different hypotheses. Thus we conclude that some hypotheses with high area overlay more than one smaller hypothesis. In other words we can say that most of the hypotheses are involved in more than one pair.

For the reduction process we rule that the hypothesis with the higher matching score remains while the other hypothesis of this pair is removed. Since the removed hypothesis often is a partner in more than one pair, for some hypotheses there will be no other correct "inside" partner. That is in practice that the number of hypotheses pairs may decrease for more than one pair. We want to illustrate this fact based on Fig. 2: We obtain a reduction of 67 hypotheses (Fig. 2a and d) if we successively remove the "inside" partner with the lower score. Considering the hypotheses that are used to create correct "inside" pairs (Fig. 2b) and their remaining partners (Fig. 2c), gives a reduction by 95 hypotheses.

From Fig. 2a and d we can see that the total number of hypotheses is only reduced by about 10%. Since this reduction does not significantly simplify the hypothesis verification process we investigate in the relation "overlapping" (see Sect. 3.2)

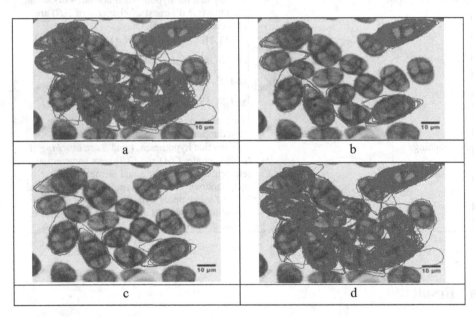

**Fig. 2.** (a) Hypothesized Matches (734 Hypotheses). (b) Hypothesis Pairs that fulfils the Relation "inside" (138 Pairs based on 136 Hypotheses). (c) Remaining Hypotheses after removing the Hypotheses with lower Score (41 Remaining Partners). (d) Remaining Hypotheses after applying the "Inside"- Criterion (667 Hypotheses).

## 4.2    The Relationship "Overlapping"

In this Section, we concentrate on the relationship "overlapping". Although we could not significantly reduce the number of hypotheses by applying the relationship "inside" we work with the reduced number of hypotheses (see Fig. 2d). For presentation purposes we first only consider the 41 remaining hypotheses of the "inside" pairs which are shown in Fig. 3 (compare to Fig. 2c). At the end of this Section we extend our investigations to the whole set of hypotheses.

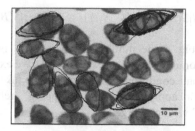

**Fig. 3.** Basic hypotheses.

Note that the hypotheses are concentrated in some regions of the image (see Fig. 3). It seems that many hypotheses are slightly transformed (shifted or twisted) with respect to other hypotheses. This means that the intersection area of two overlapped hypotheses $A$ and $B$ has approximately the same size as the area of the hypothesis $A$ and the hypothesis $B$ respectively. We express this fact with the condition (6):

$$|S(A) \cap S(B)| \geq t|S(A)| \quad \text{AND} \quad |S(A) \cap S(B)| \geq t|S(B)|, \quad t \in [0, 1] \qquad (6)$$

We restrict the size of the intersection area with respect to the size of the hypothesis area by the value $t$. If two hypotheses fulfil the Condition (6) we say they overlap each other. As well as in the discussion of the relationship inside, we assume that the best match for an object has the highest score. From two overlapping hypotheses we therefore remove the hypothesis with the lower matching score. In the first part of Fig. 4 the remaining hypotheses are given when we use the overlapping condition (6) and varying the thresholds $t$ of the minimal common hypotheses area.

Since the condition (6) mainly combines hypotheses with similar size, we replace the "AND" with "OR" (7). Then we repeat the test with the new condition (7). The results are given in the second part of Fig. 4.

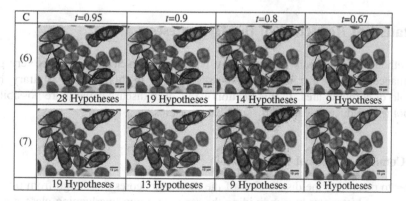

**Fig. 4.** Applying the Relationship "Overlapping" with different conditions and degrees of common area to some selected Hypotheses (C = Condition).

$$|S(A) \cap S(B)| \geq t|S(A)| \quad \text{OR} \quad |S(A) \cap S(B)| \geq t|S(B)|, \quad t \in [0,1] \quad (7)$$

As we can see from Fig. 4 the number of hypotheses is as fewer as lower the threshold of the minimal common hypotheses area is. As we expected more hypotheses are removed with the second condition (7). Therefore we take the condition (7) using the threshold $t = 0.67$ for applying the "overlapping" relation to all remaining hypotheses of the "inside" relation (Fig. 5).

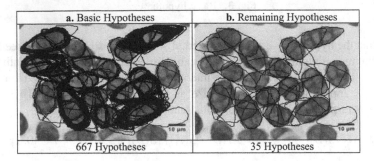

**Fig. 5.** Applying the Relationship "Overlapping" with condition (7) and the Threshold t = 0.67 after the Relationship "Inside".

From Fig. 5 we can see that the applied criterion reduces the given hypotheses to about 5%. In order to improve the performance of the reduction process we used the described rule also to all hypothesized matches. We obtain the same result as if we remove some hypotheses with the "inside" relation. Therefore we conclude that the first step in each hypothesis verification process should be the reduction of hypotheses using "overlapping" relation defined with condition (7) and the common area threshold $t = 0.67$. From each pair of overlapped hypotheses the hypothesis with the lower matching score is removed.

## 5    Statistical Reduction of the Hypotheses

The method for hypothesis reduction that is described in Sect. 3.2 shows very good performance. One of the main weaknesses of this method is the arbitrary fixed threshold of common area. In this Section we want to discuss some more possibilities of hypothesis reduction based on statistical measures.

### 5.1    Common Statistical Measures

In order to determine the distribution of the matching scores that range from 0.8 to 1, we generate a histogram by subdividing the range into non-overlapping classes with a class width of 0.05. The number of hypotheses within each class is displayed in Fig. 6.

From the Histogram in Fig. 6 we cannot conclude to any distinct distribution because we only investigate in a part of the score space. Nevertheless we can determine the mean $\mu_s$ and the standard deviation $\sigma_s$ of all scores. In order to optimize the threshold we develop a criterion for removing some hypotheses based on the mean and the standard deviation (8):

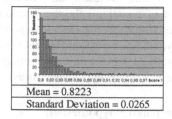

| Mean = 0.8223 |
| Standard Deviation = 0.0265 |

**Fig. 6.** Histogram, mean and standard deviation of the hypothesis scores.

$$s_i < \mu_s + f \cdot \sigma_s \Rightarrow \text{delete hypothesis } i, \quad i = 1,\ldots,h \qquad (8)$$

The number of hypotheses is denoted by $h$. Figure 7 shows the remaining hypothesized objects applying different factors $f$ to the 734 hypotheses shown in Fig. 1c.

From Fig. 7 we can see that the number of hypotheses can be reduced significantly if the threshold for the matching score is increased. Since we expect high scores if the model matches an object very well, we will obtain as more hypotheses for one object as better the model matches the object. The results reported in Fig. 7 verify this assumption.

| a. Hypotheses which Score mean > (f=0) | b. Remaining Hypotheses applying f = 1 | c. Remaining Hypotheses applying f = 2 | d. Remaining Hypotheses applying f = 3 |
|---|---|---|---|
| 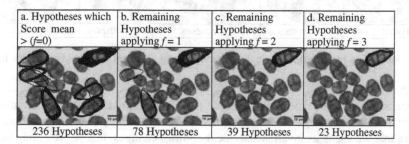 | | | |
| 236 Hypotheses | 78 Hypotheses | 39 Hypotheses | 23 Hypotheses |

**Fig. 7.** Remaining hypotheses applying condition (9) with different Factors $f$. (Color figure online)

Remember that the matcher tolerates object occlusion and touching until 20% if the score threshold is 0.8. The described method for hypotheses reduction increases the threshold for accepting a match as a hypothesis. In the consequence it removes also hypotheses of objects which are occluded or touched. If we want to consider also such objects, we must not reduce the hypothesized matches based on Eq. (8).

## 5.2    Hypotheses Reduction by the Evaluation of the Local Score

For each hypothesis we can store, during the matching process, the number of model pixels that have the same local contrast as the corresponding image pixel. In the following we denote this number with $c_{same}$. Remember that the model and the image pixel have the same local contrast if their dot product is positive. Otherwise their local contrast is inversed. In order to measure the quality of the hypothesis we determine, with respect to the number $n$ of model pixels, the fraction of contour pixels of the hypothesis that have the same contrast as the model. Depending on the global contrast between the model and the hypotheses we should accept very high and very low values of this fraction. Given a threshold $t$ the minimal fraction of the same contrast for acceptance, we determine the remaining hypotheses based on Condition (9):

$$t < \frac{c_{i,same}}{n} < (1 - t) \Rightarrow \text{remove hypothesis } i, \quad t \in [0, 0.5], \ i = 1, \ldots, h \qquad (9)$$

Figure shows the remaining hypotheses using different values of the threshold $t$.

It is strange that hypotheses which seem to match the object well are earlier removed than hypotheses which include some background. Figure 8 shows this phenomenon for the hypotheses of two selected objects in the image based on the threshold $t = 0.25$. Each image in Fig. 8 shows a labelled hypothesis whereas red parts display negative local contrast and blue parts positive local contrast. Below each image the relative fraction of negative local contrast is given (blue parts of the contour).

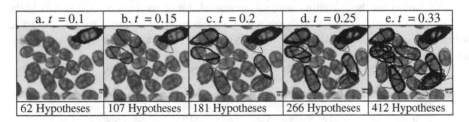

| a. $t = 0.1$ | b. $t = 0.15$ | c. $t = 0.2$ | d. $t = 0.25$ | e. $t = 0.33$ |
|---|---|---|---|---|
| 62 Hypotheses | 107 Hypotheses | 181 Hypotheses | 266 Hypotheses | 412 Hypotheses |

**Fig. 8.** Remaining hypotheses applying condition (9) with different Thresholds $t$. (Color figure online)

Because of these unexpected results we ask if the approach of the differentiation between positive and negative local contrast is correct. The power of the similarity measure that we use is its invariance to local contrast changes, that is, we eliminate the influence of the sign of the local score by summing up the absolute amount of the local scores. On the other side we exactly evaluate the sign if we calculate the score of binary contrast changes (see Fig. 9).

Figure 10 shows the same hypotheses as Fig. 9, but now pixels which local score higher than 0.9 or lower than −0.9 are marked blue. The other parts are red. Below the images the relative faction of pixels with high local score (>0.9) is given.

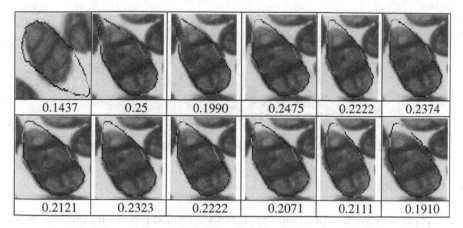

**Fig. 9.** Results for binary contrast changes for different hypotheses.

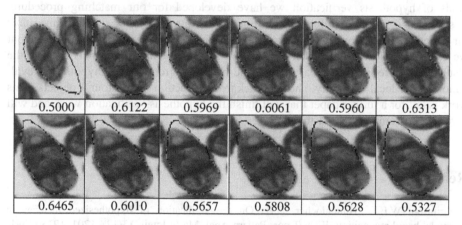

**Fig. 10.** Results for local score higher than 0.9 and lower than −0.9. (Color figure online)

Note that the fraction of high local scores depends on the threshold that defines a pixel with "*high local score*". As higher as the threshold is as lower is the value of this fraction.

Similar to the experiment we carried out for the binary contour based on contrast changes (see Fig. 8) we now create the binary contour for all hypotheses as described above by thresholding the local scores at the value 0.9. Then we remove the hypotheses which relative fraction of high local scores is lower than (a) 0.6 (b) 0.66 and (c) 0.75. The results are given in Fig. 11.

From Fig. 11 we can see that this method is another possibility to reduce the number of hypotheses. Since the hypotheses around the object in the right upper corner have high similarity scores most of these hypotheses remain.

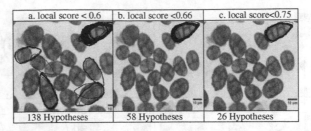

**Fig. 11.** Results for different local scores. (Color figure online)

## 6  Conclusions

In this paper we have described our hypothesis verification process for our case-based reasoning shape-object matching procedure. We have described the hypothesis-generation process in brief and the problems concerned with it. Then we described the kinds of hypothesis verification we have developed for our matching procedure. Results are given for the different rules. Finally, we introduce some statistical measures for hypothesis reduction and give results. The final results show good performance but we can still think of some other verification measures that will further improve the results. These verification measures will be based on grouping the hypotheses, evaluation of the local similarity, and the background fraction of the found objects. This work is left for a further paper that will finish the hypothesis-verification work and will give a good summary about the work.

## References

1. Grimson, W.E.L., Huttenlocher, D.P.: On the verification of hypothesized matches in model-based recognition. IEEE Trans. Pattern Anal. Mach. Intell. **13**(12), 1201–1213 (1991)
2. Jurie, F.: Hypothesis verification in model-based object recognition with a Gaussian Error Model. In: Proceedings of the European Conference on Computer Vision, Freiburg, Germany, pp. 642–656 (1998)
3. Katartzis, A., Sahli, H., Nyssen, E., Cornelis, J.: Detection of buildings from a single airborne image using a Markov Random Field Model. In: Proceedings of IEEE International Geoscience and Remote Sensing Symposium (IGARSS 2001), Sydney, Australia (2001)
4. Leibe, V.,Leonardis, A., Schiele, B.: Combined object categorization and segmentation with an implicit shape model. In: ECCV 2004 Workshop on Statistical Learning in Computer Vision, Prague, May 2004
5. Perner, P., Bühring, A.: Case-based object recognition. In: Funk, P., González Calero, P.A. (eds.) ECCBR 2004. LNCS (LNAI), vol. 3155, pp. 375–388. Springer, Heidelberg (2004). doi:10.1007/978-3-540-28631-8_28
6. Perner, P., Jänichen, S., Perner, H.: Case-based object recognition for airborne fungi recognition. J. Artif. Intell. Med. AIM **36**(2), 137–157 (2006). Special Issue on CBR
7. Shahrokni, A., Vacchetti, L., Lepetit, V., Fua, P.: Polyhedral object detection and pose estimation for augmented reality applications (2002)

# Template-Based Pattern Matching in Two-Dimensional Arrays

Yo-Sub Han[1] and Daniel Průša[2,3(✉)]

[1] Department of Computer Science, Yonsei University,
50 Yonsei-Ro, Seodaemun-Gu, Seoul 120-749, Republic of Korea
emmous@yonsei.ac.kr
[2] Faculty of Electrical Engineering, Czech Technical University,
Karlovo náměstí 13, 121 35 Prague 2, Czech Republic
prusapa1@fel.cvut.cz
[3] Czech Institute of Informatics, Robotics and Cybernetics,
Czech Technical University, Zikova 1903/4, 166 36 Prague 6, Czech Republic

**Abstract.** We propose a framework for pattern matching in two-dimensional arrays of symbols where the patterns are described by an extended version of the regular matrix grammar and the size of desired matches is prescribed. We then demonstrate how to reformulate the 2D pattern matching as the one-dimensional pattern matching (string pattern matching), and study the efficiency of the string pattern matching algorithm based on pattern complexity with respect to two finite automaton models: (1) the classical finite automaton and (2) the finite automaton equipped with two scanning heads placed in a fixed distance. We also identify several subclasses of the considered templates for which the framework yields a more efficient matching than the naive algorithm.

**Keywords:** Two-dimensional pattern matching · Matrix grammars · Pattern complexity · Finite automata · Multi-head automata

## 1 Introduction

The task of matching string patterns in a text naturally extends to higher dimensions. For example, given a two-dimensional (2D) pattern and a 2D array of symbols, the task of the exact 2D pattern matching is to detect all occurrences of this pattern in the array. Bird [6] and Baker [5] independently proposed the first efficient algorithm by reducing the problem into the one-dimensional (1D) pattern matching problem. Then later several researchers suggested different types of improved algorithms for different cases [2, 4, 9, 10, 16].

In pattern matching, another common scenario is to find matching against a set of patterns described by a suitable formalism such as regular expressions or finite automata. For instance, the problem of searching for the shortest matching substring described by a regular expression has a very efficient implementation and a wide applicability [1]. We can also easily identify many pattern searching applications specified by a certain template in the 2D setting, arising in fields

© Springer International Publishing AG 2017
V.E. Brimkov and R.P. Barneva (Eds.): IWCIA 2017, LNCS 10256, pp. 79–92, 2017.
DOI: 10.1007/978-3-319-59108-7_7

such as computer vision, robotics or data mining. However, it is not always straightforward to extend a template for string (1D) pattern matching into a template for 2D pattern matching. For example, the notion of regular expressions cannot be easily generalized for 2D arrays, especially when we intend to transfer the efficient matching algorithms for regular expressions with it.

In this paper we address this problematic scenario under the assumption that patterns are described by an extension of the 2D (regular) matrix grammar [14]. We make the 2D matching task feasible by assuming that the goal is to detect matches of a prescribed size in an input array of size $m \times n$; namely, we fix the size of desired matches to be $k \times \ell$. We say that matches are detected efficiently when the matching process is much faster than a straightforward naive approach, which goes through all subpictures of size $k \times \ell$ in the input and, for each subpicture, checks whether or not it is a match. If the subpicture checking takes linear time in the area of the subpicture, namely $\mathcal{O}(k\ell)$ time, then the whole procedure requires $\mathcal{O}(k\ell mn)$ time. We consider a matching algorithm to be efficient if it runs in time $\mathcal{O}((\log k + \log \ell)mn)$. Recently, the pattern complexity for picture languages was introduced [13]. We revisit this pattern complexity and develop the notion of *pattern complexity* for a string language. Then we classify the complexity of the studied 1D matching with respect to several subclasses of regular languages based on the new pattern complexity.

The rest of the paper is structured as follows. We briefly recall some basic notations in Sect. 2 and give examples of template usages in 2D pattern matching in Sect. 3. Then we introduce an extended regular matrix grammar and describe the matching algorithm in Sect. 4. We study in Sect. 5 complexity of the induced matching task for strings with respect to two computational models. We conclude by Sect. 6 where a future work is outlined.

## 2    Preliminaries

We use the common notation and terms on picture languages [8]. For a finite alphabet $\Sigma$, a picture $P$ over $\Sigma$ is a 2D array of symbols from $\Sigma$. If $P$ has $m$ rows and $n$ columns, it is of size $m \times n$, and we write $P \in \Sigma^{m,n}$. Rows of $P$ are indexed from 1 to $m$, columns of $P$ are indexed from 1 to $n$, $P_{i,j}$ denotes the symbol of $P$ in the $i$-th row and the $j$-th column. In graphical visualizations of pictures, position $(1,1)$ is associated with the top-left corner. The set of all pictures over $\Sigma$ is denoted by $\Sigma^{*,*}$. In addition, $\Sigma^{*,n} = \bigcup_{i=0}^{\infty} \Sigma^{i,n}$ and $\Sigma^{m,*} = \bigcup_{j=0}^{\infty} \Sigma^{m,j}$.

Let $\mathcal{A} = (Q, \Sigma, \delta, q_0, F)$ be a deterministic finite automaton (DFA), where $Q$ is a set of states, $\Sigma$ is an input alphabet, $\delta : Q \times \Sigma \to Q$ is a transition function, $q_0 \in Q$ is the initial state and $F \subseteq Q$ is a set of accepting states. The extended transition function $\hat{\delta} : Q \times \Sigma^* \to Q$ is defined by $\hat{\delta}(q, \lambda) = q$, $\hat{\delta}(q, a) = \delta(q, a)$ and $\hat{\delta}(q, aw) = \hat{\delta}(\delta(q, a), w)$ for all $a \in \Sigma$, $w \in \Sigma^*$ and $q \in Q$. The language $L(\mathcal{A})$ is a set of strings $w$ such that $\hat{\delta}(q_0, w) = f$ for an accepting state $f \in F$.

$\mathbb{N} = \{1, 2, \ldots\}$, $\mathbb{N}_0 = \mathbb{N} \cup \{0\}$ and $\mathbb{P}(S)$ is the powerset of a set $S$.

## 3   Working Examples

We can think of a template-based 2D pattern matching as a tool for image processing in digital geometry or for high-level reasoning in computer vision or robotics. An input array to be searched can be a raster image, a high-level grid structure built upon an image or a scene representation by dividing the whole area into sectors of equal size and assigning them labels by their content. We illustrate these thoughts with the following examples of possible scenarios.

- An image is obtained by scanning the surface of a component. Its area is divided into sectors, each of which is labeled as either *defective* (if it contains a defect) or *normal* (if there is no defect). The task is to locate subareas of $k \times k$ sectors with the number of defects above an acceptable threshold.
- Given a black-and-white image, search for some basic geometric shapes.
- A robot operates on a grid of sectors where each sector content is one of the following three—nothing, an obstacle, an object of interest. The robot knows the sector content. Now it is ordered to move to a location specified by "a chair located south-east of a wall in distance at most two meters".

Inspired by the scenarios, let us define the following picture languages (see also Fig. 1).

*Example 1 (Counting).* For $k \in \mathbb{N}$, let $L_{\max,k} = \{P \in \{\square,\blacksquare\} \mid |P|_\blacksquare \le k\}$, i.e., $L_{\max,k}$ consists of those pictures having at most $k$ black pixels.

*Example 2 (Digital geometry).* Let $L_{\mathrm{rect}}$ be a picture language over $\Sigma = \{\square,\blacksquare\}$ consisting of pictures where all black pixels form a boundary of a rectangle whose height and width are at least 3. Moreover, let $L_{\mathrm{diag}}$ be a picture language over $\Sigma = \{\square,\blacksquare\}$ consisting of square pictures $P$ where the main diagonal contains only black pixels, while the other pixels are white.

*Example 3 (Spatial arrangement).* Let $L_{\mathrm{sp}}$ be a picture language over $\Sigma = \{\square,\boxtimes,\blacksquare\}$ consisting of all $P \in \Sigma^{*,*}$ where $|P|_\blacksquare = |P|_\boxtimes = 1$, with $\blacksquare$ at a position $(b_x,b_y)$ and $\boxtimes$ at a position $(c_x,c_y)$, fulfilling $b_x < c_x$ and $b_y < c_y$.

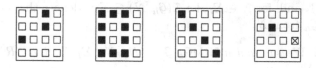

**Fig. 1.** Example pictures from $L_{\max,3}$, $L_{\mathrm{rect}}$, $L_{\mathrm{diag}}$ and $L_{\mathrm{sp}}$.

# 4    Regular Matrix Grammars with Scanning Window

One of the earliest ideas of generating pictures by grammars was to define vertical and horizontal productions (regular or context-free), and generate a picture in two phases. First, a column of symbols is produced using only vertical productions. Second, all the previously generated symbols serve as the initial nonterminals that in parallel generate rows by the horizontal productions. A picture is successfully generated only if all the produced rows are of the same length. However, it is known that such a grammar is weak [7]; for instance, it is impossible to use it to generate patterns from Example 2 since it cannot control a full neighborhood of generated fields (pixels). On the other hand, the grammar can be easily parsed and it has favorable theoretical properties.

Here we strengthen the grammar by giving it a scanning window-like mechanism allowing to synchronize the content generated at neighboring positions, which provides a control over the generated picture topology. The result is that the expressive power of the grammar is significantly increased while the simplicity of parsing is still maintained.

**Definition 4.** *A two-dimensional regular matrix grammar with scanning window of size $c$, abbreviated as* $\mathsf{2RMG}_c$ *is a tuple* $\mathcal{G} = (N_v, N_h, \Sigma_I, \Sigma, S, R^v, R^h)$, *where*

- *$N_v$ is a finite set of* vertical nonterminals,
- *$N_h$ is a finite set of* horizontal nonterminals, *with $N_v \cap N_h = \emptyset$,*
- *$\Sigma_I \subseteq N_h$ is a finite set of* intermediates,
- *$\Sigma$ is a finite set of* terminals,
- *$S \in N_v$ is a* starting symbol,
- *$R^v$ is a finite set of* vertical productions *of the form $N \to AM$ or $N \to A$ where $N, M \in N_v$ and $A \in \Sigma_I$,*
- *$R^h$ is a finite set of* horizontal productions *of the form $V \to {\_}aW$ or $V \to {\_}a$ where $V, W \in N_h$ and ${\_}a \in \Sigma^{c,1}$.*

Let $\mathcal{G}_v = (N_v, \Sigma_I, S, R^v)$ denote the regular grammar formed by the vertical productions (with the starting symbol $S$) and let $\mathcal{G}_h = (N_h, \Sigma, R^h)$ denote the regular grammar formed by the horizontal productions (without any nonterminal specified as the starting symbol). Let $L(\mathcal{G}_v) \subseteq \Sigma_I^*$ denote the set of strings generated by $\mathcal{G}_v$ and, for $N \in \Sigma_I$, let $L(\mathcal{G}_h, N) \subseteq \Sigma^{c,*}$ denote the set of pictures generated by $\mathcal{G}_h$ from $N$.

**Definition 5.** *We say that a* $\mathsf{2RMG}_c$ *$\mathcal{G} = (N_v, N_h, \Sigma_I, \Sigma, S, R^v, R^h)$ generates a picture $P \in \Sigma^{m,n}$ iff*

1. *there is $C = C_1 C_2 \cdots C_{m-c+1} \in L(\mathcal{G}_v)$, where $C_i \in \Sigma_I$,*
2. *for each $i = 1, \ldots, m - c + 1$, the subpicture of $P$ consisting of rows from $i$ to $i + c - 1$ is in $L(\mathcal{G}_h, C_i)$.*

The process of generating a picture is depicted in Fig. 2.

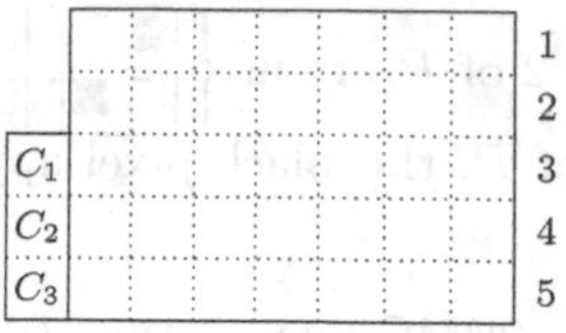

**Fig. 2.** A 2RMG$_3$ generating a picture of size $5 \times 7$. Vertical productions generate string $C_1C_2C_3$, displayed as a column. Horizontal productions generate rows $1, 2, 3$ from $C_1$, rows $2, 3, 4$ from $C_2$ and rows $3, 4, 5$ from $C_3$. Note that the content of each overlapped row must be identical in all three cases—for instance, the content of the row 3, which is overlapped by $C_1, C_2, C_3$, is identical for all the three cases.

**Corollary 6.** 2RMG$_1$ *is equivalent to the normal regular matrix grammar.*

A parser for a given 2RMG$_c$ can be constructed as follows. Let $P \in \Sigma^{m,n}$ be an input picture. For $N \in \Sigma_I$, let $\mathcal{A}^h(N)$ be a DFA accepting $L(\mathcal{G}_h, N)$. At each step, $\mathcal{A}^h(N)$ scans a column of $c$ symbols from $\Sigma$ (it has a scanning window of height $c$). Let $\mathcal{A}^h$ be the product automaton of all $\mathcal{A}^h(N)$'s for all $N \in \Sigma_I$. Apply $\mathcal{A}^h$ to process rows of $P$ from $c$ to $m$. Note that when processing row $i$, $\mathcal{A}^h$ scans also symbols in rows $i-1, \ldots, i-c+1$. Let $\Sigma_I = \{N_1, \ldots, N_k\}$ and $(q_1^i, q_2^i, \ldots, q_k^i)$ be the state entered by $\mathcal{A}^h$ at the rightmost column of row $i$. Define $\mathcal{S}_i \subseteq \Sigma_I$ to contain $N_j \in \Sigma_I$ iff $q_j^i$ is an accepting state. Let $\mathcal{A}^v$ be a DFA accepting $L(\mathcal{G}_v)$. Then, from $\mathcal{A}^v$, we construct a nondeterministic finite automaton $\mathcal{A}$ that simulates $\mathcal{A}^v$ nondeterministically by reading $\mathcal{S}_c, \mathcal{S}_{c+1}, \ldots, \mathcal{S}_m$, guessing $C_i \in \mathcal{S}_i$ in each row $i$ and simulating $\mathcal{A}^v$ over $C_c \ldots C_m$ (note that $\mathcal{A}$ rejects if $\mathcal{S}_i = \emptyset$ for some $i$).

**Lemma 7.** *Let $\Sigma$ be an alphabet and $\Theta \subseteq \Sigma^{c,d}$ be a set of pictures of size $c \times d$ for some $c, d \in \mathbb{N}$. Let $L(\Theta)$ over $\Sigma$ denote a picture language consisting of all pictures $P \in \Sigma^{*,*}$ of size at least $c \times d$ whose all subpictures of size $c \times d$ are from $\Theta$. Then, there is a 2RMG$_c$ generating $L(\Theta)$.*

*Proof.* Construct a 2RMG$_c$ $\mathcal{G}$ such that $L(\mathcal{G}_v) = \{C\}^*$ for an intermediate $C$ and $L(\mathcal{G}_h, C)$ consists of pictures of height $c$ where each subpicture of size $c \times d$ is in $\Theta$. The latter (horizontal) language over $\Sigma^{c,1}$ is regular as it is accepted by a DFA that remembers in states the lastly read subpicture $c \times d$.    □

**Lemma 8.** *The family of picture languages generated by 2RMG$_c$ is closed under union, intersection and complement.*

*Proof.* Follows from closure properties of regular languages and properties of finite automata.    □

Lemmas 7 and 8 can be used to construct a 2RMG$_2$ generating $L_{\text{diag}}$ from Example 2. For a picture $P \in \{\square, \blacksquare\}^{*,*}$ of size at least $2 \times 2$, two properties have to be checked to ensure that $P \in L_{\text{diag}}$.

- All subpictures of size $2 \times 2$ of $P$ are in $\left\{ \begin{array}{c}\blacksquare\square\\\square\blacksquare\end{array}, \begin{array}{c}\square\blacksquare\\\square\square\end{array}, \begin{array}{c}\square\square\\\blacksquare\square\end{array}, \begin{array}{c}\square\square\\\square\square\end{array} \right\}$.
- In the first and last row of $P$, the black pixel appears only at the first and last position, respectively.

We can similarly construct a 2RMG$_2$ generating $L_{\text{rect}}$. To generate $L_{\max,k}$, it suffices to construct a 2RMG$_1$ where the set of intermediates equals $\Sigma_I = \{C_0, C_1, \ldots, C_k\}$ and, for each $i$, the horizontal productions generate from $C_i$ strings $u$ over $\{\blacksquare, \square\}$ such that $|u|_\blacksquare = i$. Let $\delta(C_i) = i$. Then, the vertical productions are designed to generate strings $v \in \Sigma_I^*$ where $\sum_{i=1}^{|v|} \delta(v_i) \le k$.

As for $L_{\text{sp}}$, it can be expressed as the intersection $L_{\text{sp}} = L_{\text{sp}}^{(x)} \cap L_{\text{sp}}^{(y)}$ where $L_{\text{sp}}^{(x)}$ or $L_{\text{sp}}^{(y)}$ consists of pictures in which pixel $\blacksquare$ is at a position $(b_x, b_y)$ and pixel $\boxtimes$ is at a position $(c_x, c_y)$ fulfilling $b_x < c_x$ or $b_y < c_y$, respectively. $L_{\text{sp}}^{(y)}$ is generated by a 2RMG$_1$. $L_{\text{sp}}^{(x)}$ is generated by a variant of 2RMG$_1$ which first generates a row of intermediates and then columns of the resulting picture. A usage of this "transposed" 2RMG$_1$ could be integrated in the matching algorithm which follows next, but we do not give details on this due to the limited space.

Let $\mathcal{G} = (N_v, N_h, \Sigma_I, \Sigma, S, R^v, R^h)$ be a 2RMG$_c$. Denote $\Gamma = \Sigma^{c,1}$ and $\Delta = \mathbb{P}(\Sigma_I)$. For $N \in \Sigma_I$, $k, \ell \in \mathbb{N}$, $k \ge c$, let $\mathcal{M}_\ell^h(N)$ be a DFA accepting $\Gamma^*(\Gamma^\ell \cap L(\mathcal{G}_h, N))$ (i.e., strings whose suffix of length $\ell$ is in $L(\mathcal{G}_h, N)$) and $\mathcal{M}_k^v$ be a DFA accepting $\Delta^*(\Delta^{k-c+1} \cap L(\mathcal{A}))$ where $\mathcal{A}$ is the automaton from the parsing algorithm description. For a DFA $\mathcal{M}$, let $|\mathcal{M}|$ denote the number of states of $\mathcal{M}$. Assume also, that $c$ is a small constant (we saw that $c \le 2$ is sufficient for the working examples).

**Theorem 9 (Matching algorithm).** *Given a* 2RMG$_c$ *$\mathcal{G}$, an input $P \in \Sigma^{m,n}$ and $k, \ell \in \mathbb{N}$, $k \le m$, $\ell \le n$, there is an algorithm detecting all subpictures of $P$ that belong to $L(\mathcal{G}) \cap \Sigma^{k,\ell}$ in*

$$\mathcal{O}\left( mn\left( \log |\mathcal{M}_k^v| + \sum_{N \in \Sigma_I} \log |\mathcal{M}_\ell^h(N)| \right) \right)$$

*time.*

*Proof.* We use an auxiliary 2D array $T$ of size $m \times n$ where $T_{i,j}$ denotes its field at position $(i, j)$. The matching procedure resembles the parsing algorithm. The input $P$ is first scanned row by row. When passing through an $i$-th row $(i \ge c)$, rows $i, \ldots, i - c + 1$ are processed simultaneously by all automata $\mathcal{M}_\ell^h(N)$. We write to $T_{i,j}$ a subset $\mathcal{S} \subseteq \Sigma_I$ which contains $N \in \Sigma_I$ iff $\mathcal{M}_\ell^h(N)$ enters an accepting state after performing the $j$-th transition. The second phase goes through columns of $T$, skipping always $c - 1$ first symbols of each column. The automaton $\mathcal{M}_k^v$ is simulated over each column. Whenever $\mathcal{M}_k^v$ enters an accepting state at some $T_{i,j}$, it indicates that the position $(i, j)$ is the bottom-right corner of a match. The time complexity of the algorithm is determined by the number of scanned symbols, which is $\mathcal{O}(mn)$, and the time complexity of simulating a transition of each participating automaton, which is proportional to the length of states representation, i.e., to $\mathcal{O}(\log |\mathcal{M}|)$ for an automaton $\mathcal{M}$. Note that we do not analyze the time complexity of constructing the automata from $\mathcal{G}$ as this is independent of $m$ and $n$. □

The matching algorithm is of time complexity $\mathcal{O}\left(mn(\log k + \log \ell)\right)$ if $|\mathcal{M}_k^v|$ is polynomial in $k$ and, for all $N \in \Sigma_I$, $|\mathcal{M}_\ell^h(N)|$ is polynomial in $\ell$. Also note that instead of DFAs we could have used a different model to perform 1D matching in rows and columns. We develop these two observations in the next section.

# 5 String Languages with Polynomial Pattern Complexity

As the template-based 2D pattern matching reduces to 1D pattern matching, it is essential to investigate the complexity of matching strings of a fixed size. In this section we define the pattern complexity of string languages with respect to two models: DFA and a variant of two-head DFA.

## 5.1 Matching by DFA

Let $L$ be a language over $\Sigma$. We search for all occurrences of length $n$ patterns from $L$ in a string $w \in \Sigma^*$ by constructing a DFA accepting $L(n) = \Sigma^* (L \cap \Sigma^n)$. For any $L$, the language $L(n)$ is regular (since $L \cap \Sigma^n$ is a finite, hence regular, language) and is accepted by a DFA with $\mathcal{O}(|\Sigma|^n)$ states. Here we are interested in those languages $L$ for which $L(n)$ is accepted by a DFA with polynomially many states in $n$.

**Definition 10 (Pattern complexity of a string language).** *Let $L$ be a language over $\Sigma$. For each $n \in \mathbb{N}$, let $\mathcal{A}_n = (Q_n, \Sigma, \delta_n, q_0, F_n)$ be the state-minimal DFA accepting $L(n)$. We define the pattern complexity of $L$ to be a function $\sigma_L : \mathbb{N} \to \mathbb{N}$ where $\sigma_L(n) = |Q_n|$ for all $n \in \mathbb{N}$.*

*Example 11.* $L = \{au \mid u \in \{a,b\}^*\}$ has exponential pattern complexity.
    We prove it by applying the Myhill-Nerode theorem to $L(n)$. Let $u, v \in \{a,b\}^n$ where $u = u_1 \ldots u_n$, $v = v_1 \ldots v_n$ and there is $i$ such that $u_i \neq v_i$. Then, $a^{i-1}$ is a *distinguishing extension* as $|\{ua^{i-1}, va^{i-1}\} \cap L(n)| = 1$. This implies that $\sigma_L = \Omega(2^n)$ since there are $2^n$ mutually distinguishable strings of length $n$.

*Example 12.* $L = \{u \mid u \in \{a,b\}^* \wedge |u|_b \mod 2 = 0\}$ has exponential pattern complexity (apply the Myhill-Nerode theorem as in Example 11).

*Example 13.* $L_{\max,k} = \{u \mid u \in \{a,b\}^* \wedge |u|_a \leq k\}$ has polynomial pattern complexity. For a given $n$, construct a DFA representing in states $k+1$ counters $\mathcal{C}_1, \ldots, \mathcal{C}_{k+1}$ that memorize relative positions of the last $k+1$ occurrences of $a$'s in lastly read $n$ characters. Each counter ranges from 0 to $n$. If a counter $\mathcal{C}_i$ is of value 0, it means that the number of tracked $a$'s is less than $i$. One more counter, counting to $n$, is added to prevent accepting strings shorter than $n$. All this suffices to accept $L_k(n)$ and the constructed DFA has $\mathcal{O}(n^{k+2})$ states. On the other, we can prove that $\sigma_{L_{\max,k}} = \Omega(n^k)$. Consider two strings $u, v \in \{a, b\}^n$ such that $|u|_a = |v|_a = k$ and $u \neq v$. Let $i$ be the smallest $i$ for which $u_i \neq v_i$, hence, w.l.o.g., $u_i = a$ and $v_i = b$. Denote $\ell = |u_1 \ldots u_i|_a = 1 + |v_1 \ldots v_i|_a$. Then, $a^\ell b^{i-\ell}$ is a distinguishing extension for $u$ and $v$. As $|\{w \in \{a, b\}^n \mid |w|_a = k\}| = \Omega(n^k)$, by the Myhill-Nerode theorem, $\sigma_{L_{\max,k}} = \Omega(n^k)$.

*Example 14.* In connection with Lemma 7, consider the following regular language $L$ over $\Sigma$:

$$L = \{w \in \Sigma^* \mid \text{all length } d \text{ substrings of } w \text{ are in } \Theta\},$$

where $\Theta \subseteq \Sigma^{1 \times d}$ is a set of pictures of size $1 \times d$—we can regard $\Theta$ as a set of strings since all pictures of $\Theta$ are single-row pictures.

Then, $\sigma_L = \mathcal{O}(dn^2)$ since $L(n)$ can be accepted similarly as $L_{\max,0}$ from Example 13 (the number of substrings of length $d$ not in $\Theta$ must be zero; and it is needed to remember $d$ lastly read symbols).

**Proposition 15.** *Let $L_1$ and $L_2$ be two languages over $\Sigma$ with polynomial pattern complexity. Then, $L_\cup = L_1 \cup L_2$, $L_\cap = L_1 \cap L_2$ and $\overline{L} = \Sigma^* \setminus L_1$ have polynomial pattern complexity.*

*Proof.* For $n \in \mathbb{N}$, let $\mathcal{A}_1$ and $\mathcal{A}_2$ be DFAs with polynomially many states accepting $L_1(n)$ and $L_2(n)$, respectively. We can write

$$L_\cup(n) = \Sigma^* \left((L_1 \cup L_2) \cap \Sigma^n\right) = \Sigma^* (L_1 \cap \Sigma^n) \cup \Sigma^* (L_2 \cap \Sigma^n) = L_1(n) \cup L_2(n),$$

hence $L_\cup(n)$ is accepted by the product automaton of $\mathcal{A}_1$ and $\mathcal{A}_2$. Analogously, we derive $L_\cap(n) = L_1(n) \cap L_2(n)$, meaning again that $L_\cap(n)$ is a regular language accepted by the product automaton. For the complement, it holds

$$\overline{L}(n) = \Sigma^* \left((\Sigma^* \setminus L_1) \cap \Sigma^n\right) = \Sigma^* \Sigma^n \setminus \Sigma^* (L_1 \cap \Sigma^n) = \Sigma^* \Sigma^n \setminus L_1(n).$$

Hence, $\overline{L}(n)$ consists of those strings $w \in \Sigma^*$ where $|w| \geq n$ and $w$ is rejected by $\mathcal{A}_1$. A DFA, with polynomially many states, simultaneously counting to $n$ and simulating $\mathcal{A}_1$ can be easily constructed. $\square$

On the other hand, it is not difficult to show that concatenation of two regular languages with polynomial pattern complexity may result in a regular language with exponential complexity. This can be easily demonstrated by expressing the regular language $L$ from Example 11 as $L = \{a\} \cdot \{a, b\}^*$. In addition, $L = L_3^R$ where $L_3 = \{ua \mid u \in \{a, b\}^*\}$ and $\sigma_{L_3} = \mathcal{O}(1)$.

**Corollary 16.** *There exist regular languages $L_1, L_2, L_3$ over $\Sigma$ with polynomial pattern complexity such that $L_1 L_2$ and $(L_3)^R$ are of exponential pattern complexity.*

## 5.2 Matching by $k$-gapped Two-Head DFA

In this section we show that a broader subclass of templates has an efficient matching procedure if the matching algorithm allows to access more number of (distant) fields in the input text. Namely, the algorithm keeps tracking on the symbol that leaves a fictive scanning window whose size equals the length of matches. We model this mechanism by introducing a $k$-gapped two-head deterministic finite automaton (g2h-DFA) as a finite-state device with two reading heads $h_l, h_r$ (the left and right one). The model is almost identical to the traditional finite-state model except for that there are two heads and the distance between them on the input is $k$ while reading the input[1].

Informally speaking, given an input $u = u_1 \cdots u_n \in \Sigma^*$, the $k$-gapped g2h-DFA prepends $k - 1$ #'s and $\vdash$ to $u$ and reformats $u$ to be $u^\# = \underbrace{\#\# \cdots \# \vdash u}_{k \text{ new symbols}}$,

which allows g2h-DFA to place two heads apart from each other at the distance $k$ on the input. The $\vdash$ is a delimiter separating the original input string $u$ from dummy symbols #, which are used for $h_l$ to make the desired distance $k$ between two heads from the beginning of the computation as the initial configuration becomes $[h_l]\#\# \cdots \# \vdash q_0[h_r]u_1u_2 \cdots u_n$, where $[h_l]$ and $[h_r]$ denote the corresponding positions of two heads.

Then g2h-DFA starts processing $u^\#$ from $q_0$ by reading two symbols indicted by two heads and going to the next state defined and moving two heads to the next symbols. The computation ends when g2h-DFA read the whole $u^\#$. If it ends at an accepting state, then we say that g2h-DFA accepts $u$.

**Definition 17.** *A $k$-gapped two-head DFA (g2h-DFA) is a tuple $(k, Q, \Sigma, \delta, q_0, F)$, where $k \in \mathbb{N}$ is the distance between two heads. The transition function is of the form $\delta : Q \times (\Sigma \cup \{\#, \vdash\}) \times \Sigma \to Q$. The other components are the same as for DFA.*

Given a g2h-DFA $\mathcal{A} = (k, Q, \Sigma, \delta, q_0, F)$ and an input string $u = u_1 \cdots u_n \in \Sigma^*$, the initial configuration is

$$[h_l]\#\# \cdots \# \vdash q_0[h_r]u_1u_2 \cdots u_n.$$

The general configuration computation of $\mathcal{A}$ for $u^\#$ is defined as follows: Given a current configuration

$$x[h_l]y_1y_2 \cdots y_k q_i[h_r]y_{k+1}y_{k+2} \cdots y_n,$$

where $x$ and $y = y_1y_2 \cdots y_n$ are strings over $\Sigma \cup \{\#, \vdash\}$ and $u^\# = xy$, the next configuration is

$$xy_1[h_l]y_2 \cdots y_k y_{k+1} q_j[h_r]y_{k+2} \cdots y_n,$$

if $\delta(q_i, y_1, y_{k+1}) = q_j$ in $\mathcal{A}$.

---

[1] The proposed automaton model g2h-DFA is different from the traditional two-headed finite automaton that has two read-only bidirectional heads.

We say that a configuration is *final* if it is $x'[h_l]y'q[h_r]$, where $u^{\#} = x'y'$. We say that it is *accepting* if $q \in F$—this is the case when $\mathcal{A}$ accepts $u$.

For a language $L \subseteq \Sigma^*$, recall that $L(n) = \Sigma^* (L \cap \Sigma^n)$. Since $L(n)$ is regular, there exists a state-minimal g2h-DFA $\mathcal{A}_n = (n, Q_n, \Sigma, \delta_n, F_n)$ accepting $L(n)$. Then, we define the pattern complexity $\tau_L$ of $L$ with respect to g2h-DFA to be the number of states in $\mathcal{A}$; namely, $\tau_L : \mathbb{N} \to \mathbb{N}$, where $\tau_L(n) = |Q_n|$.

An immediate outcome of the pattern complexity over g2h-DFA is an ability to implement a counter for the number of particular symbols in its fictive scanning window. This is illustrated by the following example.

*Example 18.* Consider the regular language $L_{\max,k} = \{u \mid u \in \{a,b\}^* \wedge |u|_a \leq k\}$ from Example 13. Now consider g2h-DFA $\mathcal{A} = (n, Q, \Sigma, \delta, (q_0, 0), F)$ where $Q = \{q_0, q_1, \ldots, q_n\} \times \{0,1\}$, $\Sigma = \{a, b\}$, $F = \{q_0, \ldots, q_k\} \times \{1\}$ and

$$\delta((q_i, 0), \#, a) = (q_{i+1}, 0) \quad \text{for } i < n, \quad \delta((q_i, 1), a, a) = (q_i, 1),$$
$$\delta((q_i, 0), \#, b) = (q_i, 0), \quad\quad\quad\quad\quad \delta((q_i, 1), b, b) = (q_i, 1),$$
$$\delta((q_i, 0), \vdash, a) = (q_{i+1}, 1) \quad \text{for } i < n, \quad \delta((q_i, 1), a, b) = (q_{i-1}, 1) \quad \text{for } i > 1,$$
$$\delta((q_i, 0), \vdash, b) = (q_i, 1), \quad\quad\quad\quad\quad \delta((q_i, 1), b, a) = (q_{i+1}, 1) \quad \text{for } i < n.$$

We omit all states and transitions that are not applicable to any configuration—e.g.: $\delta((q_n, 1), b, a)$. They can be defined arbitrarily. For each state $(q_i, j)$, the first component $q_i$ acts as a counter ranged from 0 to $n$, and the second component $j$ acts as a flag determining if the prefix of length $n$ has already been read. During each computation of $\mathcal{A}$, the counter—the index $i$ of the first component of the current state $(q_i, j)$—is increased/decreased if the symbol scanned by $h_r/h_l$ is $a$. Then it is easy to verify that $\mathcal{A}$ accepts $L_{\max,k}(n)$. Since the number of states in $\mathcal{A}$ is $O(n)$, it holds $\tau_{L_{\max,k}} = \mathcal{O}(n)$, which is an improvement towards $\sigma_{L_{\max,k}} = \Omega(n^k)$.

*Example 19.* For another example of the pattern complexity, consider $L = \{au \mid u \in \{a,b\}^*\}$ from Example 11. We construct a g2h-DFA $\mathcal{A} = (n - 1, Q, \Sigma, \delta, (q_0, 0), F)$ where $Q = \{q_0, q_1\} \times \{0,1\}$, $\Sigma = \{a, b\}$, $F = \{(q_0, 1)\}$ and, for $i \in \{0,1\}$,

$$\delta((q_i, 0), \#, a) = (q_{1-i}, 0), \quad\quad \delta((q_i, 1), a, a) = (q_i, 1),$$
$$\delta((q_i, 0), \#, b) = (q_i, 0), \quad\quad\quad\; \delta((q_i, 1), b, b) = (q_i, 1),$$
$$\delta((q_i, 0), \vdash, a) = (q_{1-i}, 1), \quad\quad \delta((q_i, 1), a, b) = (q_{1-i}, 1),$$
$$\delta((q_i, 0), \vdash, b) = (q_i, 1), \quad\quad\quad \delta((q_i, 1), b, a) = (q_{1-i}, 1).$$

It is straightforward to verify that $\mathcal{A}$ accepts $L(n)$ and, thus, $\tau_L = \mathcal{O}(1)$ for $L$.

Next, we identify a subclass of regular languages whose pattern complexity is constant with respect to g2h-DFA.

**Definition 20.** *We say that a DFA $\mathcal{A} = (Q, \Sigma, \delta, q_0, F)$ is* strongly Eulerian *iff for each $q \in Q$ and $a \in \Sigma$, there is a state $p \in Q$ such that $\delta(p, a) = q$.*

Strongly Eulerian DFAs can be characterized based on the structure of underlying transition graphs. For a DFA $\mathcal{A} = (Q, \Sigma, \delta, q_0, F)$ and $a \in \Sigma$, let $G(\mathcal{A}, a) = (Q, E)$ be the directed graph where the set of vertices equals $Q$ and $(p, q)$ is an edge in $E$ iff $\delta(p, a) = q$. The automaton $\mathcal{A}$ is strongly Eulerian iff, for each $a \in \Sigma$, $G(\mathcal{A}, a)$ consists of cycles—See Fig. 3 for an example. Another example of a regular language accepted by a strongly Eulerian DFA is the language from Example 12.

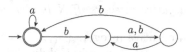

**Fig. 3.** A strongly Eulerian DFA accepting the string language described by regular expression $a^*(b((a + b)a)^*(a + b)ba^*)^*$.

Strongly Eulerian DFAs are a superset of well-known *bideterministic* automata [15]. A finite automaton $\mathcal{A}$ is bideterministic if $\mathcal{A}$ and its reversal automaton $\mathcal{A}^R$ are both deterministic. Researchers considered bideterministic automata and bideterministic languages in the context of machine learning or in coding theory in the literature [3,12]. Bideterminism also plays a crucial role in characterizing minimal DFAs in formal language theory.

**Proposition 21.** *Let $L \subseteq \Sigma^*$ be a regular language accepted by a strongly Eulerian DFA. Then, $\tau_L = \mathcal{O}(1)$.*

*Proof.* Let $L = L(\mathcal{A})$ where $\mathcal{A} = (Q, \Sigma, \delta, q_0, F)$ is a strongly Eulerian DFA. Define $\delta^{-1}(q, a)$ to be the only state $p \in Q$ such that $\delta(p, a) = q$. For $n \in \mathbb{N}$, construct a g2h-DFA $\mathcal{M} = (n, Q', \Sigma, \delta', q'_0, F')$ accepting $L(n)$ as follows. Let $w \in \Sigma^*$ be an input string of length $m$ where $m \geq n$. When processing the prefix of $w$ of length $n$, $\mathcal{M}$ simulates $\mathcal{A}$ and computes $\hat{\delta}(q, w_1 \ldots w_n)$ for each $q \in Q$ and a mapping $q \to \hat{\delta}(q, w_1 \ldots w_n)$ is stored in states of $\mathcal{M}$ at the end of the stage. The suffix $w_{n+1} \ldots w_m$ is processed during the second stage. Assume that $\mathcal{M}$ stores the mapping $q \to \hat{\delta}(q, w_i \ldots w_{n+i-1})$, its left head scans $w_{i+1}$ and the right head scans $w_{n+i}$. Then, in the next transition, $\mathcal{M}$ updates the mapping to be $q \to \hat{\delta}(q, w_{i+1} \ldots w_{n+i})$ using the formulas

$$\hat{\delta}(q, w_{i+1} \ldots w_{n+i-1}) = \hat{\delta}(\delta^{-1}(q, w_i), w_i \ldots w_{n+i-1}),$$
$$\hat{\delta}(q, w_{i+1} \ldots w_{n+i}) = \delta(\hat{\delta}(q, w_{i+1} \ldots w_{n+i-1}), w_{n+i}).$$

A state of $\mathcal{M}$ is accepting iff the stored mapping maps $q_0$ to a state from $F$ and the currently read prefix of $w$ is of length at least $n$ (i.e., $\vdash$ has already been encountered by $h_l$). To represent the mapping, $\mathcal{M}$ needs $\mathcal{O}(|Q|^{|Q|})$ states, which is a constant with respect to $n$.                                           □

As the next result we show that it is possible to efficiently detect those patterns where the number of a particular symbol occurrences corresponds to the length of a string from a unary regular language. More precisely, let $\Sigma$ be an alphabet such that $a \in \Sigma$ and $L$ be a regular language over $\{a\}$. We define the regular language $L' = \{u \in \Sigma \mid a^{|u|_a} \in L\}$. It can be equivalently expressed using the shuffle operation as $L' = L \, \Delta \, (\Sigma \setminus \{a\})^*$.

**Lemma 22.** *Let $c_1, \ldots, c_m$ be positive integers and $d = \gcd(c_1, \ldots, c_m)$. There exists a bound $B \in \mathbb{N}$ such that, for any $N \geq B$, the equation $c_1 x_1 + \cdots + c_m x_m = N$ has a solution in non-negative integers iff $N$ is divisible by $d$.*

*Proof.* For a vector of non-negative integers $(x_1, \ldots, x_m)$, $\sum_{i=1}^{m} c_i x_i$ is divisible by $d$, hence the equation has a solution only if $d$ divides $N$.

By Bézout's identity, we can write $d = \sum_{i=1}^{m} c_i z_i$ for integers $z_i$. Letting $c = \sum_{i=1}^{m} c_i$, every $N$ divisible by $d$ can be written as $N = qc + rd$ where $q, r$ are integers with $0 \leq r < \frac{c}{d}$.

If $N$ is large enough, in particular if $N \geq B = c + \frac{c^2}{d} \max_i |z_i|$, then the coefficients in $N = \sum_{i=1}^{m} (q + z_i r) c_i$ are all non-negative. □

**Proposition 23.** *Let $\Sigma$ be an alphabet, $a \in \Sigma$, $L \subseteq \{a\}^*$ be a regular language and $L' = L \, \Delta \, (\Sigma \setminus \{a\})^*$. Then, $\tau_{L'} = \mathcal{O}(n)$.*

*Proof.* By Parikh's theorem, $S = \{|u| \mid u \in L\}$ is a union of finitely many linear sets. W.l.o.g, consider that $S$ equals a linear set $\{c_0 + c_1 t_1 + \cdots + c_m t_m \mid t_1, \ldots, t_m \in \mathbb{N}_0\}$ where $c_0 \in \mathbb{N}_0$ and $c_i \in \mathbb{N}$ for all $i = 1, \ldots, m$. Let $B$ be the bound given by Lemma 22 for the equation $c_1 t_1 + \cdots + c_m t_m = N$. Moreover, let $\mathcal{S}$ be the set of all non-negative integers $N < B$ for which the equation has a non-negative solution. We then construct a g2h-DFA $\mathcal{A} = (n, Q, \Sigma, \delta, q_0, F)$ accepting $L'(n)$ as follows. It counts in states the number of $a$'s in the lastly read $n$ characters. Let the counter store a value $s$. Automaton $\mathcal{A}$ accepts iff head $h_r$ has already encountered $\vdash$, and it holds either $0 \leq s - c_0 < B$ and $s - c_0 \in \mathcal{S}$, or, $s - c_0 \geq B$ and $\gcd(c_1, \ldots, c_m)$ divides $s - c_0$. □

Note that as demonstrated in the proof of Proposition 23, a g2h-DFA may have a quite succinct representation—we construct a g2h-DFA using a constant memory regardless of $n$ (it is possible to calculate transitions as well as accepting states by a fixed set of constant-sized formulas). We believe that the problem of characterizing such g2h-DFAs should be an interesting problem.

As the last observation let us notice that g2h-DFA does not entirely solve the problem with possible exponential pattern complexity of a regular language obtained as concatenation of two regular languages of polynomial pattern complexity. This can be easily demonstrated by $L = \{a, b\} \cdot \{au \mid u \in \{a, b\}^*\}$. Using a g2h-DFA with the head distance $n - 1$ (instead of $n$) to search for matches of patterns in $L(n)$ would fix this particular case, however, there will still be another examples of the exponential increase.

# 6    Conclusion

We have presented a framework supporting to search for template-based two-dimensional patterns in two-dimensional arrays. The regular matrix grammar has been extended to a formalism powerful enough for describing patterns that occur in applications. Complexity of the proposed method has been analyzed with respect to two models considered for executing one-dimensional matching tasks for various subclasses of regular languages incorporated by the templates. We have established positive results (constant or polynomial pattern complexities) as well as negative results (exponential pattern complexities).

For future research, we see several directions. The computational model used for one-dimensional matching can be further extended to lower the pattern complexity of some regular languages. For example, more heads with varying distance can be considered.

The introduced regular matrix grammar with scanning window has its limits. It might not be powerful enough to describe some patterns of complex topology. There are two possibilities how to increase the power of the method. One option is to go beyond regular languages (and regular matrix grammars) when defining the templates. Note that the connection between regular languages and languages with polynomial pattern complexity is loose as demonstrated by our results. Other classes of string languages could establish a more tighter relation. A good candidate is the family of languages accepted by jumping finite automata [11] as these automata are related to counting symbols. The second option is to incorporate a preprocessing of the input image. Fast algorithms like depth-first search in a graph can be applied to detect topological relationships and such a preprocessed image can be passed to the matching procedure.

As the last direction, let us mention the possibility to generalize the framework to three-dimensional (or higher-dimensional) patterns as the matrix grammar naturally extends to higher dimensions and the presented matching algorithm can still be based on reduction to one-dimensional matchings for each of the original dimensions.

**Acknowledgments.** Han was supported by the Basic Science Research Program through NRF funded by MEST (2015R1D1A1A01060097) and the IITP grant funded by the Korea government (MSIP) (R0124-16-0002), and Průša was supported by the Czech Science Foundation under grant no. 15-04960S.

# References

1. Aho, A.V.: Algorithms for finding patterns in strings. In: van Leeuwen, J. (ed.) Algorithms and Complexity, Handbook of Theoretical Computer Science, vol. A, pp. 255–300. The MIT Press, Cambridge (1990)
2. Amir, A., Benson, G., Farach, M.: Alphabet independent two dimensional matching. In: Proceedings of the Twenty-fourth Annual ACM Symposium on Theory of Computing, STOC 1992, NY, USA, pp. 59–68 (1992). http://doi.acm.org/10.1145/129712.129719

3. Angluin, D.: Inference of reversible languages. J. ACM **29**(3), 741–765 (1982)
4. Baeza-Yates, R., Régnier, M.: Fast two-dimensional pattern matching. Inf. Process. Lett. **45**(1), 51–57 (1993). http://www.sciencedirect.com/science/article/pii/002001909390250D
5. Baker, T.P.: A technique for extending rapid exact-match string matching to arrays of more than one dimension. SIAM J. Comput. **7**(4), 533–541 (1978). http://dx.doi.org/10.1137/0207043
6. Bird, R.S.: Two dimensional pattern matching. Inf. Process. Lett. **6**(5), 168–170 (1977). http://dx.doi.org/10.1016/0020-0190(77)90017-5
7. Fernau, H., Paramasivan, M., Schmid, M.L., Thomas, D.G.: Scanning pictures the Boustrophedon way. In: Barneva, R.P., Bhattacharya, B.B., Brimkov, V.E. (eds.) IWCIA 2015. LNCS, vol. 9448, pp. 202–216. Springer, Cham (2015). doi:10.1007/978-3-319-26145-4_15
8. Giammarresi, D., Restivo, A.: Two-dimensional languages. In: Rozenberg, G., Salomaa, A. (eds.) Handbook of Formal Languages, vol. 3, pp. 215–267. Springer, New York (1997)
9. Kärkkäinen, J., Ukkonen, E.: Two and higher dimensional pattern matching in optimal expected time. In: Sleator, D.D. (ed.) Proceedings of the Fifth Annual ACM-SIAM Symposium on Discrete Algorithms, Arlington, Virginia, 23–25, pp. 715–723. ACM/SIAM (1994). http://dl.acm.org/citation.cfm?id=314464.314680
10. Karp, R.M., Rabin, M.O.: Efficient randomized pattern-matching algorithms. IBM J. Res. Dev. **31**(2), 249–260 (1987). http://dx.doi.org/10.1147/rd.312.0249
11. Meduna, A., Zemek, P.: Jumping finite automata. Int. J. Found. Comput. Sci. **23**(7), 1555–1578 (2012). http://www.fit.vutbr.cz/research/view_pub.php.cs?id=9795
12. Pin, J.-E.: On reversible automata. In: Simon, I. (ed.) LATIN 1992. LNCS, vol. 583, pp. 401–416. Springer, Heidelberg (1992). doi:10.1007/BFb0023844
13. Průša, D.: Complexity of sets of two-dimensional patterns. In: Han, Y.-S., Salomaa, K. (eds.) CIAA 2016. LNCS, vol. 9705, pp. 236–247. Springer, Cham (2016). doi:10.1007/978-3-319-40946-7_20
14. Siromoney, G., Siromoney, R., Krithivasan, K.: Abstract families of matrices and picture languages. Comput. Graph. Image Process. **1**(3), 284–307 (1972). http://www.sciencedirect.com/science/article/pii/S0146664X72800194
15. Tamm, H., Ukkonen, E.: Bideterministic automata and minimal representations of regular languages. Theoret. Comput. Sci. **328**(1–2), 135–149 (2004)
16. Zhu, R.F., Takaoka, T.: A technique for two-dimensional pattern matching. ACM Commun. **32**(9), 1110–1120 (1989). http://doi.acm.org/10.1145/66451.66459

# Construction of Persistent Voronoi Diagram on 3D Digital Plane

Ranita Biswas[1]([⊠]) and Partha Bhowmick[2]

[1] Department of Computer Science and Engineering,
Indian Institute of Technology, Roorkee, India
biswas.ranita@gmail.com
[2] Department of Computer Science and Engineering,
Indian Institute of Technology, Kharagpur, India
bhowmick@gmail.com

**Abstract.** Different distance metrics produce Voronoi diagrams with different properties. It is a well-known that on the (real) 2D plane or even on any 3D plane, a Voronoi diagram (VD) based on the Euclidean distance metric produces convex Voronoi regions. In this paper, we first show that this metric produces a persistent VD on the 2D digital plane, as it comprises *digitally convex* Voronoi regions and hence correctly approximates the corresponding VD on the 2D real plane. Next, we show that on a 3D digital plane $D$, the Euclidean metric spanning over its voxel set does not guarantee a digital VD which is persistent with the real-space VD. As a solution, we introduce a novel concept of *functional-plane-convexity*, which is ensured by the Euclidean metric spanning over the *pedal set* of $D$. Necessary proofs and some visual result have been provided to adjudge the merit and usefulness of the proposed concept.

**Keywords:** Digital Voronoi diagram · 3D digital plane · Distance metric · Digital convexity · Digital geometry

## 1 Introduction

Voronoi diagram in 2D and in 3D real spaces is a well-researched topic in computational geometry [2,3]. Historically, the concept dates back to the mid-19th century, since it finds potential applications ranging from modeling cells and bone micro-architecture in biology to estimating the reserves of valuable minerals and materials in mining. Voronoi diagrams are also used in designing visual arts and in numerous other applications in image processing, computer vision, and graphics. The Euclidean Delaunay triangulation, which is the dual combinatorial structure of Voronoi diagram in Euclidean space, has a bagful of applications in scientific computing and mesh generation, especially in terrain modeling.

A Voronoi diagram (VD) is a partitioning of a space into regions based on distance from a specific set of points as input, which are called *seeds* (also called *sites* or *generators*). For each seed there is a corresponding region consisting of all points closer to that seed than to any other (which forms the conventional

© Springer International Publishing AG 2017
V.E. Brimkov and R.P. Barneva (Eds.): IWCIA 2017, LNCS 10256, pp. 93–104, 2017.
DOI: 10.1007/978-3-319-59108-7_8

closest-seed Voronoi diagram). These regions are called *Voronoi cells* or *Voronoi regions*. For a given distance metric $d$, the Voronoi region $R_i$ corresponding to a seed $p_i$ $(1 \leqslant i \leqslant n)$ can be defined as follows.

$$R_i = \{q \in \mathbb{R}^2 \mid d(q, p_i) \leqslant d(q, p_j) \ \forall \ j = 1, 2, \ldots, n\} \tag{1}$$

Depending on the requirement, a specific distance measure or metric is chosen to create the VD in real space. Euclidean distance is the most commonly used metric in practice, as it connects the VD with real-life scenarios. Use of Euclidean metric produces Voronoi regions that are convex polygons in shape. Other metrics such as Manhattan distance or Mahalanobis distance produce Voronoi diagrams with different nature of the Voronoi regions, e.g., non-convex or with complex boundaries. Euclidean distance gives convex Voronoi regions because the distance travel is uniform in every direction, which is not the case with other distance measures or metrics.

A Voronoi diagram using Euclidean distance measure is called *Euclidean Voronoi diagram*. Such diagrams on the 2D digital plane can be produced by following a similar method as the generation of Euclidean Voronoi diagram in 2D real plane. On the 2D digital plane, the set of seed points are 2D integer points or pixels, and the region $R_i$ corresponding to seed $p_i$ can be defined simply by constraining the point $q$ in Eq. 1 to be a point in $\mathbb{Z}^2$. Hence, a simple pixel-coloring approach can be used on 2D digital plane to color each pixel with the color of its closest seed point (incorporating some consistent tie-breaking rule). However, as shown in [10], this leads to *debris pixels* due to the presence of *sliver polygons*, which happens when the corresponding real Voronoi regions possess very sharp corners. Hence, a parallel breadth-first-search algorithm starting from each seed point and incrementally growing each of the Voronoi regions until the boundaries touch each other, is a more effective method to produce the Euclidean Voronoi regions on the 2D digital plane [10]. In fact, the 2D digital Euclidean Voronoi regions are dual of the corresponding Delaunay triangulation, as shown in [10].

Unlike in the 2D digital plane, Euclidean distance metric does not easily fit into the topological space of 3D Voronoi diagram (VD) in the voxel space. To show this, we bring in the concept of persistence of a digital VD with its real counterpart. For this, we first show in Sect. 2 that Euclidean Voronoi regions on the 2D digital plane are always *digitally convex*. Owing to this, they closely approximate the Voronoi regions on the corresponding real plane and approach the real-plane Voronoi regions with increasing resolution of the underlying grid. Thus in 2D, a digital VD becomes persistent with the real VD.

Next, we show in Sect. 3 that the region-growing strategy, as described in [10], suffers from lack of persistence while constructing a VD on a 3D digital plane using the following equation.

$$R_i = \{q \in \mathbb{Z}^3 \mid d(q, p_i) \leqslant d(q, p_j) \ \forall \ j = 1, 2, \ldots, n\} \tag{2}$$

where, $R_i$ is the Voronoi region corresponding to the seed $p_i$, which is a 3D integer point or voxel. By '3D digital plane' we mean the thinnest digital plane (also known as 'naive plane'), which is *2-minimal* (Sect. 2).

We have investigated the reason behind the failure of Euclidean metric in the voxel space in producing a persistent VD, which is reported in this paper. To circumvent the problem, we propose a simple alteration of the Euclidean distance measure, which is commensurable with correctly constructing a VD with *FP-convex* regions on a 3D digital plane. The notion of FP-convex region is newly introduced in this paper and its significance in the context of digital VD on 3D planes is explained in Sect. 4. The property of FP-convexity of VD regions ensures that by increasing the grid resolution to a sufficiently large value, the digital VD constructed on a 3D digital plane is persistent, as it can be made to approach the real-space VD on the corresponding real plane.

## 2   Preliminaries

In this section, we explain the basic definitions and terminologies from digital geometry which are relevant in the context of our work. For two (real or integer) points $p(i, j, k)$ and $p'(i', j', k')$, we define the distance between them along each coordinate axis. For the coordinate $w \in \{`x`,`y`,`z`\}$, it is given by

$$d_w(p, p') = \begin{cases} |i - i'| & \text{if } w = `x` \\ |j - j'| & \text{if } w = `y` \\ |k - k'| & \text{if } w = `z`. \end{cases}$$

The inter-point distances define the respective $x$-, $y$-, and $z$-distances between a point $p(i, j, k)$ and a (real) surface $\Gamma$, which can be generalized as follows.

$$d_w(p, \Gamma) = \begin{cases} \min\{d_w(p, p') : p' \in \Gamma_w(p)\} & \text{if } \Gamma_w(p) \neq \emptyset \\ \infty & \text{otherwise} \end{cases}$$

where, $\Gamma_w(p) = \{p' \in \Gamma : d_v(p, p') = 0 \; \forall v \in \{`x`,`y`,`z`\} \setminus \{w\}\}$.

The above definitions are used to define the isothetic distance between two points, or between a point and a surface. Between two points $p(i, j, k)$ and $p'(i', j', k')$, isothetic distance is taken as the Chebyshev distance or Minkowski norm [15], given by

$$d_\infty(p, p') = \max\{d_x(p, p'), d_y(p, p'), d_z(p, p')\}.$$

Between a point $p(i, j, k)$ and a surface $\Gamma$, it is defined as

$$d_\perp(p, \Gamma) = \min\{d_x(p, \Gamma), d_y(p, \Gamma), d_z(p, \Gamma)\}.$$

In 2D Euclidean space, the integer points are termed as pixels and visualized as unit squares (2-cell) centered on integer points. When represented using unit squares, two pixels are said to be 1-adjacent if they share an edge (1-cell) and 0-adjacent if they share a vertex (0-cell). A 1-path (0-path) is a sequence of pixels where each pair of consecutive pixels are 1-adjacent (0-adjacent). A finite set of pixels (say, $R$) is 1-connected (0-connected) if a 1-path (0-path) exists in $R$ between any two pixels of $R$.

In 3D space, objects are represented by isothetic polyhedra composed by unit cubes (voxels) defined by the integer grid. A voxel is an integer point in 3D space, and equivalently, a 3-cell [15]. Two distinct voxels are said to be 0-adjacent if they share a vertex (0-cell), 1-adjacent if they share an edge (1-cell), and 2-adjacent if they share a face (2-cell). The 0-, 1-, and 2-neighborhood notations adopted by us in our work correspond respectively to the classical 26-, 18-, and 6-neighborhood notations used in [12].

For $l \in \{0, 1, 2\}$, an *l-path* in a 3D discrete object $A$ (or the discrete space $\mathbb{Z}^3$) is a sequence of voxels from $A$ such that every two consecutive voxels are *l*-adjacent. The object $A$ is said to be *l-connected* if there is an *l*-path in $A$ connecting any two voxels of $A$. An *l-component* is a maximal *l*-connected subset of $A$. Let $D$ be a subset of a discrete object $A$. If $A \setminus D$ is not *l*-connected, then the set $D$ is *l-separating* in $A$. Let $D$ be an *l*-separating discrete object in $A$ such that $A \setminus D$ has exactly two *l*-components. A 3-cell $c \in D$ is said to be *l-simple* w.r.t. $D$ if $D \setminus \{c\}$ is *l*-separating in $A$. An *l*-separating discrete object in $A$ is *l-minimal* if it does not contain any *l*-simple 3-cell w.r.t. $A$.

Let $A \in \mathbb{Z}^3$ be a discrete object and $A'$ be its projection on a real plane $P$. If there exists a bijection between $A$ and $A'$, then the plane $P$ is said to be a functional plane of $A$. For our work, in particular, we say that a coordinate plane, say, $xy$, is functional for $A$, if for every voxel $v = (x_0, y_0, z_0) \in A$ there is no other voxel in $A$ with the same first two coordinates. For example, $A = \{(2, 5, 3), (2, 6, 3), (3, 5, 3)\}$ is a discrete 3D object. Projecting $A$ on the coordinate planes gives us the 2D sets as follows: $\{(2, 5), (2, 6), (3, 5)\}$ in $xy$-plane, $\{(5, 3), (6, 3)\}$ in $yz$-plane, and $\{(3, 2), (3, 3)\}$ in $zx$-plane. As a bijection exists here between $A$ and its projection on the $xy$-plane, it becomes the functional plane of $A$.

## 2.1   3D Digital Plane

Digital plane is a well-researched topic in the subject of digital geometry [9, 15]. Standardized and analytical definitions of different classes of digital plane can be found in several papers [1, 6–9, 11, 14, 15]. The analytical equation of a digital plane having *thickness* $\omega$ and centered on the real plane $ax + by + cz = \mu$ is given by

$$\mu - \tfrac{\omega}{2} \leqslant ax + by + cz < \mu + \tfrac{\omega}{2}. \tag{3}$$

For other related details, we refer to [15]. Without loss of generality, we consider $\mu = 0$. Therefore, the *2-minimal digital plane* (henceforth, simply called 'digital plane') centered on the real plane $ax + by + cz = 0$ admits the following characterization [1, 9].

$$\frac{-\max(|a|, |b|, |c|)}{2} \leqslant ai + bj + ck < \frac{\max(|a|, |b|, |c|)}{2} \tag{4}$$

A digital plane always has at least one functional plane (FP) that can be obtained by removing the coordinate for which the absolute value of the coefficient is the highest. For example, for a plane $ax + by + cz = 0$, if $|c|$ is greater than both $|a|$

and $|b|$, then $xy$-plane is the FP corresponding to $ax + by + cz = 0$. As shown in [4], the isothetic distance (minimum of the axis-parallel distances) of any voxel of the digital plane is at most $\frac{1}{2}$ from the corresponding real plane.

## 3   2D Digital Voronoi Diagram

Several algorithms can be found in the literature for efficient generation of Voronoi diagrams on 2D grid. A general approach using incremental growing from the seed points can be seen in [10]. In this approach, parallel breadth-first-search is executed from the seeds until the consecutive region boundaries touch each other. There are GPU-based algorithms as well, e.g., the jump flooding algorithm in [16], which can be used to efficiently generate Voronoi diagram or distance transform on 2D grid.

Generation of Euclidean VD on 2D digital plane has, however, certain algorithmic challenges. One such challenge lies in handling *debris*, which splits a VD region into multiple connected components. As discussed in [10], a naive coloring-based algorithm using Eq. 1 on 2D digital plane for assigning the nearest-seed color to each pixel may create debris. The debris effect is more pronounced with occurrence of 'sliver' (long and sharp corner) in a VD region. Figure 1(a) shows such an instance where the sliver-containing real Voronoi region is shown using a green boundary and the corresponding digital Voronoi region is shown using pink pixels. Notice that due to the occurrence of debris points, we get two connected components here for a single Voronoi region.

A region growing algorithm solves this problem by restricting the growing of a region when it hits the points from the boundaries of the consecutive regions. However, in this way, we are letting the debris points be engulfed in a different region than where it belonged by Eq. 1. The digital VD obtained by region growing algorithm is thus not persistent with the real VD, wherefore the correspondence between the real and the digital VDs is lost. A solution to avoid occurrence of debris points and to simultaneously maintain the correspondence with the real VD is to increase the grid resolution to a sufficiently large value.

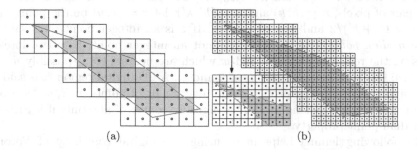

(a)                                    (b)

**Fig. 1.** (a) Occurrence of debris point due to presence of 'sliver' in the real polygon. (b) Increasing the resolution of the grid solves the problem.

As can be seen in Fig. 1(b), increasing grid resolution leads to a single component for the same Voronoi region. Therefore, henceforth we assume that we are working with a sufficiently high-resolution grid so that the slivers and resultant debris do not occur.

## 3.1 Convexity of Digital Regions

The concept of convexity of a region or a polygon in 2D real plane is quite unambiguous; however, it is not so in 2D or 3D digital space. The commonly used convexity notion in 2D digital space is *hv-convexity* or horizontal-vertical convexity. A 2D digital region $R$ (a set of pixels) is said to be *h-convex* or horizontal-convex if each row of $R$ is 1-connected. Similarly, it is *v-convex* or vertical-convex if each column is 1-connected. If $R$ is both h-convex and v-convex, then it is called *hv-convex*. The notion of *ortho-convexity* defined for 3D digital object (a set of voxels) is similar to the notion of hv-convexity in 2D. A 3D digital object $R$ is *ortho-convex* or orthogonally convex when its intersection with any plane parallel to one of the coordinate planes is either empty or an hv-convex object. In other words, each row, each column, and each stack of voxels in an ortho-convex object is 2-connected.

There is another notion of convexity in digital space, which more closely resembles the convexity in real space than hv-convexity or ortho-convexity. This convexity is known as *digital convexity* and defined as follows.

**Definition 1 (Digitally Convex [13]).** *A 2D digital region $R$ is digitally convex if and only if there does not exist any pixel $p$ which belongs to the convex hull of $R$ but not in $R$.*

It has been shown in [13] that a digital region $R$ is (digitally) convex if and only if any two points of $R$ are connected by a digital straight line segment in $R$. It can be realized that, when we increase the resolution of the underlying grid, a digitally convex region tends to a real convex region, which is not the case for hv-convex region or ortho-convex regions e.g. an 'L'-shaped hv-convex region.

From [13], we know that digital convexity satisfies *median-point property*, and it is a necessary and sufficient property to make a region digitally convex. For a pair of pixels, $u = (h, k)$ and $w = (h', k')$, let $z = (x, y)$ be the point such that $x = (h + h')/2$ and $y = (k + k')/2$. If $z$ is an integer point, it is said to be *the median point* of $u$ and $w$. If $z$ is not an integer point, then two integer points on the real line joining $u$ and $w$ which are nearest to $z$ (possibly $u$ and $w$) are said to be *the median points* of $u$ and $w$. A 2D digital region $R$ is said to be satisfying median-point property if for every pair of points in $R$, at least one of the median points belong to $R$. $R$ is digitally convex if and only if it satisfies the median-point property.

The following lemma helps in ensuring the digital convexity of Voronoi regions generated using specific distance metrics.

**Lemma 1.** *Inner pixel cover of a convex region on the 2D real plane is always digitally convex.*

*Proof.* The inner pixel cover $I$ of a real convex region $C$ includes all the pixels that lie inside $C$. Assume, there is a pair of points $u$ and $w$ in $I$ such that none of their median point(s) is included in $I$. This means the pair $(u, w)$ violates the median-point property. However, as $C$ is convex, and $u$ and $w$ both lie inside $C$, the real line segment joining $u$ and $w$ completely lies inside $C$; hence, any integer point that lies on this real line segment must lie inside $C$. Therefore, the median point(s) of $u$ and $w$ is included in $I$, which contradicts our assumption. Hence, the proof.                                                                            □

As mentioned earlier, an Euclidean VD on the 2D digital plane can be produced using Eq. 1 by restricting the domain of seed points to $\mathbb{Z}^2$. We now introduce the following theorem on the property of Euclidean VD constructed on the 2D digital plane.

**Theorem 1.** *An Euclidean Voronoi diagram on the 2D digital plane always comprises digitally convex Voronoi regions.*

*Proof.* On the 2D digital plane, each Voronoi region $R_i$ of an Euclidean VD is a 1-connected set of integer points that satisfy Eq. 1. Hence, $R_i$ is basically the inner pixel cover of the corresponding Euclidean Voronoi region on the 2D real plane. By Lemma 1, the inner pixel cover of a real convex region is always digitally convex. Hence, the proof.                                                                            □

The property of digital convexity of each Voronoi region ensures that when we have a grid of sufficiently high resolution, the digital VD tends to the real VD, and hence it is persistent.

# 4  Voronoi Diagram on 3D Digital Plane

As discussed in Sect. 1, Euclidean distance is the perfect metric for computing convex Voronoi regions on a real plane where the movements are not restricted to any direction. Euclidean metric also produces digitally convex Voronoi regions on 2D digital plane, as we have shown in Sect. 3. However, when the input is a 3D digital plane, we need to be cautious about the selection of the distance metric to make the resultant VD closely approximate the Euclidean VD on the corresponding 3D real plane. The following example gives an intuitive idea. Assume there are $n$ seed pixels on the $xy$-plane. We can assign $z$-coordinates on these seeds to lift them in 3D space such that all these $n$ 'lifted seeds' now belong to some 3D digital plane. We can do it in many possible ways and hence can get many such digital planes. For some of them, the use of inter-voxel Euclidean metric would produce Voronoi diagrams that are not persistent with their corresponding diagrams on the real planes. To show this, we introduce here a measure of convexity for Voronoi regions of an Euclidean VD on 3D digital planes.

**Definition 2 (FP-convex).** *A subset of a 3D digital plane is FP-convex if its projection on the functional plane is digitally convex.*

The above definition uses the fact that each voxel from the 3D digital plane relates to a single pixel on the projection on the functional plane. Therefore, the correspondence between an FP-convex set and a digitally convex can be made from 3D to 2D. As FP-convexity of a voxel set depends entirely on the digital convexity of its projection on FP, this convexity property does not get affected with changing orientation of the 3D plane. Further, due to the correlation of FP-convexity with digital convexity, an FP-convex region on the 3D digital plane becomes persistent with a real convex region on its corresponding 3D real plane when we move towards higher grid resolution.

Our objective is to construct Euclidean Voronoi diagram on a 3D digital plane so that it tends to the real Voronoi diagram with finer and finer grid resolution. For this, we define a *persistent Euclidean Voronoi diagram* as the one for which its comprising Voronoi regions remain FP-convex irrespective of the orientation of the corresponding real plane in 3D space. In order to ensure this, we need an appropriate distance metric. In this section, we show the implications of inter-voxel Euclidean metric on creating the Voronoi regions on 3D digital planes and how it fails to render a persistent VD. As a practical and effective solution, we propose a variation of the inter-voxel Euclidean metric, namely *inter-pedal Euclidean metric*, which is guaranteed to create persistent VD on a 3D digital plane.

## 4.1  Inter-voxel Euclidean Metric

The inter-voxel Euclidean metric directly uses Eq. 2 on the voxels of the 3D digital plane. As we have already mentioned, it can be seen from the results in Fig. 2(a), the generated Voronoi regions are not FP-convex. Figure 2(b) shows projections of the generated Voronoi regions on the functional plane to highlight the fact that the produced regions in the projection are not digitally convex.

As we have proved in Sect. 3, the Voronoi regions produced in 2D using the Euclidean distance between pixel pairs as the distance metric are always digitally convex. However, as we can see it is not the case for the 3D counterpart. The genuine reason for this lies in the fact that the voxels of a 3D digital plane do not lie exactly on the corresponding 3D real plane unless the real plane is a special case e.g. parallel to some coordinate plane. Whereas, in case of 2D plane, the pixels and their real counterparts are the same set of points.

As discussed in Sect. 2.1, the voxels of a 3D digital plane lie within an isothetic distance of $\frac{1}{2}$ from the real plane. Therefore, in a general case of a plane, the inter-voxel Euclidean distance does not necessarily represent the Euclidean distance between their corresponding real points. We name this corresponding real point of a voxel as the *pedal point*. More precisely, the pedal point of a voxel is the isothetically nearest point on the underlying real plane. Let $p_i$ and $p_j$ be two seed voxels on the 3D digital plane and their respective pedal points on the real plane be $p_i^\perp$ and $p_j^\perp$. Now, when we measure the distance of an arbitrary voxel $q$ on the 3D digital plane from these two seeds, it could be a case that $d(q, p_i) < d(q, p_j)$, but $d(q^\perp, p_i^\perp) > d(q^\perp, p_j^\perp)$, where $q^\perp$ is the pedal point of $q$. This leads to assigning some voxels along the borders of the Voronoi regions wrongly to one region instead of the other and thus makes the regions FP-non-convex.

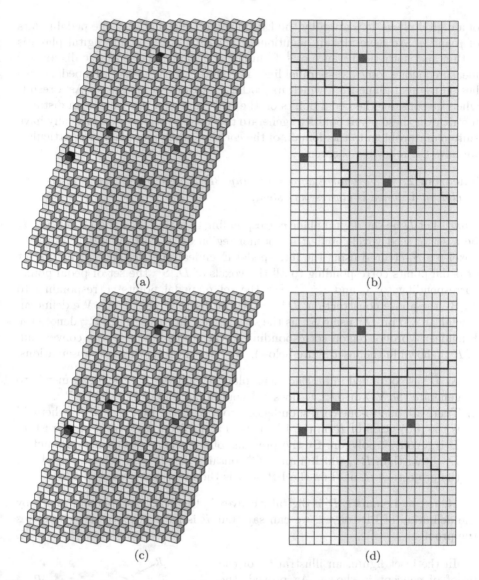

(a)

(b)

(c)

(d)

**Fig. 2.** Voronoi diagrams on a digital plane and their functional-plane projections using (a, b) Inter-voxel and (c, d) Inter-pedal Euclidean distances. Notice that in (b) the yellow region is not digitally convex, which indicates (a) is not persistent. (Color figure online)

## 4.2   Inter-pedal Euclidean Metric

As the Euclidean Voronoi regions generated on 3D digital plane are not FP-convex, we propose the generation of Persistent Euclidean VD using inter-pedal Euclidean metric. Pedal Euclidean distance is defined between two voxels $p$ and $q$

of a digital plane. It is given by the Euclidean distance between the pedal points of $p$ and $q$. We have already mentioned that each voxel of the digital plane is within and isothetic distance of $\frac{1}{2}$ from the real plane, and isothetic distance is measured along some axis-parallel line. Therefore, for each voxel, the pedal point lies on the real plane from which its isothetic distance is measured. As a result, when we consider the pedal points on the real plane and use Euclidean distance metric over these pedal points to generate the Voronoi regions, we basically have shifted our problem to the domain of the corresponding real plane. In particular, we have the following theorem.

**Theorem 2.** *A Voronoi diagram on any 3D digital plane using inter-pedal Euclidean distance is always persistent.*

*Proof.* Let $D$ be the digital plane corresponding to a 3D real plane $P$. Let $S \subset D$ be a set of seed voxels, and $R$ a Voronoi region on $D$ corresponding to a seed voxel $p \in S$, created using the inter-pedal Euclidean metric. Let $D^{\perp}$ be the set of pedal points corresponding to all the voxels of $D$, $S^{\perp}$ the set of pedal points corresponding to $S$, and $R^{\perp} \subset D^{\perp}$ the set of pedal points corresponding to $R$. By the above construction, $D^{\perp}$ is a set of discrete real points. We define an Euclidean VD on $D^{\perp}$ using $S^{\perp}$ as the set of seeds. In this VD, let $R^{\perp}$ denote the Euclidean Voronoi region corresponding to the seed $p^{\perp}$, and $C$ the convex hull of $R^{\perp}$ (shown in the inset figure below). We have the following two observations:

1. As $D^{\perp}$ is contained in an Euclidean plane $P$ and we use Euclidean metric to generate the VD on $D^{\perp}$, $C$ does not contain any point from $D^{\perp} \setminus R^{\perp}$.
2. There are one-to-one correspondences from $R$ to $R^{\perp}$ and also to $R'$ where $R'$ is the projection of $R$ on the FP of $D$. The projection $C'$ of $C$ on the FP of $D$ is a convex polygon. By the previous observation, $C$ contains no point in $D^{\perp}$ other than $R^{\perp}$, which implies $C'$ contains just $R'$ and no other pixel. So, the projection $R'$ of $R$ on the FP of $D$ is the inner pixel cover of $C'$.

Hence, by Lemma 1, $R'$ is digitally convex in the 2D digital plane. Finally, by the definition of FP-convex, we can say that $R$ is FP-convex, which completes the proof.    □

In the inset figure, an illustration of the proof of concept is shown. As proved, the regions obtained by inter-pedal Euclidean distance gives us FP-convex Voronoi regions. An example is shown in Fig. 2(c, d). Figure 3 shows Persistent Euclidean VD on a 3D digital plane at three different levels of resolution. It indicates how FP-convex Voronoi regions tend towards real Voronoi regions with finer and finer grid.

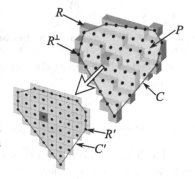

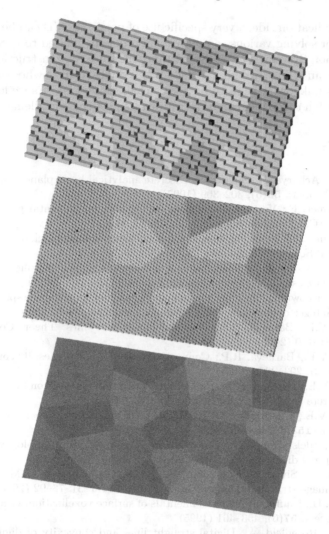

**Fig. 3.** Persistent Euclidean VD on a 3D digital plane at 3 different levels of resolution.

## 5   Conclusion

We have shown how the Euclidean distance metric defined over the pedal set of a 3D digital plane can be used for construction of a persistent Voronoi diagram on the plane. Naturally, this poses the feasibility of the proposed technique for construction of VD with similar persistence when the underlying real surface of the digital object is non-linear. For example, for a 2-minimal digital sphere [5], construction of VD would involve greater challenges and would require an appropriate convexity measure in the digital space in order to establish the persistence of the VD with its real counterpart on a real sphere.

On the application side, a very specific use of persistent VD can be related to 3D terrains for solving various computational problems related to GIS (Geography Information System). A suitable distance measure for construction of well-defined VD on an arbitrary digital surface, e.g., a digital terrain whose underlying real surface is unknown, seems to be a very challenging task. As we foresee, the work presented in this paper can be forwarded to meet these challenges in future.

# References

1. Andres, E., Acharya, R., Sibata, C.: Discrete analytical hyperplanes. Graph. Models Image Process. **59**(5), 302–309 (1997)
2. Aurenhammer, F.: Voronoi diagrams–a survey of a fundamental geometric data structure. ACM Comput. Surv. **23**(3), 345–405 (1991)
3. Aurenhammer, F., Klein, R., Lee, D.: Voronoi Diagrams and Delaunay Triangulations. World Scientific, Singapore (2013)
4. Biswas, R., Bhowmick, P.: On different topological classes of spherical geodesic paths and circles in $\mathbb{Z}^3$. Theor. Comput. Sci. **605**, 146–163 (2015)
5. Biswas, R., Bhowmick, P.: From prima quadraginta octant to lattice sphere through primitive integer operations. Theor. Comput. Sci. **624**, 56–72 (2016)
6. Brimkov, V.E., Barneva, R.P.: Graceful planes and lines. Theor. Comput. Sci. **283**(1), 151–170 (2002)
7. Brimkov, V.E., Barneva, R.P.: Connectivity of discrete planes. Theor. Comput. Sci. **319**(1–3), 203–227 (2004)
8. Brimkov, V.E., Barneva, R.P.: Plane digitization and related combinatorial problems. Discrete Appl. Math. **147**(2–3), 169–186 (2005)
9. Brimkov, V.E., Coeurjolly, D., Klette, R.: Digital planarity–a review. Discrete Appl. Math. **155**(4), 468–495 (2007)
10. Cao, T.T., Edelsbrunner, H., Tan, T.S.: Triangulations from topologically correct digital Voronoi diagrams. Comput. Geom. **48**(7), 507–519 (2015)
11. Coeurjolly, D., Sivignon, I., Dupont, F., Feschet, F., Chassery, J.M.: On digital plane preimage structure. Discrete Appl. Math. **151**(1–3), 78–92 (2005)
12. Cohen-Or, D., Kaufman, A.: Fundamentals of surface voxelization. Graph. Models Image Process. **57**(6), 453–461 (1995)
13. Kim, C.E., Rosenfeld, A.: Digital straight lines and convexity of digital regions. IEEE Trans. Pattern Anal. Mach. Intell. **4**(2), 149–153 (1982)
14. Klette, R., Stojmenović, I., Žunić, J.: A parametrization of digital planes by least square fits and generalizations. Graph. Models Image Process. **58**, 295–300 (1996)
15. Klette, R., Rosenfeld, A.: Digital Geometry: Geometric Methods for Digital Picture Analysis. Morgan Kaufmann, San Francisco (2004)
16. Rong, G., Tan, T.S.: Jump flooding in GPU with applications to Voronoi diagram and distance transform. In: Proceedings of the 2006 Symposium on Interactive 3D Graphics and Games, pp. 109–116 (2006)

# Extension of a One-Dimensional Convexity Measure to Two Dimensions

Sara Brunetti[1], Péter Balázs[2(✉)], and Péter Bodnár[2]

[1] Dipartimento di Ingegneria dell'Informazione e Scienze Matematiche,
Via Roma, 56, 53100 Siena, Italy
sara.brunetti@unisi.it

[2] Department of Image Processing and Computer Graphics,
University of Szeged, Árpád tér 2., Szeged 6720, Hungary
{pbalazs,bodnaar}@inf.u-szeged.hu

**Abstract.** In this paper we propose a new idea to design a measure for shape descriptors based on the concept of Q-convexity. The new measure extends the directional convexity measure defined in [2] to a two-dimensional convexity measure. The derived shape descriptors have the following features: (1) their values range from 0 to 1; (2) their values equal 1 if and only if the binary image is Q-convex; (3) they are invariant by reflection and point symmetry; (4) their computation can be easily and efficiently implemented.

**Keywords:** Shape descriptor · Convexity measure · Q-convexity

## 1 Introduction

Shape representation is a current topic in digital image analysis, for example, for object recognition and classification. The approaches for handling the problem consist in the design of new shape descriptors and measures for descriptors sensitive to distinguish the shapes but robust to noise. There are several methods used for describing shapes. Sometimes they provide a unified approach that can be applied to determine a variety of shape measures, but more often they are specific to a single aspect of shape. Over the years, measures for descriptors based on convexity have been developed: Area based measures form one popular category [4,15,16], while boundary-based ones [17] are also frequently used. Other methods use simplification of the contour [11] or a probabilistic approach [13,14] to solve the problem.

An alternative to "total" convexity studied in discrete geometry, and especially in discrete tomographic reconstruction is the horizontal and vertical convexity (or shortly, hv-convexity), arising inherently from the pixel-based representation of the digital image (see, e.g., [3,7,8]). A measure of horizontal (or vertical) convexity was introduced in [2], showing also that the aggregation of the measure in two dimensions can be a difficult task. In [1] the authors proposed an immediate two-dimensional convexity measure based on the concepts

V.E. Brimkov and R.P. Barneva (Eds.): IWCIA 2017, LNCS 10256, pp. 105–116, 2017.
DOI: 10.1007/978-3-319-59108-7_9

of Q-convexity [5,6] and exploiting the geometrical properties of salient points [9,10]. In this paper, we present an alternative new convexity measure based on the concept of Q-convexity that extends the directional convexity in [2] to a two-dimensional convexity measure. This new measure differs from the measures in [1] because it does not employ salient points, but uses quantitative information derived directly by the definition of Q-convexity (see Sect. 3). As a result, it is very easy to compute. We show how the measure can be normalized in two ways to obtain two shape descriptors having the following features: (1) their values range from 0 to 1; (2) their values equal 1 if and only if the binary image is Q-convex; (3) they are invariant by reflection and point symmetry; (4) their computation can be easily and efficiently implemented. We show with some experiments that the descriptors correctly incorporate the notion of Q-convexity. Finally, we briefly discuss sensitivity and robustness to noise for them.

## 2    Notation and Definitions

In this section we introduce the necessary notation and definitions. A *binary image* is a digital image containing just black (also called as object or foreground) and white (background) pixels. A binary image of size $m \times n$ (where $m, n \in \mathbb{Z}$) can also be represented by a binary matrix $F = (f_{ij})_{m \times n}$ where value 1 (respectively, value 0) indicates that the color of the corresponding pixel is black (respectively, white). $F$ is called *horizontally* (respectively, *vertically*) *convex* if its 1's (or black pixels) follow consecutively in each row (respectively, in each column). We also say that each row (column) is convex.

Let us denote the vector of *row* and *column sums* of the image $F$ by $H = (h_1, \ldots, h_m)$ and $V = (v_1, \ldots, v_n)$, respectively, where

$$h_i = \sum_{j=1}^{n} f_{ij} \ (i = 1, \ldots, m) \quad \text{and} \quad v_j = \sum_{i=1}^{m} f_{ij} \ (j = 1, \ldots, n). \tag{1}$$

Figure 1 shows a binary image $F$ with row and column sums $H = (1, 3, 3, 1, 3, 2)$, and $V = (1, 4, 3, 2, 1, 1, 1)$, respectively, and its matrix representation. $F$ is horizontally convex but not vertically convex.

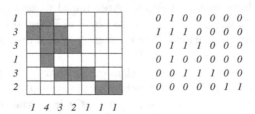

| 1 | | 0 1 0 0 0 0 0 |
| 3 | | 1 1 1 0 0 0 0 |
| 3 | | 0 1 1 1 0 0 0 |
| 1 | | 0 1 0 0 0 0 0 |
| 3 | | 0 0 1 1 1 0 0 |
| 2 | | 0 0 0 0 0 1 1 |

       1 4 3 2 1 1 1

**Fig. 1.** A binary image with its horizontal and vertical projections, and its matrix representation. The binary image is horizontally convex but not vertically convex.

The rows and columns of a binary image can be represented by using *run-length encoding* (see, e.g., [12]) which likely results in a more compact description of them (especially, if there are relatively few, but preferably long runs of identical values in the data). A *1-token* is a token of 1s and a *0-token* is a token of 0s. The *length* of a given token is the number of occurrences of the same value in that particular token. For example the row 00111000000111111 can be encoded by $0^2 1^3 0^6 1^6$, where the superscripts represent the length of each token (counters). The *length* of the row (column) is the total number of bits present in that row (column). In our example the length of the row is 17.

## 2.1   Measuring Non-convexity of a Single Row or Column

In [2] the basic idea of the definition of the directional measure is the following. First, consider only the horizontal direction. Let us recall that a row is convex if all its 1's are consecutive, otherwise 0's may separate any two 1's. Then, consider all the pairs of items 1's on the same row and the line segments connecting them. To compute the *non-convexity* of a row $R$, we split it into a sequence of 1-tokens and 0-tokens. Leading and trailing 0-tokens do not contribute to the measure, thus hereafter we shall omit them. The rest of the row can be encoded as $R = 1^{k_1} 0^{l_1} 1^{k_2} 0^{l_2} \ldots 1^{k_n}$, where $n$ is the number of 1-tokens and $k_1, l_1, k_2, l_2, \ldots, k_n > 0$. Trivially, taking two 1's from the same 1-token, the line segment connecting them will not contain any 0's and hence will not contribute to the non-convexity measure. Now, let us take two arbitrary 1s from different 1-tokens, say the $i$th and $j$th, such that $i < j$. The contribution to non-convexity of 0's in between is given by the sum of the lengths of the 0-token in between:

$$\sum_{t=i}^{j-1} l_t. \tag{2}$$

For two different 1-tokens ($i$th and $j$th), we can form $k_i k_j$ possible pairs of 1s, by picking one from each. The contribution of these 1-token pairs is

$$k_i k_j \sum_{t=i}^{j-1} l_t. \tag{3}$$

Finally, to get the contributions for the entire row $R$ sum up (3) for all possible combinations of 1-token pairs:

$$\varphi_h(R) = \sum_{1 \le i < j \le n} k_i k_j \sum_{t=i}^{j-1} l_t. \tag{4}$$

The value $\varphi_h(R)$ actually indicates the horizontal *non-convexity* of $R$, the higher $\varphi_h(R)$ is, the horizontally "less convex" $R$ is. Figure 2 shows an example of a binary image and the calculation of the horizontal non-convexity value of one particular row.

In [2] the authors proved that the non-convexity of every row can be normalized by

$$\hat{\varphi}_h(R) = \frac{\varphi_h(R)}{(n/3)^3}, \tag{5}$$

where $n$ is the length of the row, and cumulating (5) for all the rows of the matrix $F$,

$$\Phi_h(F) = \frac{\sum_{i=1}^m \hat{\varphi}_h(R_i)}{m} \tag{6}$$

is the normalized non-horizontal convexity measure. Finally to map horizontal non-convexity into horizontal convexity, they simply adopt

$$\Psi_h = 1 - \Phi_h. \tag{7}$$

Naturally, the same argument can be repeated for the columns of the image yielding the vertical convexity measure $\Psi_v$.

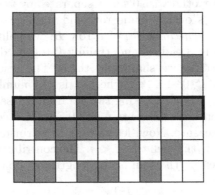

**Fig. 2.** The non-convexity of the highlighted row $R = 110100111$ is $\varphi_h(R) = k_1 k_2 l_1 + k_1 k_3 (l_1 + l_2) + k_2 k_3 l_3 = 2 \cdot 1 \cdot 1 + 2 \cdot 3 \cdot 3 + 1 \cdot 3 \cdot 2 = 26$.

Another viewpoint to think about this is to calculate the contribution of every entry 0 in between any two different 1-tokens. If a 0 entry belongs to the $t$-th 0-token (of length $l_t$), then its contribution is given by: $(k_1 + \ldots + k_t)(k_{t+1} + \ldots + k_n)$. Then, for all the 0's entries in the same token, we get

$$l_t(k_1 + \ldots + k_t)(k_{t+1} + \ldots + k_n), \tag{8}$$

and by summing (8) for all 0-token:

$$\sum_{t=1}^{n-1} l_t(k_1 + \ldots + k_t)(k_{t+1} + \ldots + k_n). \tag{9}$$

Finally we may rewrite (9) as follows:

$$\sum_{t=1}^{n-1} l_t \sum_{i=1}^{t} k_i \sum_{j=t+1}^{n} k_j. \tag{10}$$

It is easy to see that (4) is equal to (10).

## 3    New Q-Convexity Measure

Let $F = (f_{ij})_{m \times n}$ be an $m \times n$ binary matrix. Each position $(i, j)$ determines the following four quadrants (submatrices):

$$Z_0(i, j) = \{(l, k) : 1 \leq l \leq i, \ 1 \leq k \leq j\},$$
$$Z_1(i, j) = \{(l, k) : i \leq l \leq m, \ 1 \leq k \leq j\},$$
$$Z_2(i, j) = \{(l, k) : i \leq l \leq m, \ j \leq k \leq n\},$$
$$Z_3(i, j) = \{(l, k) : 1 \leq l \leq i, \ j \leq k \leq n\}.$$

Let us denote the number of object points of $F$ in $Z_p(i, j)$ by $n_p(i, j)$, for $p = 0, \ldots, 3$, i.e.

$$n_p(i, j) = card(Z_p(i, j) \cap \{(i, j) : f_{ij} = 1\}) \ (p = 0, \ldots, 3). \tag{11}$$

**Definition 1.** A binary matrix $F$ is Q-convex if for each $(i, j)$ $(n_0(i, j) > 0 \wedge n_1(i, j) > 0 \wedge n_2(i, j) > 0 \wedge n_3(i, j) > 0)$ implies $f_{ij} = 1$.

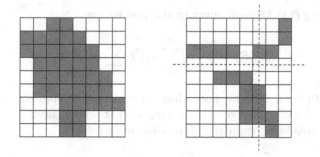

**Fig. 3.** A Q-convex image (left) and a non Q-convex image (right). The four quadrants around the position $(4, 6)$ are illustrated: $Z_0(4, 6)$ left-top, $Z_1(4, 6)$ left-bottom, $Z_2(4, 6)$ right-bottom, $Z_3(4, 6)$ right-top. The 4-th row and 6-th column (marked by dashed lines) are in common between consecutive zones.

If $F$ is not Q-convex, then there exists a position $(i, j)$ violating the Q-convexity property, i.e. $n_p(i, j) > 0$ for all $p = 0, \ldots, 3$ and $f_{ij} = 0$. Note that if $F$ is not horizontally or vertically convex, then it is not $Q$-convex. Figure 3 illustrates the definition of Q-convexity: the binary image on the right is not Q-convex because $f_{46} = 0$ but $Z_p(4, 6)$ contains 1's items, for all $p = 0, 1, 2, 3$. In the figure we have $n_0(4, 6) = 5$, $n_1(4, 6) = 9$, $n_2(4, 6) = 5$, $n_3(4, 6) = 4$.

We define the non-Q-convexity measure as the sum of the contributions of non-Q-convexity measure of each 0 entry of $F$. Formally,

$$\varphi_Q(i, j) = n_0(i, j)n_1(i, j)n_2(i, j)n_3(i, j)(1 - f_{ij}), \tag{12}$$

where $(i, j)$ is an arbitrary position of $F$, and

$$\varphi_Q(F) = \sum_{ij} \varphi_Q(i, j). \tag{13}$$

For example, $\varphi_Q(4, 6) = n_0(4, 6)n_1(4, 6)n_2(4, 6)n_3(4, 6) = 5 \cdot 9 \cdot 5 \cdot 4 = 900$.

*Remark 1.* If $f_{ij} = 1$, then $\varphi_Q(i, j) = 0$. Moreover, if $f_{ij} = 0$, but there exists $n_p(i, j) = 0$, then $\varphi_Q(i, j) = 0$. Therefore, $F$ is $Q$-convex if and only if $\varphi_Q(F) = 0$.

*Remark 2.* By definition, the measure is invariant by reflection and by point symmetry.

### 3.1   Connection with the Directional Convexity

Suppose that $F$ is constituted of just one row $R = 1^{k_1}0^{l_1}1^{k_2}0^{l_2} \ldots 1^{k_n}$. In this case $n_0(i, j) = n_1(i, j) = k_1 + \ldots + k_t$, $n_2(i, j) = n_3(i, j) = k_{t+1} + \ldots + k_n$, and so we have

$$\varphi_Q(i, j) = (\sum_{i=1}^{t} k_i)^2 (\sum_{j=t+1}^{n} k_j)^2, \tag{14}$$

and by summing (14) for each entry in the row we get

$$\varphi_Q(R) = \sum_{t=1}^{n-1} l_t (\sum_{i=1}^{t} k_i)^2 (\sum_{j=t+1}^{n} k_j)^2. \tag{15}$$

Comparing (10) with (15) we note that an exponent 2 appears in the latter one. Roughly speaking, the reason for this is that we consider "regions" (two dimensions) instead of "boundary" (one dimension).

## 4   Normalization

A desirable property for a measure is that it ranges in $[0, 1]$. In this section we show how to normalize the new measure.

*Property 1.* By definition, $Z_0(l, k) \subseteq Z_0(i, j)$ if $l \leq i$ and $k \leq j$, and hence $n_0(l, k) \leq n_0(i, j)$ with $l \leq i$ and $k \leq j$. Analogous relations hold for $Z_1$, $Z_2$, $Z_3$ and for $n_1$, $n_2$, $n_3$ accordingly.

Recall from (1) that for the row and columns sums of $F$ $h_l = \sum_{k=1}^{n} f_{lk}$ with $l = 1, \ldots, m$ and $v_k = \sum_{l=1}^{m} f_{lk}$ with $k = 1, \ldots, n$. Moreover, denote the horizontal and vertical partial sums by

$$H_p = \sum_{l=1}^{p} h_l \ (p = 1, \ldots, m) \quad \text{and} \quad V_r = \sum_{k=1}^{r} v_k \ (r = 1, \ldots, n). \tag{16}$$

Clearly, $H_m = V_n = \alpha$, where $\alpha$ is the total number of object pixels.

*Property 2.* For fixed row $i$, we use $v_k[1 \ldots i]$ to denote the sum limited to $i$, i.e. $\sum_{l=1}^{i} f_{lk}$. Thus, by Eqs. (1), (11), and (16) there follow the relations:

$$n_0(i, k) + n_3(i, k) - v_k[1 \ldots i] = H_i$$

and

$$n_1(i, k) + n_2(i, k) - v_k[i \ldots m] = \alpha - H_{i-1},$$

for $k = 1, \ldots, n$.

For fixed column $j$, we use $h_l[1 \ldots j]$ to denote the sum limited to $j$, i.e. $\sum_{k=1}^{j} f_{lk}$. Analogously, we have:

$$n_0(l, j) + n_1(l, j) - h_l[1 \ldots j] = V_j$$

and

$$n_2(l, j) + n_3(l, j) - h_l[j \ldots n] = \alpha - V_{j-1},$$

for $l = 1, \ldots, m$.

By Properties 1 and 2 there follows:

*Property 3*

$$n_0(i, j) + n_1(i, j) + n_2(i, j) + n_3(i, j) = \alpha + h_i + v_j + f_{ij}$$

for $i = 1, \ldots, m$ and $j = 1, \ldots, n$.

We need the following two lemmas:

**Lemma 1.** *Let $x, y$ be real numbers such that $x + y = p$, where $p$ is a constant. Then, the expression $xy$ has a maximum at $x = y = p/2$.*

*Proof.* By derivative of $xy = g(x) = x(p - x)$, we get $g'(x) = 0$ for $x = p/2$, and it is a maximum.

**Lemma 2.** *Let $x, y, z, w$ be real numbers such that $x + y + z + w = p$, where $p$ is a constant. Then, the expression $xyzw$ has a maximum at $x = y = w = z = p/4$.*

*Proof.* The product $xyzw$ is maximal iff $xy$ is maximal and $zw$ is maximal. Rewrite $x + y = p - (w + z) = p_1$ and $w + z = p - (x + y) = p_2$. By Lemma 1 $xy$ is maximal if $x = y = p_1/2$ and $zw$ is maximal if $w = z = p_2/2$. Since $x + y + z + w = p$, we get $2x + 2w = p$ and hence $w = p/2 - x$. Therefore $xyzw = x^2(p/2-x)^2 = g(x)$ and by derivative we get $g'(x) = 4x^2 - 3px + p^2/2 = 0$ for $x = p/4$ and it is a maximum.

Now are we able to derive an upper bound to the value $\varphi_Q(i, j)$ for the position $(i, j)$ violating the Q-convexity.

**Proposition 1.** *Let $f_{ij}$ be a 0 entry of a binary matrix $F$, and $h_i$ and $v_j$ be the $i$-th row and $j$-th column sums. Then, $\varphi_Q(i, j) \leq ((\alpha + h_i + v_j)/4)^4$.*

$$1\ 1\ 1$$
$$1\ 0\ 1$$
$$1\ 1\ 1$$

**Fig. 4.** A binary image $F$ with $\varphi_Q(F) = \varphi_Q(2,2) = 3^4 = (\frac{8+2+2}{4})^4 = 81$.

*Proof.* By Property 3 we have that $n_0(i,j) + n_1(i,j) + n_2(i,j) + n_3(i,j) = \alpha + h_i + v_j = p$. By Lemma 2 $\varphi_Q(i,j) = n_0(i,j)n_1(i,j)n_2(i,j)n_3(i,j)$ is maximal for $n_0(i,j) = n_1(i,j) = n_2(i,j) = n_3(i,j) = p/4$, then $\varphi_Q(i,j) = (p/4)^4$. The upper bound follows.

Note that the bound is tight (see Fig. 4, for example).

In the light of Proposition 1, we may normalize the measure $\varphi_Q(F)$ by normalizing each single contribution $\varphi_Q(i,j)$, i.e.,

$$\hat{\varphi}_Q(i,j) = \frac{\varphi_Q(i,j)}{(\frac{\alpha+h_i+v_j}{4})^4}, \tag{17}$$

and

$$\hat{\varphi}_Q(F) = \frac{\sum_{(i,j)\in\bar{F}} \hat{\varphi}_Q(i,j)}{card(\bar{F})}, \tag{18}$$

where $\bar{F}$ is constituted by the 0 entries of $F$ violating the Q-convexity (if $\bar{F} = \emptyset$ then we simply assign $\hat{\varphi}_Q(F) = 0$). Finally, we map the non-Q-convexity measure into a Q-convexity measure simply by

$$\Psi_Q(F) = 1 - \hat{\varphi}_Q(F). \tag{19}$$

We can also make the measure independent from $\alpha$ and its row and column sums as follows. Since $\alpha \leq mn$, $h_i \leq n, v_j \leq m$, we get that $\varphi_Q(i,j) < (\frac{mn+m+n-3}{4})^4$, and hence

$$\hat{\varphi}_Q^{mn}(i,j) = \frac{\varphi_Q(i,j)}{(\frac{mn+m+n-3}{4})^4}. \tag{20}$$

(The reason for $-3$ is that if $f_{ij}$ contributes, it is 0, and so the first three inequalities are strict). Therefore, $\varphi_Q(F)$ is normalized to

$$\hat{\varphi}_Q^{mn}(F) = \frac{\sum_{ij} \hat{\varphi}_Q^{mn}(i,j)}{mn}, \tag{21}$$

and finally,

$$\Psi_Q^{mn}(F) = 1 - \hat{\varphi}_Q^{mn}(F). \tag{22}$$

*Remark 3.* Both measures assign 1 to Q-convex images by Remark 1. On the other hand, for instance, $\Psi_Q$ assigns 0, while $\Psi_Q^{mn}$ assigns 8/9 to the image in Fig. 4.

## 4.1 Implementation

The measures can be efficiently implemented in linear time in the size of the image. Indeed, by Property 1 we can count the number of 1's in $F$ for $Z_p(f_{ij})$, for each $(i, j)$ in linear time, and store them in a matrix for any $p = \{1, 2, 3, 4\}$. Then, $\varphi_Q(i, j)$ can be computed in constant time for any $(i, j)$. Normalization is straightforward.

## 5 Experiments

We investigated the new measures on some images to show their behavior and their robustness in case of noise. We considered at first a chessboard image and a stripe image of sizes $50 \times 50$ (see Fig. 5). We report on the Q-convexity measures $\Psi_Q$, $\Psi_Q^{mn}$, and on the non-convexity measure $\varphi_Q$ as a reference. As expected $\varphi_Q$ assigns a smaller value to the second image and, accordingly, both $\Psi_Q$, $\Psi_Q^{mn}$ assign a greater value to it, since it is horizontally convex but not vertically convex. Notice that in [2] the authors discussed these two examples showing that two simple aggregations of the directional convexity did not behave correctly. For comparison, the horizontal ($\Psi_h$) and vertical ($\Psi_v$) directional convexity values are also presented (see (7) for the definition).

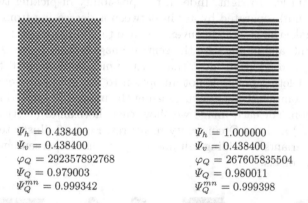

$\Psi_h = 0.438400$     $\Psi_h = 1.000000$
$\Psi_v = 0.438400$     $\Psi_v = 0.438400$
$\varphi_Q = 292357892768$     $\varphi_Q = 267605835504$
$\Psi_Q = 0.979003$     $\Psi_Q = 0.980011$
$\Psi_Q^{mn} = 0.999342$     $\Psi_Q^{mn} = 0.999398$

**Fig. 5.** A chessboard pattern (left) and a stripe pattern (right).

Secondly, we took three $50 \times 50$ images representing four square regions separated by a cross pattern as illustrated in Fig. 6 with different sizes for the black pixels. We report the values of the different measures for each image. We may note two main differences:

- $\Psi_Q^{mn}$ tends to overestimate the Q-convexity by assigning values close to 1, whereas $\Psi_Q$ assigns values closer to 0. This is true in general, because by definition $\Psi_Q^{mn}$ is normalized with respect to the size of the image itself.

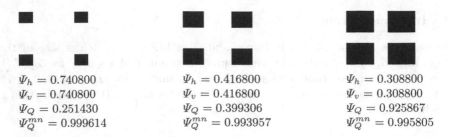

$\Psi_h = 0.740800$    $\Psi_h = 0.416800$    $\Psi_h = 0.308800$
$\Psi_v = 0.740800$    $\Psi_v = 0.416800$    $\Psi_v = 0.308800$
$\Psi_Q = 0.251430$    $\Psi_Q = 0.399306$    $\Psi_Q = 0.925867$
$\Psi_Q^{mn} = 0.999614$    $\Psi_Q^{mn} = 0.993957$    $\Psi_Q^{mn} = 0.995805$

**Fig. 6.** $50 \times 50$ images representing four square regions separated by a cross pattern with different sizes for the black pixels: $10 \times 10$ (left), $15 \times 15$ (middle) and $20 \times 20$ (right).

- By definition, $\Psi_Q^{mn}$ follows opposite behavior of $\varphi_Q$ (normalization is obtained by dividing by constants), whereas $\Psi_Q$ may have a different behavior depending on $\alpha$, and the row and column sums of the image.

In this case, $\Psi_Q$ assigns increasing values to the images, whereas $\Psi_Q^{mn}$ assigns the smallest value to the second image. It is also clearly observable, that the horizontal and vertical directional convexity measures do not take into account the two-dimensional structure of the image. They assign decreasing values to the images, from left to right. Indeed, the possibility of picking two 1 s in the same row (column) separated by a 0 in between is decreasing from left to right. However, to make the image Q-convex we have to add more points to the image from left to right, and in any case the central image is farer to be Q-convex than the right image. In addition, note that all the measures assign 0 to the empty image so that deleting points is not an option to achieve Q-convexity.

Finally, we computed our measures using the images illustrated in Figs. 7 and 8 for comparison. For each image we show their horizontal (vertical) convexity values (as in [2]) and the Q-convexity measures. Figure 8 illustrates an original image and its variants by adding salt and pepper noise in percentage of 5%, 10%, and 20% of the pixels.

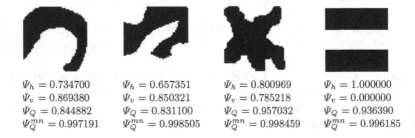

$\Psi_h = 0.734700$    $\Psi_h = 0.657351$    $\Psi_h = 0.800969$    $\Psi_h = 1.000000$
$\Psi_v = 0.869380$    $\Psi_v = 0.850321$    $\Psi_v = 0.785218$    $\Psi_v = 0.000000$
$\Psi_Q = 0.844882$    $\Psi_Q = 0.831100$    $\Psi_Q = 0.957032$    $\Psi_Q = 0.936390$
$\Psi_Q^{mn} = 0.997191$    $\Psi_Q^{mn} = 0.998505$    $\Psi_Q^{mn} = 0.998459$    $\Psi_Q^{mn} = 0.996185$

**Fig. 7.** Example binary images of size $50 \times 50$, with horizontal ($\Psi_h$), vertical ($\Psi_v$) convexity and Q-convexity ($\Psi_Q$, $\Psi_Q^{mn}$) shown.

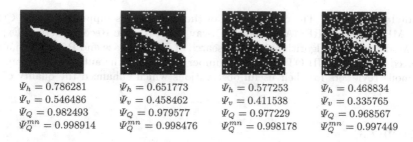

$\Psi_h = 0.786281$    $\Psi_h = 0.651773$    $\Psi_h = 0.577253$    $\Psi_h = 0.468834$
$\Psi_v = 0.546486$    $\Psi_v = 0.458462$    $\Psi_v = 0.411538$    $\Psi_v = 0.335765$
$\Psi_Q = 0.982493$    $\Psi_Q = 0.979577$    $\Psi_Q = 0.977229$    $\Psi_Q = 0.968567$
$\Psi_Q^{mn} = 0.998914$   $\Psi_Q^{mn} = 0.998476$   $\Psi_Q^{mn} = 0.998178$   $\Psi_Q^{mn} = 0.997449$

**Fig. 8.** Same image without, and with 5%, 10%, and 20% noise. For each image we show horizontal ($\Psi_h$), vertical ($\Psi_v$) convexity and Q-convexity ($\Psi_Q$, $\Psi_Q^{mn}$) measures.

An immediate observation is that for all presented images $\Psi_Q^{mn}$ lies between 0.99 and 1. This seems to be a weakness of this measure, which can be avoided by renormalizing the values into the interval $[0, 1]$. For this, the minimal possible value of $\Psi_Q^{mn}$ should be identified, which is among our further aims. However, for further comparison let us assume that the minimal value of $\Psi_Q^{mn}$ is $\min \Psi_Q^{mn} = 8/9$ (the value we observed for Fig. 4) and let

$$\widetilde{\Psi}_Q^{mn} = \beta(\Psi_Q^{mn} - \min \Psi_Q^{mn}), \qquad (23)$$

where $\beta = 9$ since $\max \Psi_Q^{mn} - \min \Psi_Q^{mn} = 1/9$, which stretches the values of $\Psi_Q^{mn}$ onto the interval $[0, 1]$, and thus provides an exact comparison of $\Psi_Q$ and $\Psi_Q^{mn}$. Denoting the images of Fig. 8 by $F_0$, $F_5$, $F_{10}$, and $F_{20}$, from left to right respectively, we get $\widetilde{\Psi}_Q^{mn}(F_0) = 0.990226$, $\widetilde{\Psi}_Q^{mn}(F_5) = 0.986284$, $\widetilde{\Psi}_Q^{mn}(F_{10}) = 0.983602$, and $\widetilde{\Psi}_Q^{mn}(F_{20}) = 0.977041$. Comparing $\widetilde{\Psi}_Q^{mn}(F_5) - \widetilde{\Psi}_Q^{mn}(F_0) = 0.003942$, $\widetilde{\Psi}_Q^{mn}(F_{10}) - \widetilde{\Psi}_Q^{mn}(F_0) = 0.006624$, and $\widetilde{\Psi}_Q^{mn}(F_{20}) - \widetilde{\Psi}_Q^{mn}(F_0) = 0.013185$ to $\Psi_Q(F_5) - \Psi_Q(F_0) = 0.002916$, $\widetilde{\Psi}_Q^{mn}(F_{10}) - \widetilde{\Psi}_Q^{mn}(F_0) = 0.005264$, and $\widetilde{\Psi}_Q^{mn}(F_{20}) - \widetilde{\Psi}_Q^{mn}(F_0) = 0.013926$, respectively, we may note that $\Psi_Q$ is more robust, while $\Psi_Q^{mn}$ is more sensitive to a moderate amount of noise (5% and 10%), but this, of course, needs a broader study.

## 6 Conclusions

In this paper, we presented a new idea to define shape descriptors based on the concept of Q-convexity. This measure is quantitative and it is an extension of the directional convexity measure proposed in [2]. This study shows some potential of these shape descriptors since they correctly incorporate the convexity along the considered directions, and in particular $\Psi_Q$ seems to be more robust, whereas $\Psi_Q^{mn}$ to be more sensitive to noise. Moreover they can be computed efficiently and are invariant by reflection and point symmetry. Further work should be done to deeply investigate normalization and to conduct experiments for object recognition and classification.

**Acknowledgements.** The collaboration of the authors was supported by the COST Action MP1207 "EXTREMA: Enhanced X-ray Tomographic Reconstruction: Experiment, Modeling, and Algorithms". The research of Péter Balázs and Péter Bodnár was supported by the NKFIH OTKA [grant number K112998]. The authors also thank the anonymous reviewers for their useful observations which enhanced the quality of the paper.

# References

1. Balázs, P., Brunetti, S.: A measure of Q-Convexity. In: Normand, N., Guédon, J., Autrusseau, F. (eds.) DGCI 2016. LNCS, vol. 9647, pp. 219–230. Springer, Cham (2016). doi:10.1007/978-3-319-32360-2_17
2. Balázs, P., Ozsvár, Z., Tasi, T.S., Nyúl, L.G.: A measure of directional convexity inspired by binary tomography. Fundamenta Informaticae **141**(2–3), 151–167 (2015)
3. Barcucci, E., Del Lungo, A., Nivat, M., Pinzani, R.: Medians of polyominoes: a property for the reconstruction. Int. J. Imag. Syst. Technol. **9**, 69–77 (1998)
4. Boxter, L.: Computing deviations from convexity in polygons. Pattern Recogn. Lett. **14**, 163–167 (1993)
5. Brunetti, S., Daurat, A.: An algorithm reconstructing convex lattice sets. Theor. Comput. Sci. **304**(1–3), 35–57 (2003)
6. Brunetti, S., Daurat, A.: Reconstruction of convex lattice sets from tomographic projections in quartic time. Theor. Comput. Sci. **406**(1–2), 55–62 (2008)
7. Brunetti, S., Del Lungo, A., Del Ristoro, F., Kuba, A., Nivat, M.: Reconstruction of 4- and 8-connected convex discrete sets from row and column projections. Linear Algebra Appl. **339**, 37–57 (2001)
8. Chrobak, M., Dürr, C.: Reconstructing hv-convex polyominoes from orthogonal projections. Inform. Process. Lett. **69**(6), 283–289 (1999)
9. Daurat, A.: Salient points of Q-convex sets. Int. J. Pattern Recogn. Artif. Intell. **15**, 1023–1030 (2001)
10. Daurat, A., Nivat, M.: Salient and reentrant points of discrete sets. Electron. Notes Discrete Math. **12**, 208–219 (2003)
11. Latecki, L.J., Lakamper, R.: Convexity rule for shape decomposition based on discrete contour evolution. Comput. Vis. Image Understand. **73**(3), 441–454 (1999)
12. Nelson, M.R.: The Data Compression Book. M&T Books, Redwood City (1991)
13. Rahtu, E., Salo, M., Heikkila, J.: A new convexity measure based on a probabilistic interpretation of images. IEEE Trans. Pattern Anal. **28**(9), 1501–1512 (2006)
14. Rosin, P.L., Zunic, J.: Probabilistic convexity measure. IET Image Process. **1**(2), 182–188 (2007)
15. Sonka, M., Hlavac, V., Boyle, R.: Image Processing, Analysis, and Machine Vision, 3rd edn. Thomson Learning, Toronto (2008)
16. Stern, H.: Polygonal entropy: a convexity measure. Pattern Recogn. Lett. **10**, 229–235 (1998)
17. Zunic, J., Rosin, P.L.: A new convexity measure for polygons. IEEE T. Pattern Anal. **26**(7), 923–934 (2004)

# Algorithms for Stable Matching
# and Clustering in a Grid

David Eppstein, Michael T. Goodrich, and Nil Mamano$^{(\boxtimes)}$

Department of Computer Science, University of California, Irvine, USA
{eppstein,nmamano}@uci.edu, goodrich@acm.org

**Abstract.** We study a discrete version of a geometric stable marriage problem originally proposed in a continuous setting by Hoffman, Holroyd, and Peres, in which points in the plane are stably matched to cluster centers, as prioritized by their distances, so that each cluster center is apportioned a set of points of equal area. We show that, for a discretization of the problem to an $n \times n$ grid of pixels with $k$ centers, the problem can be solved in time $O(n^2 \log^5 n)$, and we experiment with two slower but more practical algorithms and a hybrid method that switches from one of these algorithms to the other to gain greater efficiency than either algorithm alone. We also show how to combine geometric stable matchings with a $k$-means clustering algorithm, so as to provide a geometric political-districting algorithm that views distance in economic terms, and we experiment with weighted versions of stable $k$-means in order to improve the connectivity of the resulting clusters.

# 1 Introduction

A long line of research considers algorithms on objects embedded in $n \times n$ grids, including problems in computational geometry (e.g., see [1,2,8,17,19,26,28,29]), graph drawing (e.g., see [5,10,14,30]), geographic information systems (e.g., see [13]), and geometric image processing (e.g., see [9,11,15,20]). Continuing this line, we consider in this paper the problem of matching grid points (which we view as *pixels*) to $k$ *center* points in the grid. Pixels have a preference for centers closer to them, and centers prefer closer pixels as well. The goal is to match every center to an equal number of pixels and for the matching to be *stable*, meaning that no two elements prefer each other to their specified matches. For example, the centers could be facilities, such as polling places, fire stations, or post offices, that have assigned jurisdictions and equal operational capacities (in terms of how many pixels they can serve). Rather than optimizing some computationally challenging global quality criterion based on distance or area, we seek an assignment of pixels to centers that is locally stable. Figure 1 illustrates a solution to this *stable grid matching* problem for a $900 \times 900$ grid and 100 random centers. Note that some centers are matched to disconnected regions.

Stable grid matching is a special case of the classic *stable matching problem* [18], which was originally described in terms of arranging marriages between $N$ heterosexual men and women in a closed community. In this case, stability

© Springer International Publishing AG 2017
V.E. Brimkov and R.P. Barneva (Eds.): IWCIA 2017, LNCS 10256, pp. 117–131, 2017.
DOI: 10.1007/978-3-319-59108-7_10

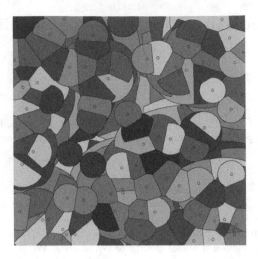

**Fig. 1.** An example solution to the *stable grid matching* problem for a $900 \times 900$ grid and 100 centers distributed randomly. Pixels of the same color are assigned to the same center. (Color figure online)

means that no man-woman pair prefers each other to their assigned mates, which is necessary (and more important than, e.g., total utility) to prevent extramarital affairs. The Gale-Shapley algorithm [18] finds a stable matching for arbitrary preferences in $O(N^2)$ time. For stable grid matching in an $n \times n$ grid this would give a running time of $O(n^4)$, since each "man" would correspond to a pixel and each "woman" would correspond to one of $\lceil n^2/k \rceil$ copies of a center. As we show, the geometric structure of the stable grid matching problem allows for significantly more efficient solutions.

We also study the effect of integrating a stable matching with a *k-means* clustering method, which alternates between assigning points to cluster centers and moving cluster centers to better represent their assigned points. Using stable matching for the assignment stage of this method allows us to fix the size of the clusters (for instance, to be all equally sized), which might be advantageous in some applications.

*Prior Related Work.* As mentioned above, there is considerable prior research on algorithms involving objects embedded in an $n \times n$ grid. The stable grid matching problem that we study can be viewed as a grid-restricted version of the classic "post office" problem of Knuth [27], where one wishes to identify each point in the plane with its closest of $k$ post offices, with the added restriction that the region assigned to each post office must have the same area. The continuous version of the stable grid matching problem, which deals with points in $\mathbb{R}^2$ instead of discrete pixels, was studied by Hoffman et al. [21]. They showed that there is a unique solution, and there is a simple numerical method to find it: Start growing a circle from each center at the same time, all growing at the same speed. When a yet-unmatched point is reached by a circle, it is assigned to the corresponding center. When a center reaches its quota (its region covers $1/k$ of the area of the square), its circle halts. (Note that if the halting condition is removed, we obtain the Voronoi diagram of the centers instead, as in the well-known solution to Knuth's post office problem, e.g., see [3].) Due to its continuous, numerical nature, Hoffman et al. did not analyze the running time of their method; hence, there is motivation to study the grid-based version of this problem.

With respect to the related problem of $k$-means clustering, we are interested in a grid-based version of this problem as well, which has been studied extensively in non-grid discrete contexts (e.g., see [22,24]). In the continuous version of this problem, one is interested in partitioning a geometric region into subregions that all have the same area (e.g., see [6]). One of the motivations for such partitions is in *political districting*, for which there is additional related prior work (e.g., see [32]). The goal of political districting is to partition a territory into regions (districts) which all have roughly the same population size and are "compact", which informally means that their shape should be connected and resemble a circle rather than an octopus [32]. Ricca et al. [31] adapted the concept of Voronoi regions to the discrete setting in order to use them for political districting. Voronoi regions ensured good compactness but poor population balance, however. Thus, there is motivation for a clustering algorithm based on the use of stable matchings, since such partitions enforce the property that all regions have the same size (at the possible cost of connectivity). Finding a scheme that guarantees both size equality and compactness is an open problem of interest.

*Problem Definition.* In the *stable grid matching problem*, we are given a square $n \times n$ grid and $k$ points called *centers* within the grid. The lattice points are called *pixels* or *sites*. Sites implicitly rank the centers in increasing order of distance, and centers similarly implicitly rank pixels in increasing order by distance. A *matching* is a mapping from sites to centers. The goal is to find a matching with the following two properties (see Fig. 2, left column):

1. The *region* of each center (the set of sites assigned to it) must have the same size up to roundoff errors. The *quota* of a center is the number of sites that must be in its region. If $n^2$ is a multiple of $k$, then all the quotas are $n^2/k$. Otherwise, some centers are allowed one extra site.
2. The matching must be *stable*. A matching is not stable when a pair of sites $(p_1, p_2)$ is assigned to centers $c_1$ and $c_2$ such that $p_1$ prefers (i.e., according to some metric is closer to) $c_2$ over $c_1$ and $c_2$ prefers $p_1$ over $p_2$. This is unstable because $p_1$ and $c_2$ prefer each other to their current matches.

*Combining $k$-means with Stable Assignment.* The $k$-means clustering method is to partition a data set (which, in our case, is an $n \times n$ grid) into $k$ regions, based on a simple iterative refinement algorithm (which is called the *k-means algorithm* or *Lloyd's algorithm*, e.g., see [24]): We begin by choosing $k$ points, called cluster centers, randomly in the space. Then, we iteratively repeat the following two phases: (1) *assignment* step: each object is assigned to its closest center, and (2) *update* step: each center is moved to the centroid of the objects assigned to it.

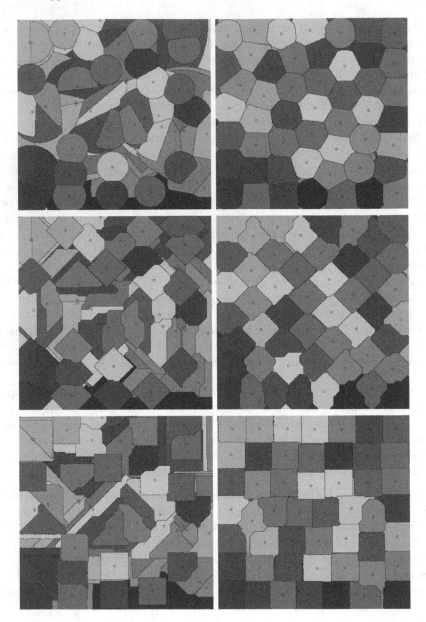

**Fig. 2.** Left: stable matching in a 300 × 300 grid with the same 50 random centers for the Euclidean (top), Manhattan (center), and Chebyshev (bottom) metrics. Right: result of the stable k-means algorithm with unweighted centroids for each metric.

Lloyd's algorithm converges to a (locally optimal) partition that minimizes the sum of the squared distances from each object to its assigned center [24]. In this paper, we propose a variation, which we call *stable k-means*, where the assignment step is replaced by a stable matching between objects and centers, so as to achieve the additional property that the regions all have equal area (to within roundoff errors). Intuitively, the goal is to implement Lloyd's algorithm with stable grid matching so as to improve the compactness of the regions while preserving equal-sized clusters.

We have found through experimentation that, although the stable $k$-means method succeeds in improving compactness, centers can sometimes stop moving while we are executing Lloyd's algorithm before their regions became completely connected (e.g., see Fig. 2). Thus, we introduce in this paper an additional heuristic, where we use weighted centroids, which are more sensitive to the outlying parts of their region. The usual centroid of a set of points $S$ is defined as $(\sum_{q\in S} q)/|S|$, where the points are regarded as two-dimensional vectors so that the sum makes sense. Instead, we can compute a weighted centroid as $(\sum_{q\in S} w_q q)/(\sum_{q\in S} w_q)$. A natural choice to use for the weight $w_q$ of a point $q$ assigned to the region of the center $c$ is the distance from $q$ to $c$ raised to some exponent $p$ that we can choose, $d(q,c)^p$. The larger $p$ is, the more sensitive the weighted centroids are to outliers. When $p = 0$, we get the usual centroid. When $p \to +\infty$, we get the circumcenter of the region, and when $p \to -\infty$ we get the current center.

*Contributions.* In this paper, we provide the following results:

- The stable grid matching problem, for a grid of $n \times n$ pixels with $k$ centers, can be solved by a randomized algorithm with expected running time $O(n^2 \log^5 n)$. Since an $n \times n$ grid has $\Theta(n^2)$ pixels, this quasilinear bound improves the $O(n^4)$ time of the Gale-Shapley algorithm. However, this algorithm uses intricate data structures that make it challenging to implement in practice.
- Given the pragmatic challenges of the above-mentioned quasilinear-time algorithm, we provide two alternative algorithms, a "circle-growing algorithm" and a "distance-sorting" method, both of which are simple to implement and have running times of $O(n^2 k)$.
- We provide an experimental analysis of these two practical algorithms, where we observe that the circle-growing algorithm is more efficient at finding low-distance matched pairs, while the distance-sorting based method is more efficient when pairs are farther apart. Therefore, we show that it is advantageous to switch from one algorithm to the other partway through the matching process, potentially achieving running times with a sublinear dependence on $k$. We experiment with the optimal cutoff for switching between these two algorithms.
- We also provide the results of experiments to test the connectivity of the clusters obtained by our stable $k$-means algorithm, with weighted variants for finding centroids. Our experiments support the conclusion that no choice of a weight exponent $p$ will always result in total connectivity. Nevertheless, our experiments provide evidence that the best results come from the range $-0.8 \le p \le 0.4$. Empirically, more highly negative values of $p$ tend to make

the algorithm converge slowly or fail to converge, while more highly positive values of $p$ lead to oscillations in the center placement. See the full version of the paper for additional figures of these cases.

## 2  Algorithms

Our stable grid matching algorithms start with an empty matching and add center–site pairs to it. Given a partial matching, we say a site is *available* if it has not been matched yet, and a center is *available* if the size of its region is smaller than its quota. A center–site pair is available if both the center and site are available, and it is a *closest available pair* if it is available and the distance from the center to the site is minimum among all available pairs. It is simple to prove that if an algorithm starts with an empty matching and only adds closest available pairs to it until it is complete, the resulting matching is stable.

### 2.1  Circle-Growing Algorithm

In this section we describe our main practical algorithm, the *circle-growing algorithm*, which mimics the continuous construction from [21]. First, we obtain the list of all the lattice points with coordinates ranging from $-n$ to $n$ sorted by distance to the origin. The resulting list $P$ emulates a circle growing from the origin. When initializing $P$, we can gain a factor of eight savings in space by sorting and storing only the points in the triangle $\triangle(0,0)(0,n)(n,n)$. The remaining points can be obtained by symmetry: if $p = (x,y)$ is a point in the triangle, the eight points with coordinates of the form $(\pm x, \pm y)$ and $(\pm y, \pm x)$ are at the same distance from the origin as $p$. Moreover, in applications where we find multiple stable grid matchings, such as in the stable $k$-means method, we need only initialize $P$ once. The way we use $P$ depends on the type of centers we consider.

*Integer Centers.* In this case we can use the fact that if we relocate the points in $P$ relative to a center, then they are in the order in which a circle growing from that center would reach them. To respect that all the circles grow at the same rate, we iterate through the points in $P$ in order. For each point $p$, we relocate it relative to each center $c$ to form the site $p + c$ (the order of the centers does not matter). We add to the matching any available center–site pair $(c, p + c)$. We iterate through $P$ until the matching is complete.

We require $O(n^2)$ space and $O(n^2 \log n)$ time to sort the points in $P$. For the Euclidean metric instead of using distances to sort $P$ we can use squared distances, which take integer values between $0$ and $2n^2$. Then, we can use an integer sorting algorithm such as counting sort to sort in $O(n^2)$ time [12, Chap. 8.2]. Since each point in $P$ results in up to $O(k)$ center–site pairs, we need $O(n^2 k)$ time to iterate through $P$.

*Real Centers (Algorithm 1).* If centers have real coordinates, we cannot translate the points in $P$ relative to the centers, because $p + c$ is not necessarily a lattice point. The workaround is to associate each center $c$ to its closest lattice point $p_c$. Let $\delta$ be the maximum distance $d(c, p_c)$ among all centers. Then, the center–site pairs "generated" by each point $p$ in $P$ have the form $(c, p + p_c)$ and their distances can vary between $d(p, O) - \delta$ and $d(p, O) + \delta$ (where $O$ denotes the origin, $(0, 0)$). Consequently, the distances of pairs generated by points $p_i, p_j$ in $P$ with $i < j$ may intertwine, but only if $d(p_j, O) - \delta \leq d(p_i, O) + \delta$. The points in $P$ after $p_i$ whose pairs might intertwine with those of $p_i$ form an annulus centered at $O$ with small radius $d(p_i, O)$ and big radius $d(p_i, O) + 2\delta$.

Since $\delta$ is a constant (for the Euclidean metric, $\delta \leq \sqrt{2}/4$), it can be derived from the Gauss circle problem that such an annulus contains $O(d(p_i, O)) = O(n)$ points.

---

**Algorithm 1.** Circle growing algorithm for $k$ real centers on an $n \times n$ grid.

---

Set all sites as unmatched.
Set the quota of the first $n^2 \bmod k$ centers to $\lceil n^2/k \rceil$.
Set the quota of the remaining centers to $\lfloor n^2/k \rfloor$.
Let $P = $ list of points $(x, y)$ such that $-n < x, y < n$.
Sort $P$ by nondecreasing distance to $(0, 0)$.
For each center $c$, let $p_c = (\text{round}(c_x), \text{round}(c_y))$.
Let $\delta = \max\{\text{dist}(c, p_c)\}$ among all centers.
$j \leftarrow 1$
**while** the matching is not complete **do**
    $L \leftarrow$ empty list
    $i \leftarrow \min(j + n, |P|)$
    **for all** $p \in P_j, \ldots, P_i$ **do**    ▷ Add to $L$ pairs generated by points in the next chunk
        **for all** centers $c$ with quota $> 0$ **do**
            $s \leftarrow p + p_c$
            **if** $0 \leq s_x, s_y < n$ and $s$ is still available **then**
                Add $(c, s)$ to $L$.
    Let $d = \max\{\text{dist}(c, s)\}$ among all pairs $(c, s) \in L$.
    **for all** $p \in P_{i+1}, \ldots, P_{|P|}$ **do**    ▷ Add to $L$ pairs closer than pairs already in $L$
        **if** $\text{dist}(p, O) > \text{dist}(P_i, O) + 2\delta$ **then**
            **break**
        **for all** centers $c$ with quota $> 0$ **do**
            $s \leftarrow p + p_c$
            **if** $0 \leq s_x, s_y < n$ and $s$ is still available and $\text{dist}(c, s) \leq d$ **then**
                Add $(c, s)$ to $L$.
    Sort $L$ by nondecreasing center–site distance.
    **for all** $(c, s) \in L$ **do**
        **if** $c$ and $s$ are available **then**
            Match $s$ and $c$.
            Reduce the quota of $c$ by 1.
    $j \leftarrow i + 1$

---

The algorithm processes the points in $P$ in chunks of $n$ at a time, adding available center–site pairs generated by points in the chunk (or points after it, as we will see) to the matching in order by distance. The invariant is that after a chunk is processed, its points do not generate any more available pairs, and we can move on to the next one until the matching is complete. To do this, for each chunk we construct the list $L$ of all the pairs generated by its points. Let $d$ be the maximum distance among these pairs. If $p_i$ is the last point in the chunk, the points in $P$ from $p_{i+1}$ up to the last point at distance to the origin at most $d(p_i, O) + 2\delta$ can generate pairs with distance less than $d$. We add any such pair to $L$. We have to check $O(n)$ additional points, so $L$ still has size $O(kn)$. We sort all these pairs and consider them in order, adding any available pair to the matching. Since each chunk has size $n$, there will be $O(n)$ chunks. Each one requires sorting a list of $O(kn)$ pairs, which requires $O(kn \log n)$ time (since $k \leq n^2$) and $O(kn)$ space. In total, we need $O(n^2 k \log n)$ time and $O(n^2 + nk)$ space.

## 2.2   Distance-Sorting Methods

Unless the centers are clustered together, the circle-growing algorithm finds many available pairs in the early iterations. However, it reaches a point in which most circles overlap. Even if the centers are randomly distributed, in the typical case a large fraction of centers have "far outliers", sites which belong to their region but are arbitrarily far because all the area in between is claimed by other centers. Consequently, many centers have to scan a large fraction of the square. At some point, thus, it is convenient to switch to a different algorithm that can find the closest available pairs quickly. In this section, let $m$ and $k \leq m$ denote, respectively, the number of available sites and centers after a matching has been partially completed.

*Pair Sort Algorithm.* This algorithm simply sorts all the center–site pairs by distance and considers them in order, adding any available pair to the matching until it is complete. This algorithm is convenient when we can use integer sorting techniques, as in the case of the Euclidean metric and integer centers. Then, it requires $O(mk)$ time and space.

While the pair sort algorithm has a big memory requirement to be used starting with an empty matching, used after the circle-growing algorithm has matched a large fraction of sites results in improved performance.

*Pair Heap Algorithm.* When centers have real coordinates, sorting all the pairs takes $O(mk \log m)$ time, but we can do better. We find for each site $s$ its closest center $c_s$, and build a min-heap with all the center–site pairs of the form $(c_s, s)$ using $d(c_s, s)$ as key. Clearly, the top of the heap is a closest available pair. We can iteratively extract and match the top of the heap until one of the centers becomes unavailable. When a center $c$ becomes unavailable, all the pairs in the heap containing $c$ become unavailable. At this point, there are two possibilities:

*Eager update.* We find the new closest available center of all the sites that had $c$ as closest center and rebuild the heap from scratch so that it again contains one pair for each available site and its closest available center.

*Lazy update.* We proceed as usual until we actually extract a pair $(c_s, s)$ with an unavailable center. Then, we find the new closest available center only for $s$, and reinsert the new pair in the heap.

In both cases, we repeat the process until the matching is complete.

We have not addressed yet how to find the closest center to a site. For this, we can use a nearest neighbor (NN) data structure that supports deletions. Such a data structure maintains a set of points and is able to answer *nearest neighbor queries*, which provide a query point $q$ and ask for the point in the set closest to $q$. For the pair heap algorithm, we initialize the NN data structure with the set of centers and delete them as they become unavailable.

Since we need deletions we can use a *dynamic* NN data structure, i.e., with support for insertions as well as deletions. The simplest NN algorithm is a linear search, and a dynamic data structure based on it has $O(k)$ time per query and $O(1)$ time per update. The best known complexity of a dynamic NN data structure is $O(\log^5 k)$ amortized time per operation [7,25].

Given that we know all the query points for our NN data structure ahead of time (the sites), we can build for each site $s$ an array $A_s$ with all the centers sorted by distance to $s$. Then, the closest center to a site $s$ is $A_s[i_s]$, where $i_s$ is the index of the first available center in $A_s$. When a center is deleted we simply mark it. When we get a query for the closest center to a site $s$, we search $A_s$ until we find an unmarked center. We can start the search from the index of the center returned in the last query for $s$. This data structure requires $O(mk)$ space and has a $O(mk \log k)$ initialization cost to sort all the arrays. The interesting property is that if we do $O(k)$ queries for a given site $s$, we require $O(k)$ time for all of them, as in total we traverse $A_s$ only once. We call this data structure *presort*, although it is not strictly a NN data structure because it knows the query points ahead of time.

In the pair heap algorithm, we can combine eager and lazy updates with any NN data structure. In any case, the running time is influenced by $\alpha$, the sum among all centers $c$ of the number of sites that had $c$ as closest center when $c$ became unavailable. In the worst case $\alpha = O(km)$, but assuming that each center is equally likely to be the closest center to each site, the expected value of $\alpha$ is $O(m)$. In the full version of the paper we test the value of $\alpha$ empirically in several different settings, and in every case we find $\alpha < 10\,m$.

With eager updates in total we have to initialize the NN data structure, perform $m$ *extract-min* operations, $O(m + \alpha)$ NN queries, $k$ NN deletions, and rebuild the heap $k$ times. Thus, the running time is $O(P(k, m) + m \log m + (m + \alpha)Q(k) + kD(k) + km)$, where $P(k, m)$ is the cost of initializing the NN data structure of choice with $k$ points (and $m$ query points, in the case of the presort data structure), and $Q(k)$ and $D(k)$ are the costs of queries and deletions, respectively. With lazy updates, instead of rebuilding the heap we have $O(\alpha)$ extra *insert* and *extract-min* heap operations, which requires $O(\alpha \log m)$ time.

For real centers, the best worst-case bound is with eager deletions and the presort NN data structure. In that case, we have that the NN queries take $O(km)$ for any $\alpha$, so the total running time is $O(mk \log k + m \log m)$. If we assume that $\alpha = O(m)$, then the best time is with lazy deletions and the NN data structure from [7,25]. The running time with this heuristic assumption is $O(m \log^5 k + m \log m)$.

### 2.3    Bichromatic Closest Pairs and Nearest Neighbor Chains

We now describe a less-practical solution based on bichromatic closest pairs which achieves the best theoretical running time that we have been able to prove. A bichromatic closest pair (BCP) data structure maintains a set of points, each colored red or blue, and is able to answer queries asking for the closest pair of different color.

The stable grid matching problem can be solved with a BCP data structure that supports deletions, either on its own or after the circle-growing algorithm. We first initialize the data structure with the available sites and centers as blue and red points, respectively. Then, we repeatedly find and match the closest pair, remove the site, and remove the center if it becomes unavailable. The running time is $O(P(m) + mQ(m) + mD(m))$, where $P(m), Q(m)$, and $D(m)$ are the initialization, query, and deletion costs, respectively, for the BCP data structure of choice containing $m$ blue points and $k \leq m$ red points.

Eppstein [16] proposed a fully dynamic BCP data structure that uses an auxiliary dynamic NN data structure. Using it, the sequence of operations required to solve the stable grid matching problem takes $O(mT(m) \log^2 m)$ time, where $T(m)$ is the cost per operation of the NN data structure. In particular, combining this with the dynamic nearest neighbor data structure of Chan [7] and Kaplan et al. [25] gives a total time bound of $O(n^2 \log^7 n)$ for this problem.

To improve this, we observe that (with a suitable tie-breaking rule to ensure that no two distances are equal) it is not necessary to find the bichromatic closest pair in each step: it suffices, instead, to find a mutual nearest neighbor pair: a pixel and a center that are closer to each other than to any other pixel or center. The reason is twofold. First, in the algorithm that repeatedly finds and removes closest pairs, every pair $(c, p)$ of mutual nearest neighbors eventually becomes a closest pair, because until they do, nothing else that the algorithm does can change the fact that they are mutual nearest neighbors. So $(c, p)$ will eventually become matched by the algorithm. Second, if we find a pair $(c, p)$ that will eventually become matched (such as a mutual nearest neighbor pair), it is safe to match them early; doing so cannot affect the correctness of the rest of the algorithm.

To find these, we may adapt the *nearest-neighbor chain algorithm* from the theory of hierarchical clustering [4,23] which uses a stack to repeatedly find pairs of mutual nearest neighbors at a cost of $O(1)$ nearest neighbor queries per pair. In more detail, the algorithm is as follows.

1. Initialize two dynamic nearest neighbor structures for the pixels and centers, and an empty stack $S$.

2. Repeat the following steps until all pixels have been matched:
   (a) If $S$ is empty, push an arbitrary point (either a pixel or a center) onto $S$.
   (b) Let $p$ be the point at the top of $S$, and use the nearest neighbor data structure to find the nearest point $q$ of the opposite color to $p$.
   (c) If $q$ is not already on $S$, push it onto $S$. Otherwise, $q$ must be the second-from-top point on $S$, and is a mutual nearest neighbor with $p$. Pop $p$ and $q$, match them to each other, and remove one or both of $p$ or $q$ from the nearest neighbor data structure (always remove the pixel, and remove the center if it becomes unavailable).

Note that in step 2. (c) $q$ must be second-from-top because we have a cycle of (non-mutual) nearest neighbors starting with $p \to q$ and then up the stack back to $p$. At each step along this cycle, the distance decreases or stays equal. But it cannot decrease, because there would be no way to increase back again, and nothing but $q \to p$ can be equal to $p \to q$, because we are using a tie-breaking rule. So the cycle has length two and $q$ is second-from-top.

Each step that pushes a new point onto $S$ can be charged against a later pop operation and its associated matched pixel, so the number of repetitions is $O(n^2)$. This algorithm gives us the following theorem.

**Theorem 1.** *The stable grid matching problem can be solved in $O(n^2)$ operations of a dynamic nearest neighbor data structure. In particular, with the structure of Chan [7] and Kaplan et al. [25], the time is $O(n^2 \log^5 n)$.*

## 3   Experiments

*Datasets.* Table 1 summarizes the parameters used in the different experiments. We use the following labels for the algorithms: $CG$ the circle-growing algorithm alone, and $PS$ and $PH$ for the combination of CG and the pair sort and pair heap algorithms, respectively. Moreover, for the pair heap algorithm we consider the following variations: eager/presort ($PH_{E,P}$), eager/linear search ($PH_{E,L}$), lazy/presort ($PH_{L,P}$), and lazy/linear search ($PH_{L,L}$).

We focus on the Euclidean metric, but in the full version of the paper we also consider the Manhattan and Chebyshev metrics. The parameter $n$ is the length of the side of the square grid, and $k$ is the number of centers. In all the experiments, the centers are chosen uniformly and independently at random. Moreover, every data point is the average of 10 runs, each starting with different centers.

The cutoff is the parameter used to determine when to switch from the circle-growing algorithm to a different one. We define it as a ratio between the number of available pairs and the number of pairs already considered by the circle-growing algorithm.

The algorithms were implemented in C++ (gcc version 4.8.2) and the interface in Qt. The experiments were executed by a Intel(R) Core(TM) CPU i7-3537U 2.00 GHz with 4 GB of RAM, on Windows 10.

**Table 1.** Summary of parameters used in the experiments section.

| Experiment | Algorithms | Metric | $n$ | $k$ | Cutoff |
|---|---|---|---|---|---|
| Exec. time (Fig. 3) | All | $L_2$ | varies | $10n$ | 0.15 |
| Cutoff (Fig. 4) | $CG, PH_{L,L}$ | $L_2$ | 1000 | varies | varies |

*Algorithm Comparison.* Figure 3 contains a comparison of all the algorithms. Pair heap is generally better than pair sort, even for integer distances where it has a higher theoretical complexity. Among pair heap variations, lazy/linear is the best for both types of centers. In general lazy updates perform better, but eager/presort is also a strong combination because they synergize: eager updates require more NN queries in exchange for less extract-min heap operations, and the presort data structure has fast NN queries.

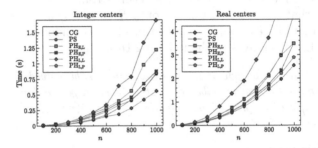

**Fig. 3.** Execution time of the various algorithms for integer (left) and real (right) centers. For all the methods but $CG$, the cutoff is 0.15. Each data point is the average of 10 runs with $10n$ randomly distributed centers and the $L_2$ metric.

*Optimal Cutoff.* When combining the circle-growing algorithm with another algorithm, the efficiency of the combination depends on the cutoff used to switch between both. If we switch too soon, we don't exploit the good behavior of the circle-growing algorithm when circles are still mostly disjoint. If we switch too late, the circle-growing algorithm slows down as it grows the circles in every direction just to reach some outlying region.

Figure 4 illustrates the role of the cutoff. It shows that most of the execution time of the circle-growing algorithm is spent with the very few last available pairs, so even a really small cutoff prompts a substantial improvement. After that, the additional time spent in the pair heap algorithm slightly beats the savings in the circle-growing algorithm, resulting in a steady increase of the total running time.

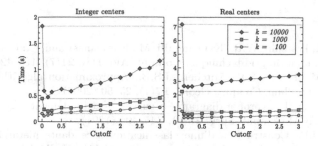

**Fig. 4.** Execution time of the circle-growing algorithm for integer (left) and real (right) centers, combined with the pair heap algorithm with lazy updates and a linear search NN data structure. The dotted lines denote the running time of the circle-growing algorithm alone, i.e., with cutoff 0. Each data point is the average of 10 runs with randomly distributed centers, $n = 1000$, and the $L_2$ metric.

## 4  Discussion

We have defined the stable grid matching problem, developed efficient theoretical algorithms and practical implementations of slower but simpler algorithms for this problem, and used our implementation to test different strategies for center placement in $k$-means like stable clustering algorithms. However, this work leaves several open questions:

- For which $n$ and $k$ does the stable grid matching problem have a placement of centers for which all clusters are connected, and how can such centers be found?
- Can the worst-case running time of our theoretical $O(n^2 \log^5 n)$-time algorithm be improved? Is it possible to achieve similar runtimes without going through fully-dynamic bichromatic closest pair data structures?
- Can we obtain practical algorithms whose runtime has lower worst-case dependence on $k$ than our $O(n^2 k)$-time circle-growing and distance-sorting methods?
- Our bichromatic closest pair and distance-sorting algorithms can be made to work for arbitrary point sets (not just pixels) but the circle-growing method assumes that the points form a grid, and its time analysis depends on the fact that the grid is a fat polygon (so that the area of each circle is proportional to the number of grid points that it covers) and that testing whether a point belongs to the grid is trivial. Can this method be extended to pixelated versions of more complicated polygons?
- How efficiently can we perform similar distance-based stable matching problems for graph shortest path distances instead of geometric distances? Can additional structure (such as the structures found in real-world road networks) help speed up this computation?

# References

1. Akman, V., Franklin, W.R., Kankanhalli, M., Narayanaswami, C.: Geometric computing and uniform grid technique. Comput.-Aid. Des. **21**(7), 410–420 (1989)
2. Arkin, E.M., Fekete, S.P., Mitchell, J.S.B.: Approximation algorithms for lawn mowing and milling. Comput. Geom. **17**(1), 25–50 (2000)
3. Aurenhammer, F.: Voronoi diagrams–a survey of a fundamental geometric data structure. ACM Comput. Surv. **23**(3), 345–405 (1991)
4. Benzécri, J.P.: Construction d'une classification ascendante hiérarchique par la recherche en chaîne des voisins réciproques. Les Cahiers de l'Analyse des Données **7**(2), 209–218 (1982). http://www.numdam.org/item?id=CAD_1982__7_2_209_0
5. Biedl, T., Bläsius, T., Niedermann, B., Nöllenburg, M., Prutkin, R., Rutter, I.: Using ILP/SAT to determine pathwidth, visibility representations, and other grid-based graph drawings. In: Wismath, S., Wolff, A. (eds.) 21st International Symposium on Graph Drawing (GD), pp. 460–471 (2013)
6. Böhringer, K.F., Donald, R.B., Halperin, D.: On the area bisectors of a polygon. Discrete Comput. Geom. **22**(2), 269–285 (1999)
7. Chan, T.M.: A dynamic data structure for 3-d convex hulls and 2-d nearest neighbor queries. J. ACM **57**(3), 16: 1–16: 15 (2010)
8. Chan, T.M., Patrascu, M.: Transdichotomous results in computational geometry, i: point location in sublogarithmic time. SIAM J. Comput. **39**(2), 703–729 (2009)
9. Chandran, S., Kim, S.K., Mount, D.M.: Parallel computational geometry of rectangles. Algorithmica **7**(1), 25–49 (1992)
10. Chrobak, M., Nakano, S.: Minimum-width grid drawings of plane graphs. Comput. Geom. **11**(1), 29–54 (1998)
11. Chun, J., Korman, M., Nöllenburg, M., Tokuyama, T.: Consistent digital rays. Discrete Comput. Geom. **42**(3), 359–378 (2009)
12. Cormen, T.H., Stein, C., Rivest, R.L., Leiserson, C.E.: Introduction to Algorithms, 2nd edn. McGraw-Hill Higher Education, Boston (2001)
13. De Floriani, L., Puppo, E., Magillo, P.: Applications of computational geometry to geographic information systems. In: Handbook of Computational Geometry, pp. 333–388 (1999)
14. De Fraysseix, H., Pach, J., Pollack, R.: How to draw a planar graph on a grid. Combinatorica **10**(1), 41–51 (1990)
15. Dehne, F., Pham, Q.T., Stojmenović, I.: Optimal visibility algorithms for binary images on the hypercube. Int. J. Parallel Programm. **19**(3), 213–224 (1990)
16. Eppstein, D.: Dynamic Euclidean minimum spanning trees and extrema of binary functions. Discrete Comput. Geom. **13**(1), 111–122 (1995)
17. Fang, T.P., Piegl, L.A.: Delaunay triangulation using a uniform grid. IEEE Comput. Graph. Appl. **13**(3), 36–47 (1993)
18. Gale, D., Shapley, L.S.: College admissions and the stability of marriage. Am. Math. Monthly **69**(1), 9–15 (1962)
19. Greene, D.H., Yao, F.F.: Finite-resolution computational geometry. In: 27th IEEE Symposium on Foundations of Computer Science (FOCS), pp. 143–152 (1986)
20. Hartley, R., Zisserman, A.: Multiple View Geometry in Computer Vision. Cambridge University Press, New York (2003)
21. Hoffman, C., Holroyd, A.E., Peres, Y.: A stable marriage of Poisson and Lebesgue. Ann. Probab. **34**(4), 1241–1272 (2006)
22. Jain, A.K., Murty, M.N., Flynn, P.J.: Data clustering: a review. ACM Comput. Surv. **31**(3), 264–323 (1999)

23. Juan, J.: Programme de classification hiérarchique par l'algorithme de la recherche en chaîne des voisins réciproques. Les Cahiers de l'Analyse des Données **7**(2), 219–225 (1982). http://www.numdam.org/item?id=CAD_1982__7_2_219_0
24. Kanungo, T., Mount, D.M., Netanyahu, N.S., Piatko, C.D., Silverman, R., Wu, A.Y.: An efficient $k$-means clustering algorithm: analysis and implementation. IEEE Trans. Pattern Anal. Mach. Intell. **24**(7), 881–892 (2002)
25. Kaplan, H., Mulzer, W., Roditty, L., Seiferth, P., Sharir, M.: Dynamic planar Voronoi diagrams for general distance functions and their algorithmic applications. Electronic preprint arxiv:1604.03654 (2016)
26. Keil, J.M.: Computational geometry on an integer grid. Ph.D. thesis, University of British Columbia (1980)
27. Knuth, D.E.: The Art of Computer Programming Sorting and Searching. Pearson Education, Reading (1998)
28. Overmars, M.H.: Computational geometry on a grid an overview. In: Earnshaw R.A. (eds.) Theoretical Foundations of Computer Graphics and CAD. NATO ASI Series (Series F: Computer and Systems Sciences), vol. 40, pp. 167–184. Springer, Heidelberg (1988)
29. Overmars, M.H.: Efficient data structures for range searching on a grid. J. Algorithms **9**(2), 254–275 (1988)
30. Rahman, M.S., Nakano, S., Nishizeki, T.: Rectangular grid drawings of plane graphs. Comput. Geom. **10**(3), 203–220 (1998)
31. Ricca, F., Scozzari, A., Simeone, B.: Weighted Voronoi region algorithms for political districting. Math. Comput. Modell. **48**(9–10), 1468–1477 (2008)
32. Solbrig, M.: Mathematical Aspects of Gerrymandering. Master's thesis, University of Washington (2013). https://digital.lib.washington.edu/researchworks/handle/1773/24334

# A Relational Generalization
# of the Khalimsky Topology

Josef Šlapal[✉]

Institute of Mathematics, Brno University of Technology, Brno, Czech Republic
slapal@fme.vutbr.cz

**Abstract.** We discuss certain $n$-ary relations ($n > 1$ an integer) and
show that each of them induces a connectedness on its underlying set.
Of these $n$-ary relations, we study a particular one on the digital plane $\mathbb{Z}^2$
for every integer $n > 1$. As the main result, for each of the $n$-ary relations
studied, we prove a digital analogue of the Jordan curve theorem for the
induced connectedness. It follows that these $n$-ary relations may be used
as convenient structures on the digital plane for the study of geometric
properties of digital images. For $n = 2$, such a structure coincides with
the (specialization order of the) Khalimsky topology and, for $n > 2$, it
allows for a variety of Jordan curves richer than that provided by the
Khalimsky topology.

## 1 Introduction

A crucial problem of digital topology, a theory that was founded for the study of
geometric and topological properties of digital images, is to provide the digital
plane $\mathbb{Z}^2$ with a convenient structure for such a study (cf. [9,10]). The classical,
graph theoretic, approach to digital topology is based on using the 4-adjacency
and 8-adjacency graphs for structuring $\mathbb{Z}^2$ (see [14,15]). Unfortunately, neither
4-adjacency nor 8-adjacency graph alone allows for an analogue of the Jordan
curve theorem (cf. [8]) so that a combination of the two adjacency graphs has
to be used. Despite this drawback, the classical approach to digital topology has
been used to solve numerous problems of digital image processing (see, e.g., [1])
and create a great deal of useful graphic software.

To eliminate the above drawback of the classical approach to digital topology,
a new, purely topological approach was proposed in [5] which utilizes a convenient topology for structuring the digital plane, namely the Khalimsky topology.
The convenience of the Khalimsky topology for structuring the digital plane was
shown in [5] by proving an analogue of the Jordan curve theorem for the topology (recall that the classical Jordan curve theorem states that a simple closed
curve in the Euclidean plane separates this plane into exactly two connected
components). The topological approach was then developed by many authors -
see, e.g., [4,6,7,11–13,17,18].

Since the Khalimsky topology is an Alexandroff $T_0$-topology, it is uniquely
determined by a partial order on $\mathbb{Z}^2$, the so-called specialization order of the

© Springer International Publishing AG 2017
V.E. Brimkov and R.P. Barneva (Eds.): IWCIA 2017, LNCS 10256, pp. 132–141, 2017.
DOI: 10.1007/978-3-319-59108-7_11

topology. The connectedness in the Khalimsky space then coincides with the connectedness in the underlying (simple) graph of the specialization order of the Khalimsky topology. Thus, when studying the connectedness of digital images with respect to the Khalimsky topology, this graph, rather than the Khalimsky topology itself, may be used for structuring the digital plane. A disadvantage of this approach is that Jordan curves in the (specialization order of the) Khalimsky topology may never turn at the acute angle $\frac{\pi}{4}$. It would, therefore, be useful to find some new, more convenient structures on $\mathbb{Z}^2$ that would allow Jordan curves to turn, at some points, to form the acute angle $\frac{\pi}{4}$. In the present note, to obtain such a convenient structure, we generalize the specialization order of the Khalimsky topology, hence a binary relation on $\mathbb{Z}^2$, by using certain $n$-ary relations on $\mathbb{Z}^2$ ($n > 1$ an integer). We will define a connectedness induced by these relations and will prove a digital Jordan curve theorem for this connectedness. Thus, the $n$-ary relations provide convenient structures on the digital plane for the study of the geometric properties of digital images that are related to boundaries because boundaries of objects in digital images are represented by digital Jordan curves.

## 2   Preliminaries

Throughout the paper, non-negative integers are considered to be finite ordinals and they are called, as usual, natural numbers. Thus, given a natural number $n > 0$, $(x_i | i < n)$ will denote the finite sequence $(x_0, x_1, ..., x_{n-1})$ and $(x_i | i \leq n)$ the finite sequence $(x_0, x_1, ..., x_n)$. These finite sequences will often be treated as sets, namely the sets $\{x_i;\ i < n\} = \{x_0, x_1, ..., x_{n-1}\}$ and $\{x_i;\ i \leq n\} = \{x_0, x_1, ..., x_n\}$, respectively.

We will work with some basic graph-theoretic concepts only - we refer to [2] for them. By a *graph* $G = (V, E)$, we understand an undirected simple graph without loops where $V \neq \emptyset$ is the *vertex* set of $G$ and $E \subseteq \{\{x, y\};\ x, y \in V,\ x \neq y\}$ is the set of *edges* in $G$. We will say that $G$ is a graph *on* $V$. Two vertices $x, y \in V$ are said to be *adjacent* (to each other) if $\{x, y\} \in E$. Recall that a *path* in $G$ is a (finite) sequence of pairwise different vertices (i.e., elements of $V$) such that every pair of consecutive vertices is adjacent. A (finite) sequence $(x_0, x_1, ..., x_n)$ of vertices of $G$ with $n > 2$ is called a *circle* in $G$ if $(x_i | i < n)$ is a path in $G$ and $x_0 = x_n$. A subset $A \subseteq V$ is *connected* in $G$ if any two points $x, y \in A$ may be joined by a path contained in $A$ (i.e., there is a path $(x_i | i \leq n)$ with $x_0 = x$, $x_n = y$ and $\{x_i | i \leq n\} \subseteq A$). A subset $A \subseteq V$ is said to be a *component* of $G$ if it is a maximal (with respect to set inclusion) connected subset of $V$. A circle $C$ in a graph $G$ is said to be a *simple closed curve* if, for every vertex $z \in C$, $C$ contains precisely two vertices adjacent to $z$. A simple closed curve $J$ in a graph with the vertex set $V$ is called a *Jordan curve* if it separates the set $V$ into precisely two components, i.e., if the induced subgraph $V - J$ has exactly two components.

Recall that, given a directed graph (i.e., a set with a binary relation) $D$, its *underlying graph* is the (undirected) graph obtained by just ignoring the direction of the edges in $D$.

For every point $(x, y) \in \mathbb{Z}^2$, we denote by $A_4(x, y)$ and $A_8(x, y)$ the sets of all points that are 4-adjacent and 8-adjacent to $(x, y)$, respectively. Thus, $A_4(x, y) = \{(x + i, y + j); \; i, j \in \{-1, 0, 1\}, \; ij = 0, \; i + j \neq 0\}$ and $A_8(x, y) = A_4(x, y) \cup \{(x + i, y + j); \; i, j \in \{-1, 1\}\}$. The graphs $(\mathbb{Z}^2, A_4)$ and $(\mathbb{Z}^2, A_8)$ are called the *4-adjacency graph* and *8-adjacency graph*, respectively.

In digital image processing, the 4-adjacency and 8-adjacency graphs are the most frequently used structures on the digital plane. But, since the late 1980's, another structure on $\mathbb{Z}^2$ has been used too, namely the Khalimsky topology [5]. It is the product of two copies of the topology on $\mathbb{Z}$ given by the subbase $\{\{2k - 1, 2k, 2k + 1\}; \; k \in \mathbb{Z}\}$ (for the basic concepts of general topology see [3]). Recall that, given a topology $\mathcal{T}$ on a set $X$, the *specialization preorder* of $\mathcal{T}$ is the preorder $\leq$ on $X$ defined by $x \leq y \Leftrightarrow x \in \overline{\{y\}}$ for all $x, y \in X$ (where $\overline{A}$ denotes the closure operator with respect to $\mathcal{T}$). Since the Khalimsky topology is $T_0$ (i.e., for all $t, z \in \mathbb{Z}^2$, $t \in \overline{\{z\}}$ and $z \in \overline{\{t\}}$ imply $t = z$), its specialization preorder is a (partial) order on $\mathbb{Z}^2$. And, since the Khalimsky topology is an Alexandroff topology (i.e., for all $A \subseteq \mathbb{Z}^2$, $\overline{A} = \bigcup_{z \in A} \overline{\{z\}}$), it is uniquely determined by its specialization order.

The specialization order of the Khalimsky topology coincides with the binary relation $\leq$ on $\mathbb{Z}^2$ given as follows:

For any $(x, y), (z, t) \in \mathbb{Z}^2$, $(x, y) \leq (z, t)$ if and only if

$(x, y) = (z, t)$ or
$x, y$ are even and $(z, t) \in A_8(x, y)$ or
$x$ is even, $y$ is odd, $z = x + i$ where $i \in \{-1, 1\}$, and $t = y$ or
$x$ is odd, $y$ is even, $z = x$, and $t = y + i$ where $i \in \{-1, 1\}$.

A portion of the specialization order $\leq$ of the Khalimsky topology is demonstrated in Fig. 1 by a directed graph with the vertex set $\mathbb{Z}^2$ where an oriented edge from a point $p$ to a point $q$ means that $q \leq p$.

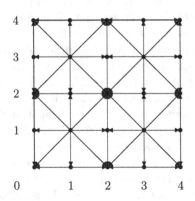

**Fig. 1.** A portion of the specialization order of the Khalimsky topology.

The underlying graph of the specialization order of the Khalimsky topology coincides with the *connectedness graph* of the topology, i.e., the graph with the vertex set $\mathbb{Z}^2$ in which two points are adjacent if and only if they are different and constitute a connected subset of the Khalimsky space. It may easily be seen that the connectedness in the Khalimsky space coincides with the connectedness in the connectedness graph of the Khalimsky topology, i.e., in the underlying graph of the specialization order the topology.

The famous Jordan curve theorem proved for the Khalimsky topology in [4] may be formulated as follows:

**Theorem 1.** *In the underlying graph of the specialization order of the Khalimsky topology, every simple closed curve with at least four points is a Jordan curve.*

It is readily verified that a simple closed curve (and thus also a Jordan curve) in the underlying graph of the specialization order of the Khalimsky topology may never turn at the acute angle $\frac{\pi}{4}$. It could therefore be useful to replace the specialization order of the Khalimsky topology with some more convenient structure (relation on $\mathbb{Z}^2$) that would allow Jordan curves to turn at the acute angle $\frac{\pi}{4}$ at some points. And this is what we will do in the next section.

## 3   Plain Relations and Induced Connectedness

Recall that, given a natural number $n > 0$ and a set $X$, an *n-ary relation* on $X$ is a subset $R \subseteq X^n$. Thus, the elements of $R$ are finite sequences (ordered $n$-tuples) $(x_0, x_1, ..., x_{n-1}) = (x_i | i < n)$ consisting of elements of $X$ (for the basic properties of $n$-ary relations see [16]). In the sequel, to eliminate the trivial case $n = 1$, we will restrict our considerations to $n > 1$.

**Definition 1.** An $n$-ary relation $R$ on a set $X$ is said to be *plain* if, for any $g, h \in R$, $g \neq h$ implies card $(g \cap h) \leq 1$.

**Definition 2.** Let $R$ be a plain $n$-ary relation on a set $X$ and $m$ a natural number. A sequence $C = (c_k | k \leq m)$ of elements of $X$ is called an *R-walk* if the following two conditions are satisfied:

I. For every non-negative integer $k_0 < m$, there exist $(x_i | i < n) \in R$ and $i_0 < n - 1$ such that $\{c_{k_0}, c_{k_0+1}\} = \{x_{i_0}, x_{i_0+1}\}$.

II. Every $(x_i | i < n) \in R$ satisfies the following two conditions:

   (i) if there exist $k_0 < m$ and $i_0 < n-1$ such that $c_{k_0} = x_{i_0}$ and $c_{k_0+1} = x_{i_0+1}$, then $k_0 \geq i_0$ and $c_{k_0-j} = x_{i_0-j}$ for all $j = 1, 2, ..., i_0$,

   (ii) if there exist $k_0 < m$ and $i_0 < n-1$ such that $c_{k_0} = x_{i_0+1}$ and $c_{k_0+1} = x_{i_0}$, then $k_0 \leq m - i_0 - 1$ and $c_{k_0+j} = x_{i_0-j+1}$ for all $j = 2, 3, ..., i_0 + 1$.

An $R$-walk $(c_k | k \leq m)$ with the property that $m \geq 2$ and $c_i = c_j \Leftrightarrow \{i, j\} = \{0, m\}$ is said to be an *R-circle*.

Observe that, if $(c_0, c_1, ..., c_n)$ is an $R$-walk, then $(c_n, c_{n-1}, ..., c_0)$ is an $R$-walk, too (so that $R$-walks are closed under reversion) and, if $(d_k|k \leq p)$ and $(e_k|k \leq q)$ are $R$-walks with $d_p = e_0$, then, putting $f_k = d_k$ for all $k \leq p$ and $f_k = e_{k-p}$ for all $k$ with $p \leq k \leq p+q$, we get an $R$-walk $(f_k|k \leq p+q)$ (so that $R$-walks are closed under composition).

Given a plain $n$-ary relation $R$ on a set $X$, a subset $Y \subseteq X$ is said to be $R$-*connected* if, for every pair $a, b \in Y$, there is an $R$-walk $(c_k|k \leq m)$ such that $c_0 = a$, $c_m = b$ and $c_i \in Y$ for all $i \in \{0, 1, ..., m\}$. A maximal (with respect to set inclusion) $R$-connected subset of $X$ is called an $R$-*component* of $X$.

**Definition 3.** Let $R$ be a plain $n$-ary relation on a set $X$. A nonempty, finite and $R$-connected subset $J$ of $X$ is said to be an $R$-*simple closed curve* if every element $(x_i|i < n) \in R$ with $\{x_0, x_1\} \subseteq J$ satisfies $\{x_i|i < n\} \subseteq J$ and every point $z \in J$ fulfills one of the following two conditions:

(1) There are exactly two elements $(x_i|i < n) \in R$ satisfying both $\{x_i|i < n\} \subseteq J$ and $z \in \{x_0, x_{n-1}\}$ and there is no element $(y_i|i < n) \in R$ satisfying both $\{y_i|i < n\} \subseteq J$ and $z = y_i$ for some $i \in \{1, 2, ..., n-1\}$.
(2) There is exactly one element $(y_i|i < n) \in R$ satisfying both $\{y_i|i < n\} \subseteq J$ and $z = y_i$ for some $i \in \{1, 2, ..., n-2\}$ and there is no element $(x_i|i < n) \in R$ satisfying both $\{x_i|i < n\} \subseteq J$ and $z \in \{x_0, x_{n-1}\}$.

Clearly, every $R$-simple closed curve is an $R$-circle.

**Definition 4.** Let $R$ be a plain $n$-ary relation on a set $X$. An $R$-simple closed curve $J$ is called an $R$-*Jordan curve* if the subset $X - J \subseteq X$ consists (i.e., is the union) of precisely two $R$-components.

*Remark 1.* In the Euclidean plane $\mathbb{R}^2$, every Jordan curve $J$ is a minimal separator of $\mathbb{R}^2$, i.e., the subset $\mathbb{R}^2 - (J - \{p\}) \subseteq \mathbb{R}^2$ is connected for every point $p \in J$. This is not true for $R$-Jordan curves, which means that an $R$-Jordan curve $J$ may have a point $p \in J$ such that the subset $\mathbb{Z}^2 - (J - \{p\}) \subseteq \mathbb{Z}^2$ is not $R$-connected.

From now on, for every natural number $n > 1$, $R_n$ will denote the plain $n$-ary relation on $\mathbb{Z}^2$ given as follows: For every $((x_i, y_i)|i < n)$ such that $(x_i, y_i) \in \mathbb{Z}^2$ for every $i < n$, $((x_i, y_i)|i < n) \in R_n$ if and only if one of the following eight conditions is satisfied:

(1) $x_0 = x_1 = ... = x_{n-1}$ and there is $k \in \mathbb{Z}$ such that $y_i = (2k+1)(n-1) + i$ for all $i < n$,
(2) $x_0 = x_1 = ... = x_{n-1}$ and there is $k \in \mathbb{Z}$ such that $y_i = (2k+1)(n-1) - i$ for all $i < n$,
(3) $y_0 = y_1 = ... = y_{n-1}$ and there is $l \in \mathbb{Z}$ such that $x_i = (2l+1)(n-1) + i$ for all $i < n$,
(4) $y_0 = y_1 = ... = y_{n-1}$ and there is $l \in \mathbb{Z}$ such that $x_i = (2l+1)(n-1) - i$ for all $i < n$,

(5) there is $k \in \mathbb{Z}$ such that $x_i = (2k+1)(n-1)+i$ for all $i < n$ and there is $l \in \mathbb{Z}$ such that $y_i = (2l+1)(n-1)+i$ for all $i < n$,

(6) there is $k \in \mathbb{Z}$ such that $x_i = (2k+1)(n-1)+i$ for all $i < n$ and there is $l \in \mathbb{Z}$ such that $y_i = (2l+1)(n-1)-i$ for all $i < n$,

(7) there is $k \in \mathbb{Z}$ such that $x_i = (2k+1)(n-1)-i$ for all $i < n$ and there is $l \in \mathbb{Z}$ such that $y_i = (2l+1)(n-1)+i$ for all $i < n$,

(8) there is $k \in \mathbb{Z}$ such that $x_i = (2k+1)(n-1)-i$ for all $i < n$ and there is $l \in \mathbb{Z}$ such that $y_i = (2l+1)(n-1)-i$ for all $i < n$.

A portion of $R_n$ is demonstrated in Fig. 2. The ordered $n$-tuples belonging to $R_n$ are represented by arrows oriented from first to last terms. Between any pair of neighboring parallel horizontal or vertical arrows (having the same orientation), there are $n-2$ more parallel arrows with the same orientation that are not displayed in order to make the Figure transparent.

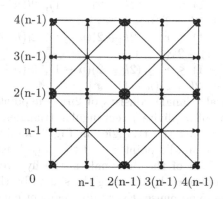

Fig. 2. A portion of $R_n$.

It may easily be seen that $R_2$ coincides with the specialization order of the Khalimsky topology. Thus, Theorem 1 is a Jordan curve theorem for $R_2$. We will prove a Jordan curve theorem for every $R_n$ with $n > 2$.

In Fig. 3, (a section of) a graph on $\mathbb{Z}^2$ is demonstrated but only the vertices $(2k(n-1), 2l(n-1))$, $k, l \in \mathbb{Z}$, are marked out. Thus, on every edge (denoted by a line segment), there are $2n-1$ vertices that are not displayed.

**Theorem 2.** *If $n > 2$, then every circle in the graph demonstrated in Fig. 3 that turns only at some of the marked out points $(2k(n-1), 2l(n-1))$, $k, l \in \mathbb{Z}$, is an $R_n$-Jordan curve.*

*Proof.* For every point $z = ((2k+1)(n-1), (2l+1)(n-1))$, $k, l \in \mathbb{Z}$, each of the following four subsets of $\mathbb{Z}^2$ will be called an $n$-*fundamental triangle* (given by $z$):

$$\{(r,s) \in \mathbb{Z}^2;\ 2k(n-1) \leq r \leq (2k+2)(n-1),\ 2l(n-1) \leq s \leq (2l+2)(n-1),\ s \leq r + 2l(n-1) - 2k(n-1)\},$$

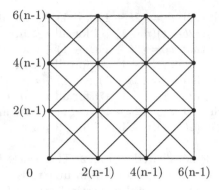

**Fig. 3.** $R_n$-Jordan curves.

$\{(r,s) \in \mathbb{Z}^2;\ 2k(n-1) \leq r \leq (2k+2)(n-1),\ 2l(n-1) \leq s \leq (2l+2)$
$(n-1),\ s \geq 2l(n-1) + (2k+2)(n-1) - r\}$,

$\{(r,s) \in \mathbb{Z}^2;\ 2k(n-1) \leq r \leq (2k+2)(n-1),\ 2l(n-1) \leq s \leq (2l+2)$
$(n-1),\ s \geq r + 2l(n-1) - 2k(n-1)\}$,

$\{(r,s) \in \mathbb{Z}^2;\ 2k(n-1) \leq r \leq (2k+2)(n-1),\ 2l(n-1) \leq s \leq (2l+2)$
$(n-1),\ s \leq 2l(n-1) + (2k+2)(n-1) - r\}$.

Every $n$-fundamental triangle consists of $2n^2 - n$ points and forms a segment having the shape of a (digital) rectangular triangle. The $n$-fundamental triangles given by $z$ are just the triangles in Fig. 3 obtained by dividing the square segment with the middle point $z$ and the edge length $2(n-1)$ by one of the two diagonals. Each of the diagonals is the hypotenuse of the two $n$-fundamental triangles obtained by dividing the square by the diagonal and $z$ is the middle point of the hypotenuse. Every line segment constituting an edge of an $n$-fundamental triangle consists of precisely $2n - 1$ points. Clearly, the edges of any $n$-fundamental triangle form an $R_n$-simple closed curve. We will show that every $n$-fundamental triangle is $R_n$-connected and so is every set obtained from an $n$-fundamental triangle by subtracting some of its edges.

Let $z = ((2k+1)(n-1), (2l+1)(n-1))$, $k, l \in \mathbb{Z}$, be a point and consider the $n$-fundamental triangle $T = \{(r,s) \in \mathbb{Z}^2;\ 2k(n-1) \leq r \leq (2k+2)(n-1),\ 2l(n-1) \leq s \leq (2l+2)(n-1),\ y \leq x + 2l(n-1) - 2k(n-1)\}$. Then $T$ is the (digital) triangle $ABC$ with the vertices $A = (2k(n-1), 2l(n-1))$, $B = ((2k+2)(n-1), 2l(n-1))$, $C = ((2k+2)(n-1), (2l+2)(n-1))$. For every $u \in \mathbb{Z}$, $(2k+1)(n-1) \leq u \leq (2k+2)(n-1)$, the sequence $G_u = ((u,y)|2l(n-1) \leq y \leq u + 2(l-k)(n-1))$ is an $R_n$-walk (contained in $T$), so that $G_u$ is an $R_n$-connected set. Similarly, for every $v \in \mathbb{Z}$, $2l(n-1) \leq v \leq (2l+1)(n-1)$, the sequence $H_v = ((x,v)|v + 2(k-l)(n-1) \leq x \leq (2k+2)(n-1))$ is an $R_n$-walk (contained in $T$), so that $H_v$ is an $R_n$-connected set. We clearly have $T = \bigcup\{G_u;\ (2k+1)(n-1) \leq u \leq (2k+2)(n-1)\} \cup \bigcup\{H_v;\ 2l(n-1) \leq v \leq (2l+1)(n-1)\}$. It may easily be seen that $G_u \cap H_v \neq \emptyset$ whenever $(2k+1)(n-1) \leq u \leq (2k+2)(n-1)$ and $2l(n-1) \leq v \leq (2l+1)(n-1)$. For every natural number $i < 2n$, we put

$$S_i = \begin{cases} G_{(2k+1)(n-1)+\frac{i}{2}} & \text{if } i \text{ is even,} \\ H_{2l(n-1)+\frac{i-1}{2}} & \text{if } i \text{ is odd.} \end{cases}$$

Then $(S_i | i < 2n)$ is a sequence with the property that its members with even indices form the sequence $(G_u | (2k+1)(n-1) \leq u \leq (2k+2)(n-1))$ and those with odd indices form the sequence $(H_v | 2l(n-1) \leq v \leq (2l+1)(n-1))$. Hence, $\bigcup\{S_i | i < 2n\} = \bigcup\{G_u; (2k+1)(n-1) \leq u \leq (2k+2)(n-1)\} \cup \bigcup\{H_v; 2l(n-1) \leq v \leq (2l+1)(n-1)\}$ and every pair of consecutive members of $(S_i | i < 2n)$ has a non-empty intersection. Thus, since $T = \bigcup\{S_i | i < 2n\}$, $T$ is $R_n$-connected. For each of the other three $n$-fundamental triangles given by $z$, the proof is analogous, and the same is true also for every set obtained from an $n$-fundamental triangle (given by $z$) by subtracting some of its edges.

We will say that a (finite or infinite) sequence $S$ of $n$-fundamental triangles is a tiling sequence if the members of $S$ are pairwise different and every member of $S$, excluding the first one, has an edge in common with at least one of its predecessors. Given a tiling sequence $S$ of $n$-fundamental triangles, we denote by $S'$ the sequence obtained from $S$ by subtracting, from every member of the sequence, all its edges that are not shared with any other member of the sequence. By the firs part of the proof, for every tiling sequence $S$ of $n$-fundamental triangles, the set $\bigcup\{T; T \in S\}$ is $R_n$-connected and the same is true for the set $\bigcup\{T; T \in S'\}$.

Let $J$ be an $R_n$-simple closed curve. Then $J$ constitutes the border of a polygon $S_F \subseteq \mathbb{Z}^2$ consisting of $n$-fundamental triangles. More precisely, $S_F$ is the union of some $n$-fundamental triangles such that any pair of them is disjoint or meets in just one edge in common. Let $U$ be a tiling sequence of the $n$-fundamental triangles contained in $S_F$. Since $S_F$ is finite, $U$ is finite, too, and we have $S_F = \bigcup\{T; T \in U\}$. As every $n$-fundamental triangle $T \in U$ is $R_n$-connected, so is also $S_F$. Similarly, $U'$ is a finite sequence with $S_F - J = \bigcup\{T; T \in U'\}$ and, since every member of $U'$ is $R_n$-connected (by the first part of the proof), $S_F - J$ is connected, too.

Further, let $V$ be a tiling sequence of $n$-fundamental triangles which are not contained in $S_F$. Since the complement of $S_F$ in $\mathbb{Z}^2$ is infinite, $V$ is infinite, too. Put $S_I = \bigcup\{T; T \in V\}$. As every $n$-fundamental triangle $T \in V$ is $\mathcal{B}^2$-connected, so is also $S_I$. Similarly, $V'$ is a finite sequence with $S_I - J = \bigcup\{T; T \in V'\}$ and, since every member of $V'$ is connected (by the first part of the proof), $S_I - J$ is connected, too.

It may easily be seen that every $R_n$-walk $C = (z_i | i \leq k)$, $k > 0$ a natural number, connecting a point of $S_F - J$ with a point of $S_I - J$ meets $J$ (i.e., meets an edge of an $n$-fundamental triangle which is contained in $J$). Therefore, the set $\mathbb{Z}^2 - J = (S_F - J) \cup (S_I - J)$ is not $R_n$-connected. We have shown that $S_F - J$ and $S_I - J$ are $R_n$-components of $\mathbb{Z}^2 - J$, $S_F - J$ finite and $S_I - J$ infinite, with $S_F$ and $S_I$ $R_n$-connected. The proof is complete.

The circles in the graph demonstrated in Fig. 3 that do not turn at any point $((2k+1)(n-1), (2l+1)(n-1))$, $k, l \in \mathbb{Z}$, which are $R$-Jordan curves by Theorem 2, provide a rich enough variety of circles to be used for representing borders of objects in digital images. The advantage of the circles over the Jordan

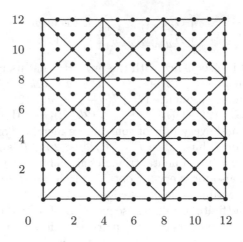

**Fig. 4.** $R_3$-Jordan curves.

curves in the Khalimsky topology is that they may turn at the acute angle $\frac{\pi}{4}$ at some points.

*Example 1.* Every circle in the graph demonstrated in Fig. 4 that does not turn at any point $(4k + 2, 4l + 2)$, $k, l \in \mathbb{Z}$, is an $R_3$-Jordan curve by Theorem 2. Thus, for example, the triangle with vertices $(0,0), (8,0), (4,4)$ is an $R_3$-Jordan curve but not an $R_2$-Jordan curve. For this triangle to become an $R_2$-Jordan curve, we have to delete the points $(0,0), (1,0), (7,0), (8,0)$. But this will cause a considerable deformation of the triangle.

## 4   Conclusions

We have shown that every plain $n$-nary relation induces connectedness on its underlying set. This connectedness may be used to define the concepts of a simple closed curve and a Jordan curve in the underlying set of a given plain $n$-ary relation. We introduced and discussed a particular plain $n$-nary relation on the digital plane $\mathbb{Z}^2$ for every natural number $n > 1$ and showed that the connectedness induced by each of these relations allows for a digital analogue of the Jordan curve theorem. Thus, we have shown that the $n$-ary relations introduced provide convenient structures on the digital plane for the study of digital images. While for $n = 2$ this structure coincides with the Khalimsky topology, for $n > 2$ the structures have the advantage over the Khalimsky topology that they allow the Jordan curves to turn at the acute angle $\frac{\pi}{4}$ at some points. Since Jordan curves represent borders of objects in digital images, the structures on $\mathbb{Z}^2$ provided by the $n$-ary relations discussed may be used in digital image processing for solving problems related to boundaries, such as pattern recognition, boundary detection, contour filling, data compression, etc.

**Acknowledgement.** This work was supported by the Brno University of Technology Specific Research Program, project no. FSI-S-17-4464.

# References

1. Brimkov, V.E., Klette, R.: Border and surface tracing - theoretical foundations. IEEE Trans. Pattern Anal. Mach. Intell. **30**, 577–590 (2008)
2. Bondy, J.A., Murty, U.S.R.: Graph Theory. Graduate Texts in Mathematics. Springer, Heidelberg (2008)
3. Engelking, R.: General Topology. Państwowe Wydawnictwo Naukowe, Warszawa (1977)
4. Han, S.-E.: Compression of Khalimsky topological spaces. Filomat **26**, 1101–1114 (2012)
5. Khalimsky, E.D., Kopperman, R., Meyer, P.R.: Computer graphics and connected topologies on finite ordered sets. Topology Appl. **36**, 1–17 (1990)
6. Khalimsky, E.D., Kopperman, R., Meyer, P.R.: Boundaries in digital plane. J. Appl. Math. Stochast. Anal. **3**, 27–55 (1990)
7. Kiselman, C.O.: Digital Jordan curve theorems. In: Borgefors, G., Nyström, I., Baja, G.S. (eds.) DGCI 2000. LNCS, vol. 1953, pp. 46–56. Springer, Heidelberg (2000). doi:10.1007/3-540-44438-6_5
8. Kong, T.Y., Kopperman, R., Meyer, P.R.: A topological approach to digital topology. Am. Math. Monthly **98**, 902–917 (1991)
9. Kong, T.Y., Roscoe, W.: A theory of binary digital pictures. Comput. Vision Graphics Image Process. **32**, 221–243 (1985)
10. Kong, T.Y., Rosenfeld, A.: Digital topology: introduction and survey. Comput. Vis. Graph. Image Process. **48**, 357–393 (1989)
11. Kopperman, R., Meyer, P.R., Wilson, R.G.: A Jordan surface theorem for three-dimensional digital space. Discr. Comput. Geom. **6**, 155–161 (1991)
12. Melin, E.: Digital surfaces and boundaries in Khalimsky spaces. J. Math. Imaging and Vision **28**, 169–177 (2007)
13. Melin, E.: Continuous digitization in Khalimsky spaces. J. Approx. Theory **150**, 96–116 (2008)
14. Rosenfeld, A.: Connectivity in digital pictures. J. Assoc. Comput. Math. **17**, 146–160 (1970)
15. Rosenfeld, A.: Digital topology. Amer. Math. Monthly **86**, 621–630 (1979)
16. Šlapal, J.: Cardinal arithmetics of general relational systems. Publ. Math. Debrecen **18**, 39–48 (1991)
17. Šlapal, J.: Convenient closure operators on $\mathbb{Z}^2$. In: Wiederhold, P., Barneva, R.P. (eds.) IWCIA 2009. LNCS, vol. 5852, pp. 425–436. Springer, Heidelberg (2009). doi:10.1007/978-3-642-10210-3_33
18. Šlapal, J.: Topological structuring of the digital plane. Discr. Math. Theor. Comput. Sci. **15**, 425–436 (2013)

# Toward Parallel Computation of Dense Homotopy Skeletons for nD Digital Objects

Pedro Real[1], Fernando Diaz-del-Rio[2(✉)], and Darian Onchis[3,4]

[1] Institute of Mathematics, University of Seville,
Avda. Reina Mercedes s/n, 41012 Seville, Spain
real@us.es
[2] Computer Architecture and Technology Department,
University of Seville, Avda. Reina Mercedes s/n, 41012 Seville, Spain
fdiaz@us.es
[3] Faculty of Mathematics, Univ. of Vienna,
Oskar-Morgenstern-Platz 1, 1090 Vienna, Austria
[4] Faculty of Mathematics and Computer Science,
West Univ. of Timisoara, Bulevardul Vasile Parvan 4,
300223 Timisoara, Romania

**Abstract.** An appropriate generalization of the classical notion of abstract cell complex, called primal-dual abstract cell complex (pACC for short) is the combinatorial notion used here for modeling and analyzing the topology of $nD$ digital objects and images. Let $D \subset I$ be a set of n-xels (ROI) and $I$ be a n-dimensional digital image. We design a theoretical parallel algorithm for constructing a topologically meaningful asymmetric pACC $HSF(D)$, called Homological Spanning Forest of $D$ (HSF of $D$, for short) starting from a canonical symmetric pACC associated to $I$ and based on the application of elementary homotopy operations to activate the pACC processing units. From this HSF-graph representation of $D$, it is possible to derive complete homology and homotopy information of it. The preprocessing procedure of computing $HSF(I)$ is thoroughly discussed. In this way, a significant advance in understanding how the efficient HSF framework for parallel topological computation of $2D$ digital images developed in [2] can be generalized to higher dimension is made.

**Keywords:** Computational topology · nD digital image · Primal-dual abstract cell complex · Parallelism · Homological Spanning Forest · Homotopy operation

## 1 Introduction

The problem of developing a topologically consistent framework for efficient parallel topological analysis and recognition of $n$-dimensional digital objects is nowadays a major challenge. Intimately associated to this problem, we encounter the issue to find a suitable representation model from which the extraction of

© Springer International Publishing AG 2017
V.E. Brimkov and R.P. Barneva (Eds.): IWCIA 2017, LNCS 10256, pp. 142–155, 2017.
DOI: 10.1007/978-3-319-59108-7_12

topological features and characteristics of the object can be as fast and the most complete as possible. A successful strategy for achieving these goals is to "cellularize" the images. A primal-dual abstract cell complex [2] (or, pACCs for short), an appropriate generalization of the notion of abstract cell complex [8,9] for describing bitopological spaces, efficiently encodes local topological (incidences between cells, working at sub-$n$-xel level) information of the digital object in order to be promoted to global consistent topological information. We are mainly interested in information related to "homology holes", which are abstract generalizations at any dimension of the intuitive notion of curve bounding an arc or surface bounding a volume [7]. Classically, the different homology holes of a complex are obtained via linear algebra algorithms based on diagonalization of incidence matrices to Smith Normal Form [17]. The technique employed here for parallel processing is based on building asymmetric pACCs from symmetric ones. The asymmetric and non-redundant output pACCs resulting from our framework encompass the hierarchical graph notion of Homological Spanning Forest (HSF, for short) developed in [10,11,14]. Roughly speaking, an HSF of a digital object is a flexible topological model described by a kind of dense topological skeleton inside the object. Figure 1 shows two different HSFs of the same 2D digital object. The inclusion of an optimal vector field over each tree installed "inside the object" allows us not only counting the different homological holes of dimension 0 (connected components or CCs for short) and dimension 1 but also to removing them via cutting or filling. Moreover, if we retain the vicinity relations between these HSF graphs, we can reach homotopy-based representations of 2D digital images like the adjacency tree of a binary image or the region-adjacency-graph of a grey-level image [13].

In this paper, we design a theoretical parallel algorithm for computing an HSF-structure of a nD-digital object. Let us emphasize that: (a) the HSF-approach can be considered as a Morse-based pre-homology computation method

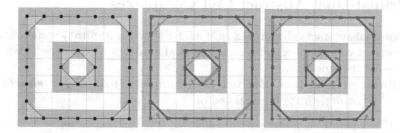

**Fig. 1.** (Left) ROI consisting in the set of black pixels. The implicit cellularization of the ROI -using 8-adjacency and being the 0-cells the square physical pixels- is superimposed; (Center) Visualization of an HSF of the ROI. The two trees spanning 0-cells (in red) of the ROI mean that it has two 8-CCs. The yellow "trees" -derived from the optimal vector field linking the rest of 1-cells with the set of 2-cells of the ROI-containing a 1-cell marked with a thick yellow segment determine two one-dimensional homological holes of the ROI or, equivalently, two 4-CCs of the background; (Right) Another possible HSF. (Color figure online)

(e.g. [3,4]) in the sense that a discrete vector field is "optimally" installed over the pACC. Its novelty lies in dealing with this issue as a pure combinatorial optimization problem in a fully parallel way over a scenario subdivided space and substituting the classical vector field language of homology by that of the new dynamic notion of crack (called link in [2]); (b) the theoretical time complexity of the parallel algorithm of [2] for computing an HSF structure of a binary digital $n \times m$ image is approximately logarithmic (precisely, $O(log(n + m))$). It seems that its generalization to nD image context can be done without excessive cost in complexity; (c) another strength of this framework is its potentiality to generate new topological representation models of nD objects and images involving homological holes (not only of dimension zero) and topologically strong relationships between them (for instance, generalizing to nD the notions of adjacency tree or RAG 2D models).

A flowchart of this nD-HSF algorithm is shown in Fig. 2.

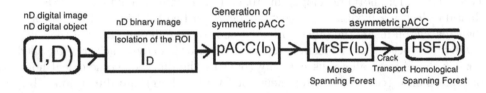

**Fig. 2.** Workflow of nD-HSF Algorithm.

In what follows, after a section of technical definitions related to the concept of primal-dual abstract cell complex, we formally describe the different stages of the previous theoretical algorithm.

## 2  Primal-Dual Abstract Cell Complexes

A primal-dual abstract cell complex (pACC, for short) is a suitable generalization of an abstract cell complex and a combinatorial model of a geometric subdivided object as bitopological spaces.

*A finite primal-dual abstract cell complex (pACC for short)* $C = (C, {}^{C}B^p, {}^{C}B^d, dim_p^C, dim_d^C)$ is composed of:

- $C \bigcup \{\emptyset\}$, where $C$ is a finite set of cells and $\emptyset$ is the empty set.
- two dimension functions: *(primal dimension)* $dim_p^C : C \rightarrow \{0, 1, 2, \ldots, \ell_p\}$ and *(dual dimension)* $dim_d^C : C \rightarrow \{0, 1, 2, \ldots, \ell_d\}$, where $\ell_p, \ell_d \in \mathbb{N} \cup \{0\}$. The set $C_i^p$ (resp. $C_i^d$) is the set of cells such that their primal (resp. dual) dimension is $i$.
- two bounding maps: (primal bounding map) a graded function ${}^{C}B^p = \{{}^{C}B_i^p\}_i$, such that ${}^{C}B_i^p : C_i^p \times C_{i+1}^p \rightarrow \mathbb{N} \cup \{0\}$ ($\forall 0 \le i \le \ell_p - 1$) and (dual bounding map) a graded function ${}^{C}B^d = \{{}^{C}B_i^d\}_i$, such that ${}^{C}B_i^d : C_i^d \times C_{i+1}^d \rightarrow \mathbb{N} \cup \{0\}$, $\forall 0 \le i \le \ell_d - 1$. We extend the respective definitions of ${}^{C}B^p$ and ${}^{C}B^d$ to

$C \times C$ by simply assigning value zero to the rest of ordered pairs of cells not belonging to the original domains.

The set of values the bounding maps takes on as output is the semi-ring $\mathbb{N} \cup \{0\}$. Of course, it is possible to change the images of the bounding maps to a ring (like $\mathbb{Z}$) or to a field (like $\mathbb{Q}$ o $\mathbb{R}$).

The pACC C is called *uniquely dimensional* if its primal and dual dimensions both depend on *a unique dimension* function $dms : C \rightarrow \{0, 1, 2, \ldots, \ell\}$, being $\ell = \ell_p = \ell_d$. $\ell$ is called *the dimension of* C. In fact, $dim_p = dms$ and $dim_d = \ell - dms$. Let us denote the set of cells $C_i^p$ of primal dimension $i$ simply by $C_i$ and an $i$-cell means a primal $i$-cell. A uniquely dimensional pACC C is called *symmetric* if $^{C}B_i^p(c, c') = {}^{C}B_i^d(c', c)$, $\forall 0 \leq i \leq \ell$ and $\forall c, c' \in C$. In this case, the bounding maps $^{C}B^p$ and $^{C}B^d$ are respectively denoted by $^{C}B$ and $^{C}B^{-1}$.

From now on, to simplify the notation, we drop the subindex $i$ (corresponding to primal dimension) and the superindex C (corresponding to the ACC name) from the dimension and bounding maps, unless otherwise specified.

Given two cells $c'$ and $c''$ of C, we say that the ordered pair $(c', c'')$ is an $(i, i + 1)$ *primal (resp. dual) vector* $(i = 0, 1, \ldots)$ of the pACC C if *its primal (resp. dual) multiplicity* $B^p(c', c'') \neq 0$ (resp. if $B^d(c', c'') \neq 0$), being $c' \in C_i^p$ (resp. $c' \in C_i^d$). The cell $c'$ is called the tail and $c''$ is the head of the primal (resp. dual) vector $(c', c'')$. We say that the set $\{c', c''\}$ is an $(i, i + 1)$ *primal (resp. dual) incidence set* of the pACC if $B^p(c', c'') \neq 0$ or $B^p(c'', c') \neq 0$ (resp. if $B^d(c', c'') \neq 0$ or $B^d(c'', c') \neq 0$), being $c'$ or $c''$ a cell of $C_i^p$ (resp. $C_i^d$).

Given a pACC $C = (C, B^p, B^d, dim_p, dim_d)$, let us define a sub-pACC $D = (D, {}^{D}B^p, {}^{D}B^d, dim_p, dim_d)$ of C as a new pACC with $D \subset C$ whose: (a) primal and dual dimension functions agree with those of C restricted to $D$; (b) the primal (resp. dual) bounding map satisfies that if $^{D}B^p(c', c'') = q \neq 0$ (resp. $^{D}B^d(c', c'') = q \neq 0$), then $B^p(c', c'') \geq q$ (resp. $B^d(c', c'') \geq q$). If $^{D}B^p = B^p|_{D \times D}$ and $^{D}B^d = B^d|_{D \times D}$, the sub-pACC D of C is called *complete*.

The complete sub-pACC $St^p(c, C)$ (resp. $St^d(c, C)$) of C, consisting of $c$ and all elements $c'$ in C, such that $B^p(c, c') \neq 0$ (resp. $B^d(c, c') \neq 0$) is called the *primal (resp. dual) open star of $c$ in* C. It is exactly the same as the smallest primal (resp. dual) neighborhood of $c$ in C [9]. If C is an uniquely dimensional symmetric pACC, so are $St^p(c, C)$ and $St^d(c, C)$.

Any pACC can be expressed as *a node-arc weighted graph*. The *incidence graph* $G(C)$ associated to a pACC C is the graph such that its nodes are the different cells of C and an edge $\{c, c'\}$ of this graph is either a primal or dual incidence set of the pACC or both. If C is symmetric, we propose as label for an edge $\{c', c''\}$ $(c' \in C_i, c'' \in C_{i+1})$ of $G(C)$, the ordered pair $(B^p(c', c''), B^d(c'', c'))$. As weight for a node $c \in C$, we choose the number $dms(c)$.

A *primal (resp. dual) crack* associated to the $(i, i + 1)$-primal (resp. dual) vector $(c, c')$ is the set $crk^p(c, c')$ (resp. $crk^d(c, c')$) of triplets $(c, c', c'')$, for all the cells $c''$ such that $(c', c'')$ is a dual (resp. primal) vector. A crack $crack(c, c')$ can be considered as an uniquely dimensional asymmetric sub-pACC of C. For example, for a primal crack $crk^p(c, c')$, its bounding functions $\bar{B}^p$ and $\bar{B}^d$ satisfy an "ortogonality" condition: for all the triplets $(c, c', c'')$ of $crk^p(c, c')$, $\bar{B}^p(c, c') =$

$B^p(c,c') \neq 0$, $\bar{B}^p(c',c'') = 0$, $\bar{B}^d(c',c'') = B^d(c',c'') \neq 0$, $\bar{B}^d(c,c') = 0$. Let us note that the crack notion is an extension of the term link in [2].

A geometric cell complex $\mathcal{K}$ can be represented by a uniquely dimensional symmetric pACC $\mathsf{K} = (K, B, B^{-1}, dms, \ell - dms)$, such that $B(c',c'') \in \{0,1\}$, $\forall (c',c'') \in K \times K$. In fact, the primal and dual bounding relation maps can automatically be obtained from the complete set of incidences between cells of $\mathsf{K}$ which differ in one dimension and the dimension map $dms$ of $\mathsf{K}$ agrees with the dimension function of the cell complex $\mathcal{K}$.

Finally, let us note that both bounding graded functions $\{B_i^p\}_i$ and $\{B_j^d\}_j$ of a pACC $\mathsf{C} = (C, B^p, B^d, dim_p, dim_d)$ can be extended to $C \times C$ in an asymmetric, irreflexive and transitive way without difficulty, giving raise to two different (primal and dual) classical ACCs associated to the pACC $\mathsf{C}$. Due to the fact that every finite topological space with the T0-separation property is isomorphic to an abstract cellular complex [9], *a pACC can be interpreted as a finite bitopological space*. The primal and dual ACC of a uniquely dimensional symmetric pACC can be deduced one from each other by simply reversing the order of the factors in the bounding relations.

## 3   pACC Homotopy Computation

First, we succinctly describe here the distinct steps of the theoretical nD-HSF Algorithm (whose flowchart is (2)). The rest of this section is devoted to understand the concept of elementary homotopy operation and the sequential algorithm computing an HSF of a pACC.

(a) **Input data:** The pair $(I, D)$. The nD digital image $I : \{1, \ldots, m_1\} \times \{1, \ldots, m_2\} \times \ldots \times \{1, \ldots, m_n\} \to \{0, 1, \ldots, 2^c - 1\}$ is represented by a $m_1 \times m_2 \times \ldots \times m_n$ $(m_1, m_2, \ldots, m_n, c \in \mathbb{N})$ integer-valued matrix. The digital object $D$, called region-of-interest (or ROI, for short), is formed by a set of pixels (represented by their corresponding (row,column) coordinates) of $I$. In fact, in order to avoid the mathematical ill-posed problems of the segmentation and noise, which are ubiquitous in the area of Digital Imagery, $I$ is a pre-segmented digital image, and $D$ is a region of this previous segmentation.

(b) **Extraction of the ROI:** From $I$, we "isolate" the ROI $D$ by means of new digital binary image $I_D$ of the same dimension than $I$. The set of black pixels (numbered by 1's) of $I_D$ is exactly $D$.

(c) **Generation of topological pACCs:** In this phase, we compute two kinds of pACCs in this order: (a) first, symmetric pACCs, modeling in a redundant way the connectivity (incidence) information of $D$ and $I$; (b) finally, asymmetric pACCs, which are non-redundant sub-pACCs of the previous ones, specifying a kind of dense homotopy graph-skeleton of them.

Some key notions for understanding our topological scaffolding are those of primal and dual pACC-homotopy operations. Given a uniquely $n$-dimensional symmetric pACC $\mathsf{C} = (C, B, B^{-1}, dms, n - dms)$ and a primal vector $(c, c')$, then *the primal pACC-homotopy operation* $Op^p(\overrightarrow{(c,c')}(\mathsf{C}))$ is a new symmetric pACC $(C\backslash\{\mathbf{c},\mathbf{c'}\}, \tilde{B}, \tilde{B}^{-1}, dms, n - dms)$, such that the new bounding function $\tilde{B}$ is defined by:

- $\forall \bar{c} \in St^d(\mathbf{c'}, \mathsf{C})\backslash\{\mathbf{c}\}$, $\forall \bar{c'} \in St^p(\mathbf{c}, \mathsf{C})\backslash\{\mathbf{c'}\}$,

$$\tilde{B}(\bar{c}, \bar{c'}) = B(\bar{c}, \bar{c'}) + B(\bar{c}, \mathbf{c'})B^{-1}(\mathbf{c'}, \mathbf{c})B(\mathbf{c}, \bar{c'});$$

- for the rest of pairs of cells $(c, c')$, $\tilde{B}(c, c') = B(c, c')$

Analogously, we can define elementary dual pACC-homotopy operations. We emphasize that such kind of operations is not, in general, a map of pACCs (that is, a map of sets compatible with the dimensions and bounding relations), but it can be considered as a function $Op^p(\overrightarrow{(c,c')}(\mathsf{C})) : pACC \times pACC \rightarrow pACC$. For example, considering the primal crack pACC $crk(c, c')$, we can construct a primal pACC-homotopy operation $Op^p(crk(c, c'), \mathsf{C})$ providing us the same resulting pACC than $Op^p(\overrightarrow{(c,c')}(\mathsf{C}))$.

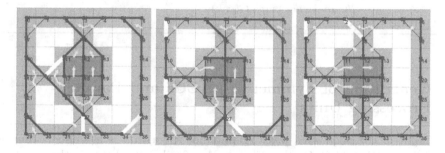

**Fig. 3.** Three different possible HSF outputs of Algorithm primal-HSF applied to a 2D digital object $X$ of black pixels, depending of the concrete ordered list of cells of $X$ chosen for sequential processing.

Now, we are able to design a sequential computational method for computing an HSF of the pACC $pACC(I_D)$, based on an appropriate reduction of cells via primal homotopy operations.

The output of the previous algorithm consists of a set of asymmetric pACCs $\{\mathcal{F}_{k-1,k}\}_{k=1}^n$ and a minimal pACC $\mathcal{H}$ formed by a set of isolated cells of different primal dimension. Figure 3 shows some outputs of the algorithm for 2D objects. The cells of H are called *critical cells*. These data can be reorganized and interpreted in terms of a set $HSF(\mathsf{C})$ of connected sub-graphs spanning the set of cells of C, called *Homological Spanning Forest associated to* C. In fact, these graphs can not be trees in dimension higher than two but we use this name because they appear as a suitable generalization to higher dimension of the notion of the spanning forest as a tool for labeling connected components of a graph [6]. Let us limit ourselves to say that the importance to save this combinatorial homology information of nD digital objects in terms of cracks and graphs primarily lies in its capacity of creating robust topological models involving homological holes of the objects and strong homology (incidence) relations between them.

For a better understanding, we only work the three-dimensional case in the rest of sections. The nD case is completely analogous.

---

**Algorithm 1.** (Sequential pACC-Homology Algorithm)

---

**Input:** A uniquely dimensional symmetric pACC $\mathsf{C} := \{C, {}^C B, {}^C B^{-1}, dms, n \backslash dms\}$
*A list of all the cells of* $\mathsf{C}$ *ordered by primal dimension*      $c_1^0 \ldots c_{\ell_0}^0, c_1^1, \ldots, c_{\ell_1}^1, \ldots c_1^n,$
$\ldots, c_{\ell_n}^n$ *such that* $dim_p(c_j^k) = k, \forall k, j.$
  1: $\mathsf{H} \leftarrow \mathsf{C}$
  2: **for** $k = 1$ to $n$ **do**
  3:     $\mathcal{F}_{(k-1,k)} \leftarrow \emptyset$
  4:     $crk \leftarrow \emptyset$
  5:     **for** $j = 1$ to $\ell_k$ **do**
  6:        **if** $\exists \bar{c} \in St^d(c_j^k, \mathsf{H}) / {}^{\mathsf{H}} B(\bar{c}, c_j^k) = 1$ **then**
  7:          $\mathsf{H} \leftarrow Op^p(crk(\bar{c}, c_j^k), \mathsf{H})$
  8:          $crk \leftarrow crk \bigoplus \{crk(\bar{c}, c_j^k)\};$
  9:     $\mathcal{F}_{(k-1,k)} \leftarrow$ *the incidence graph* $G(crk)$
10: **Output:** $((\mathcal{F}_{(0,1)}, \ldots, \mathcal{F}_{(n-1,n)}), \mathcal{H})$

---

## 4    Generation of Symmetric pACCs and Parallel Processing Units

The input of the Sequential pACC-Homology Algorithm is a uniquely dimensional symmetric pACC. On the other hand, a fundamental step in the workflow of nD-HSF Algorithm (Fig. 2) is the generation of such objects. Apart from building these initial pACCs, we also create the parallel processing units of our framework.

The scenario in which we need to "embed" the digital image $I_D$ is a uniquely dimensional symmetric pACC intimately associated to the contractible set of cells denoted by $Cell(I_D)$. $Cell(I_D)$ only depends on the dimensions of $I_D$ and can be constructed in a straightforward way. The 0-cells are the voxels (elements of the matrix) of $I_D$ (black or whites), the 1-cells are given by the set of two 6-adjacent voxels ($x$-frame, $y$-frame or $z$-frame adjacent), 2-cells are given by sets of four mutually 6-adjacent voxels and, 3-cells are given by sets of eight mutually 6-adjacent voxels. Thus, a dimension function $dms : Cell(I_D) \rightarrow \{0, 1, 2, 3\}$ is well-defined in this way. In order to create topological coordinates (automatically detecting incidences between cells) preserving the initial coordinate system ($row, colum, depth$) existing for the voxels of $I_D$, we use the following geometric realization for the cells of $Cell(I_D)$: (a) 0-cells are points in $\mathbb{R}^3$ with natural-value coordinates; (b) a 1-cell is represented at sub-voxel level by the coordinates of the barycenter of the segment determined by its corresponding pair of voxels, (c) a 2-cell is represented at sub-voxel level by the coordinates of the barycenter of the square formed by the 4-uple of voxels barycenters; (d) a 3-cell is represented at sub-voxel level by the coordinates of the barycenter of the cube formed by its corresponding 8-uple of voxels. For instance, a 1-cell is specified by topological coordinates of the type $(x_1, x_2, x_3)$, where two value of them are natural numbers and the third is a natural number minus $\frac{1}{2}$ (for example, $x_3$). The geometric boundary of this 1-cell which is formed by the set of two 0-cells $\{(x_1, x_2, x_3 - \frac{1}{2}), (x_1, x_2, x_3 + \frac{1}{2})\}$ completely describes the dual bounding relation of the 1-cell. Its geometric coboundary, formed by the set of four 2-cells $\{(x_1 \pm \frac{1}{2}, x_2, x_3), (x_1, x_2 \pm \frac{1}{2}, x_3 + \frac{1}{2})\}$ fully specifies its primal bounding relation. Then, it is straightforward to construct the uniquely dimensional symmetric pACC $pACC(I_D) = (Cell(I_D), B_{I_D}, B_{I_D}^{-1}, dim_p^{I_D}, dim_d^{I_D})$. Notice that $pACC(I_D) = pACC(I)$, and, in consequence, $pACC(I_D)$ is independent of $D$. We can also define another uniquely dimensional symmetric sub-pACC $pACC(D)$ of $pACC(I_D)$, being $Cell(D)$ its set of cells. $Cell(D)$ is the *topological hull of the set of black voxels* $D$ within $I_D$, which means that the 0-cells of $Cell(D)$ are the black voxels of $I_D$ and its $i$-cells $c$ ($i = 1, 2, 3$) can be recursively defined in terms of $(i - 1)$-cells by imposing that $St^d(c) \subset Cell(D)$.

Any node ($i$-cell) $(x, y, z)$ of the incidence graph $G(pACC(I_D))$ has the number $color(x, y, z)$ as weight. The function $color : Cell(I_D) \to \{0, \frac{1}{2}, 1\}$ is defined as follows: (a) for a 0-cell, it is the voxel value in $I_D$; (b) for an $i$-cell $c$ with $i \geq 1$, if all the values of the color function over the 0-cells of $c$ is 0 (resp. 1), then $color(c)$ is 0 (resp. is 1). In another case, $color(c) = \frac{1}{2}$.

For creating the parallel processing units, the idea is to establish a regular partition of the $Cell(I_D)$ into *cellular units* $Cell(x, y, z)$. There are as many cellular units as voxels the image has (equivalently, as 0-cells the $pACC(I_D)$ has). The *cellular unit* $Cell_8(x, y, z)$ associated to the voxel of topological coordinates $(x, y, z)$ is the set $\{(x, y, z), (x + \frac{1}{2}, y, z), (x, y + \frac{1}{2}, z), (x, y, z + \frac{1}{2}), (x + \frac{1}{2}, y + \frac{1}{2}, z), (x + \frac{1}{2}, y, z + \frac{1}{2}), (x, y + \frac{1}{2}, z + \frac{1}{2}), (x + \frac{1}{2}, y + \frac{1}{2}, z + \frac{1}{2})\}$ (one 0-cell, three 1-cells, three 2-cells, one 3-cell). Considered as an uniquely dimensional asymmetric sub-pACC of $pACC(I_D)$, *the processing unit* $PE(x, y, z)$ is defined as the sum of pACCs $\bigoplus_{(c', c'') \in U} crk^p(c', c'')$, where $U = Cell_8(x, y, z) \times Cell_8(x, y, z)$. Its underlying set of cells involves 27 cells which belong to the topological hull generated by the cells $(x, y, z), (x+1, y, z), (x, y+1, z), (x, y, z+1), (x+1, y+1, z), (x+1, y, z+1), (x, y+1, z+1)$ and $(x+1, y+1, z+1)$. The number of primal vectors (see Fig. 4) involved in $PE(x, y, z)$ is twelve (three $(0, 1)$ vectors, six $(1, 2)$ vectors and three $(2, 3)$ vectors).

## 5   Generation of MrSFs

The next step in the Algorithm nD-HSF is the parallel building of an HSF of the initial geometric symmetric pACC $pACC(I_D)$. This particular asymmetric pACC $MrSF(I_D)$ is called Morse Spanning Forest (MrSF for short). An MrSF has the property that the set of its elementary primal cracks applied in some order in a sequential process of reduction based on primal homotopy operations provides a final pACC consisting in only one 0-cell (critical cell). In this way, a MrSF for $I_D$ is seen as a kind of "dense combinatorial skeleton" of the contractible cell complex $Cell(I_D)$. This notion has been already developed in [12] making use exclusively of homological arguments. Finally, the last process of the pipeline of Fig. 2, called *crack transport*, consists in a "homotopy optimization" of $MrSF(I_D)$ in order to get another MrSF, denoted by $HSF(I_D)$, such that its restriction to $Cell(D)$ is a true HSF $HSF(D)$ of $D$. This optimization is done by suitably "transporting" cracks of the $MrSF(I_D)$, with the objective to maximize the number of its primal bounding relations between cells of $pACC(D)$. We focus here in the parallel algorithmic techniques for MrSF construction; the crack transport step of the algorithm will be studied in detail elsewhere.

A *Morse Spanning Forest* for a three dimensional digital image $I$ of dimension $m_1 \times m_2 \times m_3$ is any output $((\mathcal{F}_{(0,1)}, \mathcal{F}_{(1,2)}, \mathcal{F}_{(2,3)}), \mathcal{H})$ of Sequential pACC-Homology Algorithm applied to $pACC(I)$. It is not difficult to prove that any MrSF has only one $(0,1)$-tree.

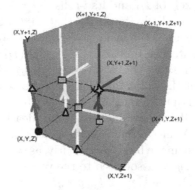

**Fig. 4.** An activation state (local MrSF rule: direction $+Y$) of the processing unit $PE(x, y, z)$ showing its eight active cells, primal and dual activation vectors and associated cracks. The 0-cell $(x, y, z)$ is drawn with a circle, the 1-cells with triangles, the 2-cells with squares and the 3-cell with a star. The active primal vectors are drawn with an arrow and using different colors depending on its dimension.

Our algorithm of MrSF generation is divided into two main steps: (a) building a MrSF at local (voxel's neighborhood) level by means of a process of activation of processing units; (b) building the MrSF at global level, specifying the membership of any cell to the corresponding tree of the MrSF. Afterwards, we can proceed to the Final HSF determination via crack transports.

**(a) MrSF building at local level: Activation of processing units.** There are nine possible activation states for any $\mathsf{PE}(x, y, z)$, each one associated to a particular configuration of four disjoint primal vectors (called primal activation vectors) involving cells of $Cell_8(x, y, z)$. The sum of the crack pACCs of $\mathsf{PE}(x, y, z)$ associated to these primal activation vectors fully defines the corresponding activation state.

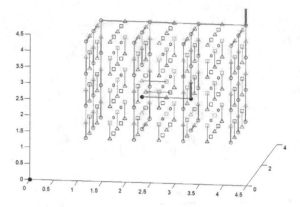

**Fig. 5.** A (4,3,4) binary 3D image showing active primal $(0, 1)$-vectors (red and green colors) and dual $(1, 0)$-vectors (black thin vectors) of the MrSF. Thicker vectors indicate possible critical 0-cells. (Color figure online)

For activating in parallel all the processing units of $pACC(I_D)$, we can use *local MrSF rules*. For each $\mathsf{PE}(x, y, z)$, we choose an activation's state depending of giving preference to some order in the principal directions or the particular configuration of the color function of the cells in $\mathsf{PE}(x, y, z)$ (Fig. 4).

In our current implementation of the algorithm of MrSF generation: (a) the local MrSF rules are first defined for the lowest dimension cells and then progressively extended to higher dimension; (b) we give preference to $+Z$ direction, then to $+Y$, and finally to $+X$.

Once the primal $(0, 1)$-vector of the $\mathsf{PE}(x, y, z)$ is activated, the two primal $(1, 2)$-vectors and the $(2, 3)$-vector are activated following the same direction of the first one. This implies that only one 1-cell of $Cells(x, y, z)$ belong to the $(0, 1)$-tree of the MrSF, and the other two 1-cells reside in the $(1, 2)$-tree. Figure 5 shows an example of the primal $(0, 1)$ and $(1, 2)$ vectors for a binary 3D image that contains two black voxels in the center.

The above MrSF arrangement is one the many possible configurations. Its main advantage is that it can be computed in a fully parallel manner for each voxel. Other possibilities can be exploited, but the parallelism feature should be preserved if we would want to process real 3D images in an efficient way.

**(b) Global MrSF construction.** Once a local MrSF has been defined it is necessary to introduce global relations between the cells of the whole MrSF. This process can be done in a similar way to that of [2]. That algorithm was much easier since it was written only for two dimensional images. Nevertheless, the idea is the same: to label each cell of the incidence graph (forest) $G(MrSF)$ of the MrSF, according to its membership to some connected subgraph (tree) of $G(MrSF)$. At the end of this process, the different connected components of $G(MrSF)$ must have been labeled (Fig. 6).

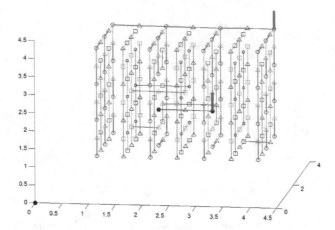

**Fig. 6.** The same (4,3,4) binary 3D image of Fig. 5 with the complete MrSF. Thicker links indicate potential critical 0-cells.

**(c) Final HSF determination via crack transports.** This final step of the nD-HSF Algorithm is aimed to minimize the number of critical cells. This would produce the final HSF. As an example, the trees of Fig. 7(Left) would transform into that of Fig. 7(Right). A graphical explanation of this process from the lower dimensional MrSF trees to the higher ones is the following. Firstly, the $(0, 1)$-crack marked as 'D' is transported to the right inferior crack in the $(0, 1)$-tree. Secondly, the crack 'C' is laid to its left to continue the closing of the 0-1 tree. Transports of 'C' and 'D' supposes the cancellation of one critical 0-cell and one critical 1-cell. In fact, these cells should be detected as false critical cells in the initial MrSF. This transport process is really a pairing of critical cells of different dimensions going through the corresponding tree. Finally, cracks 'A' and 'B' must be also transported so as to "close" properly the 1–2 tree. This yields to an equivalent set of trees, which composed the HSF for the ROI. Obviously this final HSF indicates that the ROI contains only one critical 0-cell being the

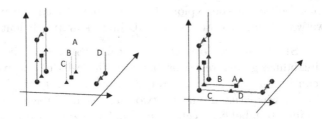

**Fig. 7.** (Left) A ROI (composed of 6 voxels in 'L' shape) that contains only one CC, and whose MrSF presents two separated 0–1 trees. Cracks that go out from the ROI indicate possible critical cells. (Right) The same 3D image that contains only one 0–1 tree after the necessary transports that complete the HSF.

representative of the CC (connected component). The correct computation of final HSF will yield to the homology of any CC inside a digital image. Some examples are shown in the next section.

## 6    Examples of Homological Magnitudes of Several Shapes Obtained Through 3-Dimensional HSFs

The topological nature of 3D digital images are much richer than that of 2-D images. Attending exclusively to homology groups, apart from cavities and connected components of a digital object (somewhat comparable to holes and connected components in 2D imagery context), tunnels appear in 3D. In a nut-shell, each critical cell of any dimension is in direct relationship with a different homology generator. Figure 8 and 9 shows different shapes and their correspond-ing critical cells (those belonging to a crack of a MrSF "going out" of the ROI). To ease the viewing of these figures, only ROIs are represented and axes are not drawn. Cells belonging to the black ROI have been filled. These results are summarized in Table 1. Table 1 shows the results of the different simple shapes of Figs. 8 to 9 and their critical cells. Excepting Fig. 9 Left (due to its false critical cells), the number of critical 0-cells agree with the number of CCs, the number of

**Table 1.** Results of the different simple shapes of Figs. 8 and 9 and their critical cells

| Shapes | # Critical 0-cells | # Critical 1-cells | # Critical 2-cells |
|---|---|---|---|
| Two perpendicular rings with contact | 1 | 2 | 0 |
| Two perpendicular crossing rings | 2 | 2 | 0 |
| An empty polyhedron (showing its MrSF) | 1 | 2 | 3 |
| An emptypolyhedron (showing its HSF) | 1 | 2 | 0 |

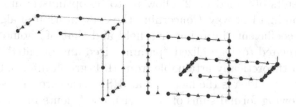

**Fig. 8.** Left: Two perpendicular $3 \times 3$ rings with contact resulting in two critical 1-cells (inferior right corner and superior left corner), representative of its two tunnels, and one critical 0-cell (upper right corner), representative of the CC. Right: Two perpendicular crossing $3 \times 3$ rings resulting in two critical 1-cells (inferior left corners), representative of two tunnels, and two critical 0-cells (upper right corners), representative of the two CCs.

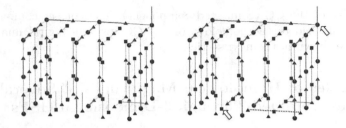

**Fig. 9.** Left: A MrSF of an empty $3 \times 3 \times 3$ polyhedron. There is one critical 0-cell (upper right corner), representative of the CC. In addition, three critical 2-cells and two 3 critical 1-cells (all of them in the inferior side) have appeared. Two pairs of them are false critical cells. Right: After proper transports (marked with thicker dotted lines), the HSF of the same empty polyhedron yields to only one critical 2-cell (representative of the cavity), and the same critical 0-cell. Arrows indicates the position of these resultant critical cells.

critical 1-cells indicates the number of tunnels and the number of critical 2-cells represents the number of cavities.

## 7   Conclusions

Based on the notions of primal-dual abstract cell complex and homotopy operation, and generalizing to higher dimension the work developed in [2], a theoretical algorithm for computing combinatorial homology structures, called HSFs of $nD$ digital objects, has been sketched. Focusing in a topological pre-processing step, called Morse Spanning Forest generation, we set a fully parallel algorithm for determining a kind of dense topological skeleton associated to the image scenario within which the digital object is embedded. Both to analyze the efficiency of the procedure and to advance in increasing the degree of understanding on HSF or pACC homology computation of digital objects, an unpretentious implementation done in Matlab is used for experimentation. Although a theoretical complexity study of the parallel algorithm has not yet been carried out, the encouraging results obtained in [2] allow us to be optimistic in computing the HSF information in a fast way. Concerning the computation of algebraic homology holes with coefficients in a ring or a field and that of "homotopy holes" of objects (those related to generalized "parametrized and oriented closed curves" [7]), they sound theoretically attainable from HSF-graph information. An argument supporting this idea is the fact that an HSF structure can be algebraically interpreted (allowing formal sums of cells with coefficients in some ground ring or field) as an operator controlling a chain homotopy equivalence between an object and its homology [1,5,12,15,16]. Finally, the possibility to detect homological hole relationships (like adjacency or "to be surrounded by" between path connected components in 2D) in an HSF allows holding high expectations in achieving functional implementations of parallel algorithms of topological pattern recognition based on HSF information.

**Acknowledgments.** This work has been supported by the Spanish research projects (supported by the Ministerio de Economía y Competitividad and FEDER funds) COFNET (Event-based Cognitive Visual and Auditory Sensory Fusion, TEC2016-77785-P) and TOP4COG (Topological Recognition of 4D Digital Images via HSF model, MTM2016-81030-P (AEI/FEDER,UE)). The last co-author gratefully acknowledges the support of the Austrian Science Fund FWF-P27516.

# References

1. Berciano, A., Molina-Abril, H., Real, P.: Searching high order invariants in computer imagery. Appl. Algebra Eng. Commun. Comput. **23**(1–2), 17–28 (2012)
2. Díaz-del-Río, F., Real, P., Onchis, D.: A parallel homological spanning forest framework for 2D topological image analysis. Pattern Recogn. Lett. **83**, 49–58 (2016)
3. Harker, S., Mischaikow, K., Mrozek, M., Nanda, V., Wagner, H., Juda, M., Dlotko, P.: The efficiency of a homology algorithm based on discrete Morse theory and coreductions. Image-A: Appl. Math. Image Eng. **1**(1), 41–48 (2010)
4. Floriani, L., Fugacci, U., Iuricich, F.: Homological shape analysis through discrete morse theory. In: Breuß, M., Bruckstein, A., Maragos, P., Wuhrer, S. (eds.) Perspectives in Shape Analysis. MV, pp. 187–209. Springer, Cham (2016). doi:10.1007/978-3-319-24726-7_9
5. González-Díaz, R., Real, P.: On the cohomology of 3D digital images. Discrete Appl. Math. **147**(2), 245–263 (2005)
6. Hopcroft, J., Tarjan, R.: Algorithm 447: efficient algorithms for graph manipulation. Commun. ACM **16**(6), 372–378 (1973)
7. Hurewicz, W.: Homology and homotopy theory. In Proceedings of the International Mathematical Congress of 1950, p. 344. University of Toronto Press (1952)
8. Klette, R.: Cell complexes through time. In International Symposium on Optical Science and Technology. International Society for Optics and Photonics, pp. 134–145 (2000)
9. Kovalevsky, V.: Finite topology as applied to image analysis. Comput. Vis. Graph. Image Process. **46**, 141–161 (1989)
10. Molina-Abril, H., Real, P., Nakamura, A., Klette, R.: Connectivity calculus of fractal polyhedrons. Pattern Recogn. **48**(4), 1150–1160 (2015)
11. Molina-Abril, H., Real, P.: Homological spanning forest framework for 2D image analysis. Ann. Math. Artif. Intell. **64**(4), 385–409 (2012)
12. Molina-Abril, H., Real, P.: Homological optimality in Discrete Morse Theory through chain homotopies. Pattern Recogn. Lett. **11**, 1501–1506 (2012)
13. Pavlidis, T.: Algorithms for Graphics and Image Processing. Springer Science and Business Media, Heidelberg (1977)
14. Real, P., Molina-Abril, H., Gonzalez-Lorenzo, A., Bac, A., Mari, J.L.: Searching combinatorial optimality using graph-based homology information. Appl. Algebra Eng. Comm. Comp. **26**(1–2), 103–120 (2015)
15. Real, P.: Homological perturbation theory and associativity. Homology, Homotopy Appl. **2**(1), 51–88 (2000)
16. Real, P.: An algorithm computing homotopy groups. Math. Comput. Simul. **42**(4–6), 461–465 (1996)
17. Veblen, O.: Analisis Situs, vol. 5. A.M.S. Publications, Providence (1931)

# Polynomial Time Algorithm for Inferring Subclasses of Parallel Internal Column Contextual Array Languages

Abhisek Midya[1]([✉]), D.G. Thomas[2], Alok Kumar Pani[3], Saleem Malik[1], and Shaleen Bhatnagar[1]

[1] Information Technology, Alliance University, Bangalore 562106, India
abhisekmidyacse@gmail.com, baronsaleem@gmail.com,
shaleenbhatnagar@gmail.com
[2] Department of Mathematics, Madras Christian College, Chennai 600059, India
dgthomasmcc@yahoo.com
[3] Computer Science and Engineering, Christ University Faculty of Engineering,
Bangalore 560074, India
alok.kumar@christuniversity.in

**Abstract.** In [2,16] a new method of description of pictures of digitized rectangular arrays is introduced based on contextual grammars, called parallel internal contextual array grammars. In this paper, we pay our attention on parallel internal column contextual array grammars and observe that the languages generated by these grammars are not inferable from positive data only. We define two subclasses of parallel internal column contextual array languages, namely, k-uniform and strictly parallel internal column contextual languages which are incomparable and not disjoint classes and provide identification algorithms to learn these classes.

**Keywords:** Parallel internal column contextual array grammars · k-uniform · Identification in the limit from positive data

## 1 Introduction

In theoretical computer science, formal language theory is one of the fundamental areas. This study has its origin in Chomskian grammars. Contextual grammars which are different from Chomskian grammars, have been studied in [3,13] by formal language theorists, as they provide novel insight into a number of issues central to formal language theory. In a total contextual grammar, a context is adjoined depending on the whole current string. Two special cases of total contextual grammars, namely internal and external are very natural and have been extensively investigated. (External) Contextual grammars are introduced by S. Marcus in 1969 [13] with a linguistic motivation in mind. An external contextual grammar generates a language starting from a finite set of strings (the base) and iteratively adjoining to its contexts outside the current string.

© Springer International Publishing AG 2017
V.E. Brimkov and R.P. Barneva (Eds.): IWCIA 2017, LNCS 10256, pp. 156–169, 2017.
DOI: 10.1007/978-3-319-59108-7_13

In other families of contextual grammars, such as internal contextual grammars [13], the contexts are adjoined inside the current string.

There has been a great interest in adapting the techniques of formal string language theory for developing methods to study the problem of picture generation and description, where pictures are considered as connected, digitized finite arrays in the two-dimensional plane [15]. Recently, extensions of string contextual grammars to array structures and hyper graphs have been made in [1,2,6–8,11,12,14,16].

On the other hand, Grammatical Inference refers to the method of inferring a grammar (and possibly a target language) from data. Data can be text or informant. The difference between text and informant is that a text gives only positive examples (all strings do belong to the same language) where informant is both positive and negative examples. A learning procedure is an algorithm which is executed on a never-ending stream of inputs. The inputs are grammatical strings/arrays, taken from a target language which is in a known class of languages. The task is to identify a grammar that generates the target language. At each point in the process, any string is given as an input to the algorithm. After each input the algorithm produces a guess at the grammar which is eventually correct and could be unaltered when additional inputs are given. This model of learning is Gold's model of identification in the limit from positive data [5]. It is proved that no super finite language(it contains all finite languages and at least one infinite language) can be learn-able in the limit from positive examples. Hence, regular, context free, context sensitive grammars are not learn-able in the limit from positive examples only.

In this paper, we have introduced two subclasses of parallel internal column contextual array grammar, called, strictly parallel internal column contextual array grammar (SPICCAG), k-uniform parallel internal column contextual array grammar (k-UPICCAG) in order to find out identification algorithms. Our learning strategy is based on Gold's model.

## 2    Definition and Examples

If $V$ is a finite alphabet, then $V^*$ is the set of all strings including the empty string $\lambda$. An image or a picture over $V$ is a rectangular $m \times n$ array of elements of $V$ or in short $[a_{ij}]_{m \times n}$, the set of all images including the empty array $\Lambda$ is denoted by $V^{**}$. A picture language or two dimensional language over $V$ is a subset of $V^{**}$. In this paper $\Lambda$ denotes any empty array. The notion of column concatenation is as follows: if $X$ and $Y$ are two arrays where

$$X = \begin{bmatrix} a_{1,j} & \cdots & a_{1,k} \\ a_{2,j} & \cdots & a_{2,k} \\ \cdots & \cdots & \cdots \\ a_{l,j} & \cdots & a_{l,k} \end{bmatrix}, Y = \begin{bmatrix} b_{1,m} & \cdots & b_{1,n} \\ b_{2,m} & \cdots & b_{2,n} \\ \cdots & \cdots & \cdots \\ b_{l,m} & \cdots & b_{l,n} \end{bmatrix} then, X\Phi Y = \begin{bmatrix} a_{1,j} & \cdots & a_{1,k} & b_{1,m} & \cdots & b_{1,n} \\ a_{2,j} & \cdots & a_{2,k} & b_{2,m} & \cdots & b_{2,n} \\ \cdots & \cdots & \cdots & \cdots & \cdots & \cdots \\ a_{l,j} & \cdots & a_{l,k} & b_{l,m} & \cdots & b_{l,n} \end{bmatrix}$$

If $L_1, L_2$ are two picture languages over an alphabet $\Sigma$, the column concatenation $L_1\Phi L_2$ of $L_1, L_2$ is defined by $L_1\Phi L_2 = \{X\Phi Y \mid X \in L_1, Y \in L_2\}$. If $X$ is an array, the set of all subarrays of $X$ is denoted by $sub(X)$. We now recall the notion of column array context [2,16].

**Definition 1.** *Let $V$ be an alphabet. A column array context $c$ over $V$ is of the form*

$$c = [\begin{smallmatrix} u_1 \\ u_2 \end{smallmatrix}] \psi [\begin{smallmatrix} v_1 \\ v_2 \end{smallmatrix}]$$

$\in V^{**}\psi V^{**}$ *where $u_1, u_2$ are arrays of sizes $1 \times p$, and $v_1, v_2$ are arrays of sizes $1 \times q$, for some $p, q \geq 1$ and $\psi$ is a special symbol not in $V$.*

The next definition deals with parallel internal column contextual operation.

**Definition 2.** *Let $V$ be an alphabet, $C$ be a finite subset of $V^{**}\psi V^{**}$ whose elements are the column array contexts and $\varphi : V^{**} \to 2^C$ be mapping, called choice mapping.*

*For an array* $X = \begin{bmatrix} a_{1,j} & \cdots & a_{1,k} \\ a_{2,j} & \cdots & a_{2,k} \\ \cdots & \cdots & \cdots \\ a_{l,j} & \cdots & a_{l,k} \end{bmatrix}$,

$j \leq k, a_{ij} \in V$, *we define* $\hat{\varphi} : V^{**} \to 2^{V^{**}\psi V^{**}}$ *such that $L\psi R \in \hat{\varphi}[X]$, where*

$$L = \begin{bmatrix} u_1 \\ u_2 \\ \vdots \\ u_l \end{bmatrix}, R = \begin{bmatrix} v_1 \\ v_2 \\ \vdots \\ v_l \end{bmatrix},$$

*and*

$$c_i = [\begin{smallmatrix} u_i \\ u_{i+1} \end{smallmatrix}] \psi [\begin{smallmatrix} v_i \\ v_{i+1} \end{smallmatrix}] \in \varphi [\begin{smallmatrix} a_{i,j} & \cdots a_{i,k} \\ a_{i+1,j} & \cdots a_{i+1,k} \end{smallmatrix}],$$

*with $c_i \in C, (1 \leq i \leq l - 1)$, not all need to be distinct.*

*Given an array $X = [a_{ij}]$ of size $m \times n$, $a_{ij} \in V, X = X_1 \Phi X_2 \Phi X_3$ where*

$$X_1 = \begin{bmatrix} a_{1,1} & \cdots & a_{1,p-1} \\ a_{2,1} & \cdots & a_{2,p-1} \\ \vdots & \vdots & \vdots \\ a_{m1} & \cdots & a_{m,p-1} \end{bmatrix}, X_2 = \begin{bmatrix} a_{1,p} & \cdots & a_{1,q} \\ a_{2,p} & \cdots & a_{2,q} \\ \vdots & \vdots & \vdots \\ a_{m,p} & \cdots & a_{m,q} \end{bmatrix}, X_3 = \begin{bmatrix} a_{1,q+1} & \cdots & a_{1,n} \\ a_{2,q+1} & \cdots & a_{2,n} \\ \vdots & \vdots & \vdots \\ a_{m,q+1} & \cdots & a_{m,n} \end{bmatrix}$$

*and $1 \leq p \leq q \leq n$, we write $X \Rightarrow Y$ if $Y = X_1 \Phi L \Phi X_2 \Phi R \Phi X_3$ such that $L\psi R \in \hat{\varphi}[X_2]$. Here $L$ and $R$ are called left and right contexts respectively. We say that $Y$ is obtained from $X$ by parallel internal column contextual operation $(\Rightarrow_{in})$.*

Now we consider the notion of parallel internal column contextual array grammar [2, 16].

**Definition 3.** *A parallel internal column contextual array grammar is an ordered system $G = (V, A, C, \varphi)$ where $V$ is an alphabet, $A$ is a finite subset of $V^{**}$ called the axiom set, $C$ is a finite subset of $V^{**}\psi V^{**}$ called column array contexts, $\varphi : V^{**} \to 2^C$ is the choice mapping which performs the parallel internal column contextual operation. When $\varphi$ is omitted we call $G$ as a parallel internal contextual array grammar without choice.*

*For any $X, Y \in V^{**}, X \Rightarrow Y$ if and only if $X = X_1 \Phi X_2 \Phi X_3, Y = X_1 \Phi L \Phi X_2 \Phi R \Phi X_3$ with $L\psi R \in \hat{\varphi}[X_2]$. We denote by $\Rightarrow^*$ the reflexive transitive closure of $\Rightarrow_{in}$. Then the parallel internal column contextual array language generated by the parallel internal column contextual array grammar $G$ is defined as the set $L_{in}(G) = \{Y \in V^{**}/\exists X \in A \text{ such that } X \Rightarrow^* Y\}$.*

# 3 Subclasses of Parallel Internal Column Contextual Array Grammars

In this paper our main focus is on designing an identification algorithm to infer parallel internal column contextual array grammar. According to Gold model [5], no superfinite class of languages is inferable from positive data only. A class of languages that consists of all finite languages and atleast one infinite language, is called a super finite class of languages.

**Proposition 1.** *The class of parallel internal column contextual array languages (PICCAL), is not inferable from positive data only.*

*Proof.* In the case of string languages, the class of internal contextual languages, is not inferable from positive data only [4]. From this fact, we can conclude Theorem 1.

As we know that the class $(PICCAL)$ is not inferable from positive data only, it is natural to look for subclasses of these languages which can be identified in the limit from positive data only. We now define strictly parallel internal column contextual array grammar $(SPICCAG)$ and k-uniform parallel internal column contextual array grammar $(k - UPICCAG)$.

**Definition 4.** *A strictly parallel internal column contextual array grammar $(SPICCAG)$ is a 6 tuple $G = (V, X, C, \varphi, P, A)$ where*

- *$V$ is the alphabet.*
- *$X$ is a finite subset of $V^{**}$, called selector set and $C$ is a finite subset of $V^{**}\psi V^{**}$, called context set.*
- *$\varphi : V^{**} \to 2^C$ is a choice mapping.*
- *$P$ is a finite set of parallel internal column contextual rules of the form, $\varphi[x_i] = L_i\psi R_i$ where $L_i, R_i \in C$ are the ith left and right context of ith selector $x_i \in X$, $L_i, R_i$ have same number of rows.*
- *$first[L_i] \neq first[R_i]$ where $first[W]$ denotes the first column of $W$ and $L_i$ is not a subarray of $R_i$ and vice versa.*
- *$A$ is a finite subset of $V^{**}$, called the axiom set.*
- *for each selector, there is exactly one rule.*

*The language generated by strictly parallel internal column contextual array grammar $(SPICCAG)$ is called a strictly parallel internal column contextual array language $(SPICCAL)$ which is $L_{sin}(G) = \{Y \in V^{**} \mid Q \Rightarrow^* Y, Q \in A\}$.*

## 3.1 Example

Let $G = (V, X, C, \varphi, P, A)$ be a strictly parallel internal column contextual array grammar $(SPICCAG)$ where $V = \{a, b\}$,

$$X = \left\{ \begin{bmatrix} a & b \\ b & a \end{bmatrix}, \begin{bmatrix} a & b \\ a & b \end{bmatrix}, \begin{bmatrix} b & a \\ b & a \end{bmatrix} \right\}, C = \left\{ \begin{bmatrix} a \\ b \end{bmatrix} \psi \begin{bmatrix} b \\ a \end{bmatrix}, \begin{bmatrix} a \\ a \end{bmatrix} \psi \begin{bmatrix} b \\ b \end{bmatrix}, \begin{bmatrix} b \\ b \end{bmatrix} \psi \begin{bmatrix} a \\ a \end{bmatrix} \right\}$$

$\varphi$ is a choice mapping

$$P = \left\{ \varphi \begin{bmatrix} a & b \\ b & a \end{bmatrix} = \begin{bmatrix} a \\ b \end{bmatrix} \psi \begin{bmatrix} b \\ a \end{bmatrix}, \varphi \begin{bmatrix} a & b \\ a & b \end{bmatrix} = \begin{bmatrix} a \\ a \end{bmatrix} \psi \begin{bmatrix} b \\ b \end{bmatrix}, \varphi \begin{bmatrix} b & a \\ b & a \end{bmatrix} = \begin{bmatrix} b \\ b \end{bmatrix} \psi \begin{bmatrix} a \\ a \end{bmatrix} \right\},$$

$$A = \left\{ Q = \begin{bmatrix} a & a & b & b \\ a & a & b & b \\ b & b & a & a \\ b & b & a & a \end{bmatrix} \right\}$$

Here for each rule $\varphi[x_i] = L_i \psi R_i$, $first(L_i) \neq first(R_i), i \geq 1$, so it does satisfy Definition 4. Clearly, $L_{sin}(G) = \left\{ \begin{bmatrix} (a^n \ b^n)_m \\ (b^n \ a^n)_m \end{bmatrix} \mid n \geq 2, m = 2 \right\}$ Here $a^n = aaa...a$

(n times) and $a_m = \begin{matrix} a \\ \vdots \\ a \end{matrix}$, m rows are there. A simple derivation of a member of

$L_{sin}(G)$ is as follows,

$$Q = \begin{bmatrix} a & a & b & b \\ a & a & b & b \\ b & b & a & a \\ b & b & a & a \end{bmatrix} \Rightarrow \begin{bmatrix} a & a & a & b & b & b \\ a & a & a & b & b & b \\ b & b & b & a & a & a \\ b & b & b & a & a & a \end{bmatrix} = \begin{bmatrix} (a^3 \ b^3)_2 \\ (b^3 \ a^3)_2 \end{bmatrix} \in L_{sin}(G).$$

**Definition 5.** *A k-uniform parallel internal column contextual array grammar is a 6-tuple $(k - UPICCAG)$, $k \geq 1, G = (V, X, C, \varphi, P, A)$ where*

- *V is the alphabet.*
- *X is a finite subset of $V^{**}$, called selector set and C is a subset of $V^{**} \psi V^{**}$, called context set.*
- *$\varphi : V^{**} \to 2^C$ is a choice mapping.*
- *P is a finite set of parallel internal column contextual array rules of the following form, $\varphi[x_i] = L_i \psi R_i$ where $L_i, R_i \in C$ are the ith left and right context of ith selector $x_i \in X$, $L_i, R_i$ have same number of rows.*
- *A is the finite subset of $V^{**}$, called axiom set. Each member of A is an axiom which contains mk number of columns, for some $m \geq 1$ and we put the following restrictions.*

*If the rule is $\varphi[x] = L\psi R$ then,*

- *$|x| = |L| = |R| = k$, where $|W|$ denotes the number of columns in an array W.*
- *for each selector, there is exactly one rule.*

*The language generated by $k - UPICCAG$ is called a k-uniform parallel internal column contextual array language(k-UPICCAL) which is $L_{k-uin}(G) = \{Y \in V^{**} \mid Q \Rightarrow^* Y, Q \in A\}$.*

### 3.2    Example of 2-UPICCAG

$G = (V, X, C, \varphi, P, A)$ is a 2-UPICCAG where $V = \{a, b\}$,

$$X = \left\{ \begin{bmatrix} a & b \\ b & b \end{bmatrix}, \begin{bmatrix} b & b \\ a & b \end{bmatrix} \right\}, C = \left\{ \begin{bmatrix} a & b \\ b & a \end{bmatrix} \psi \begin{bmatrix} a & b \\ b & a \end{bmatrix}, \begin{bmatrix} b & a \\ a & b \end{bmatrix} \psi \begin{bmatrix} b & a \\ a & b \end{bmatrix} \right\},$$

$\varphi$ is a choice mapping,

$$P = \left\{ \varphi \begin{bmatrix} a & b \\ b & b \end{bmatrix} = \begin{bmatrix} a & b \\ b & a \end{bmatrix} \psi \begin{bmatrix} a & b \\ b & a \end{bmatrix}, \varphi \begin{bmatrix} b & b \\ a & b \end{bmatrix} = \begin{bmatrix} b & a \\ a & b \end{bmatrix} \psi \begin{bmatrix} b & a \\ a & b \end{bmatrix} \right\},$$

$A = \left\{ Q = \begin{bmatrix} a & b & a & b \\ b & b & b & b \\ a & b & a & b \end{bmatrix} \right\}$. Here, $|x| = |L| = |R| = 2, m = 2$, number of columns in $A = mk = 4$. So it satisfies Definition 5. Clearly

$$L_{k-uin}(G) = \left\{ A, \begin{bmatrix} (ab)^{n-1} \ ab \ (ab)^{n-1} \ ab \\ (ba)^{n-1} \ bb \ (ba)^{n-1} \ bb \\ (ab)^{n-1} \ ab \ (ab)^{n-1} \ ab \end{bmatrix} \mid n \geq 2 \right\}$$

For instance, $\begin{bmatrix} a & b & a & b \\ b & b & b & b \\ a & b & a & b \end{bmatrix} \Rightarrow \begin{bmatrix} a & b & a & b & a & b & a & b \\ b & a & b & b & b & b & a & b & b \\ a & b & a & b & a & b & a & b \end{bmatrix} \in L_{k-uin}(G)$.

Now, if we consider a = black box and b = white box, we get a nice rectangular picture.

**Theorem 1.** $\mathcal{L}_{SPICCAG}$ *is incomparable with* $\mathcal{L}_{K-UPICCAG}$ *and not disjoint.*

*Proof.* We prove this theorem using following lemmas whose proofs are omitted. □

**Lemma 1.** $\mathcal{L}_{SPICCAG} - \mathcal{L}_{K-UPICCAG} \neq \phi$

**Lemma 2.** $\mathcal{L}_{K-UPICCAG} - \mathcal{L}_{SPICCAG} \neq \phi$

**Lemma 3.** $\mathcal{L}_{K-UPICCAG} \cap \mathcal{L}_{SPICCAG} \neq \phi$

## 4 Identification of Subclasses of Parallel Internal Column Contextual Array Languages

In this section, we propose an algorithm to infer $SPICCAG$ from positive data only. We recall the notion of an insertion rule. The insertion operation is first considered by Haussler in [9] and based on the operation, insertion systems are introduced by L. Kari in [10]. Informally, if a string $\alpha$ is inserted between two parts $w_1$ and $w_2$ of a string $w_1w_2$ to get $w_1\alpha w_2$, we call the operation as insertion.

This algorithm takes finite sequences of positive examples in the different time interval or all together. Our goal is to find out $SPICCAG$ G, such that $IP \subseteq L(G)$ where $IP$ is the input set of arrays. The algorithm works in the following way. After receiving the first set of arrays as an input, based on the size(actually based on number of columns), firstly the algorithm determines the axiom, then it defines 2D insertion rules in order to find out context and selector from input example. After that, insertion rules are converted into 1-sided[1] contextual rules which will be a guess about the unknown grammar. Then we will convert 1-sided contextual rule into 2-sided contextual rule to take care of over generalization. Then updates with new contextual rules if the next input array cannot be generated by the existing contextual rules. All the guessing will be done in a flexible way in the sense that the correction is done at every instance. Finally we will find the parallel internal column contextual rules according to Definition 2.

In this paper we consider single axiom $A$ and finite selector set. Now, we present our algorithm with a description for better understanding.

---

[1] In an 1-sided contextual rule either left context is $\Lambda$ or right context is $\Lambda$.

# 5   Pseudocode of Our Algorithm

---

1: $axiom \leftarrow Find - Smallest(IPS)$
2: $inser \leftarrow Generate - Inser(axiom, IP_i)$
3: $1 - Sided - Contextual - Rule \leftarrow \{\}$
4: $1 - Sided - Correct - Rule \leftarrow \{\}$
5: $2 - Sided - Correct - Rule \leftarrow \{\}$
6: $Parallel - Rule \leftarrow \{\}$
7: $Table \leftarrow \sqcap$
8: $1 - Sided - Contextual - Rule.push[Convert - into - Contextual - Rule(inser)]$
9: $IPS \leftarrow Remove(IPS, IP_i)$
10: **for** $(1 - Sided - Contextual - Rule_i \in \{1 - Sided - Contextual - Rule\})$ **do**
11:     **for** $(IP_i \in \{IPS\})$ **do**
12:         $S \leftarrow Check - Contextual - Rule(1 - Sided - Contextual - Rule_i, IP_i)$
13:         **if** $S = 1$ **then**
14:             $1 - Sided - Correct - Rule.push[1 - Sided - Contextual - Rule_i]$
15:         **if** $S = 0$ **then**
16:             $1 - Sided - Correct - Rule.push[Correction - Contextual - Rule(1 - Sided - Contextual - Rule_i, IP_i)]$
17: **for** $(1 - Sided - Correct - Rule_i \in \{1 - Sided - Correct - Rule\})$ **do**
18:     **for** $(IP_i \in \{IPS\})$ **do**
19:         $Table.insert[Find - Nof - App - of - EachRule - in - EachMember(1 - Sided - Correct - Rule_i, IP_i)]$
20: **if** $TableRow_i = TableRow_j$ **then**
21:     $2 - Sided - Correct - Rule.push[Merge(1 - Sided - Correct - Rule_i, 1 - Sided - Correct - Rule_j)]$
22: **for** $(2 - Sided - Correct - RULE_i \in \{2 - Sided - Correct - Rule\})$ **do**
23:     $Parallel - Rule.push(2 - Sided - Correct - Rule_i)$

---

In the next few subsections we will explain all the steps of our pseudocode in detail.

## 5.1   Finding Axiom - Pseudocode-Step: 1

**axiom** $\leftarrow$ **Find** $-$ **Smallest(IPS):** It finds the smallest array from the IPS (input set). The output of the function will be considered as an axiom.

In order to find out the axiom, the number of columns of each array is evaluated, the array with the smallest number of columns, will be considered as the axiom. Also a new input array will be compared with the existing axiom based on the number of columns, and the smaller one will be considered as an axiom and Let the single axiom be denoted by $A$.

## 5.2  Defining Insertion Rule and Converting It into Contextual Rule - Pseudocode-Step: 2, 8, 9

- insr ← **Generate – Inser(axiom, IP$_i$)**: It generates the insertion rule from axiom and member of input set $(IP_i)$. The output of the function will be stored in $insr$ as an insertion rule.
- 1 – Sided – Contextual – Rule.push ← [**Convert – into – Contextual – Rule(inser)**]: It converts $insr$ into $1 - Sided - Contextual - Rule$ and store that.
- **IPS ← Remove(IPS, IP$_i$)**: It removes the current input member $IP_i$ from $IPS$.
- We now shortly describe about the intuitive idea of the parts 1–4. We try to identify the selectors from the axiom and contexts from examining input.
- Let the format of 2D insertion rule be $LIR$ where $L, I, R \in V^{++}$ are left context, inserted portion, and right context respectively. Axiom and examining array are respectively

$$A = \begin{bmatrix} a_{1,1} & \cdots & a_{1,n} \\ a_{2,1} & \cdots & a_{2,n} \\ \vdots & \vdots & \vdots \\ a_{m,1} & \cdots & a_{m,n} \end{bmatrix}, \ E = \begin{bmatrix} a_{1,1} & \cdots & a_{1,p} \\ a_{2,1} & \cdots & a_{2,p} \\ \vdots & \vdots & \vdots \\ a_{m,1} & \cdots & a_{m,p} \end{bmatrix}$$

Let the initial insertion rule be $LIR$ and from the axiom we can have the following consideration:

**Part 1:**
$$L = \begin{bmatrix} a_{1,1} \\ a_{2,1} \\ \vdots \\ a_{m,1} \end{bmatrix}, \ R = \begin{bmatrix} a_{1,2} & \cdots & a_{1,n} \\ a_{2,2} & \cdots & a_{2,n} \\ \vdots & \vdots & \vdots \\ a_{m,2} & \cdots & a_{m,n} \end{bmatrix}$$

Check whether any $I = [I_{i,j}]_{m \times r}$ where $r \leq p$ exists with $LIR \in sub(E)$ or not. If yes then fix that $I = [I_{i,j}]_{m \times r}$ and go to part 3, else go to part 2.

**Part 2:** Remove the last column of the right context R and the rule becomes $LIR$ where
$$L = \begin{bmatrix} a_{1,1} \\ a_{2,1} \\ \vdots \\ a_{m,1} \end{bmatrix}, \ R = \begin{bmatrix} a_{1,2} & \cdots & a_{1,n-1} \\ a_{2,2} & \cdots & a_{2,n-1} \\ \cdots & \cdots & \cdots \\ a_{m,2} & \cdots & a_{m,n-1} \end{bmatrix}$$

Check whether any $I = [I_{i,j}]_{m \times r}$ where $r \leq p$ exists with $LIR \in sub(E)$ or not. If yes then fix that $I = [I_{i,j}]_{m \times r}$ and go to part 3, else go to part 2 recursively, until $L = \begin{bmatrix} a_{1,1} \\ a_{2,1} \\ \vdots \\ a_{m,1} \end{bmatrix}, \ R = \begin{bmatrix} a_{1,2} \\ a_{2,2} \\ \vdots \\ a_{m,2} \end{bmatrix}$, and then go to part 4.

**Part 3 - Conversion of 2D insertion rule into 1 sided 2D contextual rule:** Here $L^{IN}, I^{IN}, R^{IN}$ are left context, inserted portion, and right context for insertion rule respectively. On the other hand, $L^{IC}, x^{IC}, R^{IC}$ are left context, selector, and right context for internal contextual rule respectively.

$(LIR)^{IN} \rightarrow (\hat{\varphi}[x] = L\psi R)^{IC}$ where $x^{IC} = L^{IN}, L^{IC} = \Lambda, R^{IC} = I^{IN}$. Once we get a selector and associated context with it, we have the following conditions for each 2D insertion rule:

- **Condition 1:** If $(|L|+|I|+|R|)^{IN} = |E|$, it implies that on this current axiom $A$, only one rule has been applied and we obtain the rule.
- **Condition 2:** If $(|L| + |R|)^{IN} \leq |A|$, then we remove $L^{IN}$ from axiom $A$, and obtain a new temporary axiom, also consider $R^{IN}$ as a $L^{IN}$ for the next insertion rule. Also we remove $(LI)^{IN}$ as a subarray from the examining input $E$ and obtain a new temporary input. Now we continue our procedure with this temporary axiom and temporary examing input in the same way.
- **Condition 3:** If $(|L| + |I| + |R|)^{IN} \leq |E|$ but $(|L| + |R|)^{IN} = |A|$, then it can be understood that some part of the examining input is still left to scan, and that is considered directly as the left context $L^{IC}$ of the first selector $x^{IC}_{first}$ or right context $R^{IC}$ of the last selector $x^{IC}_{last}$. We define new rule internal contextual rule.
- $(\hat{\varphi}[x] = L\psi R)_{new}$ where $L_{new} = L^{IC}, R_{new} = \Lambda, x_{new} = x^{IC}_{first}$, another rule can be $(\hat{\varphi}[x] = L\psi R)_{new}$ where $L_{new} = \Lambda, R_{new} = R^{IC}, x_{new} = x^{IC}_{last}$. It should be noted that these particular rules will not be considered for updation and correction.

**Part 4:** At that moment, existing first column of $R$ will be concatenated with existing $L$.

$$L = \begin{bmatrix} a_{1,1} & a_{1,2} \\ a_{2,1} & a_{2,2} \\ \vdots & \vdots \\ a_{m,1} & a_{m,2} \end{bmatrix}, R = \begin{bmatrix} a_{1,3} & \cdots & a_{1,n} \\ a_{2,3} & \cdots & a_{2,n} \\ \vdots & \vdots & \vdots \\ a_{m,3} & \cdots & a_{m,n} \end{bmatrix}, \text{ go to part 1 until } L = \begin{bmatrix} a_{1,1} & \cdots & a_{1,n} \\ a_{2,1} & \cdots & a_{2,n} \\ \vdots & \vdots & \vdots \\ a_{m,1} & \cdots & a_{m,n} \end{bmatrix},$$

in that case defining insertion rule is not possible. We may need to define insertion rule with the current examining array, if we are still unable to define insertion rule, then we will conclude that the choosen axiom is wrong. It is a negative example as we are dealing with single axiom.

So in this section, we get the selectors from axiom and contexts from examining input. Later on for new input, we may need to guess (next section).

### 5.3 Making Correction and Updating Rules - Pseudocode-Step: 10–16

- **S ← Check − Contextual − Rule(1 − Sided − Contextual − Rule$_i$, IP$_i$):** It checks the correctness of $1 - Sided - Contextual - Rule_i$ for $IP_i$. If $S$ is true then the correct $1 - Sided - Contextual - Rule_i$ will be pushed onto set $\{1 - Sided - Correct - Rule\}$ or it goes for correction.
- **Correction − Contextual − Rule(1 − Sided − Contextual − Rule$_i$, IP$_i$):** In that case we need to go for correction of the rule in such a way so that our new corrected rule can take care of new inputs and as well as previous inputs.
- Let the initial rule be $\hat{\varphi}[x_i] = L_i \psi R_i$ where $L_i, R_i$ are ith left and right context of the ith selector $x_i$. Here $x_{i+1}$ is also introduced because we will make the correction using $x_{i+1}$.

**Proposition 2.** *In case of correction, we deal with only 1-sided contextual rules where left context is always empty and selector is not the last one. (see condition*

*3 of Subsect. 5.2) We will try to find the rule as a subarray from the examining input.*

Let the examining input be $E = \begin{bmatrix} a_{i,1} & \cdots & a_{i,p} \\ a_{i+1,1} & \cdots & a_{i+1,p} \\ \vdots & \vdots & \vdots \\ a_{m,1} & \cdots & a_{m,p} \end{bmatrix}$. We can represent the examining input in the following format $E = P\Phi x_i \Phi Q \Phi \Phi x_{i+1} \Phi Z$. where $P, Z$ are the rest of the part of string and they can be empty also, $Q$ is the inserted subarray portion. Now we present the examining input in 2D form.

$$E = P\Phi \begin{bmatrix} a_{i,k} & \cdots & a_{i,\alpha} \\ a_{i+1,k} & \cdots & a_{i+1,\alpha} \\ \vdots & \vdots & \vdots \\ a_{m,k} & \cdots & a_{m,\alpha} \end{bmatrix} \Phi Q \Phi \begin{bmatrix} a_{i,j} & \cdots & a_{i,\beta} \\ a_{i+1,j} & \cdots & a_{i+1,\beta} \\ \vdots & \vdots & \vdots \\ a_{m,j} & \cdots & a_{m,\beta} \end{bmatrix} \Phi Z$$

Now we need to check the contexts. $R$ must be matched with $Q$. $R = R_i \Phi R_{i+1} \Phi...\Phi R_w$ where $1 \leq i \leq w$, and $R_i$ presents the ith column of array. $Q = Q_i \Phi Q_{i+1} \Phi...\Phi Q_z$ where $1 \leq i \leq z$, and $Q_i$ presents the ith column of array.

Here we are making an analysis to find out the partially equal part (as a prefix/suffix) between $R_1 \Phi R_2 \Phi...\Phi R_w$ and $Q_1 \Phi Q_2 \Phi...\Phi Q_z$ and we have shown the correction part for one rule, in the same way can make the correction for other rules. In Theorems 3 and 4, we obtain the common-prefix and common-suffix part between $R$ and $Q$.

**Theorem 2.** *If the analysis starts with equality such that $Q_1 = R_1, Q_2 = R_2 \Phi..\Phi Q_f = R_s$, and $Q_{f+1} \neq R_{s+1}$ or $f = z$ or $s = w$, then we can have four different types of errors which are stated in terms of following lemmas.*

**Lemma 4.** *If $(f = z$ and $s = w)$ then it implies that matching is correct, so no need to make any correction for this rule and the rule is correct.*

**Lemma 5.** *If $(f = z$ and $s < w)$ then we infer the following two new rules.*

– $Rule_{i'} : \hat{\varphi}[x_{i'}] = L_{i'} \psi R_{i'}$ where $R_{i'} = Q_1 \Phi Q_2 \Phi...\Phi Q_f, L_{i'} = \Lambda, x_{i'} = x_i$.
– $Rule_{(i+1)'} : \hat{\varphi}[x_{(i+1)'}] = L_{(i+1)'} \psi [R_{(i+1)'}$ where $L_{(i+1)'} = R_{s+1} \Phi R_{s+2} \Phi... \Phi R_w, R_{(i+1)'} = \Lambda, x_{(i+1)'} = x_{(i+1)}$.

**Lemma 6.** *If $(f < z$ and $s = w)$ then we infer the following two new rules.*

– $Rule_{i'} : \hat{\varphi}[x_{i'}] = L_{i'} \psi R_{i'}$ where $R_{i'} = R_1 \Phi..\Phi R_w, L_{i'} = \Lambda, x_{i'} = x_i$.
– $Rule_{(i+1)'} : \hat{\varphi}[x_{(i+1)'}] = L_{(i+1)'} \psi R_{(i+1)'}$ where $L_{(i+1)'} = Q_{f+1} \Phi Q_{f+2} \Phi...\Phi Q_z$, $R_{(i+1)'} = \Lambda, x_{(i+1)'} = x_{(i+1)}$.

**Lemma 7.** *If $(f < z$ and $s < w)$ then we infer the following three new rules.*

– $Rule_{i'} : \hat{\varphi}x_{i'} = L_{i'} \psi R_{i'}$ where $R_{i'} = Q_1 \Phi Q_2 \Phi...\Phi Q_f, L_{i'} = \Lambda, x_{i'} = x_i$.
– $Rule_{(i+1)'} : \hat{\varphi}[x]_{(i+1)'} = L_{(i+1)'} \psi R_{(i+1)'}$ where $L_{(i+1)'} = R_{s+1} \Phi..\Phi R_w, R_{(i+1)'} = \Lambda, x_{(i+1)'} = x_{i+1}$.
– $Rule_{(i+2)'} : \hat{\varphi}[x]_{i+2} = L_{(i+2)'} \psi R_{(i+2)'}$ where $L_{(i+2)'} = Q_{f+1} \Phi...\Phi Q_z$, $R_{(i+2)'} = \Lambda, x_{(i+2)'} = x_{i+1}$.

**Theorem 3.** *If the analysis starts with inequality such that $Q_1 \neq R_1$, but $Q_z = R_w, Q_{z-1} = R_{w-1}\Phi...\Phi Q_f = R_s$, and $Q_{f-1} \neq R_{s-1}$ then we can have three different types of errors which can be seen in the following lemmas.*

**Lemma 8.** *If $(s = 1, f > 1)$ then we infer the following two new rules.*

- $Rule_{i'} : \hat{\varphi}[x_{i'}] = L_{i'}\psi R_{i'}$ where $L_{i'} = R_1 \Phi R_2 \Phi...\Phi R_w, R_{i'} = \Lambda, x_{i'} = x_{i+1}$.
- $\hat{\varphi}[x_{(i+1)'}] = L_{(i+1)'}\psi R_{(i+1)'}$ where $R_{(i+1)'} = Q_1 \Phi Q_2 \Phi...\Phi Q_{f-1}, L_{(i+1)'} = \Lambda, x_{(i+1)'} = x_i$.

**Lemma 9.** *If $(s > 1)$ then we infer the following three new rules. $Rule_{i'}$ : $\hat{\varphi}[x_{i'}] = L_{i'}\psi R_{i'}$ where $L_{i'} = R_s \Phi R_{s+1}\Phi...\Phi R_w, R_{i'} = \Lambda, x_{i'} = x_{i+1}$. $Rule_{(i+1)'}$ : $\hat{\varphi}[x_{(i+1)'}] = L_{(i+1)'}\psi R_{(i+1)'}$ where $R_{(i+1)'} = Q_1 \Phi Q_2 \Phi...\Phi Q_{f-1}, L_{(i+1)'} = \Lambda, x_{(i+1)'} = x_i$. $Rule_{(i+2)'}$ : $\hat{\varphi}[x_{(i+2)}]' = L_{(i+2)'}\psi R_{(i+2)'}$ where $L_{(i+2)'} = \Lambda, R_{(i+2)'} = R_1 \Phi R_2 \Phi...\Phi R_{s-1}, x_{(i+2)'} = x_i$.*

**Lemma 10.** *If $Q_z \neq R_w$ then we infer the following two new rules.*

- $Rule_{i'} : \hat{\varphi}[x_{i'}] = L_{i'}\psi R_{i'}$ where $R_{i'} = R_1 \Phi R_2 \Phi...\Phi R_w, L_{i'} = \Lambda, x_{i'} = x_i$.
- $Rule_{(i+1)'}$ : $\hat{\varphi}[x_{(i+1)'}] = L_{(i+1)'}\psi R_{(i+1)'}$ where $R_{(i+1)'} = Q_1 \Phi Q_2 \Phi...\Phi Q_z, L_{(i+1)'} = \Lambda, x_{(i+1)'} = x_i$.

In this section, we must notice that we have different rules with same selectors. According to Definitions 4 and 5, for each selector there must be one rule. As we are inferring 1-sided contextual rule, it does not satisfy our Definitions 4 and 5. In the next section we will convert 1-sided contextual rule into 2-sided contextual rule in order to take care of over generalization and Definitions 4 and 5.

### 5.4  Controlling over Generalization - Pseudocode-Step: 17–21

- **Table.insert[Find – Nof – App – of – EachRule – in – EachMember (1 – Sided – Correct – Rule$_i$, IP$_i$)]:** It finds out the application of each rule on each member of the input and insert that record into the table.
- **2–Sided – Correct – Rule.push[Merge(1 – Sided – Correct – Rule$_i$, 1 – Sided – Correct – Rule$_j$)]:** In this case if we find that ith row (*Table Row$_i$*) and jth row (*TableRow$_j$*) is same then we merge these two rules (1 − *Sided − Correct − Rule$_i$*, 1 − *Sided − Correct − Rule$_j$*) and store as a 2 − *Sided − Correct − Rule*.
- In this section we determine the number of applications of each rule to generate the given input set. It will be presented in table. We put priority in applying rules where left context is empty and context is smaller in size. If it is found that without using any rule we can generate the full input set then we can ignore that rule.
- Actually all the rules are 1-sided where left contexts or right contexts are empty that generate more elements. Thus, to control this over generalization, we check that how many times each rule is applied in each member of the input set. Rules which are applied equal number of times in each member, those can be merged into one rule based on condition (discussed in Lemmas 11 and 12).

– Also in this way we satisfy our required condition for $SPICCAG$ (Definition 4), that is, for each selector atmost one rule is applicable.

**Lemma 11.** *If consecutive selectors are* $x_i, x_j$ *with* $(j - i) = 1$ *and left contexts(right contexts) are empty in a set of rule then we can get 1-sided or 2-sided internal contextual rule after merging them.*

*Proof.* Let $x_i, x_j$ denote ith and jth selector, $R_i, R_j$ be ith and jth right context and $L_i, L_j$ are ith and jth left context.

– **case 1:** If $x_i, x_j$ are such that $(j - i) = 1$ and if $R_i = R_j = \Lambda$ then rule becomes $\hat{\varphi}[x_i] = L_i\psi R_i$ where $R_i = L_j$.
– **case 2:** If $x_i, x_j$ are such that $(j - i) = 1$ and if $L_i = L_j = \Lambda$ then the rule becomes $\hat{\varphi}[x_i] = L_i\psi R_i$ where $L_j = R_i, x_i = x_j$.

**Lemma 12.** *If consecutive selectors are* $x_i, x_j$ *with* $(j - i) = 1$ *and left contexts of ith rule and right context of jth rule are empty then we can get 1-sided internal contextual rule after merging them.*

*Proof.* Let $x_i, x_j$ denote ith and jth selector, $R_i, R_j$ are left contexts of ith rule and right context of jth rule respectively.

If $x_i, x_j$ are such that $(j - i) = 1$ and if $L_i = R_j = \Lambda$ then the rule becomes $\hat{\varphi}[x_i] = L_i\psi R_i$ where $R_i = R_i\Phi R_j$

## 5.5  Parallalization Contextual Array Rules - Pseudocode-Step: 22, 23

– **Parallel – Rule.push(2 – Sided – Correct – Rule$_i$):** It converts the $2 - Sided - Correct - Rule_i$ into parallel rule and push onto set $\{Parallel - Rule\}$. If we get a rule $\hat{\varphi}[x] = L\psi R$ where

$$x = \begin{bmatrix} a_{i,k} & \cdots & a_{i,\alpha} \\ a_{i+1,k} & \cdots & a_{i+1,\alpha} \\ \vdots & \vdots & \vdots \\ a_{m,k} & \cdots & a_{m,\alpha} \end{bmatrix}, L = \begin{bmatrix} a_{i,j} & \cdots & a_{i,k-1} \\ a_{i+1,j} & \cdots & a_{i+1,k-1} \\ \vdots & \vdots & \vdots \\ a_{m,j} & \cdots & a_{m,k-1} \end{bmatrix}, R = \begin{bmatrix} a_{i,\alpha+1} & \cdots & a_{i,n} \\ a_{i+1,\alpha+1} & \cdots & a_{i+1,n} \\ \vdots & \vdots & \vdots \\ a_{m,\alpha+1} & \cdots & a_{m,n} \end{bmatrix}$$

According to Definition 2, we can have $(m-1)$ parallel rules $\varphi[Px_i] = PL_i\psi PR_i$ where $Px_i, PL_i, PR_i$ are respectively selector, left context, right context.

$$Px_i = \begin{bmatrix} a_{i,k} & \cdots & a_{i,\alpha} \\ a_{i+1,k} & \cdots & a_{i+1,\alpha} \end{bmatrix}, PL_i = \begin{bmatrix} a_{i,j} & \cdots & a_{i,k-1} \\ a_{i+1,j} & \cdots & a_{i+1,k-1} \end{bmatrix}, PR_i = \begin{bmatrix} a_{i,\alpha+1} & \cdots & a_{i,n} \\ a_{i+1,\alpha+1} & \cdots & a_{i+1,n} \end{bmatrix}$$

where $1 \le i \le m - 1$.

*Remark 1.* The above algorithm can also be used to identify a $k - UPICCAG$. A modification required in the algorithm is that, $k$ is also given along with the positive presentation as an input to the algorithm.

In this case, at the time of defining insertion rule (Sect. 5.2), we need to focus on the size of selectors and contexts in terms of number of columns as $k$ is given as an input. Defining insertion rule should be done in the following way, $LIR \in sub(E)$ where $|I| = |L| = |R| = k$ and also $|A| = mk, L, I, R \in V^{++}$.

# 6 Correctness of the Algorithm and Characteristic Sample

The correctness of the algorithm can be noticed in view of the fact that the specific properties of the subclasses considered allow the positive examples. The correctness of the algorithm can be seen by considering a characteristic sample for a target language. Also it can be seen that the algorithm runs in polynomial time in the sum of the size of the examples given. (discussed in Sect. 7). The correctness of the algorithm, can be seen by considering a characteristic sample for a target $SPICCAL$. Let $L$ be an $SPICCAL$. A finite set $IPS$ is called a characteristic sample of $L$ if and only if $L$ is the smallest $SPICCAL$ containing $IPS$.

# 7 Running Time Complexity of Our Algorithm

In this section we show the running time of our algorithm to infer the column contextual rules.

**Theorem 4.** *The running time complexity of the given pseudocode in Sect. 5, is polynomial in the size of the input set, that is, $SumofSize(IPS)$ where $IPS = \{IP_i, IP_{i+1}, ..., IP_k\}$.*

*Proof.* proof is omitted.

# 8 Conclusion and Future Work

In this paper we present a polynomial time algorithm to infer subclasses of parallel internal column contextual array languages from positive examples only. Here we deal with only column contextual rules. In the form of future direction of this work, we can deal with column and row contextual rules together, that is, parallel internal array contextual languages.

# References

1. Chandra, H., Martin-Vide, C., Subramanian, K.G., Van, D.L., Wang, P.S.P.: Parallel contextual array grammars and trajectories. In: Chen, C.H., Wang, P.S.P. (eds.) Handbook of Pattern Recognition and Computer Vision, 3rd edn., pp. 55-70 (2004)
2. Chandra, H., Subramanian, K.G., Thomas, D.G.: Parallel contextual array grammars and languages. Electron. Notes Discrete Math. **12**, 106–117 (2003)
3. Ehrenfeucht, A., Paun, G., Rozenberg, G.: Contextual grammars and formal languages. In: Rozenberg, G., Salomaa, A. (eds.) Handbook of Formal Language, vol. 2, pp. 237–293 (1997)
4. Emerald, J.D., Subramanian, K.G., Thomas, D.G.: Inferring subclasses of contextual languages. In: Oliveira, A.L. (ed.) ICGI 2000. LNCS, vol. 1891, pp. 65–74. Springer, Heidelberg (2000). doi:10.1007/978-3-540-45257-7_6

5. Gold, E.M.: Language identification in the limit. Inf. Control **10**, 447–474 (1967)
6. Fernau, H., Freund, R., Holzer, M.: Representations of recursively enumerable array languages by contextual array grammars. Fundamenta Informatica **64**, 159–170 (2005)
7. Fernau, H., Freund, R., Siromoney, R., Subramanian, K.G.: Contextual array grammars with matrix and regular control. In: Câmpeanu, C., Manea, F., Shallit, J. (eds.) DCFS 2016. LNCS, vol. 9777, pp. 98–110. Springer, Cham (2016). doi:10.1007/978-3-319-41114-9_8
8. Fernau, H., Freund, R., Siromoney, R., Subramanian, K.G.: Non-isometric contextual array grammars with regular control and local selectors. In: Durand-Lose, J., Nagy, B. (eds.) MCU 2015. LNCS, vol. 9288, pp. 61–78. Springer, Cham (2015). doi:10.1007/978-3-319-23111-2_5
9. Haussler, D.: Insertion and iterated insertion as operations on formal languages. Ph.D. Thesis, University of Colorado, Boulder (1982)
10. Kari, L.: Contextual insertions/deletions and computability. Inf. Comput. **1**, 47–61 (1996)
11. Krithivasan, K., Balan, M.S., Rama, R.: Array contextual grammars. In: Martin-Vide, C., Paun, G. (eds.) Recent Topics in Mathematical and Computational Linguistics, pp. 154-168 (2000)
12. Lalitha, D., Rangarajan, K., Thomas, D.G.: Petri net generating hexagonal arrays. In: Aggarwal, J.K., Barneva, R.P., Brimkov, V.E., Koroutchev, K.N., Korutcheva, E.R. (eds.) IWCIA 2011. LNCS, vol. 6636, pp. 235–247. Springer, Heidelberg (2011). doi:10.1007/978-3-642-21073-0_22
13. Marcus, S.: Contextual grammars. Revue Roumane de Mathematiques Pures et Appliques **14**(10), 1525–1534 (1969)
14. Rama, R., Smitha, T.A.: Some results on array contextual grammars. Int. J. Pattern Recogn. Artif. Intell. **14**, 537–550 (2000)
15. Rosenfield, A., Siromoney, R.: Picture languages - a survey. Lang. Design **1**, 229–245 (1993)
16. Subramanian, K.G., Van, D.L., Chandra, P.H., Quyen, N.D.: Array grammars with contextual operations. Fundamenta Informaticae **83**, 1–18 (2008)

# Parallel Contextual Array Insertion Deletion P System

S. James Immanuel[1]([⊠]), D.G. Thomas[1], Robinson Thamburaj[1], and Atulya K. Nagar[2]

[1] Department of Mathematics, Madras Christian College,
Tambaram, Chennai 600059, India
james_imch@yahoo.co.in, dgthomasmcc@yahoo.com, robin.mcc@gmail.com
[2] Department of Mathematics and Computer Science,
Liverpool Hope University, Liverpool, UK
nagara@hope.ac.uk

**Abstract.** We introduce a new P system model called as parallel contextual array insertion deletion P system, based on the modified row and column contextual rules of parallel contextual array grammar. We can generate a family of two-dimensional picture languages using this P system. We discuss some properties of this P system and find its generating power by comparing this new family of languages with that of certain other well known families of two-dimensional picture languages.

**Keywords:** P system · Rectangular array · Parallel contextual array

## 1 Introduction

One of the extensions of string language theory is two-dimensional languages. There has been a continued interest in adapting the techniques of formal string language theory for developing methods to study the problem of picture generation and description, where pictures are considered as connected, digitized finite arrays in the two-dimensional plane. The literature on array grammars and array acceptors has steadily grown over the past several years.

Rosenfeld [16,17] has investigated isometric array genration, pointing out the need for array rewriting rules for picture languages. In an array grammar, the idea is to have rewriting rules that allow replacement of a subarray of a picture with another subarray, thus generalizing the Chomskian string grammars to arrays. Siromoney et al. [18] proposed a simple generative model, called two-dimensional matrix grammar, to describe digital pictures viewed as rectangular arrays of terminals. Motivated by the need to generate picture languages that cannot be generated by two-dimensional matrix grammars, Siromoney et al. [19] introduced array models, generalizing the notion of rewriting rules in which the catenation of strings is extended to row and column catenation of arrays.

While the study of formal language theory has its origins in Chomskian grammars, another class of grammars, called contextual grammars was introduced by

© Springer International Publishing AG 2017
V.E. Brimkov and R.P. Barneva (Eds.): IWCIA 2017, LNCS 10256, pp. 170–183, 2017.
DOI: 10.1007/978-3-319-59108-7_14

S. Marcus in 1969 [12]. Contextual grammars have been intensively investigated by formal language theorists, as they offer novel insight into a number of issues central to formal language theory [3,14]. A contextual grammar produces a language by starting from a given finite set of strings and adding, iteratively, pairs of strings (called as contexts), associated to sets of words (called selectors) to the string already obtained. Extension of these grammars to 2-dimensional array structures has been attempted in [6,8,10]. In [10] a model of array contextual grammars is introduced. But this model is different from parallel contextual array grammars in [6,8]. In [10], instead of a finite set of contextual rules, a language of arrays which may be an infinite set is used for choosing the contexts. In the parallel contextual array grammars, row as well as column contexts are allowed and the contextual rules are finite. The array contextual style introduced recently in [8] is a modified contextual style of [6].

A P system or membrane system, which was introduced by Paun [13,15], evolves in parallel; at each step all objects, which can evolve should evolve. A computation starts from an initial configuration of a system, defined by a membrane structure with objects and evolutions rule in each membrane, and terminates when no further rule can be applied. In P systems with string objects one uses the Chomskian way of rewriting for computations. In [11] the contextual way of handling string objects in P systems has been considered and that the contextual P systems are found to be more powerful than ordinary string contextual grammars and its variants. Extending the string rewriting P systems to arrays, Ceterchi et al. introduced array P systems of the isometric variety using context-free type of rules [1]. Henceforth, several P system models for generating arrays, both isometric and non-isometric variety, have been considered in the literature (for example [2,7,20]).

Picture languages and array grammars are important parts of image processing. Based on the analogy to Chomskian string grammars and languages there was a belief that these two-dimensional languages will be useful in pattern recognition etc. For this we would need efficient generating tools and also efficient parsing tools for the generated language families. This is the motivation behind this paper. In this paper we introduce new P system models, called as parallel contextual array insertion deletion P system, based on the modified contextual style [8] of external and internal parallel contextual array grammars considered in [6]. In Sect. 2, we give some prerequisites. In Sect. 3, we define parallel contextual array insertion deletion P system and give an example. We also consider another P system model involving only the insertion operation. In Sect. 4, some properties for the families of languages generated by both parallel contextual array insertion deletion P system and parallel contextual insertion P system are discussed. In Sect. 5, we compare the family of languages generated by the new P system model with that of certain other well known families of two-dimensional picture languages like $LOC$, REC and Families of Siromoney matrix languages and thus bring out their generative powers. In Sect. 6, we conclude the article with a brief remark.

## 2  Preliminaries

In this section we recall some notions related to formal language theory, array grammars and parallel contextual array grammars.

Let $V$ be a finite alphabet, $V^*$ is the set of words over $V$ including the empty word $\lambda$. $V^+ = V^* - \{\lambda\}$. For $w \in V^*$ and $a \in V$, $|w|_a$ denotes the number of occurrences of $a$ in $w$. An array consists of finitely many symbols from $V$ that are arranged as rows and columns in some particular order and is written in the form, $A = \begin{bmatrix} a_{11} & \cdots & a_{1n} \\ \vdots & \ddots & \vdots \\ a_{m1} & \cdots & a_{mn} \end{bmatrix}$ or in short $A = [a_{ij}]_{m \times n}$, for all $a_{ij} \in V$, $i = 1, 2, \ldots, m$ and $j = 1, 2, \ldots, n$. The set of all arrays over $V$ is denoted by $V^{**}$ which also includes the empty array $\Lambda$ (zero rows and zero columns). $V^{++} = V^{**} - \{\Lambda\}$. For $a \in V$, $|A|_a$ denotes the number of occurrences of $a$ in $A$. The column concatenation of $A = \begin{bmatrix} a_{11} & \cdots & a_{1p} \\ \vdots & \ddots & \vdots \\ a_{m1} & \cdots & a_{mp} \end{bmatrix}$, and $B = \begin{bmatrix} b_{11} & \cdots & b_{1q} \\ \vdots & \ddots & \vdots \\ b_{n1} & \cdots & b_{nq} \end{bmatrix}$, defined only when $m = n$, is given by $A \oplus B = \begin{bmatrix} a_{11} & \cdots & a_{1p} & b_{11} & \cdots & b_{1q} \\ \vdots & \ddots & \vdots & \vdots & \ddots & \vdots \\ a_{m1} & \cdots & a_{mp} & b_{n1} & \cdots & b_{nq} \end{bmatrix}$. As $1 \times n$-dimensional arrays can be easily interpreted as words of length $n$ (and vice versa), we will then write their column catenation by juxtaposition (as usual). Similarly, the row concatenation, defined only when $p = q$, is given by $A \ominus B = \begin{bmatrix} a_{11} & \cdots & a_{1p} \\ \vdots & \ddots & \vdots \\ a_{m1} & \cdots & a_{mp} \\ b_{11} & \cdots & b_{1q} \\ \vdots & \ddots & \vdots \\ b_{n1} & \cdots & b_{nq} \end{bmatrix}$. The empty array acts as the identity for column and row catenation of arrays of arbitrary dimensions.

**Definition 1.** *A* Phase-structure matrix grammar (Context-sensitive matrix grammar (CSMG) Context-free matrix grammar (CFMG), Right-linear matrix grammar(RLMG)) *is defined by a 7-tuple* $G = (V_h, V_v, \Sigma_I, \Sigma, S, HR, VR)$, *where:* $V_h$ *is a finite set of* horizontal nonterminals; $V_v$ *is a finite set of* vertical nonterminals; $\Sigma_I \subseteq V_v$ *is a finite set of* intermediates; $\Sigma$ *is a finite set of* terminals; $S \in V_h$ *is a* starting symbol; $HR$ *is a finite set of* horizontal phase-structure (context-sensitive, context-free, right-linear) rules; $VR$ *is a finite set of* vertical right-linear rules. *For more information, we can refer to [18].*

**Definition 2.** *Let* $V$ *be a finite alphabet. A* two-dimensional language $L \subseteq V^*$ *is local if there exists a finite set of* $\Theta$ *of tiles over the alphabet* $V \cup \{\#\}$ *such that* $L = \{p \in V^{**} | B_{2,2}(\hat{p}) \subseteq \Theta\}$.

*Given a language* $L$, *we can consider the set* $\Theta$ *as the set of all possible blocks of size (2, 2) of pictures that belong to* $L$ *(when considered with the frame of* $\#$ *symbols). The language* $L$ *is local if, given such a set* $\Theta$, *we can exactly retrieve the language* $L$. *We call the set* $\Theta$ *a representation by tiles for the local language* $L$ *and write* $L = L(\Theta)$.

*The family of local picture languages will be denoted by* LOC.

**Definition 3.** *A* two-dimensional language *is "Tiling Recognizable" (REC) if it can be obtained as a projection of a local picture language.*

We can refer to [4,5] for further details about $LOC$ and $REC$ languages.

**Definition 4.** *Let $V$ be a finite alphabet. A column array context over $V$ is of the form, $c = \$_c \left[\begin{smallmatrix} u_1 \\ u_2 \end{smallmatrix}\right] \$_c \in \$_c V^{**}\$_c$, $u_1, u_2$ are of size $1 \times p$, $p \geq 1$ and $\$_c$ is a special symbol not in $V$.*

*A row array contexts over $V$ is of the form, $r = \$_r \left[ u_1\ u_2 \right] \$_r \in \$_r V^{**}\$_r$, $u_1, u_2$ are of size $p \times 1$, $p \geq 1$ and $\$_r$ is a special symbol not in $V$.*

**Definition 5.** *The parallel column contextual insertion operation is defined as follows: Let $V$ be an alphabet, $C$ be a finite subset of $\$_c V^{**}\$_c$ whose elements are the column array contexts and $\varphi_c^I : (V^{**}, V^{**}) \to 2^C$ be a choice mapping. For arrays, $A = \begin{smallmatrix} a_{1j} & \cdots & a_{1(k-1)} \\ \vdots & \ddots & \vdots \\ a_{mj} & \cdots & a_{m(k-1)} \end{smallmatrix}, B = \begin{smallmatrix} a_{1k} & \cdots & a_{1(l-1)} \\ \vdots & \ddots & \vdots \\ a_{mk} & \cdots & a_{m(l-1)} \end{smallmatrix}, j < k < l, a_{ij} \in V$, we define $\hat\varphi_c^I : (V^{**}, V^{**}) \to \$_c V^{**}\$_c$ such that, $\$_c I_c \$_c \in \hat\varphi_c^I(A, B)$, $I_c = \left[\begin{smallmatrix} u_1 \\ u_2 \\ \vdots \\ u_m \end{smallmatrix}\right]$ if $c_i = \$_c \left[\begin{smallmatrix} u_i \\ u_{i+1} \end{smallmatrix}\right] \$_c \in \varphi_c^I \left(\begin{smallmatrix} a_{ij} & \cdots & a_{i(k-1)} \\ a_{(i+1)j} & \cdots & a_{(i+1)(k-1)} \end{smallmatrix}, \begin{smallmatrix} a_{ik} & \cdots & a_{i(l-1)} \\ a_{(i+1)k} & \cdots & a_{(i+1)(l-1)} \end{smallmatrix}\right)$, $c_i \in C, 1 \leq i \leq m - 1$, not all need to be distinct.*

*Given an array $X = [a_{ij}]_{m \times n}, a_{ij} \in V$ such that $X = X_1 \textcircled{D} A \textcircled{D} B \textcircled{D} X_2$,*

$$X_1 = \begin{smallmatrix} a_{11} & \cdots & a_{1(j-1)} \\ \vdots & \ddots & \vdots \\ a_{m1} & \cdots & a_{m(j-1)} \end{smallmatrix}, A = \begin{smallmatrix} a_{1j} & \cdots & a_{1(k-1)} \\ \vdots & \ddots & \vdots \\ a_{mj} & \cdots & a_{m(k-1)} \end{smallmatrix}, B = \begin{smallmatrix} a_{1(k+p)} & \cdots & a_{1(l-1)} \\ \vdots & \ddots & \vdots \\ a_{m(k+p)} & \cdots & a_{m(l-1)} \end{smallmatrix}, X_2 = \begin{smallmatrix} a_{1l} & \cdots & a_{1n} \\ \vdots & \ddots & \vdots \\ a_{ml} & \cdots & a_{mn} \end{smallmatrix},$$

$1 \leq j \leq k < l \leq n+1$ *(or)* $1 \leq j < k \leq l \leq n+1$, *we write $X \Rightarrow_i Y$ if $Y = X_1 \textcircled{D} A \textcircled{D} I_c \textcircled{D} B \textcircled{D} X_2$, such that $\$_c I_c \$_c \in \hat\varphi_c^I(A, B)$. $I_c$ is called as the inserted column context. We say that $Y$ is obtained from $X$ by parallel column contextual insertion operation. The following 4 special cases for $X = X_1 \textcircled{D} A \textcircled{D} B \textcircled{D} X_2$ is also considered,*

1. *For $j = 1$ we have $X_1 = \Lambda$.*
2. *For $j = k$, we have $A = \Lambda$. If $j = k = 1$, then $X_1 = \Lambda$ and $A = \Lambda$.*
3. *For $k = l$, we have $B = \Lambda$.*
4. *For $l = n + 1$, we have $X_2 = \Lambda$. If $k = l = n + 1$, then $B = \Lambda$ and $X_2 = \Lambda$.*

*The case $j = k = l$ is not considered for parallel column contextual insertion operation.*

Similarly we can define parallel row contextual insertion operation also.

**Definition 6.** *The parallel column contextual deletion operation is defined as follows: Let $V$ be an alphabet, $C$ be a finite subset of $\$_c V^{**}\$_c$ whose elements are the column array contexts and $\varphi_c^D : (V^{**}, V^{**}) \to 2^C$ be a choice mapping. For arrays*

$$A = \begin{smallmatrix} a_{1j} & \cdots & a_{1(k-1)} \\ \vdots & \ddots & \vdots \\ a_{mj} & \cdots & a_{m(k-1)} \end{smallmatrix}, B = \begin{smallmatrix} a_{1(k-p)} & \cdots & a_{1(l-1)} \\ \vdots & \ddots & \vdots \\ a_{m(k-p)} & \cdots & a_{m(l-1)} \end{smallmatrix}, j < k < l, a_{ij} \in V$$

*we define $\hat\varphi_c^D : (V^{**}, V^{**}) \to \$_c V^{**}\$_c$ such that,*

$$\$_c D_c \$_c \in \hat\varphi_c^I(A), \ D_c = \left[\begin{smallmatrix} u_1 \\ u_2 \\ \vdots \\ u_m \end{smallmatrix}\right] \text{ if}$$

$$c_i = \$_c \left[ \begin{smallmatrix} u_i \\ u_{i+1} \end{smallmatrix} \right] \$_c \in \varphi_c^I \left( \begin{smallmatrix} a_{ij} & \cdots & a_{i(k-1)} & a_{i(k+p)} & \cdots & a_{i(l-1)} \\ a_{(i+1)j} & \cdots & a_{(i+1)(k-1)}, & a_{(i+1)(k+p)} & \cdots & a_{(i+1)(l-1)} \end{smallmatrix} \right)$$

$c_i \in C, 1 \le i \le m-1$, *not all need to be distinct.*

*Given an array* $X = [a_{ij}]_{m \times n}, a_{ij} \in V$ *such that* $X = X_1 \oplus A \oplus D_c \oplus B \oplus X_2$,

$$X_1 = \begin{smallmatrix} a_{11} & \cdots & a_{1(j-1)} \\ \vdots & \ddots & \vdots \\ a_{m1} & \cdots & a_{m(j-1)} \end{smallmatrix}, A = \begin{smallmatrix} a_{1j} & \cdots & a_{1(k-1)} \\ \vdots & \ddots & \vdots \\ a_{mj} & \cdots & a_{m(k-1)} \end{smallmatrix}, B = \begin{smallmatrix} a_{1(k+p)} & \cdots & a_{1(l-1)} \\ \vdots & \ddots & \vdots \\ a_{m(k+p)} & \cdots & a_{m(l-1)} \end{smallmatrix}, X_2 = \begin{smallmatrix} a_{1l} & \cdots & a_{1n} \\ \vdots & \ddots & \vdots \\ a_{ml} & \cdots & a_{mn} \end{smallmatrix},$$

$1 \le j \le k < l \le n+1$, *we write* $X \Rightarrow_d Y$ *if* $Y = X_1 \oplus A \oplus B \oplus X_2$, *such that* $\$_c D_c \$_c \in \hat{\varphi}_c^D(A, B)$. $D_c$ *is called as the deleted column context. We say that* $Y$ *is obtained from* $X$ *by parallel column contextual deletion operation. The following 4 special cases for* $X = X_1 \oplus A \oplus D_c \oplus B \oplus X_2$ *are to be considered,*

1. *For* $j = 1$ *we have* $X_1 = \Lambda$.
2. *For* $j = k$, *we have* $A = \Lambda$. *If* $j = k = 1$, *then* $X_1 = \Lambda$ *and* $A = \Lambda$.
3. *For* $k + p = l$, *we have* $B = \Lambda$.
4. *For* $l = n+1$, *we have* $X_2 = \Lambda$. *If* $k+p = l = n+1$, *then* $B = \Lambda$ *and* $X_2 = \Lambda$.

Similarly we can define parallel row contextual deletion operation also.

# 3   Parallel Contextual Array Insertion Deletion P Systems

In this section we define parallel contextual array insertion deletion P system and give an example.

**Definition 7.** *A parallel contextual array insertion deletion P system is a construct,*

$$\prod = (V, T, \mu, C, R, (M_1, I_1, D_1), \ldots, (M_h, I_h, D_h), \varphi_c^I, \varphi_r^I, \varphi_c^D, \varphi_r^D, i_0)$$

*where,*

$V$ *is the finite nonempty set of symbols called alphabet;*

$T \subseteq V$ *is the output alphabet;*

$\mu$ *is the membrane structure with* $h$ *membranes or regions;*

$C$ *is the finite subset of* $\$_c V^{**} \$_c$ *called column array contexts;*

$R$ *is the finite subset of* $\$_r V^{**} \$_r$ *called row array contexts;*

$M_i$ *is the finite set of arrays over* $V$ *called as axioms, each associated with the regions of* $\mu$;

$\varphi_c^I : (V^{**}, V^{**}) \to C$ *is the choice mapping performing parallel column contextual insertion operations;*

$\varphi_r^I : (V^{**}, V^{**}) \to R$ *is the choice mapping performing parallel row contextual insertion operations;*

$\varphi_c^D : (V^{**}, V^{**}) \to C$ *is the choice mapping performing parallel column contextual deletion operations;*

$\varphi_r^D : (V^{**}, V^{**}) \to R$ is the choice mapping performing parallel row contextual deletion operations;

$$I_i = \emptyset \ (or) \ \left\{ \left( \left\{ \varphi_c^I(A_i, B_i) = \$_c \left[ {}_{u_{i+1}}^{u_i} \right] \$_c \middle| i = 1, 2, \ldots, m-1 \right\}, \alpha \right) \right\}$$

$$A_i = \left[ \begin{matrix} a_{ij} & \cdots & a_{i(k-1)} \\ a_{(i+1)j} & \cdots & a_{(i+1)(k-1)} \end{matrix} \right], B_i = \left[ \begin{matrix} a_{ik} & \cdots & a_{i(l-1)} \\ a_{(i+1)k} & \cdots & a_{(i+1)(l-1)} \end{matrix} \right], 1 \leq j \leq k < l \leq n+1$$

$(or)$ $1 \leq j < k \leq l \leq n+1$, $\alpha \in \{here, out, in_t\}$, $u_i$ and $u_{i+1}$ are of size $1 \times p$ with $p \geq 1$.

$(or)$

$$\left\{ \left( \left\{ \varphi_r^I(C_i, E_i) = \$_r \left[ u_i \ u_{i+1} \right] \$_r \middle| i = 1, 2, \ldots, n-1 \right\}, \alpha \right) \right\}$$

$$C_i = \left[ \begin{matrix} a_{ji} & a_{j(i+1)} \\ \vdots & \vdots \\ a_{(k-1)i} & a_{(k-1)(i+1)} \end{matrix} \right], E_i = \left[ \begin{matrix} a_{ki} & a_{k(i+1)} \\ \vdots & \vdots \\ a_{(l-1)i} & a_{(l-1)(i+1)} \end{matrix} \right], 1 \leq j \leq k < l \leq m+1$$

$(or)$ $1 \leq j < k \leq l \leq m+1$, $\alpha \in \{here, out, in_t\}$, $u_i$ and $u_{i+1}$ are of size $p \times 1$ with $p \geq 1$.

$$D_i = \emptyset \ (or) \ \left\{ \left( \left\{ \varphi_c^D(A_i, B_i) = \$_c \left[ {}_{u_{i+1}}^{u_i} \right] \$_c \middle| i = 1, 2, \ldots, m-1 \right\}, \alpha \right) \right\}$$

$$A_i = \left[ \begin{matrix} a_{ij} & \cdots & a_{i(k-1)} \\ a_{(i+1)j} & \cdots & a_{(i+1)(k-1)} \end{matrix} \right], B_i = \left[ \begin{matrix} a_{i(k+p)} & \cdots & a_{i(l-1)} \\ a_{(i+1)(k+p)} & \cdots & a_{(i+1)(l-1)} \end{matrix} \right], 1 \leq j \leq k < l \leq n+$$

$1$, $\alpha \in \{here, out, in_t\}$, $u_i$ and $u_{i+1}$ are of size $1 \times p$ with $p \geq 1$.

$(or)$

$$\left\{ \left( \left\{ \varphi_r^D(C_i, E_i) = \$_r \left[ u_i \ u_{i+1} \right] \$_r \middle| i = 1, 2, \ldots, n-1 \right\}, \alpha \right) \right\}$$

$$C_i = \left[ \begin{matrix} a_{ji} & a_{j(i+1)} \\ \vdots & \vdots \\ a_{(k-1)i} & a_{(k-1)(i+1)} \end{matrix} \right], E_i = \left[ \begin{matrix} a_{(k+p)i} & a_{(k+p)(i+1)} \\ \vdots & \vdots \\ a_{(l-1)i} & a_{(l-1)(i+1)} \end{matrix} \right] 1 \leq j \leq k < l \leq m+1,$$

$\alpha \in \{here, out, in_t\}$, $u_i$ and $u_{i+1}$ are of size $p \times 1$ with $p \geq 1$.

$i_0$ is the output membrane

The direct derivation with respect to $\prod$ is a binary relation $\Rightarrow$ on $V^{**}, T^{**}$ and is defined as $X \Rightarrow_{i,d} Y$, where $X \in V^{**}, Y \in T^{**}$ if and only if, $X = X_1 \oplus A \oplus B \oplus X_2$, $Y = X_1 \oplus A \oplus I_c \oplus B \oplus X_2$ or $X = X_3 \ominus A \ominus B \ominus X_4$, $Y = X_3 \ominus A \ominus I_r \ominus B \ominus X_4$ for some $X_1, X_2, X_3, X_4 \in V^{**}$ and $I_c, I_r$ are inserted column and row contexts obtained by using the insertion rules based on the parallel column or row contextual insertion operations according to the choice mappings. (or) $X = X_1 \oplus A \oplus D_c \oplus B \oplus X_2, Y = X_1 \oplus A \oplus B \oplus X_2$ or $X = X_3 \ominus A \ominus D_r \ominus B \ominus X_4$, $Y = X_3 \ominus A \ominus B \ominus X_4$ for some $X_1, X_2, X_3, X_4 \in V^{**}$ and $D_c, D_r$ are deleted column and row contexts obtained by using the deletion rules based on the parallel column or row contextual deletion operations according to the choice mappings.

The initial configuration of the system consists of the membrane structure with $h$ membranes labelled $1, 2, \ldots, h$ where the outermost membrane being the skin membrane is labelled as $1$, which also acts as our output membrane. Using the insertion or deletion rules $I_i$ or $D_i$ based on the choice mapping $\varphi_i$ present in the region $i$ we do the step by step computation. The array we obtain after each computation is placed in the membrane indicated by $\alpha$. If we choose $\alpha$ to be 'here', it means that the resulting array remains in the same membrane. If we choose $\alpha$ to be 'out', it means that the resulting array is sent out of the current membrane and enters the immediate outer membrane. If that outer membrane happens to be the skin membrane and if no further computation is possible, we say that the resulting array is present in the language generated by this P system. If we choose $\alpha$ to be '$in_t$', it means that the resulting array is sent to the membrane labelled $t$. When there is no rule applicable to the choice array obtained after the last computation we say that the computation is successful and it halts. A successful computation depending on $\alpha$ may result in an array being sent out to the skin membrane. All the arrays with symbols over T collected in the skin membrane is the language generated by the parallel contextual array insertion deletion P system $\prod$ and is denoted by PCAIDP ($\prod$). The family of all array languages PCAIDP ($\prod$) generated by parallel contextual array insertion deletion P system with at most $h$ membranes is denoted by PCAIDP$_h$.

We also consider P system which involves only the insertion operation i.e., $\prod = (T, \mu, C, R, (M_1, I_1), \ldots, (M_h, I_h), \varphi_c^I, \varphi_r^I, i_0)$. The language generated by this P system is denoted by PCAIP ($\prod$). The family of all array languages generated by parallel contextual array insertion P system with at most $h$ membranes is denoted by PCAIP$_h$.

*Example 1.* We consider an example for a PCAIDP ($\prod$),
$$\prod = (V, T, \mu, C, R, (M_1, I_1, D_1), (M_2, I_2, D_2), \varphi_c^I, \varphi_c^D, \varphi_r^I, \varphi_r^D, 1)$$

where,

$V = \{\bullet, X, Y\}$

$T = \{\bullet, X\}$

$\mu = [_1[_2]_2]_1$

$C = \left\{ \$_c \left[\begin{smallmatrix} X & Y \\ \bullet & Y \end{smallmatrix}\right] \$_c, \quad \$_c \left[\begin{smallmatrix} \bullet & Y \\ \bullet & Y \end{smallmatrix}\right] \$_c, \quad \$_c \left[\begin{smallmatrix} \bullet & Y \\ X & Y \end{smallmatrix}\right] \$_c, \$_c \left[\begin{smallmatrix} Y & X \\ Y & \bullet \end{smallmatrix}\right] \$_c, \quad \$_c \left[\begin{smallmatrix} Y & \bullet \\ Y & \bullet \end{smallmatrix}\right] \$_c, \right.$
$\left. \$_c \left[\begin{smallmatrix} Y & \bullet \\ Y & X \end{smallmatrix}\right] \$_c, \$_c \left[\begin{smallmatrix} Y \\ Y \end{smallmatrix}\right] \$_c \right\}$

$R = \left\{ \$_r \left[\begin{smallmatrix} Y & Y \\ \bullet & \bullet \end{smallmatrix}\right] \$_r, \quad \$_r \left[\begin{smallmatrix} Y & Y \\ \bullet & Y \end{smallmatrix}\right] \$_r, \quad \$_r \left[\begin{smallmatrix} Y & Y \\ Y & X \end{smallmatrix}\right] \$_r, \quad \$_r \left[\begin{smallmatrix} X & Y \\ X & Y \end{smallmatrix}\right] \$_r, \quad \$_r \left[\begin{smallmatrix} Y & Y \\ Y & \bullet \end{smallmatrix}\right] \$_r, \right.$
$\left. \$_r \left[\begin{smallmatrix} \bullet & \bullet \\ Y & Y \end{smallmatrix}\right] \$_r, \$_r \left[\begin{smallmatrix} \bullet & Y \\ Y & Y \end{smallmatrix}\right] \$_r, \$_r \left[\begin{smallmatrix} Y & X \\ Y & Y \end{smallmatrix}\right] \$_r, \$_r \left[\begin{smallmatrix} X & Y \\ Y & Y \end{smallmatrix}\right] \$_r, \$_r \left[\begin{smallmatrix} Y & \bullet \\ Y & Y \end{smallmatrix}\right] \$_r \right\}$

$M_1 = \emptyset$

$M_2 = \left\{ \left[\begin{smallmatrix} X & X & X \\ \bullet & X & \bullet \\ X & X & X \end{smallmatrix}\right] \right\}$

$I_1 = \emptyset$

$D_1 = \left\{ \left( \left\{ \varphi_r^D [X X, \bullet \bullet] = \$_r [Y Y] \$_r, \quad \varphi_r^D [X X, \bullet X] = \$_r [Y Y] \$_r, \right. \right. \right.$
$\left. \varphi_r^D [X X, X \bullet] = \$_r [Y Y] \$_r \right\}, here), \left( \left\{ \varphi_r^D [\bullet \bullet, X X,] = \$_r [Y Y] \$_r, \right. \right.$
$\left. \left. \varphi_r^D [\bullet X, X X] = \$_r [Y Y] \$_r, \varphi_r^D [X \bullet, X X] = \$_r [Y Y] \$_r \right\}, here \right) \right\}$

$I_2 = \left\{ \left( \left\{ \varphi_c^I [\begin{smallmatrix} X \\ \bullet \end{smallmatrix}, \begin{smallmatrix} X \\ X \end{smallmatrix}] = \$_c [\begin{smallmatrix} X & Y \\ \bullet & Y \end{smallmatrix}] \$_c, \quad \varphi_c^I [\begin{smallmatrix} \bullet \\ \bullet \end{smallmatrix}, \begin{smallmatrix} X \\ X \end{smallmatrix}] = \$_c [\begin{smallmatrix} \bullet & Y \\ \bullet & Y \end{smallmatrix}] \$_c, \quad \varphi_c^I [\begin{smallmatrix} \bullet \\ X \end{smallmatrix}, \begin{smallmatrix} X \\ X \end{smallmatrix}] = \right. \right. \right.$
$\left. \$_c [\begin{smallmatrix} \bullet & Y \\ X & Y \end{smallmatrix}] \$_c \right\}, here), \left( \left\{ \varphi_c^I [\begin{smallmatrix} X \\ X \end{smallmatrix}, \begin{smallmatrix} X \\ \bullet \end{smallmatrix}] = \$_c [\begin{smallmatrix} Y & X \\ Y & \bullet \end{smallmatrix}] \$_c, \quad \varphi_c^I [\begin{smallmatrix} X \\ X \end{smallmatrix}, \begin{smallmatrix} \bullet \\ \bullet \end{smallmatrix}] = \$_c [\begin{smallmatrix} Y & \bullet \\ Y & \bullet \end{smallmatrix}] \$_c, \right.$
$\varphi_c^I [\begin{smallmatrix} X \\ X \end{smallmatrix}, \begin{smallmatrix} \bullet \\ X \end{smallmatrix}] = \$_c [\begin{smallmatrix} Y & \bullet \\ Y & X \end{smallmatrix}] \$_c \right\}, here), \left( \left\{ \varphi_r^I [X X, \bullet \bullet] = \$_r [\begin{smallmatrix} Y & Y \\ \bullet & \bullet \end{smallmatrix}] \$_r, \right. \right.$
$\varphi_r^I [X Y, \bullet Y] = \$_r [\begin{smallmatrix} Y & Y \\ \bullet & Y \end{smallmatrix}] \$_r, \quad \varphi_r^I [Y X, Y X] = \$_r [\begin{smallmatrix} Y & Y \\ Y & X \end{smallmatrix}] \$_r, \quad \varphi_r^I [X Y, X Y] =$
$\$_r [\begin{smallmatrix} Y & Y \\ X & Y \end{smallmatrix}] \$_r, \quad \varphi_r^I [Y X, Y \bullet] = \$_r [\begin{smallmatrix} Y & Y \\ Y & \bullet \end{smallmatrix}] \$_r \right\}, here), \left( \left\{ \varphi_r^I [\bullet \bullet, X X] = \right. \right.$

$$\$_r\left[\begin{smallmatrix}\bullet&\bullet\\Y&Y\end{smallmatrix}\right]\$_r,\quad \varphi_r^I[\bullet\,Y,\,X\,Y]\;=\;\$_r\left[\begin{smallmatrix}\bullet&Y\\Y&Y\end{smallmatrix}\right]\$_r,\quad \varphi_r^I[Y\,X,\,Y\,X]\;=\;\$_r\left[\begin{smallmatrix}Y&X\\Y&Y\end{smallmatrix}\right]\$_r,$$

$$\varphi_r^I[X\,Y,\,X\,Y]=\$_r\left[\begin{smallmatrix}X&Y\\Y&Y\end{smallmatrix}\right]\$_r,\;\varphi_r^I[Y\,\bullet,\,Y\,X]=\$_r\left[\begin{smallmatrix}Y&\bullet\\Y&Y\end{smallmatrix}\right]\$_r\Big\},here\Big)\Big\}$$

---

$$D_2\;=\;\Big\{\Big(\Big\{\varphi_c^D\left[\begin{smallmatrix}X\\Y\end{smallmatrix},\begin{smallmatrix}X\\Y\end{smallmatrix}\right]\;=\;\$_c\left[\begin{smallmatrix}Y\\Y\end{smallmatrix}\right]\$_c,\;\;\varphi_c^D\left[\begin{smallmatrix}Y\\\bullet\end{smallmatrix},\begin{smallmatrix}\bullet\\X\end{smallmatrix}\right]\;=\;\$_c\left[\begin{smallmatrix}Y\\Y\end{smallmatrix}\right]\$_c,\;\;\varphi_c^D\left[\begin{smallmatrix}\bullet\\Y\end{smallmatrix},\begin{smallmatrix}\bullet\\X\end{smallmatrix}\right]\;=$$

$$\$_c\left[\begin{smallmatrix}Y\\Y\end{smallmatrix}\right]\$_c,\;\;\varphi_c^D\left[\begin{smallmatrix}\bullet\\Y\end{smallmatrix},\begin{smallmatrix}X\\Y\end{smallmatrix}\right]\;=\;\$_c\left[\begin{smallmatrix}Y\\Y\end{smallmatrix}\right]\$_c\Big\},here\Big),\Big(\Big\{\varphi_c^D\left[\begin{smallmatrix}X\\Y\end{smallmatrix},\begin{smallmatrix}X\\Y\end{smallmatrix}\right]\;=\;\$_c\left[\begin{smallmatrix}Y\\Y\end{smallmatrix}\right]\$_c,$$

$$\varphi_c^D\left[\begin{smallmatrix}Y\\X\end{smallmatrix},\begin{smallmatrix}Y\\X\end{smallmatrix}\right]\;=\;\$_c\left[\begin{smallmatrix}Y\\Y\end{smallmatrix}\right]\$_c,\varphi_c^D\left[\begin{smallmatrix}X\\X\end{smallmatrix},\begin{smallmatrix}\bullet\\\bullet\end{smallmatrix}\right]\;=\;\$_c\left[\begin{smallmatrix}Y\\Y\end{smallmatrix}\right]\$_c,\;\;\varphi_c^D\left[\begin{smallmatrix}X\\Y\end{smallmatrix},\begin{smallmatrix}\bullet\\Y\end{smallmatrix}\right]\;=\;\$_c\left[\begin{smallmatrix}Y\\Y\end{smallmatrix}\right]\$_c,$$

$$\varphi_c^D\left[\begin{smallmatrix}Y\\X\end{smallmatrix},\begin{smallmatrix}Y\\X\end{smallmatrix}\right]\quad=\quad\$_c\left[\begin{smallmatrix}Y\\Y\end{smallmatrix}\right]\$_c\Big\},here\Big),\Big(\Big\{\varphi_r^D[X\,X,\,\bullet\,\bullet]\quad=\quad\$_r[Y\,Y]\$_r,$$

$$\varphi_r^D[X\,X,\,\bullet\,X]\quad=\quad\$_r[Y\,Y]\$_r,\quad \varphi_r^D[X\,X,\,X\,\bullet]\quad=\quad\$_r[Y\,Y]\$_r\Big\},\alpha\Big),$$

$$\Big(\Big\{\varphi_r^D[\bullet\,\bullet,\,X\,X]=\$_r[Y\,Y]\$_r,\;\varphi_r^D[\bullet\,X,\,X\,X]=\$_r[Y\,Y]\$_r,\;\varphi_r^D[X\,\bullet,\,X\,X]=$$

$$\$_r[Y\,Y]\$_r\Big\},\alpha\Big)\Big\},\alpha\in\{here,out\}$$

Membrane labelled 1 i.e., the skin membrane is the output membrane.

The language generated by this parallel contextual array insertion deletion P system is,

$$L(\textstyle\prod)=\left\{\begin{matrix}X^n\;X\;X^n\\(\bullet^n\;X\;\bullet^n)_{2n-1}\\X^n\;X\;X^n\end{matrix}\;\Big|\;n\ge1\right\}$$

This language can also be generated by a parallel contextual array insertion P system,

$$\textstyle\prod\;=\;(T,\mu,C,R,(M_1,I_1),(M_2,I_2),(M_3,I_3),(M_4,I_4),(M_5,I_5),\varphi_c^I,\varphi_r^I,1)$$

where,

$$T=\{\bullet,X\}$$
$$\mu=[_1[_2[_3]_3]_2[_4[_5]_5]_4]_1$$
$$C=\left\{\$_c\left[\begin{smallmatrix}X\\\bullet\end{smallmatrix}\right]\$_c,\;\$_c\left[\begin{smallmatrix}\bullet\\\bullet\end{smallmatrix}\right]\$_c,\;\$_c\left[\begin{smallmatrix}\bullet\\X\end{smallmatrix}\right]\$_c\right\}$$
$$R=\left\{\$_r[\bullet\,\bullet]\$_r,\;\$_r[\bullet\,X]\$_r,\;\$_r[X\,\bullet]\$_r\right\}$$
$$M_1=\emptyset$$
$$M_2=\emptyset$$
$$M_3=\left\{\left[\begin{smallmatrix}X&X&X\\\bullet&X&\bullet\\X&X&X\end{smallmatrix}\right]\right\}$$
$$M_4=\emptyset$$
$$M_5=\emptyset$$
$$I_1=\emptyset$$
$$I_2\;=\;\Big\{\Big(\Big\{\varphi_c^I\left[\begin{smallmatrix}X\\X\end{smallmatrix},\begin{smallmatrix}X\\\bullet\end{smallmatrix}\right]\;=\;\$_c\left[\begin{smallmatrix}X\\\bullet\end{smallmatrix}\right]\$_c,\;\;\varphi_c^I\left[\begin{smallmatrix}X\\X\end{smallmatrix},\begin{smallmatrix}\bullet\\\bullet\end{smallmatrix}\right]\;=\;\$_c\left[\begin{smallmatrix}\bullet\\\bullet\end{smallmatrix}\right]\$_c,\;\;\varphi_c^I\left[\begin{smallmatrix}X\\X\end{smallmatrix},\begin{smallmatrix}\bullet\\X\end{smallmatrix}\right]\;=$$

$$\$_c\left[\begin{smallmatrix}\bullet\\X\end{smallmatrix}\right]\$_c\Big\},in_5\Big)\Big\}$$

$$I_3\;=\;\Big\{\Big(\Big\{\varphi_c^I\left[\begin{smallmatrix}X\\X\end{smallmatrix},\begin{smallmatrix}X\\X\end{smallmatrix}\right]\;=\;\$_c\left[\begin{smallmatrix}X\\\bullet\end{smallmatrix}\right]\$_c,\;\;\varphi_c^I\left[\begin{smallmatrix}\bullet\\\bullet\end{smallmatrix},\begin{smallmatrix}X\\X\end{smallmatrix}\right]\;=\;\$_c\left[\begin{smallmatrix}\bullet\\\bullet\end{smallmatrix}\right]\$_c,\;\;\varphi_c^I\left[\begin{smallmatrix}\bullet\\X\end{smallmatrix},\begin{smallmatrix}X\\X\end{smallmatrix}\right]\;=$$

$$\$_c\left[\begin{smallmatrix}\bullet\\X\end{smallmatrix}\right]\$_c\Big\},out\Big)\Big\}$$

$$I_4\;=\Big\{\Big(\Big\{\varphi_r^I[\bullet\,\bullet,\,X\,X]\quad=\quad\$_r[\bullet\,\bullet]\$_r,\quad\varphi_r^I[\bullet\,X,\,X\,X]\quad=\quad\$_r[\bullet\,X]\$_r,$$

$$\varphi_r^I[X\,\bullet,\,X\,X]=\$_r[X\,\bullet]\$_r\Big\},out\Big)\Big\}$$

$$I_5 = \left\{ \left( \left\{ \varphi_r^I [X\,X, \bullet\bullet] = \$_r[\bullet\bullet]\$_r, \quad \varphi_r^I[X\,X, \bullet\,X] = \$_r[\bullet\,X]\$_r, \right. \right. \right.$$
$$\left. \left. \left. \varphi_r^I[X\,X, X\,\bullet] = \$_r[X\,\bullet]\$_r \right\}, \alpha \right) \right\}, \alpha \in \{out, in_3\}$$

Membrane labelled 1 i.e., the skin membrane is the output membrane.

By replacing $X$ by ◇ and • by ◆ we can arrive at pictures like,

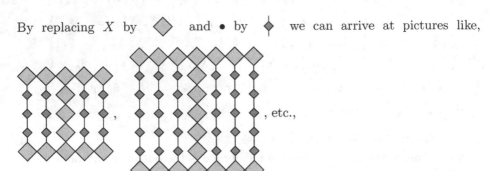

, etc.,

## 4 Properties of Parallel Contextual Array Insertion Deletion P Systems

In this section we give some properties of parallel contextual array insertion deletion P system.

**Theorem 1.** *The families $PCAIDP_h$ and $PCAIP_h$ are closed under union, column catenation and row catenation.*

The proofs are straight forward. □

**Theorem 2.** *The families $PCAIDP_h$ and $PCAIP_h$ are closed under reflection on the base and right leg, transpose and rotations by $90°, 180°, 270°$.*

The proofs are straight forward. □

## 5 Comparison Results

In this section we compare the generative power of $PCAIDP_h$ and $PCAIP_h$ along with that of $LOC$, REC, families of Siromoney matrix languages and family of languages generated by parallel contextual array insertion deletion grammar (PCAIDG) available in the literature [4,5,8,18].

**Theorem 3.** $PCAIDG \subseteq PCAIDP_2$.

*Proof.* For every parallel contextual array insertion deletion grammar, $G = (V, T, B, C, R, \varphi_c^I, \varphi_c^D, \varphi_c^D, \varphi_r^D)$ we can easily construct a parallel contextual array insertion deletion P system $\prod$ with two membranes. $\prod = (V, T, [_1[_2]_2]_1, C, R, (M_1, I_1, D_1), (M_2, I_2, D_2), \varphi_c^I, \varphi_c^D, \varphi_r^I, \varphi_r^D, 1)$ where,
$M_1 = \emptyset$

$M_2 = B$

$I_1 = \emptyset$

$I_2 = \left\{ \left( \left\{ \varphi_c^I \left[ {}^{a_1,}_{a_2,} {}^{b_1}_{b_2} \right] = \$_c \left[ {}^{u_1}_{u_2} \right] \$_c \;\middle|\; \$_c \left[ {}^{u_1}_{u_2} \right] \$_c \in \varphi_c^I \left[ {}^{a_1,}_{a_2,} {}^{b_1}_{b_2} \right] \text{ in } G \right\}, here \right) \right\}$

$\cup \left\{ \left( \left\{ \varphi_r^I \left[ a_1\, a_2,\, b_1\, b_2 \right] = \$_r \left[ u_1\, u_2 \right] \$_r \;\middle|\; \$_r \left[ u_1\, u_2 \right] \$_r \in \varphi_r^I \left[ a_1\, a_2,\, b_1\, b_2 \right] \text{ in } G \right\}, \right. \right.$

$\left. \left. here \right) \right\} \cup \left\{ \left( \left\{ \varphi_c^I \left[ \Lambda,\, {}^{a_1}_{a_2} \right] = \$_c \,\Lambda\, \$_c \;\middle|\; a_1, a_2 \in T \right\}, out \right) \right\}$

$D_1 = \emptyset$

$D_2 = \left\{ \left( \left\{ \varphi_c^D \left[ {}^{a_1,}_{a_2,} {}^{b_1}_{b_2} \right] = \$_c \left[ {}^{u_1}_{u_2} \right] \$_c \;\middle|\; \$_c \left[ {}^{u_1}_{u_2} \right] \$_c \in \varphi_c^D \left[ {}^{a_1,}_{a_2,} {}^{b_1}_{b_2} \right] \text{ in } G \right\}, here \right) \right\}$

$\cup \left\{ \left( \left\{ \varphi_r^D \left[ a_1\, a_2,\, b_1\, b_2 \right] = \$_r \left[ u_1\, u_2 \right] \$_r \;\middle|\; \$_r \left[ u_1\, u_2 \right] \$_r \in \varphi_r^D \left[ a_1\, a_2,\, b_1\, b_2 \right] \text{ in } G \right\}, \right. \right.$

$\left. here \right) \right\}$

1 is the output membrane.

Clearly we can see that $L(\prod) = L(G)$.

Hence PCAIDG $\subseteq$ PCAIDP$_2$. □

**Theorem 4.** $PCAIP_h \subset PCAIDP_h, \forall\, h \geq 2$.

*Proof.* Every parallel contextual array insertion P system, $\prod = (T, \mu, C, R, (M_1, I_1), (M_2, I_2), \varphi_c^I, \varphi_r^I, 1)$ is also a parallel contextual array insertion deletion P system, $\prod = (V, T, \mu, C, R, (M_1, I_1, D_1), (M_2, I_2, D_2), \varphi_c^I, \varphi_c^D, \varphi_r^I, \varphi_r^D, 1)$ where, $V = T$, $D_1 = D_2 = \emptyset$. Hence PCAIP$_h \subseteq$ PCAIDP$_h$. Strict inclusion follows from example 1, where the language $L(\prod)$ cannot be generated by any $PCAIP_2$, $PCAIP_3$ and $PCAIP_4$. □

**Theorem 5.** $LOC \subset PCAIDP_2$.

*Proof.* From [8] where we have $LOC \subset$ PCAIDG. From Theorem 3, we have PCAIDG $\subseteq$ PCAIDP$_2$. Hence $LOC \subset$ PCAIDP$_2$. □

**Theorem 6.** $REC \subset PCAIDP_2$.

*Proof.* From [8] where we have $REC \subset$ PCAIDG. From Theorem 3, we have PCAIDG $\subseteq$ PCAIDP$_2$. Hence $REC \subset$ PCAIDP$_2$. □

**Theorem 7.** $LOC \subset PCAIP_2$.

*Proof.* Every $LOC$ language can be generated by some parallel contextual array insertion P system with two membranes. Let $L$ be a language over $\Gamma$ in $LOC$ with a finite set of tiles, $\Theta$ such that $L = L(\Theta)$. Consider the parallel contextual array insertion P system with 2 membranes, $\prod = (T, \mu, C, R, (M_1, I_1), (M_2, I_2), \varphi_c^I, \varphi_r^I, 1)$ where,

$T = \Gamma \cup \{\#\}$,

$\mu = [_1 [_2 ]_2 ]_1$,

$C = \left\{ \$_c \,{}^e_f\, \$_c \;\middle|\; {}^{a\ b}_{c\ d},\; {}^{b\ e}_{d\ f},\; {}^{\#\ \#}_{b\ e} \in \Theta \right\}$,

$R = \left\{ \$_r \,e\, f\, \$_r \;\middle|\; {}^{a\ b}_{c\ d},\; {}^{c\ d}_{e\ f} \in \Theta \right\}$.

$M_1 = \emptyset$

$$M_2 = \left\{ \begin{smallmatrix} a & b \\ c & d \end{smallmatrix} \;\middle|\; \begin{smallmatrix} \# & \# \\ \# & a \end{smallmatrix}, \; \begin{smallmatrix} \# & \# \\ a & b \end{smallmatrix}, \; \begin{smallmatrix} \# & a \\ \# & b \end{smallmatrix}, \; \begin{smallmatrix} a & b \\ c & d \end{smallmatrix} \in \Theta \right\}$$

$$I_1 = \emptyset$$

$$I_2 = \left\{ \left( \left\{ \varphi_c^I \left[ \begin{smallmatrix} a & b \\ c & d \end{smallmatrix}, \Lambda \right] = \$_c \begin{smallmatrix} e \\ f \end{smallmatrix} \$_c \right\}, here \right) \;\middle|\; \begin{smallmatrix} a & b \\ c & d \end{smallmatrix}, \; \begin{smallmatrix} b & e \\ d & f \end{smallmatrix}, \; \begin{smallmatrix} \# & \# \\ b & e \end{smallmatrix} \in \Theta \right\} \cup$$

$$\left\{ \left( \left\{ \varphi_c^I \left[ \begin{smallmatrix} a & b \\ c & d \end{smallmatrix}, \Lambda \right] = \$_c \begin{smallmatrix} e \\ f \end{smallmatrix} \$_c \right\}, out \right) \;\middle|\; \begin{smallmatrix} a & b \\ c & d \end{smallmatrix}, \; \begin{smallmatrix} b & e \\ d & f \end{smallmatrix}, \; \begin{smallmatrix} \# & \# \\ b & e \end{smallmatrix}, \; \begin{smallmatrix} \# & \# \\ e & \# \end{smallmatrix} \in \Theta \right\} \cup$$

$$\left\{ \left( \left\{ \varphi_r^I \left[ \begin{smallmatrix} a & b \\ c & d \end{smallmatrix}, \Lambda \right] = \$_r \, e \, f \, \$_r \;\middle|\; \begin{smallmatrix} a & b \\ c & d \end{smallmatrix}, \; \begin{smallmatrix} c & d \\ e & f \end{smallmatrix}, \; \begin{smallmatrix} \# & c \\ \# & e \end{smallmatrix} \in \Theta \right\} \cup \left\{ \varphi_r^I \left[ \begin{smallmatrix} a & b \\ c & d \end{smallmatrix}, \Lambda \right] = \$_r \, e \, f \, \$_r \;\middle|\;$$

$$\begin{smallmatrix} a & b \\ c & d \end{smallmatrix}, \; \begin{smallmatrix} c & d \\ e & f \end{smallmatrix} \in \Theta \right\} \cup \left\{ \left( \left\{ \varphi_r^I \left[ \begin{smallmatrix} a & b \\ c & d \end{smallmatrix}, \Lambda \right] = \$_r \, e \, f \, \$_r \;\middle|\; \begin{smallmatrix} a & b \\ c & d \end{smallmatrix}, \; \begin{smallmatrix} c & d \\ e & f \end{smallmatrix}, \; \begin{smallmatrix} d & \# \\ f & \# \end{smallmatrix} \in \Theta \right\}, here \right) \right\} \cup$$

$$\left\{ \left( \left\{ \varphi_r^I \left[ \begin{smallmatrix} a & b \\ c & d \end{smallmatrix}, \Lambda \right] = \$_r \, e \, f \, \$_r \;\middle|\; \begin{smallmatrix} a & b \\ c & d \end{smallmatrix}, \; \begin{smallmatrix} c & d \\ e & f \end{smallmatrix}, \; \begin{smallmatrix} \# & c \\ \# & e \end{smallmatrix}, \; \begin{smallmatrix} \# & e \\ \# & \# \end{smallmatrix} \in \Theta \right\} \cup \left\{ \varphi_r^I \left[ \begin{smallmatrix} a & b \\ c & d \end{smallmatrix}, \Lambda \right] =$$

$$\$_r \, e \, f \, \$_r \;\middle|\; \begin{smallmatrix} a & b \\ c & d \end{smallmatrix}, \; \begin{smallmatrix} c & d \\ e & f \end{smallmatrix}, \; \begin{smallmatrix} e & f \\ \# & \# \end{smallmatrix} \in \Theta \right\} \cup \left\{ \left( \left\{ \varphi_r^I \left[ \begin{smallmatrix} a & b \\ c & d \end{smallmatrix}, \Lambda \right] = \$_r \, e \, f \, \$_r \;\middle|\; \begin{smallmatrix} a & b \\ c & d \end{smallmatrix}, \; \begin{smallmatrix} c & d \\ e & f \end{smallmatrix}, \; \begin{smallmatrix} e & f \\ \# & \# \end{smallmatrix}, \right.$$

$$\left. \begin{smallmatrix} f & \# \\ \# & \# \end{smallmatrix} \in \Theta \right\}, out \right) \right\} \text{ and 1 is the output membrane.}$$

Clearly, $\prod$ can generate any language in $LOC$ and hence $LOC \subseteq PCAIP_2$.

Now to prove the proper inclusion, we consider the language $L' = \{(X)_m^3 | m \geq 1\}$. This language, $L'$ is not in $LOC$ as can be seen in [4]. But this language can be generated by the parallel contextual array insertion P system, $\prod = (T, \mu, C, R, (M_1, I_1), (M_2, I_2), \varphi_c^I, \varphi_r^I, 1)$, where $T = \{X\}$, $\mu = [_1[_2]_2]_1$, $C = \emptyset$, $R = \{\$_r \, X \, X \, \$_r\}$, $M_1 = \emptyset$, $M_2 = \{ \begin{smallmatrix} X & X & X \\ X & X & X \end{smallmatrix} \}$, $I_1 = \emptyset$, $I_2 = \{(\{\varphi_r^I [X \, X, X \, X] = \$_r \, X \, X \, \$_r\}, \alpha)\}$, $\alpha \in \{here, out\}$ and 1 is the output membrane.

Hence $LOC$ is properly contained in $PCAIP_2$. $\square$

**Theorem 8.** *REC is incomparable with $PCAIP_3$ but not disjoint.*

*Proof.* The language considered in example 1 cannot be generated by any parallel contextual array insertion P system with 3 membranes, whereas it is a language in $REC$. This proves that $REC - PCAIP_3 = \emptyset$.

Now we consider the picture language $L$ consisting of arrays describing stair-cases of X's of the form,

$$\begin{smallmatrix} \bullet & \bullet & \bullet & \bullet & \bullet & \bullet & \bullet & \bullet & X \\ \bullet & \bullet & \bullet & \bullet & \bullet & \bullet & \bullet & \bullet & X \\ \bullet & \bullet & \bullet & \bullet & \bullet & \bullet & X & X & X & X \\ \bullet & \bullet & \bullet & \bullet & \bullet & \bullet & X & \bullet & \bullet & \bullet \\ \bullet & \bullet & \bullet & \bullet & X & X & X & X & \bullet & \bullet & \bullet \\ \bullet & \bullet & \bullet & \bullet & X & \bullet & \bullet & \bullet & \bullet \\ \bullet & \bullet & X & X & X & X & \bullet & \bullet & \bullet & \bullet \\ \bullet & \bullet & X & \bullet & \bullet & \bullet & \bullet & \bullet & \bullet \\ X & X & X & X & \bullet & \bullet & \bullet & \bullet & \bullet & \bullet \end{smallmatrix}$$

. This language can be generated by a parallel contextual array insertion P system with 3 membranes, $\prod = (T, \mu, C, R, (M_1, I_1), (M_2, I_2), (M_3, I_3), \varphi_c^I, \varphi_r^I, 1)$ where, $T = \{X, \bullet\}$, $M = \{ \begin{smallmatrix} \bullet & \bullet & \bullet & X \\ X & X & X & X \end{smallmatrix} \}$, $C = \{ \$_c \begin{smallmatrix} \bullet \\ \bullet \\ \bullet \end{smallmatrix} \$_c \}$, $R = \{ \$_r \, X \, X \, \$_r, \; \$_r \begin{smallmatrix} \bullet \\ \bullet \end{smallmatrix} X \, \$_r, \; \$_r \begin{smallmatrix} X \\ \bullet \end{smallmatrix} \bullet \, \$_r, \; \$_r \begin{smallmatrix} \bullet \\ \bullet \end{smallmatrix} \$_r \}$,

$$I_2 = \left\{ \left( \left\{ \varphi_r^I [\begin{smallmatrix} \bullet \\ \bullet \end{smallmatrix}, \Lambda] = \$_r \, X \, X \, \$_r, \varphi_r^I [\begin{smallmatrix} \bullet \\ \bullet \end{smallmatrix}, \Lambda] = \$_r \begin{smallmatrix} \bullet \\ \bullet \end{smallmatrix} \$_r, \varphi_r^I [\begin{smallmatrix} \bullet \\ \bullet \end{smallmatrix} X, \Lambda] = \right. \right. \right.$$

$$\$_r \begin{smallmatrix} \bullet \\ X \end{smallmatrix} X \$_r, \varphi_r^I [X \, X, \Lambda] = \$_r \begin{smallmatrix} X \\ \bullet \end{smallmatrix} \$_r, \varphi_r^I [X \, X, \Lambda] = \$_r \begin{smallmatrix} \bullet \\ \bullet \end{smallmatrix} \$_r, \varphi_r^I [X \, X, \Lambda] =$$

$$\left. \left. \$_r \begin{smallmatrix} \bullet \\ \bullet \end{smallmatrix} \$_r \right\}, \alpha \right) \right\}, \alpha \in \{out, in_3\}$$

$$I_3 = \left\{ \left( \left\{ \varphi_c^I [\Lambda, \begin{smallmatrix} \bullet \\ \bullet \end{smallmatrix}] = \$_c \begin{smallmatrix} \bullet \\ \bullet \\ \bullet \end{smallmatrix} \$_c, \varphi_c^I [\Lambda, X \, X] = \$_c \begin{smallmatrix} \bullet \\ \bullet \\ \bullet \end{smallmatrix} \$_c \right\}, out \right) \right\}$$

A kind of pumping lemma is available for the picture languages of the family $REC$ [4] called as the horizontal iteration lemma and the vertical iteration lemma. It can be seen that this necessary condition cannot be satisfied in case we assume that $L$ is in the $REC$ family. Hence $L \notin REC$. This proves that $PCAIP_3 - REC \neq \emptyset$.

Now to prove $REC \cap \text{PCAIP}_3 \neq \emptyset$, we consider the language $L'$ from the proof of Theorem 7. This language is in $REC$ as can be seen in [4].

Hence $REC$ is incomparable with $\text{PCAIP}_3$ but not disjoint.    $\square$

**Theorem 9.** $CSML \subset PCAIDP_2$.

*Proof.* From [8], we have $CSML \subset \text{PCAIDG}$. From Theorem 3, we have $\text{PCAIDG} \subseteq \text{PCAIDP}_2$. Hence the proof.    $\square$

**Theorem 10.** $CFML \subset PCAIP_3$.

*Proof.* Every $CFML$ can be generated by some parallel contextual array insertion P system $\prod = (T, [_1[_2]_2[_3]_3]_1, C, R, (M_1, I_1), (M_2, I_2), (M_3, I_3), \varphi_c^I, \varphi_r^I, 1)$ where every column insertion rule in $I_2$ is only of the form, $\left\{ \left( \left\{ \varphi_c^I \begin{bmatrix} a_1 & \cdots & a_n \\ b_1 & \cdots & b_n \end{bmatrix}, \right.\right.\right.$
$\left.\left.\left. \begin{bmatrix} a_1' & \cdots & a_n' \\ b_1' & \cdots & b_n' \end{bmatrix} = \$_c \begin{bmatrix} u_1 & \cdots & u_p \\ v_1 & \cdots & v_p \end{bmatrix} \$_c \middle| n, p \geq 1 \right\}, in_3 \right) \right\}$ in $I_2$ with $a_i, b_i, a_i', b_i', u_j, v_j \in T, i = 1, \ldots, n, j = 1, \ldots, p$ and for each such rule in $I_2$ there is a column insertion rule, $\left\{ \left( \left\{ \varphi_c^I \begin{bmatrix} a_1' & \cdots & a_n' \\ b_1' & \cdots & b_n' \end{bmatrix}, \begin{bmatrix} a_1 & \cdots & a_n \\ b_1 & \cdots & b_n \end{bmatrix} = \$_c \begin{bmatrix} u_p & \cdots & u_1 \\ v_p & \cdots & v_1 \end{bmatrix} \$_c \middle| n, p \geq 1 \right\}, \alpha \right) \right\}, \alpha \in \{in_2, out\}$ in $I_3$ and only such column insertion rules exists in $I_3$.

Also, every row insertion rule in $I_2$ is only of the form, $\left\{ \left( \left\{ \varphi_r^I \begin{bmatrix} a_1 & \cdots & a_n \\ b_1 & \cdots & b_n \end{bmatrix}, \right.\right.\right.$
$\left.\left.\left. \begin{bmatrix} a_1' & \cdots & a_n' \\ b_1' & \cdots & b_n' \end{bmatrix} = \$_r \begin{bmatrix} u_1 & \cdots & u_p \\ v_1 & \cdots & v_p \end{bmatrix} \$_r \middle| n, p \geq 1 \right\}, here \right) \right\}$ and for each such rule in $I_2$ there is the same row insertion rule in $I_3$ and only such row insertion rules exist in $I_3$.

To prove proper inclusion we consider the language $L$ in the proof of Theorem 8, which is not in $CFML$ but in $\text{PCAIP}_3$. To see that the language $L$ is not in $CFML$ we can refer [19].

Hence $CFML \subset \text{PCAIP}_3$.    $\square$

**Corollary 1.** $RML \subset PCAIP_3$.

# 6  Conclusion

In this paper, we have introduced parallel contextual array insertion P System along with parallel contextual array Insertion P system and listed some of their closure properties. We have given some comparison results with some well known families of two-dimensional picture languages available in the literature. We can exhibit more comparison results of the new models with other families of languages generated by certain other types of P systems in the literature like parallel contextual array P systems [9] and parallel array-rewriting P systems [21]. The application of this model can be seen in floor designing, kolam pattern generation, etc. We can also study various other applications and properties.

# References

1. Ceterchi, R., Mutyam, M., Păun, G., Subramanian, K.G.: Array - rewriting P systems. Nat. Comput. **2**, 229–249 (2003)
2. Dersanamiba, K.S., Krithivasan, K.: Contextual array P systems. Int. J. Comput. Math. **81**(8), 955–969 (2004)
3. Ehrenfeucht, A., Păun, G., Rozenberg, G.: Contextual grammars and formal languages. In: Rozenberg, G., Salomaa, A. (eds.) Handbook of Formal Languages: Volume 2. Linear Modeling: Background and Application, pp. 237–293. Springer, Heidelberg (1997)
4. Giammarresi, D., Restivo, A.: Two-dimensional languages. In: Rozenberg, G., Salomaa, A. (eds.) Handbook of Formal Languages: Volume 3 Beyond Words, pp. 215–267. Springer, Heidelberg (1997)
5. Giammarresi, D., Restivo, A.: Recognizable picture languages. Int. J. Pattern Recognit. Artif. Intell. **6**, 241–256 (1992)
6. Helen Chandra, P., Subramanian, K.G., Thomas, D.G.: Parallel contextual array grammars and languages. Electron. Notes Discrete Math. **12**, 106–117 (2003)
7. Fernau, H., Freund, R., Schmid, M.L., Subramanian, K.G., Wielderhold, P.: Contextual array grammars and array P systems. Ann. Math. Artif. Intell. **75**, 5–26 (2013)
8. James Immanuel, S., Thomas, D.G.: Parallel contextual array insertion deletion grammar. In: Presented in the International Conference on Theoretical Computer Science and Discrete Mathematics (ICTCSDM 2016). Kalasalingam University (2016)
9. James Immanuel, S., Thomas, D.G., Thamburaj, R., Nagar, A.K.: Parallel contextual array P systems. In: The Proceedings of Asian Conference on Membrane Computing (ACMC 2014), pp. 1–9. IEEE Xplore (2014)
10. Krithivasan, K., Balan, M.S., Rama, R.: Array contextual grammars. In: Recent Topics in Mathematical and Computational Linguistics, pp. 154–168. The Publishing house of the Romanian Academy(2000)
11. Madhu, M., Krithivasan, K.: Contextual P systems. Fund. Info. **49**, 179–189 (2002)
12. Marcus, S.: Contextual grammars. Rev. Roum. Math. Pures et Appl. **14**(10), 1525–1534 (1969)
13. Păun, G.: Computing with membranes. J. Comput. Syst. Sci. **61**, 108–143 (2000)
14. Păun, G.: Marcus Contextual Grammars. Kluwer, Dordrecht (1997)
15. Păun, G., Rozenberg, G., Salomaa, A. (eds.): The Oxford Handbook of Membrane Computing. Oxford Univ. Press, Oxford (2010)
16. Rosenfeld, A.: Isotonic grammars, parallel grammars and picture grammars. In: Michie, D., Meltzer, D. (eds.) Machine Intelligence VI, pp. 281–294. University of Edinburgh Press, Scotland (1971)
17. Rosenfeld, A.: Picture Languages: Formal Models for Picture Recognition. Academic Press, New York (1979)
18. Siromoney, G., Siromoney, R., Krithivasan, K.: Abstract families of matrices and picture languages. Comput. Graph. Image Process. **1**, 234–307 (1972)

19. Siromoney, G., Siromoney, R., Krithivasan, K.: Picture languages with array rewriting rules. Inf. Contr. **22**, 447–470 (1973)
20. Subramanian, K.G., Venkat, I., Wiederhold, P.: A P system model for contextual array languages. In: Barneva, R.P., Brimkov, V.E., Aggarwal, J.K. (eds.) IWCIA 2012. LNCS, vol. 7655, pp. 154–165. Springer, Heidelberg (2012). doi:10.1007/978-3-642-34732-0_12
21. Subramanian, K.G., Isawasan, P., Venkat, I., Pan, L.: Parallel array-rewriting P systems. Rom. J. Inf. Sci. Technol. **17**(1), 103–116 (2014)

# A 3D Curve Skeletonization Method

Nilanjana Karmakar[1]([✉]), Sharmistha Mondal[2], and Arindam Biswas[2]

[1] Department of Information Technology,
St. Thomas' College of Engineering and Technology, Kolkata, India
nilanjana.nk@gmail.com
[2] Department of Information Technology,
Indian Institute of Engineering Science and Technology, Shibpur, India
sharmistha28101990@gmail.com, barindam@gmail.com

**Abstract.** An efficient and robust technique for the determination of the 3D curve skeleton of a digital object is presented in this paper. As a preprocessing step, the 3D isothetic inner cover of the digital object is constructed. The voxels adjacent to the surface of the inner cover are represented in a topological space. The object voxels which are interior to the inner cover and satisfy certain conditions along the three coordinate planes are also expressed in another topological space. Homotopy equivalence of the topological spaces is utilized to report the 3D curve skeleton. The resultant skeleton is a single voxel thick, connected, and centered representation of the object that preserves the object topology. Accuracy of shape representation by the skeleton may be varied by using control values according to the requirement of the application. Experimental results on a wide range of objects demonstrate the efficacy and robustness of the method.

**Keywords:** 3D curve skeleton · 3D isothetic inner cover · 3D object topology · Homotopy · Attaching spaces

## 1 Introduction

The concept of skeletonization has received considerable attention from a wide range of research domains due to its potential applications in diverse areas. In two dimensions as well as in three dimensions, skeletonization has often been associated with the concept of medial axis and thinning. An object boundary has been proved to be homotopy equivalent to its medial axis [6]. Skeletonization of 3D digital objects provide important information about the object topology. An efficient skeletonization algorithm that is insensitive to object boundary complexity, preserves basic connectivity and centeredness, and facilitates object hole detection is proposed in [19]. The method involves the SS-coding that converts objects into a directed cluster graph leading to shortest path extraction and the BS-coding that generates a traditional minimum distance field. An effective sequential thinning algorithm has been presented in [18] that directly produces medial curves from 3D binary objects. The algorithm preserves topology and

© Springer International Publishing AG 2017
V.E. Brimkov and R.P. Barneva (Eds.): IWCIA 2017, LNCS 10256, pp. 184–197, 2017.
DOI: 10.1007/978-3-319-59108-7_15

exploits the local topological parameters of a digital image to extract the curve skeleton. Partitioning of an object can assist in the creation of a skeleton and any segmentation of the skeleton can infer a partitioning of the object. In [11], a volume-based shape-function called the shape-diameter-function (SDF) has been used for the purpose. These algorithms are largely insensitive to pose changes of the same object and also present similar results in analoguous parts of different objects.

The survey in [14] presents an overview of 3D shape skeletonization for both surface and curve skeletons using mesh based and voxel based representations. The survey includes definitions and properties of different types of 3D skeletons and comparison among them based on those properties like homotopy, invariance, thinness, centeredness, smoothness, etc. A taxonomy of the methods, based on dimensionality and sampling, used to compute different types of 3D skeletons is also included along with a discussion on the assumptions, advantages, and limitations of the methods. Another comprehensive and concise survey of different skeletonization algorithms has been presented in [10] where the principles, challenges, and benefits of different skeletonization algorithms have been discussed. Discussion on topology preservation, parallelization, and multi-scale skeleton approaches constitute a specialty of the survey. Various applications of skeletonization and the fundamental challenges of assessing the performance of different skeletonization algorithms have also been included. A comparison of six curve-skeletonization and four surface-skeletonization methods based on voxel models has been presented in [12]. A two-level method of comparison is carried out where firstly, the curve and surface skeletons are globally compared in terms of the standard criteria like homotopy, thinness, centeredness, smoothness, etc. followed by a detailed comparison based on resolution. A detailed visualization has been proposed here which is able to highlight small-scale centeredness differences between curve and surface skeletons.

3D skeletonization has also been attempted in the orthogonal domain. Straight skeleton of an orthogonal polyhedron is constructed by an output sensitive algorithm [7] that exploits a plane-sweep approach instead of shrinking the object boundary. The curve skeleton extraction algorithm in [13] is restricted to surface-like objects and is based on the detection of curves and junctions between different surfaces. In [17], a novel valence driven spatial median (VDSM) algorithm has been developed which eliminates crowded regions and ensures that the output skeleton is unit-width. It computes the center of a crowded region, and applies Dijkstra's shortest path algorithm to generate a unit-width curve to replace the crowded region. The 3D thinning algorithm proposed in [9] directly extracts medial lines consisting of arcs and/or curves instead of surfaces. The thinning strategy used here is the hybrid method, which is a combination of both directional and subfield methods. A variation in the form of an efficient 3D parallel thinning algorithm has been reported in [8], which produces medial surfaces. Each iteration step is composed of three parallel subiterations according to the three deletion directions. Other skeletonization algorithms where voxel connectivity is ensured or the density of the skeletal structure is controlled by a thinness parameter, etc. may be found in [2,16].

The use of vertex antipodal points for extracting 3D mesh skeletons has been reported in literature [1]. A vertex antipodal point is the diametrically opposite point that belongs to the same mesh. The set of centers of the connecting lines between each vertex and its antipodal point represents the desired skeleton of the 3D mesh. In another algorithm based on Discrete Euclidean Distance Transform [15], each interior voxel in the 3D image object is classified according to its relative distance from the object border. Recently, algorithms for centerline extraction of tubular objects based on surface normal accumulation [5] and based on Voronoi covariance measure using orthogonal planes [3] have also been proposed. Though our algorithm exploits the concept of antipodal points, they are localized to regular specific ranges along each coordinate plane. Also, unlike the above cases, the centrally located voxels in each grid range are selected to be a part of the skeleton depending upon the number of object voxels present in the grid range.

The rest of the paper is organized as follows. A few preliminary concepts related to the work is explained in Sect. 2. In Sect. 3, the 3D curve skeleton of a given digital object has been extracted by proving its homotopy equivalence to the topological space representing the 3D isothetic inner cover of the object. The paper is concluded with some experimental results in Sect. 4.

## 2    Definitions and Preliminaries

The following concepts have been used in the current work.

### 2.1    Directional Distance

In a specific grid range along a given coordinate plane, a voxel $v$ has eight neighbors, four of which are 2-adjacent and four are 1-adjacent to $v$. Directional distance between two voxels $v_1$ and $v_2$ can be defined if $v_1$ and $v_2$ are such that $v_2$ can be reached from $v_1$ following a path consisting of only 1-adjacent (only 2-adjacent) voxels. Then the number of voxels constituting the path (including $v_2$) is the directional distance between the voxels $v_1$ and $v_2$. In Fig. 1(a), the voxels from $v_1$ to $v_2$ are 1-adjacent and in Fig. 1(b), they are 2-adjacent. In both the cases the directional distance between $v_1$ and $v_2$ is 3. Note that directional distance can be calculated for only those pairs of voxels which are connected according to the above criteria. For example, in Fig. 1(a), there exists no directional distance between voxels $v_1$ and $v_3$.

### 2.2    Local Antipodal Points

Antipodal points are defined as diametrically opposite points on a sphere of any dimension. Along a given coordinate plane in the orthogonal domain, we define two voxels $v_2$ and $v_2'$ as local antipodal points w.r.t. another voxel $v_1$ if

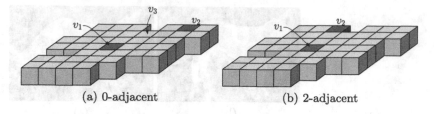

**Fig. 1.** Directional distance between voxels $v_1$ (red) and $v_2$ (blue) is 3. The consecutive voxels connecting $v_1$ and $v_2$ are (a) 1-adjacent or (b) 2-adjacent. No directional distance exists between $v_1$ and $v_3$ in (a). (Color figure online)

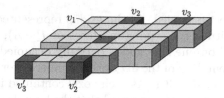

**Fig. 2.** $v_2$, $v_2'$ (blue) and $v_3$, $v_3'$ (blue) are two pairs of local antipodal points equidistant from $v_1$ (red). (Color figure online)

- $v_1$, $v_2$, and $v_2'$ lie in the same grid range,
- $v_2$ and $v_2'$ are located at equal (or almost equal) directional distance from $v_1$, and
- $v_2$ and $v_2'$ are diametrically opposite w.r.t. $v_1$.

In Fig. 2, $v_2$ and $v_2'$ are local antipodal points which are at equal directional distance from $v_1$. In this case, 2-adjacency is considered for the directional distance. $v_3$ and $v_3'$ are also local antipodal points at equal directional distance from $v_1$ where the connecting voxels are 1-adjacent. Here, $v_2$ and $v_3$ do not qualify as local antipodal points because their locations are not diametrically opposite w.r.t. $v_1$. Note that, given a coordinate plane, there may be at most four pairs of local antipodal points w.r.t. a given voxel.

## 3    Proposed Work

Let us consider a 3D digital object $A$ provided as a triangulated data set such that exactly two triangles are incident on each edge of the triangulation (Fig. 3(a)). Let the object be embedded on a 3D digital grid represented as a set of unit grid cubes (UGCs) each of length $g$. We construct the 3D isothetic inner cover $\underline{P}_{\mathbb{G}}(A)$ which is defined as the 3D polyhedron of maximum volume defined w.r.t. an underlying grid $\mathbb{G}$ having surfaces parallel to the coordinate planes and inscribing the entire object (Fig. 3(b)) [4]. A voxel $p$ is considered as an object voxel if it is intersected by one or more triangles on the object surface. If each of the voxels in a UGC are object voxels, then the UGC is called totally object-occupied. If at least one of the voxels in a UGC is a background voxel, then it

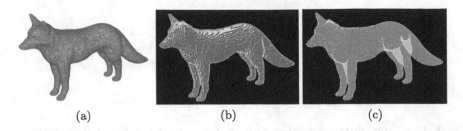

(a)                          (b)                          (c)

**Fig. 3.** A digital object Fox, its 3D isothetic inner cover, and 3D curve skeleton.

is called partially object-occupied. As the grid is represented by a set of UGCs, each UGC-face constituting the polytope faces of $\underline{P}_\mathbb{G}(A)$, is neighbored by a totally occupied UGC on one side and a partially occupied UGC on the other side. Therefore, the boundary of the 3D isothetic inner cover may be represented by a set of totally occupied UGCs. As each voxel contained in a totally occupied UGC is an object voxel, it may be assumed that the boundary of the 3D isothetic inner cover is represented by a set of object voxels. Our objective is to find a single voxel thick 3D curve skeleton of the object which is connected, centered w.r.t. the object thereby capturing the object shape, as shown in Fig. 3(c).

Let $\mathcal{X}$ be the set of object voxels $\mathbf{u_b}$ representing the boundary of $\underline{P}_\mathbb{G}(A)$. Henceforth, object voxels representing the boundary of $\underline{P}_\mathbb{G}(A)$ will be referred to as boundary voxels and those representing the interior of $\underline{P}_\mathbb{G}(A)$ will be referred to as interior voxels. Let $\Gamma_{\mathcal{X}x}$, $\Gamma_{\mathcal{X}y}$, and $\Gamma_{\mathcal{X}z}$ be three topologies defined on $\mathcal{X}$ w.r.t. the three coordinate planes $yz$-, $zx$-, and $xy$-planes. Let $\beta_{\mathcal{X}x}$, $\beta_{\mathcal{X}y}$, and $\beta_{\mathcal{X}z}$ be the bases for $\mathcal{X}$ for the corresponding topologies defined as a collection of basis elements such that

i. a basis element $\mathcal{P}_i$ consists of boundary voxels intercepted between grid values $g_i$ and $g_{i+1}$, where $0 < i < l$, where $l$ is the length of $\underline{P}_\mathbb{G}(A)$ along a given coordinate plane expressed in units of $g$.

ii. if $\exists \mathcal{P}_i, \mathcal{P}_j \in \beta_{\mathcal{X}m}$, $m \in \{x, y, z\}$, such that $\mathcal{P}_i \cap \mathcal{P}_j \neq \emptyset$, then $\exists \mathcal{P}_k \in \beta_{\mathcal{X}m}$ such that $\mathcal{P}_k \subset \mathcal{P}_i \cap \mathcal{P}_j$.

In Fig. 4, a sample set of voxels representing the basis element $\mathcal{P}_i$ in the grid range $g_i$ and $g_{i+1}$ along the $zx$-plane are shown in blue.

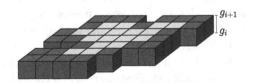

**Fig. 4.** A sample set of voxels (blue) that represents the basis element $\mathcal{P}_i$ in the grid range $g_i$ to $g_{i+1}$. $\mathcal{P}_i$ is a basis element of the basis $\beta_{\mathcal{X}y}$ that defines the topological space $\mathcal{X}$ along the $zx$-plane. (Color figure online)

Let $\mathbf{u_s}$ be an interior voxel that is at equal (or almost equal) *directional distance* from a pair of boundary voxels $\mathbf{u_b}$ and $\mathbf{u_b'}$ such that $\mathbf{u_b}$ and $\mathbf{u_b'}$ are *local antipodal points* w.r.t $\mathbf{u_s}$ within the grid range $g_i$ to $g_{i+1}$ along a given coordinate plane. If more than one such pairs exist for a $\mathbf{u_s}$ along a coordinate plane, then $\mathbf{u_s}$ is selected. In Fig. 5(a), three pairs of local antipodal points ($\{x_1, x_1'\}$, $\{x_2, x_2'\}$, and $\{x_3, x_3'\}$) (blue) exist such that the voxel $\mathbf{u_s}$ (red) is equidistant from the antipodal voxels in each pair. Hence, $\mathbf{u_s}$ is selected as shown in Fig. 5(b). Note that all the voxels connecting $\mathbf{u_s}$ to the local antipodal points are either 0-adjacent or 2-adjacent. Let $\mathcal{W}$ be the set of all selected interior voxels $\mathbf{u_s}$ and let $\Gamma_{\mathcal{W}x}$, $\Gamma_{\mathcal{W}y}$, and $\Gamma_{\mathcal{W}z}$ be three topologies defined on $\mathcal{W}$ w.r.t. the three coordinate planes. Let $\beta_{\mathcal{W}x}$, $\beta_{\mathcal{W}y}$, and $\beta_{\mathcal{W}z}$ be the bases for $\mathcal{W}$ for the corresponding topologies defined as a collection of basis elements such that

i. a basis element $\mathcal{R}_i$ consists of the selected interior voxels intercepted between grid values $g_i$ and $g_{i+1}$, where $0 < i < l$, where $l$ is the length of $\underline{P}_\mathbb{G}(A)$ along a given coordinate plane expressed in units of $g$.

ii. if $\exists \mathcal{R}_i, \mathcal{R}_j \in \beta_{\mathcal{W}m}$, $m \in \{x, y, z\}$, such that $\mathcal{R}_i \cap \mathcal{R}_j \neq \emptyset$, then $\exists \mathcal{R}_k \in \beta_{\mathcal{W}m}$ such that $\mathcal{R}_k \subset \mathcal{R}_i \cap \mathcal{R}_j$.

In Fig. 5(b), a sample set of voxels representing the basis element $\mathcal{R}_i$ in the grid range $g_i$ and $g_{i+1}$ along the $zx$-plane are shown in red. There exists more than one pair of antipodal points w.r.t. each voxel in the set.

Let $n$ be the total number of voxels that represent $\underline{P}_\mathbb{G}(A)$. Along a given coordinate plane

$$n = \sum_{i=0}^{l} n_i$$

where $n_i$ denotes the number of voxels intercepted between $g_i$ and $g_{i+1}$. Let $\mathcal{W}'$ be a topological space defined with a topology $\Gamma_{\mathcal{W}'}$. Let $w$ be an element of $\mathcal{W}$ that belongs to the basis elements $\mathcal{R}_{ix}$, $\mathcal{R}_{iy}$, and $\mathcal{R}_{iz}$ in the topologies $\Gamma_{\mathcal{W}x}$, $\Gamma_{\mathcal{W}y}$, and $\Gamma_{\mathcal{W}z}$ respectively. $w \in \mathcal{W}'$ if

$$(((w \in \mathcal{R}_{ix}) \wedge (n_{ix} = n')) \vee ((w \in \mathcal{R}_{iy}) \wedge (n_{iy} = n')) \vee ((w \in \mathcal{R}_{iz}) \wedge (n_{iz} = n'))) = 1$$

where $n_{ix}$, $n_{iy}$, and $n_{iz}$ denote the number of voxels intercepted between the grid ranges $g_i$ to $g_{i+1}$ along the three coordinate planes and $n' = min(n_{ix}, n_{iy}, n_{iz})$. The basis $\beta_{\mathcal{W}'}$ for $\mathcal{W}'$ is defined as a collection of basis elements $\mathcal{R}_i'$ that satisfy the same conditions as for the basis elements of $\beta_{\mathcal{W}}$.

The method of extracting the 3D curve skeleton may be summarized in the following steps.

1. Homotopy equivalence: Prove that $\mathcal{X}$ and $\mathcal{W}'$ are homotopy equivalent.
2. Space attachment: If $\mathcal{W}'$ is disconnected, then connect it by space attachment technique to form $\mathcal{W}''$.
3. Retraction: Define a retraction $r : \mathcal{W}'' \to \mathcal{S}$ such that $\mathcal{S}$ represents the resultant 3D curve skeleton.

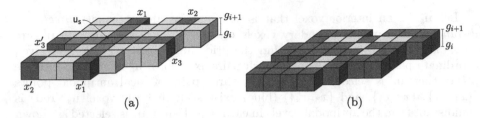

**Fig. 5.** (a) A sample case of three pairs of local antipodal points ($\{x_1, x_1'\}$, $\{x_2, x_2'\}$, and $\{x_3, x_3'\}$) (blue) such that the voxel $\mathbf{u_s}$ (red) is equidistant from the two points in each pair. (b) A sample set of voxels (red) selected according to (a) that represents the basis element $\mathcal{R}_i$ in the grid range $g_i$ to $g_{i+1}$. $\mathcal{R}_i$ is a basis element of the basis $\beta_{\mathcal{W}_y}$ that defines the topological space $\mathcal{W}$ along the $zx$-plane. (Color figure online)

### 3.1   Homotopy Equivalence

Let us consider the grid range $g_i$ to $g_{i+1}$. As defined above, an element $w \in \mathcal{R}_i'$ is equidistant (equal or nearly equal directional distance) from more than one pairs of elements $x, x' \in \mathcal{P}_i$. As elements $x$ and $x'$ represent local antipodal points, they are termed as local antipodal elements. It is observed that a pair of local antipodal elements is equidistant from a single element $w \in \mathcal{R}_i'$. Let $f : \mathcal{X} \to \mathcal{W}'$ be a function that maps to $w \in \mathcal{W}'$ one element out of each pair of local antipodal elements in $\mathcal{X}$ which are equidistant from $w$ within the grid range $g_i$ to $g_{i+1}$. Let $g : \mathcal{W}' \to \mathcal{X}$ be a function that maps $w \in \mathcal{W}'$ to one element out of each pair of local antipodal elements in $\mathcal{X}$ which are equidistant from $w$ within the grid range $g_i$ to $g_{i+1}$. Our objective is to show that the topological space $\mathcal{W}'$ is a part of the 3D curve skeleton of the object by proving that topological spaces $\mathcal{W}'$ and $\mathcal{X}$ are homotopy equivalent. If $f : \mathcal{X} \to \mathcal{W}'$ and $g : \mathcal{W}' \to \mathcal{X}$ are continuous maps such that $f \circ g$ and $g \circ f$ are homotopic to the identity map of $\mathcal{X}$ and $\mathcal{W}'$ respectively, then $\mathcal{X}$ and $\mathcal{W}'$ are homotopy equivalent.

We have the following lemma.

**Lemma 1.** $f : \mathcal{X} \to \mathcal{W}'$ and $g : \mathcal{W}' \to \mathcal{X}$ are continuous maps.

*Proof.* Let $\mathcal{P}_i$ be a basis element of $\mathcal{X}$ and let $\mathcal{R}_i'$ be a basis element of $\mathcal{W}'$. Let $x \in \mathcal{P}_i$ and $w \in \mathcal{R}_i'$ such that $f(x) = w$, where $x, x' \in \mathcal{P}_i$ are local antipodal elements equidistant from $w$. Let $\mathcal{V}$ be an open subset of $\mathcal{W}'$ such that $w \in \mathcal{V}$ and $\mathcal{V} = \bigcup_{i \in J} \mathcal{R}_i'$, where $J$ is the set of indices of the basis elements that comprise the open set $\mathcal{V}$. Therefore, $f^{-1}(\mathcal{V}) = \bigcup_{i \in J} f^{-1}(\mathcal{R}_i')$. This implies that if each set $f^{-1}(\mathcal{R}_i')$ is open, then $f^{-1}(\mathcal{V})$ is an open subset of $\mathcal{X}$. Since, $f(x) = w$ is true, $f(\mathcal{P}_i) \subset \mathcal{R}_i'$ holds. As $x \in \mathcal{P}_i$ and $\mathcal{P}_i \subset f^{-1}(\mathcal{R}_i')$ holds, $f^{-1}(\mathcal{R}_i')$ is an open set $\forall i \in J$. Since union of open sets is open, $f^{-1}(\mathcal{V})$ is an open subset of $\mathcal{X}$. Hence, $f$ is a continuous map.

Let $g(w) = x$ where $x, x' \in \mathcal{P}_i$ are local antipodal elements equidistant from $w$. Let $\mathcal{V}'$ be an open subset of $\mathcal{X}$ such that $\mathcal{V}' = \bigcup_{i \in J} \mathcal{P}_i$. Hence, $g^{-1}(\mathcal{V}') = \bigcup_{i \in J} g^{-1}(\mathcal{P}_i)$. This implies that if each set $g^{-1}(\mathcal{P}_i)$ is open, then $g^{-1}(\mathcal{V}')$ is an open subset of $\mathcal{W}'$. Since, $g(w) = x$ holds, $g(\mathcal{R}_i') \subset \mathcal{P}_i$. Therefore, $w \in \mathcal{R}_i'$ and

$\mathcal{R}'_i \subset g^{-1}(\mathcal{P}_i)$. Hence, $g^{-1}(\mathcal{P}_i)$ is an open set $\forall i \in J$. It follows that $g$ is a continuous map.    □

Let $Id_{\mathcal{W}'}$ be the identity map of $\mathcal{W}'$ and $Id_{\mathcal{X}}$ be that of $\mathcal{X}$. We have the following lemma.

**Lemma 2.** $f \circ g$ *is homotopic to* $Id_{\mathcal{W}'}$ *and* $g \circ f$ *is homotopic to* $Id_{\mathcal{X}}$.

*Proof.* Let $x \in \mathcal{X}$ and $w \in \mathcal{W}'$. Let $g(w) = \{x_1, x_2, ..., x_k\}$. This means that $w$ is equidistant from $k$ pairs of local antipodal elements $\{x_1, x'_1\}$, $\{x_2, x'_2\}$, ..., $\{x_k, x'_k\}$ belonging to $\mathcal{X}$, where $2 \leqslant k \leqslant 4$. This is justified because initially a voxel is selected for topological space $\mathcal{W}$ only if it is equidistant from more than one pair of local antipodal voxels (Sect. 3).

Let

$$f \circ g = f(g(w))$$
$$= f(x_1) \cap f(x_2) \cap ... \cap f(x_k)$$
$$= w.$$

Note that the function $f$ involves intersection operation instead of standard union operation. Since $f(g(w)) = w \; \forall \; w \in \mathcal{W}'$, we conclude that $f \circ g \simeq Id_{\mathcal{W}'}$.

Now, let $f(x) = \{w_1, w_2, w_3, ..., w_k\}$. Here, $x \in \mathcal{X}$ is mapped to $\mathcal{W}'$ using $f$ with no knowledge about the local antipodal element to $x$. Hence, $x$ may be mapped to any number of elements in $\mathcal{W}'$.

Let

$$g(w_1) = \{x_{11}, x_{12}, ..., x, ...., x_{1m}\},$$
$$g(w_2) = \{x_{21}, x_{22}, ..., x, ...., x_{2m}\},$$
$$g(w_3) = \{x_{31}, x_{32}, ..., x, ...., x_{3m}\},$$
$$\vdots$$
$$g(w_k) = \{x_{k1}, x_{k2}, ..., x, ...., x_{km}\}.$$

This indicates that $w_i$ is equidistant from $im$ number of pairs of local antipodal elements $\{x_{i1}, x'_{i1}\}$, $\{x_{i2}, x'_{i2}\}$, ..., $\{x_{im}, x'_{im}\}$, where $1 \leqslant i \leqslant k$. Note that $x$ remains common in all the cases whereas its local antipodal element is distinct for each case.

Let

$$g \circ f = g(f(x))$$
$$= g(w_1) \cap g(w_2) \cap g(w_3) \cap ... \cap g(w_k)$$
$$= x.$$

Note that the function $g$ also involves intersection operation instead of standard union operation. Also, the intersection operation involved in the function $g$ in case of $f \circ g$ and in the function $f$ in case of $g \circ f$ are trivial because the functions are imposed on a single element. Since $g(f(x)) = x \; \forall \; x \in \mathcal{X}$, we conclude that $g \circ f \simeq Id_{\mathcal{X}}$.    □

The relation between the topological spaces $\mathcal{X}$ and $\mathcal{W}'$ is established by the following theorem.

**Theorem 1.** $\mathcal{X}$ and $\mathcal{W}'$ are homotopy equivalent.

*Proof.* From Lemmas 1 and 2, $f : \mathcal{X} \to \mathcal{W}'$ and $g : \mathcal{W}' \to \mathcal{X}$ are continuous maps such that the path $f \circ g$ is homotopic to the identity map of $\mathcal{W}'$ and the path $g \circ f$ is homotopic to the identity map of $\mathcal{X}$. Since the cardinality of the bases (number of basis elements) of $\mathcal{X}$ and $\mathcal{W}'$ are equal, the path homotopies are satisfied for all $x \in \mathcal{X}$ and for all $w \in \mathcal{W}'$. Hence, the topological spaces $\mathcal{X}$ and $\mathcal{W}'$ are homotopy equivalent. □

## 3.2   Space Attachment

$\underline{P_G}(A)$ may be disconnected [4]. As the topological space $\mathcal{X}$ represents the boundary voxels of $\underline{P_G}(A)$, $\mathcal{X}$ may also be disconnected. Hence, the topological space $\mathcal{W}'$ may also be disconnected. In order to connect two components of $\mathcal{W}'$, we define a subspace of one of the components, find the closest points between the subspace and the other component, and then connect them by space attachment technique. Let $\mathcal{W}'$ be represented as $\mathcal{W}' = \mathcal{W}'_1 \cup \mathcal{W}'_2 \cup \mathcal{W}'_3 \cup ... \cup \mathcal{W}'_k$. Let us consider two topological spaces $\mathcal{W}'_i \subset \mathcal{W}'$ and $\mathcal{W}'_j \subset \mathcal{W}'$ which are components of $\mathcal{W}'$. In order to attach two topological spaces, we need to define a subspace $\mathcal{D} \subset \mathcal{W}'_i$ and identify the points in $\mathcal{D}$ with the points in $\mathcal{W}'_j$. Let $\mathcal{D}$ be a subset of $\mathcal{W}'_i$ with a topology $\Gamma_\mathcal{D}$ defined on it such that $\Gamma_\mathcal{D} = \{\mathcal{D} \cap \mathcal{V} \mid \mathcal{V} \in \Gamma_{\mathcal{W}'_i}\}$, where $\mathcal{V}$ is an open set of $\Gamma_{\mathcal{W}'_i}$. As a degenerate case, $\Gamma_\mathcal{D}$ will be equivalent to $\Gamma_{\mathcal{W}'_i}$ if $\mathcal{D} \cap \mathcal{V} = \mathcal{V} \, \forall \, \mathcal{V} \in \Gamma_{\mathcal{W}'_i}$.

Let the function $h : \mathcal{D} \to \mathcal{W}'_j$ be defined by the following method of alternate BFS (breadth first search) performed on the topological spaces $\mathcal{D}$ and $\mathcal{W}'_j$ in order to find the closest points between them. Let us consider $p_1 \in \mathcal{D}$ and $p_2 \in \mathcal{W}'_j$ and record the Euclidean distance $d(p_1, p_2)$ between them. Starting from $p_1$, BFS is performed as long as the dequeued element $p_k$ belongs to the neighborhood of $p_1$. BFS is paused and $d(p_1, p_2)$ is replaced with $d(p_k, p_2)$ if $d(p_k, p_2) < d(p_1, p_2)$. Next, a separate BFS traversal is started from $p_2$ and is continued as long as the dequeued element $p_l$ belongs to the neighborhood of $p_2$. Again, BFS is paused and $d(p_k, p_2)$ is replaced with $d(p_k, p_l)$ if $d(p_k, p_l) < d(p_k, p_2)$. The two separate BFS traversals are continued alternately until both the queues are empty. Finally, $p_k$ and $p_l$ are selected as the closest points between $\mathcal{D}$ and $\mathcal{W}'_j$.

Next, $\mathcal{W}'_i$ and $\mathcal{W}'_j$ are to be attached by space attachment technique as described below. Let the elements in the 26-neighborhood of $p_k(x, y, z)$ belonging to $\mathcal{D}$ be given by

$$N(p_k) = \{p'_k : p'_k \in \mathbb{Z}^3 \wedge p'_k \in \mathcal{D} \wedge L_1(p_k, p'_k) \in \{1, 2, 3\} \wedge L_\infty(p_k, p'_k) = 1\},$$

where $p'_k = (x', y', z')$, $L_\infty(p_k, p'_k) = \max\{|x - x'|, |y - y'|, |z - z'|\}$, and $L_1(p_k, p'_k) = |x - x'| + |y - y'| + |z - z'|$.

Let $t(i, j, k)$ denote the direction of proceeding with the space attachment where

$i, j, k \in \{-1, 0, 1\}$. Starting from $p_k$, we find $p_k'' = (x + i, y + j, z + k)$. The set of neighboring elements of $p_k''$ belonging to $\mathcal{D}$ is given by $N(p_k'')$. Next, the neighborhood $N(p_k)$ is imposed on $p_k''$ to give $N^*(p_k'')$ (explained next in the example in Fig. 6). In the process, the set of neighboring elements of $p_k''$ that are added to $\mathcal{D}$ is given by $N^*(p_k'') - (N^*(p_k'') \cap N(p_k''))$. Next, $p_k''$ is considered as $p_k$ and the procedure is continued until $p_k = p_l$ is reached. Thus, $\mathcal{W}_i'$ is topologically attached with $\mathcal{W}_j'$ by adding a new set of elements between the closest points $p_k$ and $p_l$.

For instance, in Fig. 6(a), the 26-neighborhood of $p_k(x, y, z)$ (yellow color) is highlighted in blue.

$N(p_k) = \{\{x - 1, y, z\}, \{x - 1, y, z - 1\}, \{x, y + 1, z - 1\}, \{x, y - 1, z\}, \{x + 1, y + 1, z\}, \{x + 1, y + 1, z + 1\}, \{x + 1, y - 1, z - 1\}\}$ (Fig. 6(b), orange color).

Let $t = (-1, 0, 0)$. Therefore, $p_k'' = (x - 1, y, z)$ as shown in Fig. 6(c) (magenta color). The 26-neighborhood of $p_k''$ is highlighted in green (Fig. 6(d)).

$N(p_k'') = \{\{x - 2, y, z - 1\}, \{x - 2, y - 1, z + 1\}, \{x - 1, y, z - 1\}, \{x, y + 1, z - 1\}, \{x, y, z\}, \{x, y - 1, z\}\}$ (Fig. 6(e), cyan color).

If the neighborhood of $p_k$ is imposed on $p_k''$, then $N^*(p_k'')$ is obtained by replacing $(x, y, z)$ by $(x - 1, y, z)$ in $N(p_k)$, i.e.,

$N^*(p_k'') = \{\{x - 2, y, z\}, \{x - 2, y, z - 1\}, \{x - 1, y + 1, z - 1\}, \{x - 1, y - 1, z\}, \{x, y + 1, z\}, \{x, y + 1, z + 1\}, \{x, y - 1, z - 1\}\}$, as shown by the green and cyan elements in Fig. 6(f). The set of elements added to $\mathcal{D}$ is given by $N^*(p_k'') - (N^*(p_k'') \cap N(p_k'')) = \{\{x - 2, y, z\}, \{x - 1, y + 1, z - 1\}, \{x - 1, y - 1, z\}, \{x, y + 1, z\}, \{x, y + 1, z + 1\}, \{x, y - 1, z - 1\}\}$, as shown by the green elements in Fig. 6(f).

## 3.3  Retraction

Let $\mathcal{W}''$ be the connected topological space thus obtained. Let $\mathcal{V}$ be an open set in $\mathcal{W}''$. Let $\mathcal{S} \subset \mathcal{W}''$ be a subspace of $\mathcal{W}''$ with a topology $\Gamma_\mathcal{S}$ defined on it such that $\Gamma_\mathcal{S} = \{\mathcal{S} \cap \mathcal{V} \mid \mathcal{V} \in \Gamma_{\mathcal{W}''}\}$. Our aim is to define a topological retraction from $\mathcal{W}''$ to $\mathcal{S}$. Let an element $w \in \mathcal{W}''$. Let the elements of the 26-neighborhood of $w(x, y, z)$ belonging to $\mathcal{W}''$ be given by

$$N(w) = \{w' : w' \in \mathbb{Z}^3 \wedge w' \in \mathcal{W}'' \wedge L_1(w, w') \in \{1, 2, 3\} \wedge L_\infty(w, w') = 1\},$$

where $w' = (x', y', z')$, $L_\infty(w, w') = \max\{|x - x'|, |y - y'|, |z - z'|\}$, and $L_1(w, w') = |x - x'| + |y - y'| + |z - z'|$.

The number of elements of $N(w)$ is given by $|N(w)|$. Let $|N'(w)| = 26 - |N(w)|$. $w$ is an element of the subspace $\mathcal{S}$ if any one of the following conditions is true.

i.  $|N'(w)| > T$,
ii. $0 \leqslant |N'(w)| \leqslant T$ and $\mathcal{W}'' \backslash w$ is disconnected,

where $T$ is a threshold such that $1 \leqslant T < 26$. It may be noted that if $w$ does not satisfy any of the above conditions, then it remains out of consideration while checking the next element in $\mathcal{W}''$.

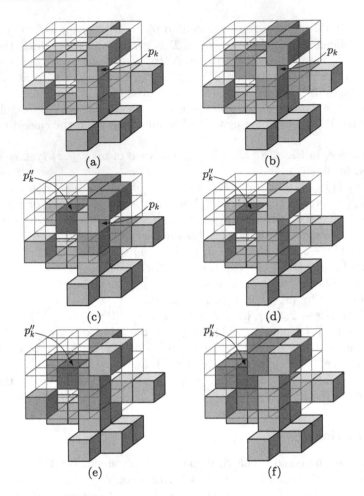

**Fig. 6.** Attaching spaces $\mathcal{W}'_i$ and $\mathcal{W}'_j$ by adding a set of elements to the subspace $\mathcal{D}$ of $\mathcal{W}'_i$. (a) 26-neighborhood of $p_k(x, y, z)$ (yellow) is highlighted in blue. (b) $N(p_k) = \{\{x-1, y, z\}, \{x-1, y, z-1\}, \{x, y+1, z-1\}, \{x, y-1, z\}, \{x+1, y+1, z\}, \{x+1, y+1, z+1\}, \{x+1, y-1, z-1\}\}$ (orange color). (c) $p''_k = (x-1, y, z)$ (magenta). (d) 26-neighborhood of $p''_k$ is highlighted in green. (e) $N(p''_k) = \{\{x-2, y, z-1\}, \{x-2, y-1, z+1\}, \{x-1, y, z-1\}, \{x, y+1, z-1\}, \{x, y, z\}, \{x, y-1, z\}\}$ (cyan color). (f) $N^*(p''_k) = \{\{x-2, y, z\}, \{x-2, y, z-1\}, \{x-1, y+1, z-1\}, \{x-1, y-1, z\}, \{x, y+1, z\}, \{x, y+1, z+1\}, \{x, y-1, z-1\}\}$ (cyan and green). Therefore, the set of elements $N^*(p''_k) - (N^*(p''_k) \cap N(p''_k)) = \{\{x-2, y, z\}, \{x-1, y+1, z-1\}, \{x-1, y-1, z\}, \{x, y+1, z+1\}, \{x, y-1, z-1\}\}$ added to $\mathcal{D}$ is shown in green color in (f). (Color figure online)

Let $r : \mathcal{W}'' \to \mathcal{S}$ maps all the elements belonging to $\mathcal{W}'' - \mathcal{S}$ to $\mathcal{S}$. That is, all the elements of $\mathcal{W}''$ that do not satisfy the above conditions are mapped to those elements which satisfies any of the conditions. Also, $r(w') = w', \forall w' \in \mathcal{S}$ because $\mathcal{S} \backslash w'$ is disconnected. Hence, $r \mid \mathcal{S} = Id_{\mathcal{W}''}$. Therefore, the function $r$

represents a retraction from $\mathcal{W}''$ to $\mathcal{S}$. Since the topological space $\mathcal{S}$ is obtained as a retraction from $\mathcal{W}''$, it is concluded that the 3D curve skeleton of a given triangulated object is reported in the topological space $\mathcal{S}$.

## 4   Experimental Results and Conclusion

The proposed algorithm has been implemented in C in Linux Fedora Release 13, Dual Intel Xeon Processor 2.8 GHz, 800 MHz FSB. The experimental results in

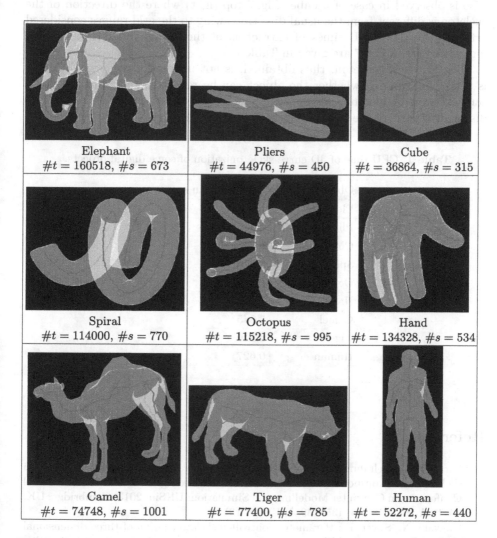

| | | |
|---|---|---|
| Elephant<br>$\#t = 160518, \#s = 673$ | Pliers<br>$\#t = 44976, \#s = 450$ | Cube<br>$\#t = 36864, \#s = 315$ |
| Spiral<br>$\#t = 114000, \#s = 770$ | Octopus<br>$\#t = 115218, \#s = 995$ | Hand<br>$\#t = 134328, \#s = 534$ |
| Camel<br>$\#t = 74748, \#s = 1001$ | Tiger<br>$\#t = 77400, \#s = 785$ | Human<br>$\#t = 52272, \#s = 440$ |

**Fig. 7.** Results for 3D curve skeletons of some digital objects. The number of voxels in the 3D isothetic inner cover ($\#t$) and that in the 3D curve skeleton ($\#s$) are given below each figure. (Color figure online)

Fig. 7 demonstrate the 3D curve skeleton of the digital objects Elephant, Pliers, Cube, Spiral, Octopus, Hand, Camel, Tiger, and Human. The number of voxels in the 3D isothetic inner cover ($\#t$) and that in the 3D curve skeleton ($\#s$) for each digital object are given below each result. The skeleton is connected, single voxel thick, and centered w.r.t. the object thereby capturing the object shape up to a high degree of accuracy. The accuracy of capturing the object shape varies with the accuracy of the 3D isothetic inner cover which may be adjusted by appropriate scaling of the object. The accuracy of the curve skeleton also varies with a variation in the grid resolution of the inner cover. A special case is observed in case of a cube (Fig. 7 top right) where the direction of the skeleton is different from the usual direction owing to the grid ranges considered orthogonally. The CPU times of extraction of the 3D curve skeleton for the digital objects in Fig. 7 are given in Table 1.

The 3D curve skeleton, thus obtained, is not unique for a given object. It is yet to be explored whether the object can be reconstructed from the resultant skeleton. A hierarchical representation of the skeleton that is suitable to distinguish parts of the object may be attempted in future.

**Table 1.** CPU time of 3D curve skeletonization of some digital objects.

| Object ($g = 2$) | CPU time (in secs.) |
|---|---|
| Elephant | 4.005 |
| Pliers | 0.965 |
| Cube | 6.045 |
| Spiral | 6.276 |
| Octopus | 2.952 |
| Hand | 1.242 |
| Camel | 5.845 |
| Tiger | 6.520 |
| Human | 0.627 |

# References

1. Farag, S., Abdelrahman, W., Creighton, D., Nahavandi, S.: Extracting 3D mesh skeletons using antipodal points locations. In: Proceedings of the 15th International Conference on Computer Modelling and Simulation: UKSim 2013, Cambridge, UK, 10–12 April 2013, pp. 135–139. IEEE (2013)
2. Gagvani, N., Silver, D.: Parameter controlled skeletonization of three dimensional objects. Department of Electrical and Computer Engineering, Rutgers University, Piscataway, NJ, Technical report CAIP-TR-216 (1997)

3. Grélard, F., Baldacci, F., Vialard, A., Domenger, J.-P.: Centerlines of tubular volumes based on orthogonal plane estimation. In: Normand, N., Guédon, J., Autrusseau, F. (eds.) DGCI 2016. LNCS, vol. 9647, pp. 427–438. Springer, Cham (2016). doi:10.1007/978-3-319-32360-2_33
4. Karmakar, N., Biswas, A., Bhowmick, P., Bhattacharya, B.B.: A combinatorial algorithm to construct 3D isothetic covers. Int. J. Comput. Math. **90**(8), 1571–1606 (2013)
5. Kerautret, B., Krähenbühl, A., Debled-Rennesson, I., Lachaud, J.-O.: 3D geometric analysis of tubular objects based on surface normal accumulation. In: Murino, V., Puppo, E. (eds.) ICIAP 2015. LNCS, vol. 9279, pp. 319–331. Springer, Cham (2015). doi:10.1007/978-3-319-23231-7_29
6. Lieutier, A.: Any open bounded subset of $\mathbb{R}^n$ has the same homotopy type as its medial axis. Comput.-Aid. Des. **36**(11), 1029–1046 (2004)
7. Martinez, J., Vigo, M., Pla-Garcia, N.: Skeleton computation of orthogonal polyhedra. Comput. Graph. Forum **30**(5), 1573–1582 (2011)
8. Palágyi, K.: A 3D 3-subiteration thinning algorithm for medial surfaces. In: Borgefors, G., Nyström, I., Baja, G.S. (eds.) DGCI 2000. LNCS, vol. 1953, pp. 406–418. Springer, Heidelberg (2000). doi:10.1007/3-540-44438-6_33
9. Palágyi, K., Kuba, A.: A hybrid thinning algorithm for 3D medical images. J. Comput. Inf. Technol. **6**(2), 149–164 (1998)
10. Saha, P.K., Borgefors, G., di Baja, G.S.: A survey on skeletonization algorithms and their applications. Pattern Recogn. Lett. **76**, 3–12 (2016)
11. Shapira, L., Shamir, A., Cohen-Or, D.: Consistent mesh partitioning and skeletonisation using the shape diameter function. Vis. Comput. **24**(4), 249 (2008)
12. Sobiecki, A., Jalba, A., Telea, A.: Comparison of curve and surface skeletonization methods for voxel shapes. Pattern Recogn. Lett. **47**, 147–156 (2014)
13. Svensson, S., Nyström, I., di Baja, G.S.: Curve skeletonization of surface-like objects in 3D images guided by voxel classification. Pattern Recogn. Lett. **23**(12), 1419–1426 (2002)
14. Tagliasacchi, A., Delame, T., Spagnuolo, M., Amenta, N., Telea, A.: 3D skeletons: a state-of-the-art report. Comput. Graph. Forum **35**(2), 573–597 (2016)
15. Tran, S., Shih, L.: Efficient 3D binary image skeletonization. In: IEEE Computational Systems Bioinformatics Conference-Workshops: CSBW 2005, Stanford, CA, 8–12 August 2005, pp. 364–372. IEEE Computer Society (2005)
16. Wang, T., Basu, A.: A note on 'A fully parallel 3D thinning algorithm and its applications'. Pattern Recogn. Lett. **28**, 501–506 (2007)
17. Wang, T., Cheng, I.: Generation of unit-width curve skeletons based on Valence Driven Spatial Median (VDSM). In: Bebis, G., Boyle, R., Parvin, B., Koracin, D., Remagnino, P., Porikli, F., Peters, J., Klosowski, J., Arns, L., Chun, Y.K., Rhyne, T.-M., Monroe, L. (eds.) ISVC 2008. LNCS, vol. 5358, pp. 1051–1060. Springer, Heidelberg (2008). doi:10.1007/978-3-540-89639-5_100
18. Wu, J., Duan, H., Zhong, Q.: A new 3D thinning algorithm extracting medial curves. In: IEEE International Conference on Intelligent Computing and Intelligent Systems: ICIS 2010, Xiamen, China, 29–31 October 2010, vol. 2, pp. 584–587. IEEE (2010)
19. Zhou, Y., Toga, A.W.: Efficient skeletonization of volumetric objects. IEEE Trans. Vis. Comput. Graph. **5**(3), 196–209 (1999)

# Inscribing Convex Polygons in Star-Shaped Objects

Nikolay M. Sirakov[1(✉)] and Nona Nikolaeva Sirakova[2]

[1] Department of Mathematics, Department of CSCI,
Texas A&M University Commerce, Commerce, TX 75428, USA
Nikolay.Sirakov@tamuc.edu
[2] Computer Science Department, University of Washington,
Seattle, WA 98105, USA
nonas@cs.washington.edu

**Abstract.** This study develops a new algorithm which automatically inscribes a convex polygon in a star shaped object $\breve{O}$. Starting at $\breve{O}$'s mass center, our active contour (E-AC) expands until it encounters the boundary of $\breve{O}$ ($\partial\breve{O}$). As a result it constructs a star-shaped polygon on $\partial\breve{O}$. We measure the Euclidean distance from $\partial\breve{O}$'s mass center to each vertex of the star-shaped polygon defined by E-AC. The distances form a distance function, whose local minima construct star-shaped polygon inscribed in $\partial\breve{O}$. Its consecutive convex triplets of vertices define a unique pair of convex polygons inscribed in $\partial\breve{O}$. The Convex Core (CC) of $\breve{O}$ is defined to be the polygon with the largest area (perimeter if the areas are equal). The CC is unique and invariant to rotation, translation and scaling. Experiments validate the new algorithm. The paper ends listing our contributions and comparing them with contemporary papers.

**Keywords:** Active contour · Distance function · Local minima · Convexity

## 1 Introduction

Automatic object decomposition into parts is a subject of interest for the computer vision community and games industry. Three major approaches are used to solve the problem. One of them is to use low level geometric properties [8,9,12], the 2nd one applies high level semantic information [2,9] and preliminary knowledge, the 3rd approach is a hybrid of the previous two [9].

Polygon decomposition methods using geometric features and mathematical concepts are described in [7,12]. Object's plausible hypotheses are used in [8] to train a continuous model to determine the likelihood that a segment is part of an object. The method in [9] integrates low level edge detection and a region growing with semantic knowledge. Such knowledge about regions and parts is also used in [2] along with a scanning window and global appearance cues.

A useful analysis on a number of shape representation methods is given by Tari in [16]. In this work the author uses the shape representation approach of

© Springer International Publishing AG 2017
V.E. Brimkov and R.P. Barneva (Eds.): IWCIA 2017, LNCS 10256, pp. 198–211, 2017.
DOI: 10.1007/978-3-319-59108-7_16

partial differential equations (PDE) and considers an object as created by a gross and peripheral (limbs-convexities, protrusions) parts. The gross part is defined as the least deformable part under a geometric (the author used "visual" [16]) transformation. The above approach is further elaborated in [17]. The authors utilized the solution of a special form of the Poisson PDE to generate "Ambrosio-Tortorelli Phase Field" [17] which provides more accurate object decomposition. To represent the objects' shape the authors designed a Randomized Hierarchy Tree. They validated the theory with a number of sophisticated object decompositions from natural images.

In the present paper we adopt the idea that objects are composed of "gross" or "core" part(s) and "peripheral" part(s), if any [16]. Since we seek the core's boundary as a convex polygon, we extend the notion of core to Convex Core (CC) following an analogy with a concept used in manifolds [4], where the CC of a hyperbolic 3-manifold is defined as the smallest convex sub-manifold.

The methods capable of inscribing convex polygons into objects have a long history [7,10–13] and are employed in object (polygons) decomposition. The latter alone has applications to database management and access, data compression, computer graphics and image processing [12]. The same source also describes a number of methods used for polygon decomposition and proves that decomposition of objects with voids is NP complete.

Recall that the convex hull (CH) of an object $\breve{O}$ is the minimal area convex polygon circumscribing $\breve{O}$ and the CH of $\breve{O}$ is unique [14]. Useful definitions and study on star-shaped sets are presented in [5]. Finding the maximum area convex polygon in a star shaped object is a subject of interest to the mathematical and computer science communities. Multiple algorithms are developed to solve the problem [7,11,12] including "finding largest potato" and/or "longest stick" [10,13]. For certain objects the solution is unique, however for others is not. Such an example is the cross (Fig. 2), where the method in [7] will detect two maximum area convex subsets. Furthermore, the problem of finding the minimum area convex polygon inscribed in another polygon is meaningless because an infinite number of such polygons are possible [1].

In this paper we develop an approach capable of inscribing a convex polygon in a given 2D star shaped object $\breve{O}$ [12]. Section 2 formulates basic notions and an expanding active contour (E-AC). In Sect. 3 we define a distance function, find its minima and use convex triplets of the minima's vertices to design the convex core (CC) of $\partial \breve{O}$. Section 4 shows that the newly defined CC is unique, invariant to rotation, scaling and translation. Section 5 validates the theory with experimental results. The paper ends deriving conclusions and comparing the new method with contemporary ones.

## 2    Background Notions

Let $\breve{O}$ denotes a closed object in $R^2$, $\partial \breve{O} = \{v_1, ..., v_n\}$ be $\breve{O}$'s boundary of $\breve{O}$. Let $r_1, ...., r_k$ be the subset of $\partial \breve{O}'s$ vertices which are concave (also called "reflex vertices" [7]). Therefore $k < n$. A concave (reflex) vertex is the vertex of an internal (for $\partial\breve{O}$) angle larger than $\pi$ radians and a convex vertex $v_i$ is a vertex of an internal angle smaller than $\pi$ (Fig. 1(d)). Consider the following four definitions:

**Definition 1.** *A point $A \in \breve{O}$ is visible from $B \in \breve{O}$ if and only if $AB \in \breve{O}$.*

**Definition 2.** *$\breve{O}$ is a convex object if and only if every pair of points in $\breve{O}$ are visible from each other.*

**Definition 3.** *$\breve{O}$ is a star shaped object if and only if there is a point $A \in \breve{O}\backslash\partial\breve{O}$ such that $Av_i \in \breve{O}$ for $i = 1, ..., n$ (all vertices $v_i$ are visible from $A$) ([7]).*

*The set of points $A$ with the above property is called kernel of $\partial\breve{O}$, and is denoted by $Kern(\partial\breve{O})$ ([7]). In [5] it is proven that the kernel is convex.*

**Definition 4.** *A chain is a set of consecutive vertices on $\partial\breve{O}$. A chain is monotone if there exists a line-l, in the plane, such that the projections of the chain vertices on $l$ follow the same order as on $\partial\breve{O}$. The polygon $\partial\breve{O}$ is monotone if there exist $l$ which divides $\partial\breve{O}$ into two monotone chains (see [12]).*

A polygon is inscribed in a circle if and only if its vertices lie on the circle. For the purpose of our study we follow this idea and introduce four new definitions.

**Definition 5.** *We say that the polygon $\Pi$ is inscribed in the polygon $\partial\breve{O}$ if and only if every $\Pi$'s vertex lies on an edge from $\partial\breve{O}$, such that $\Pi$ is entirely in $\breve{O}$.*

Let $\Pi$ be inscribed in $\partial\breve{O}$ such that $v_i \cong v$ ($v_i$ coincides with $v$), $v_i \in \partial\breve{O}$, $v \in \Pi$, then we consider that $v$ belongs to the two $\partial\breve{O}$ edges which meet in $v_i$.

**Definition 6.** *The polygon $\Pi$ is exactly inscribed in the polygon $\partial\breve{O}$ if and only if $\Pi$ is inscribed in $\partial\breve{O}$, and every $\partial\breve{O}$ edge contains exactly one $\Pi$'s vertex.*

Therefore no $\partial\breve{O}$ edge contains any edge from the exactly inscribed polygon $\Pi$ (Fig. 1). Further, it is known that in a closed polygon $\Pi$, with no self-intersections, the number of edges equals the number of vertices. Denote this number by $| \Pi |$.

**Lemma 1.** *Consider a convex polygon $\Pi$ such that $|\Pi| = n$. Then there are $\lfloor n/2 \rfloor + 1$ convex polygons, with different number of vertices, which could be exactly inscribed in $\Pi$. The polygons with the smallest number of vertices has $\lceil n/2 \rceil$, while the one with the largest number has $n$ (Fig. 1(a), (b), (c)).*

*Proof:* 1. Assume $|\Pi| = n = 2k$. Denote $\Pi$'s vertices with $v_i$, the edges with $e_i$, such that $e_i = v_i v_{i+1}$, $i = 1, ..., n$ and $v_{n+1} \cong v_1$ ($v_{n+1}$ coincides with $v_1$). Select $v_i$, $i = 1, 3, ..., 2k - 1$, denote them by $V_j \cong v_{2j-1}$ for $j = 1, .., k$. Therefore, we constructed a polygon $\Pi_k$ exactly inscribed in $\Pi$ such that $|\Pi_k| = k$, which is the smallest number of vertices to comply with the exact inscribing (Fig. 1(a)).

If we selected $V_i \in e_i$ such that $V_i \not\cong v_i$ and $V_i \not\cong v_{i+1}$ for $i = 1, ..., n$ and $v_{n+1} \cong v_1$, we receive the set of vertices $V_1, ..., V_n$ which constitute the exactly inscribed polygon $\Pi_n$ with the largest number of vertices (see Fig. 1(b)).

**Fig. 1. (a) and (b)** Hexagon and its exactly inscribed polygons with minimum and maximum number of vertices; **(c)** Pentagon and exactly inscribed polygon with minimum number of vertices; **(d)** A star shaped polygon and a non-convex exactly inscribed one; **(e)** A star shaped polygon into which no polygon could be exactly inscribed.

Consider $\Pi_k$ and remove its vertex $V_i$. We know that $V_i \cong v_{2i-1}$ and $v_{2i-1}$ is a join of $e_{2i-1}$ and $e_{2i-2}$. Select the new pints $V_{2i-1} \in e_{2i-1}$ and $V_{2i-2} \in e_{2i-2}$. Thus we have the set of points $V_1, ..., V_{i-1}, V_{2i-2}, V_{2i-1}, V_{i+1}, ..., V_k$ which constitutes an exactly inscribed polygon $\Pi_{k+1}$ such that $|\Pi_{k+1}| = k+1$.

Applying the above procedure on any point from the set $\Pi \cap \Pi_{k+1}$ we construct $\Pi_{k+2}$ such that $|\Pi_{k+2}| = k+2$. Continue the algorithm since all vertices $V_1, ..., V_k$ are exhausted we construct $\Pi_k, \Pi_{k+1}, ..., \Pi_{2k}$ exactly inscribed polygons. Follows that the total number of exactly inscribed, in $\Pi$, polygons with different number of vertices is $k+1$.

2. Assume $|\Pi| = n = 2k+1$. In this case the polygon with the smallest number of vertices is $\Pi_{k+1}$ such that $|\Pi_{k+1}| = k+1$, where $V_i \cong v_{2i-1}$ for $i = 1, ..., k$ and $V_{2k+1} \in e_{2k+1}$ (see Fig. 1(c)).

The reasoning to prove the remaining part of the odd case is analogous to the reasoning we presented above for the case with even number of vertices.◇

Lemma 1 implies that every star shaped convex polygon has multiple exactly inscribed convex polygons, with a different number of vertices. On the other hand, some star shaped polygons, with reflex points, may not have a single exactly inscribed polygon (Fig. 1(d)).

**Definition 7.** *The polygon $\Pi$ is partially on the polygon $\partial \breve{O}$ if and only if $\Pi$ is inscribed in $\partial \breve{O}$, and at least one of $\Pi$'s edges lies on a $\partial \breve{O}$ edge. The polygon $\Pi$ is entirely on the polygon $\partial \breve{O}$ if and only if each of $\Pi$'s edges lies on an edge from $\partial \breve{O}$ and $\Pi \in \breve{O}$.*

*The above definition implies that if $\Pi$ is entirely on $\partial \breve{O}$ then $\Pi \cong \partial \breve{O}$.*

**Lemma 2.** *Every star shaped polygon is monotone.*

*Proof:* Consider a star shaped object $\breve{O}$, with a boundary $\partial \breve{O}$, whose vertices are $v_i, i = 1, ..., n$. Follows from Definition 4 that we have to find a line $l$ which divides $\partial \breve{O}$ into two monotone chains. Consider a point $A \in Kern(\partial \breve{O})$. Build the lines $Av_i$ for $i = 1, ..., n$ and select a line $l \not\cong Av_i$ ($l$ does not coincide with $Av_i$) for $i = 1, ..., n$ such that $l$ intersects, inside $\breve{O}$, the maximum number of consecutive lines $Av_i$, for $i = u, ..., j$ (see Fig. 2(c)). Consider the $Av_i$ intersections with $l$ as $v_i$ projections on $l$ and denote the projections with $v_i'$ for $i = u, ..., j$.

Assume that the pair $v_i v_{i+1}$ is mapped on $l$ to a pair $v_{i+1}' v_i'$. This implies that $Av_i \cap Av_{i+1} = U$ such that $U \not\cong A$. Follows that $Av_i \cong Av_{i+1}$, which is a contradiction because the vertices $v_i$ and $v_{i+1}$ does not coincide $v_i \not\cong v_{i+1}$.

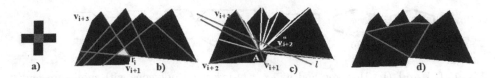

**Fig. 2.** **(a)** A cross and its kernel- the inner square; **(b)** An object $\breve{O}_{12}$ with $\partial\breve{O}_{12}$ and its $Kern(\partial\breve{O}_{12})$-the light triangle with vertex $r_i$; **(c)** $A \in Kern(\partial\breve{O}_{12})$, the lines $Av_i$, and the line $l$ splitting $\partial\breve{O}_{12}$ to two chains. **(d)** Inscribed convex polygon.

Follows that the projections $v_i'$ keep the same order on $l$ as $v_i$ do on $\partial\breve{O}$. Therefore the vertices $v_i$, for $i = u, ..., j$ are monotone and represent a chain.

The projections for the remaining consecutive vertices $v_i$, where $i = j + 1, ..., u - 1$, are defined by the intersection of $l$ with the $Av_i$ reflections about $A$ (see Fig. 2(c)). Denote these intersections with $v_i''$ for $i = j + 1, ..., u - 1$. The same reasoning as above asserts that the projections $v_i''$ keep on $l$ the same order as $v_i$ on $\partial\breve{O}$. Therefore the set of vertices $v_i$, for $i = j + 1, ..., u - 1$ is monotone and represents a chain.◊

Define an expanding active contour (E-AC) with evolution Eq. 1:

$$\mathbf{r}(q, t(u)) = e^{aq - 4a^2 t}[x(q), y(q)], \tag{1}$$

$$\mathbf{r}(q, t(0)) = R.e^{(0.1q - 10)}[x(q), y(q)]. \tag{2}$$

In Eq. 1, $x(q) = C_1 \cos(caq)$, $y(q) = C_2 \sin(caq)$, $t = (t_0 + u\partial t)$ is the time parameter, $q$ is a space parameter, $a = 0.5|\frac{\partial r(q,t)}{\partial q}|$, and $C_1, C_2, c$ are coefficients. From practice, we determined that Eq. 1 describes a circle with radius $R = C_1 = C_2$ if $c = 1000$, $a^2 t = 0.001$ and $q \in [0, \frac{2\pi}{ac}]$. The equation describes a point (pixel) if $c = 1000$ and $a^2 t = 2.5$. Follows that the curve $r(q, t)$ will evolve from a point (pixel) to a circle as $t$ decreases from the upper to the lower bound of $2.5/a^2 \geq t \geq 0.001/a^2$. Thus, to make Eq. 1 defines a point (pixel) we substitute $C_1 = C_2 = R, a = 0.1, c = 1000, t_0 = 250, u = 0$ and receive Eq. 2 where $x(q) = \cos(10^2 q)$, $y(q) = \sin(10^2 q)$.

Denote the image function by $f(x, y)$. The following boundary condition (BC) halts the AC vertices on objects boundaries:

$$\mathbf{r}(q, t(u)) = r(q, t(u) - \partial t) \text{ if } \varepsilon_2 > \frac{\partial f(r(q,t))}{\partial t} > \varepsilon_1, \tag{3}$$

where $q \in [0, \frac{2\pi}{ac}]$ and $2.5/a^2 \geq t(u) = t_0 + u\partial t \geq 0.001/a^2$ for $u = -1, -2, .....$ But, if the double inequality in Eq. 3 fails, then:

$$\mathbf{r}(q, t) \neq \mathbf{r}(q, t - \partial t). \tag{4}$$

The initial conditions (IC) defined with Eq. 2, evolution Eq. (1), BC (3) and $u \to -\infty$, define an enlarging parametric AC (E-AC), which represents Euclidean growth from a point $r(q, 250) \in \breve{O}$ toward $\partial\breve{O}$. Figure 3(a) depicts an image of a neutrophil [6] whose boundary is star shaped and was extracted by E-AC in $0.036\,s$ (Fig. 3(b)).

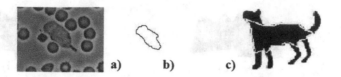

**Fig. 3.** (a) Neotrophil's video [6] 4th frame; (b) Neutrophil's boundary extracted by E-AC; (c) Cat's (original image from [3]) star shaped boundary visible from the mass center of the entire boundary. The star shaped boundary is extracted in 0.016 s.

## 3    Active Convex Core Model

Let $\breve{O}$ be a star shaped object whose boundary $\partial\breve{O}$ is a polygon with vertices $v_1, ..., v_n$ of which $r_1, ..., r_k$ are concave. Apply E-AC with IC 2 using the initial point $\aleph_0 \cong \mathbf{r}(q, t(0)) \in \breve{O}\backslash\partial\breve{O}$. At $u = -1$, $\aleph_0$ "springs" to $\mathbf{r}(q_i, t(-1))$, $q_i \in [0, \frac{2\pi}{ac}]$, $i = 1, ..., e$ vertices, which draw a closed curve. If $u = -2, -3, ...., -\infty$ then Eq. (1) evolves the curve by moving each of its vertices, on a straight line, until BC (3) halts the vertex on $\partial\breve{O}$. Define a function to measure the Euclidean distance traveled by every E-AC vertex $Q_i$ for $u = 0, -1, ..., -u_i < 0$:

$$d^E(Q_1(q_1, t(0)), Q_i(q_i, t(u_i))) = |\mathbf{r}(q, 250 - u_i\partial t)) - \mathbf{r}(q, 250)|. \qquad (5)$$

Denote the E-AC vertices, at the time they halt on $\partial\breve{O}$, by $V_i$ such that $\mathbf{r}(q_i, t(u_i)) = V_i$ for $i = 1, ..., e$ and the mass center of $\partial\breve{O}$ by $m(\partial\breve{O}) = \sum_{i=1}^n \frac{v_i}{n}$.

Assume $m(\partial\breve{O}) \in \breve{O}$, $\aleph_0 \cong m(\partial\breve{O}) \notin Kern(\partial\breve{O})$ and $\aleph_0 \in \breve{O}\backslash\partial\breve{O}$ (Fig. 2(b)). It follows from the E-AC definition that the lines $\aleph_0 V_i \in \breve{O}$ for $i = 1, ..., e$. Therefore, $\partial\breve{O}(\aleph_0) = \{V_1, ..., V_e\}$ is a star shaped polygon, partially on $\partial\breve{O}$ (Fig. 4(c), (d)), with $\aleph_0 \in Kern(\partial\breve{O}(\aleph_0))$ and mass center:

$$m(\partial\breve{O}(\aleph_0)) = \sum_{i=1}^e \frac{V_i}{e} \cong \aleph_1. \qquad (6)$$

If $\aleph_0 \cong m(\partial\breve{O}) \in Kern(\partial\breve{O})$ (Fig. 2(a)), then there exists an E-AC with $e > n$ vertices such that at time $u^*$ a subset of these vertices coincide with the $\partial\breve{O}$ vertices: $V_{i_1} \cong v_1, ..., V_{i_n} \cong v_n$. Follows that the star shaped polygon $\partial\breve{O}(\aleph_0) \cong \{V_1, ..., V_e\}$ is entirely on $\partial\breve{O}$, $\partial\breve{O}(\aleph_0) \cong \partial\breve{O}$ (Fig. 4(a), (b)) and $\aleph_0 \cong \aleph_1$. Thus, we choose $\aleph_0 \cong m(\partial\breve{O})$ as E-AC initial point and measure the distances $d^E(\aleph_0, V_i)$, $i = 1, ..., e$. Then we calculate their local minima, employing the statements:

(i) the Euclidean distances $|\aleph_0 r_i|$ to the concave vertices $r_i, i = 1, ..., k$ represent local minima;

(ii) if a $\partial\breve{O}(\aleph_0)$'s edge belongs to a convexity, then the local minimum, of the distances to the edge, equals the length of the perpendicular from $\aleph_0$ to the line defined by the edge. For this case we consider two subcases:

– if the end of the perpendicular lies on the edge, then the local minimum is the length of the perpendicular;

**Fig. 4.** $\breve{O}_{12}$ from Fig. 2. **(a)** The star shaped polygon $\partial\breve{O}_{12}(\aleph_0)$ produced by E-AC with initial point $\aleph_0 \cong A \in Kern(\partial\breve{O}_{12})$; **(b)** $\partial\breve{O}_{12}(\aleph_0)$ is entirely on $\partial\breve{O}$; **(c)** $\partial\breve{O}_{12}(\aleph_0)$ produced by E-AC with $\aleph_0 \cong A \notin Kern(\partial\breve{O}_{12})$; **(d)** $\partial\breve{O}_{12}(\aleph_0)$ is partially on $\partial\breve{O}$.

– if the end of the perpendicular does not lie on the edge, the algorithm considers that there is no minimum distance to the edge Fig. 5(b)).

**Theorem 1.** *Given a convex object $\breve{O}$. The end points of the local minima of the distances from $\aleph_0$ to $\partial\breve{O}(\aleph_0)$) define a convex polygon inscribed in $\partial\breve{O}$.*

*Proof:* Since $\breve{O}$ is convex follows that $\aleph_0 \in Kern(\breve{O})$ and $\partial\breve{O}(\aleph_0) \cong \partial\breve{O}$. Lemma 1 asserts that, in a convex polygon with $n$ edges one may inscribe a convex polygon with maximum $n$ vertices. Since $\breve{O}$ is convex, follows that the polygon $\partial\breve{O}$ has no concave (reflex) points. Therefore the local minima of the distances are represented only by the perpendiculars from $\aleph_0$ to $\partial\breve{O}$. If the end of every perpendicular to the line of every edge is inner to the edge, then the polygon, defined by the end points of the local minima (perpendiculars), is exactly inscribed in $\partial\breve{O}$. If the end of at least one local minimum (perpendicular) is outer for the edge then the polygon defined by the end points of the perpendiculars is inscribed in $\partial\breve{O}$.

We prove now that the minimum number of perpendiculars from $\aleph_0$ to the boundary $\partial\breve{O}$ is 2. Assume that $\partial\breve{O}$ is a triangle (the closed polygon with, non zero area, and the smallest number of edges). Recall that $\aleph_0$ is inner for $\partial\breve{O}$. Connect $\aleph_0$ with the $\partial\breve{O}$ vertices. $\aleph_0$ is a vertex common for three angles. At least two of them are greater than $\pi/2$, which implies that the ends of at least two perpendiculars, from $\aleph_0$, lie on $\partial\breve{O}$ edges. Follows that any polygon whose number of edges is larger than 3 will have at least two minima. $\diamond$

Theorem 1 constructs exactly inscribed convex polygon in a convex object $\breve{O}$. But if $\breve{O}$ is star shaped then the end points of the local minima of the distances $\aleph_0 V_{j_i}$ define star shaped polygon $\{V_{j_1}, ..., V_{j_w}\} \cong \Gamma(\partial\breve{O}(\aleph_0))$ inscribed in $\partial\breve{O}$. Traverse $\Gamma(\partial\breve{O}(\aleph_0))$ starting with $V_{j_i}$ and generate the consecutive triplets $V_{j_{i-1}} V_{j_i} V_{j_{i+1}}$ for $i = 1, ..., w$, such that $V_{j_0} \cong V_{j_w}$ and $V_{j_{w+1}} \cong V_{j_1}$. Therefore, a set of $w$ distinct triplets of consecutive vertices exist on $\Gamma(\partial\breve{O}(\aleph_0))$. The set is invariant according to the starting point, and we check the convexity of its triplets:

$$(x_{ji} - x_{j(i-1)})(y_{j(i+1)} - y_{ji}) < (x_{j(i+1)} - x_{ji})(y_{ji} - y_{j(i-1)}), \qquad (7)$$

If a triplet satisfies Eq. (7) then it is convex and the algorithm keeps the three points, otherwise the algorithm discards (deletes):

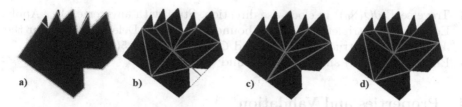

**Fig. 5.** (a) $\breve{O}_{15}$ star shaped and the E-AC generated star shaped polygon $\partial\breve{O}_{15}(\aleph_0)$ for $\aleph_0 \cong m(\partial\breve{O}_{15})$; **(b)** The star shaped polygon $\Gamma(\partial\breve{O}_{15}(\aleph_0))$; **(c)** The convex polygon $C_1(O_{15}(\aleph_0))$ extracted from $\Gamma(\partial\breve{O}_{15}(\aleph_0))$ using Eq. (7) and deleting the 1st point from every triplet of $\Gamma(\partial\breve{O}_{15}(\aleph_0))$ vertices; **(d)** The convex polygon $C_3(\breve{O}_{15}(\aleph_0))$ extracted from $\Gamma(\partial\breve{O}_{15}(\aleph_0))$ using Eq. (7) and deleting the 3rd point.

\* the first point $V_{j_{i-1}}$, in clockwise direction, and keeps the remaining two. When the set of triplets is exhausted the remaining vertices form a convex polygon $C_1(\breve{O}(\aleph_0))$ inscribed in $\partial\breve{O}$ as shown in Fig. 5(c);

\*\* the third point $V_{j_{i+1}}$, in clockwise direction, and keeps the remaining two. When the set of triplets is exhausted the remaining vertices form a convex polygon $C_3(\breve{O}(\aleph_0))$ inscribed in $\partial\breve{O}$ as shown in Fig. 5(d).

Deleting the middle point may produce an edge a part of which does not belong to $\breve{O}$. Therefore the generated polygon will be convex but will neither be inscribed in, nor will be on $\partial\breve{O}$ (partially or entirely). Therefore such a case is not in consideration.

Follows that the new algorithm inscribes, in $\partial\breve{O}$, two convex polygons: $C_1(\breve{O}(\aleph_0))$, $C_3(\breve{O}(\aleph_0))$, and calculates their areas $A_{C_1}$, $A_{C_3}$ using [15]:

$$A = \sum_{i=1}^{p+1}(x_{ji}y_{j(i+1)} - x_{j(i+1)}y_{ji}), \qquad (8)$$

where p denotes the number of polygon's vertices, and $(p+1)_{MODp} = 1$.

**Definition 8.** *Consider star shaped object $\breve{O}$ and the inscribed-convex polygons $C_1(\breve{O}(\aleph_0))$ and $C_3(\breve{O}(\aleph_0))$. The one with largest area we call Convex Core (CC) of $\breve{O}$, $CC(\breve{O})$. If $A_{C_1} = A_{C_3}$ the one with largest perimeter is considered as CC.*

The derivations made so far could be summarized in the following parametric active CC algorithm (PACCA) for star shaped object $\breve{O}$:

1. Run E-AC with initial point $\aleph_0 \cong m(\partial\breve{O})$ and find $\partial\breve{O}(\aleph_0)$ entirely on $\partial\breve{O}$ if $\aleph_0 \in Kern(\partial\breve{O})$ (Fig. 4(b)), and partially on $\partial\breve{O}$ otherwise (Figs. 5(a) and 4(d));
2. Calculate the Euclidean distances from $\aleph_0$ to the $\partial\breve{O}(\aleph_0)$ vertices;
3. Find the local minima of the distances. Use their end points to construct the polygon $\Gamma(\partial\breve{O}(\aleph_0)) \cong \{V_{j_1}, ..., V_{j_w}\}$ (Fig. 5(b));

4. Traverse $\Gamma(\partial \breve{O}(\aleph_0))$ in clockwise direction. Start with any vertex $V_{j_i}$. Apply Eq. 7 on every triplet consecutive boundary vertices. Deleting the 1st or the 3rd vertices construct $C_1(\breve{O}(\aleph_0))$ and $C_3(\breve{O}(\aleph_0))$ (Figs. 5(c), (d));
5. Employing Eq. (8) and Definition 8 find $CC(\breve{O})$.

## 4    Properties and Validation

Hereafter we prove the existence and uniqueness of the CC of a star shaped object $\breve{O}$. Recall that the CC is a closed convex polygon. In this study we consider that the closed convex polygon with the smallest number of vertices and zero area is the straight segment, which may represent CC as shown in Fig. 6(b)).

**Fig. 6.** (a) A pentagon $O_5$ and the E-AC generated polygon $\partial O_5(\aleph_0)$, which is entirely on $O_5$ for $\aleph_0 \cong m(\partial O_5)$; (b) The CC for $\partial O_5$ is the straight segment.

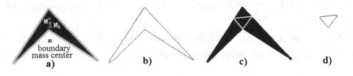

**Fig. 7.** (a) $O_M$ and the star shaped polygon $\partial O_M(\aleph_0)$ produced by E-AC with $\aleph_0 \in Kern(\partial O_M)$; (b) $\partial O_M(\aleph_0)$ alone. (c) $O_M$ along with its CC; (d) $CC(O_M)$ alone.

Denote the mass center of the $\partial \breve{O}$ vertices with $m(\partial \breve{O}) \cong \aleph_0$, and assume $\aleph_0 \in Kern(\breve{O})$. The Euclidean growth driven by E-AC starting at $\aleph_0$ exists and is unique. Follows that the star shaped polygon $\{V_1, ..., V_e\} \cong \partial \breve{O}(\aleph_0)$ exists, is unique and is entirely on $\partial \breve{O}$. Therefore $\partial \breve{O}(\aleph_0) \cong \partial \breve{O}$ (Figs. 4(b) and 6(a)).

If $\aleph_0 \in \breve{O} \backslash \partial \breve{O}$ but $\aleph_0 \notin Kern(\breve{O})$ the star shaped polygon $\partial \breve{O}(\aleph_0)$ exists, is unique and is partially on $\partial \breve{O}$ (Figs. 5(a) and 4(d)).

From the two statements above follows that if $\aleph_0 \in \breve{O} \backslash \partial \breve{O}$ the set of distances $|\aleph_0 V_1|, ...., |\aleph_0 V_e|$ exists, is unique and belongs to $\breve{O}$. Therefore the set of local minima of the distances exists and is unique. This implies that the star-shaped polygon $\Gamma(\partial \breve{O}(\aleph_0))$, inscribed in $\partial \breve{O}$, is unique.

Note, there are star shaped objects such that $\aleph_0 \notin \breve{O} \backslash \partial \breve{O}$ (Fig. 7(a)). Therefore the set of distances does not belong to $\breve{O} \backslash \partial \breve{O}$ either. In this case we select $\aleph_0 \in Kern(\breve{O})$ and $\aleph_0 \ncong m(\partial \breve{O})$. Applying E-AC with initial point $\aleph_0$ we find $\partial \breve{O}(\aleph_0)$ which exists, is unique and entirely on $\partial \breve{O}$. Using Eq. 6 we calculate $m(\partial \breve{O}(\aleph_0)) \cong \aleph_1$, which is the mass center of the E-AC vertices $\{V_i, ..., V_e\} \in \partial \breve{O}$.

Follows that $\aleph_1 \in \breve{O} \backslash \partial \breve{O}$, which implies that $|\aleph_1 V_i| \in \breve{O}$ for $i = 1, ..., e$. Thus we calculate the set of distances $|\aleph_1 V_1|, ...., |\aleph_1 V_e|$ and determine their minima in a unique way. Therefore the star-shaped polygon $\Gamma(\partial \breve{O}(\aleph_1))$ is unique.

Theorem 1 asserts that if $\partial \breve{O}$ is non-zero area convex polygon its distance function attains at least 2 local minima. Assume a star shaped $\partial \breve{O}$ having at least one concave vertex $r_1$. Follows that $|\aleph_0 r_1|$ is a local minimum. Recall, if $\aleph_0 \in \breve{O} \backslash \partial \breve{O}$ the distance function is calculated using $\aleph_0$, and $\aleph_1 \in \breve{O} \backslash \partial \breve{O}$ is utilized if $\aleph_0 \notin O \backslash \partial \breve{O}$. This is known that if a point is inner for a polygon, there is at least one perpendicular from the point to the polygon's edge such that the end of a the perpendicular lies on a edge. Follows that there are at least two minima ($w \geq 2$). Therefore the star shaped polygon $\{V_{j_1}, ..., V_{j_w}\} \cong \Gamma(\partial \breve{O}(\aleph))$ exists for both: convex and star shaped objects.

If $w = 2$ then $\Gamma(\partial \breve{O}(\aleph_0)) \cong C_1(\breve{O}(\aleph_0)) \cong C_3(\breve{O}(\aleph_0))$ and the CC is a straight segment (Fig. 6(b)). If $w = 3$ then $\Gamma(\partial \breve{O}(\aleph_0))$ is a triangle which yields $C_1(\breve{O}(\aleph_0)) \cong C_3(\breve{O}(\aleph_0))$ and the CC is a triangle (Fig. 7).

Recall, the convex polygons $C_1(O(\aleph_0))$, $C_3(O(\aleph_0))$ are generated traversing $\Gamma(\partial O(\aleph_0))$'s vertices in clockwise direction and using the convex triplets found by Eq. (7) deleting the 1st or the 3rd point in non convex triplets. We proved that $\Gamma(\partial \breve{O}(\aleph_0))$ exists and is unique. Therefore the set of consecutive triplets of its boundary vertices exists and is unique as well. Follows that $C_1(\breve{O}(\aleph_0))$, $C_3(\breve{O}(\aleph_0))$ are invariant to the starting point of $\Gamma(\partial \breve{O}(\aleph_0))$ traversal, and according to rotation in clockwise direction (Fig. 8). $C_1(O(\aleph_0))$, $C_3(O(\aleph_0))$ are invariant to traversal and rotation in counterclockwise direction as well. The statement holds because deleting the 1st (the 3rd) point of a triplet in a counterclockwise direction of traversing a boundary is the same as deleting the 3rd (1st) point in clockwise direction.

Recall that, to find the inscribed convex polygons $C_1(\breve{O}(\aleph_0))$, $C_3(\breve{O}(\aleph_0))$ we use $\Gamma(\partial \breve{O}(\aleph_0))$ if $\aleph_0 \cong m(\partial \breve{O}) \in \breve{O} \backslash \partial \breve{O}$ and $\Gamma(\partial \breve{O}(\aleph_1))$ is used if $m(\partial \breve{O}) \notin O \backslash \partial \breve{O}$. Since $\aleph_0$ and $\aleph_1$ are inner to $\partial \breve{O}$ follows that the shapes of the inscribed polygons are invariant to translation and scaling of $\breve{O}$ (Fig. 10).

So far we proved that the pair of convex polygons $C_1(\breve{O}(\aleph_0))$, $C_3(\breve{O}(\aleph_0))$, inscribed in a star-shaped object $\breve{O}$, exists is unique and invariant according to

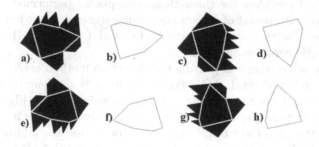

**Fig. 8.** The object $\breve{O}_{15}$ from Fig. 5, with shortened second convexity; **(a)** The object $\breve{O}_{15}$ with its CC; **(b)** the CC alone; **(c)–(h)** rotations by $-\pi/2, -\pi, -3\pi/2$.

**Fig. 9.** An object $\breve{O}_{14-4}$ with 14 vertices and 4 reflex points, symmetric according to the vertical line connecting two of them; **(a)** $\breve{O}_{14-4}$ along with $\partial\breve{O}_{14-4}(\aleph_0)$ which is entirely on $\partial\breve{O}_{14-4}$,; **(b)** $\partial\breve{O}_{14-4}(\aleph_0)$ alone; **(c)** $\breve{O}_{14-4}$ along with $C_1(\breve{O}_{14-4}(\aleph_0))$; **(d)** $C_1(\breve{O}_{14-4}(\aleph_0))$ alone; **(e)** $\breve{O}_{14-4}$ along with $C_3(\breve{O}_{14-4}(\aleph_0))$; **(f)** $C_3(\breve{O}_{14-4}(\aleph_0))$ alone.

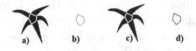

**Fig. 10. (a)** The star and its CC in an image of $1024 \times 1024$; **(b)** the CCalone; **(c)** The star and its CC in an image of $128 \times 128$; **(d)** the CC alone. Original image from [3].

rotation, translation and scaling the object $\breve{O}$. Among the two polygons we call CC of $\breve{O}$ the one with largest area. Note, there are star-shaped objects for which $C_1(\breve{O}(\aleph_0)) = C_3(\breve{O}(\aleph_0))$ but do not coincide (Fig. 9). In this case we consider that the CC is represented by a single polygon, located in two different parts of the object.

**Lemma 3.** *If $\partial\breve{O}$ is convex then $C_1(\breve{O}(\aleph_0)) \cong C_3(\breve{O}(\aleph_0))$.*

In Lemma 3 the two polygons coincide because $\Gamma(\partial\breve{O}(\aleph_0))$ is convex. Therefore, all consecutive triplets of its vertices are convex and there are no points to delete.

If $C_1(\breve{O}(\aleph_0)) \neq C_3(\breve{O}(\aleph_0))$ but $A_{C_1} = A_{C_2}$ we select as a CC the polygon with the greatest perimeter. Follows that the CC is unique.

## 5    Experimental Results

We codded in Java the Euclidean Growth E-AC model along with the CC detection algorithm. To validate the theoretical concepts we performed a number of experiments on a diverse set of images containing star-shaped objects of varying sizes and shapes. For this purpose we used a PC with CPU 2.40 GHz, 4GB RAM such that a single core was engaged.

The computation complexity of the CC detection method is $O(m * e)$, where $e$ shows the number of the E-AC vertices, while $m$ is the length, in pixels, of the largest side of the box enveloping the star-shaped object $\breve{O}$. Figure 11 shows experimental results in finding the CC of six star shaped objects in images with varying sizes. Studying these results, one may notice that the additional branch of the object in Fig. 11(j) compared to the object in (g) added a horizontal straight segment to the CC (Figs. 11(i), (l)). On the other hand the run time for the two experiments is same: 0.76 s.

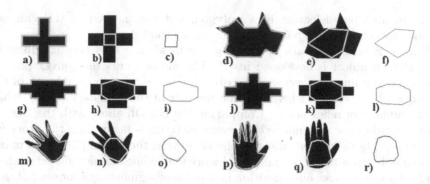

**Fig. 11.** The images in columns 1 and 4 contain six objects $\breve{O}$ along with the polygons $\partial\breve{O}(\aleph_0)$; Columns 2 and 5 show the objects $\breve{O}$ along with their CCs; Columns 3 and 6 contain the CCs alone. The images in (m) and (p) come from Tari's collection [3].

Although E-AC is capable of expanding the active contour through color images we refrain from presenting experiments with color objects because the scope of the present study is to define the geometric structure CC. Also, input must be pre-processed to remove noise, since certain types of noise can affect the shape of the polygon $\Gamma(\partial\breve{O}(\aleph_0))$ and consequently the shape of CC as well.

# 6   Conclusions

The main contribution of the paper is the development of a new algorithm for inscribing a convex polygon (named convex core (CC)) in a star shaped object. We proved that the CC of a star shaped object exists, is unique, invariant to rotation, translation and scaling. These properties show that the CC segments, in a unique way, every star-shaped object to convex parts.

The algorithm is implemented in Java along with an expanding active contour. The software was validated on a large amount of star-shaped objects including objects from a collection used by Tari [3].

An other contribution of this paper is the definition of the following new notions: *(a)* inscribed polygon in another polygon; *(b)* exactly inscribed polygon in another polygon; *(c)* a polygon partially on another polygon; *(d)* a polygon entirely on another polygon. We prove also that every star-shaped polygon is monotone.

The importance of the problem for automatic objects segmentation provoked interest among the computer vision and mathematical societies. This interest led to the development of a number of methods and algorithms capable of automatically defining maximum area convex polygons (objects) inside another polygon (object) "peeled potato," "longest stick" [10]. Compared to our new method, in accordance with the new definitions introduced in this paper, the polygons defined in [10] are partially on the original polygon, while the CC is an inscribed polygon. The method in [11] may define inscribed and even exactly inscribed maximum area parallelogram in a polygon, but the method in [11] can not inscribe any other convex polygon, while our method does.

A more general approach for finding the maximum area convex subset of a star shaped object is developed in [7]. The method cuts the convex protrusions and determines the kernel of the object. The calculation complexity of the method is $O(n + k \log k)$, where $n$ is the number of the boundary points, while $k$ is the number of reflex points. Comparing the Big Oh above with the Big Oh of our algorithm one may notice that every methods is faster than the other for certain cases. Both methods differ in the sense that the one in [7] defines maximal area polygon which is partially on the original, while our method determines inscribed polygon. Also, our algorithm is capable of segmenting a convex polygon while the one in [7] can't. Moreover, for a cross as the one in Fig. 11 our method determines a single inscribed polygon, while the method in [7] determines two - every branch is considered as a maximal subset.

A sophisticated and conceptually different method for object decomposition is given in [3,16,17]. The method applies Ambrosio-Tortorelli field generated by the solution of the Poisson PDE, and separates the object to different regions. An advantage of the method in [3,16,17] is that it segments any object with concavities, while our algorithm segments only star-shaped objects. But an advantage of our method is that it is capable of segmenting convex objects while the other one is not.

Our study continues with extending E-AC capabilities and making this tool resilient to noise. Also, we are investigating the opportunity to extend the present method and develop one capable of inscribing convex polygons in non star shaped objects using any point in $\breve{O}\backslash(\partial\breve{O})$ as initial point for the active contour.

**Acknowledgement.** Thanks to the anonymous reviewers whose notes helped us improve the paper. In loving memory of our parents and grandparents, Mariika and Metody Sirakov, for their support throughout our lives and professional development.

# References

1. Aggarwal, A., Booth, H., O'Rourke, J., Suri, S.: Finding minimal convex nested polygons. J. Inf. Comput. **83**(1), 98–110 (1989)
2. Arbelaez, P., Hariharan, P., Gu, C., Gupta, S., Bourdev, L., Malik, J.: Semantic segmentation using regions and parts. In: Proceedings of IEEE CVPR 2012, Rhode Island, 16–21 June (2012). doi:10.1109/CVPR.2012.6248077
3. Aslan, C., Tari, M.: An axis based representation for recognition. In: ICCV, pp. 1339–1346 (2005)
4. Bridgeman, M., Canary, R.D.: From the boundary of the convex core to the conformal boundary. J. Geometrica Dedicata **96**(1), 211–240 (2003)
5. Brimkov, V.E., Barneva, R.P.: Digital stars and visibility of digital objects. In: Barneva, R.P., Brimkov, V.E., Hauptman, H.A., Natal Jorge, R.M., Tavares, J.M.R.S. (eds.) CompIMAGE 2010. LNCS, vol. 6026, pp. 11–23. Springer, Heidelberg (2010). doi:10.1007/978-3-642-12712-0_2
6. Crawling neutrophil chasing a bacterium. http://www.youtube.com/watch?v=I_xh-bkiv_c. Accessed 01 Dec 2015

7. Coeurjolly, D., Chassery, J.-M.: Fast approximation of the maximum area convex subset for star-shaped polygons. In: RR-LIRIS-2004-006, CNRS (2004). http://liris.cnrs.fr/Documents/Liris-1909.pdf
8. Carreira, J., Sminchisescu, C.: CPMC: automatic object segmentation using constrained parametric min-cuts. IEEE Trans. PAMI **34**(7), 1312–1328 (2012)
9. Fan, J., Yao, D.K., Elmagarmid, A.K., Aref, W.G.: Automatic image segmentation by integrating color-edge extraction and seeded region growing. IEEE TIP **10**(10), 1454–1466 (2001)
10. Hall-Holt, O., Katz, M.J., Kumar, P., Mitchell, J.S., Sityon, B.A.: Finding large stick and potato in polygons. In: Proceedings of the 17th SIAM SODA 2006, pp. 474–483 (2006). doi:10.1145/1109557.1109610
11. Jin, K., Matulef, R.D.: Finding the maximum area paralelogram in a convex polygon. In: CCCG2011, Toronto, 10–12 August 2011 (2011). cccg.ca/PDFschedule/papers/paper3.pdf
12. Keil, J.M.: Polygon Decomposition Survey in Handbook of Computational Geometry. Elsevier Sciences Publishing, Amsterdam (2000). Sack, J.R., Urrutia, J. (eds.) ISBN: 9780444825377
13. Lien, J.-M., Amato, N.M.: Approximate convex decomposition of poligons. Comput. Geom. **35**(2006), 100–123 (2006)
14. Sirakov, N.M.: A new active convex hull model for image regions. J. Math. Imaging Vis. **26**(3), 309–325 (2006)
15. Sirakov, N.M.: Monotonic vector forces and Green's theorem for automatic area calculation. In: Proceedings of IEEE ICIP2007, San Antonio, September, vol. IV, pp. 297–300 (2007)
16. Tari, S.: Extracting parts of 2D shapes using local and global interactions simultaneously. In: Chen, C.H. (ed.) Handbook of Pattern Recognition and Computer Vision, 4th edn., pp. 283–303 (2009)
17. Tari, S., Genctav, M.: From a non-local Ambrosio-Tortorelli phase field to a randomized part hierarchy tree. JMIV **49**(1), 69–86 (2014)

# On Characterization and Decomposition of Isothetic Distance Functions for 2-Manifolds

Piyush K. Bhunre, Partha Bhowmick$^{(\boxtimes)}$, and Jayanta Mukhopadhyay

Department of Computer Science and Engineering,
Indian Institute of Technology, Kharagpur, India
kbpiyush@gmail.com, bhowmick@gmail.com, jay@cse.iitkgp.ernet.in

**Abstract.** We introduce in this paper certain interesting characterization of isothetic distance functions in the 3D space. The characterization done by us eventually leads to decomposition of an isothetic distance function for higher-order simplices to that of lower-order ones, which subsequently helps in efficient computation. We show how inter-simplex isothetic distance is a natural choice for determining an appropriate voxel size during the voxelization of a 2-manifold surface, such as the most-commonly used triangle mesh. Preliminary test result have been furnished to demonstrate its merit and aptness.

**Keywords:** Digital geometry · Digital line · Digital triangle · Discretization · Isothetic distance · Manifolds

## 1 Introduction

Discretization of a geometric object in the real space to a set of isotropic pixels or voxels is a well-studied problem in the subject of digital geometry. Different distance metrics are used for this, each having its own merits and issues while being used from one domain to another or from one application to another. Out of the most commonly used metrics, 'Euclidean distance' is a natural choice for some, owing to its easy comprehensibility and implementability both in the real and in the discrete spaces. However, it is not readily commensurable with the process of pixelization or voxelization of an object when it is subject to certain topological conditions, as we show in this paper. In fact, it is better replaced by 'isothetic distance', which is the focus of this study.

### 1.1 Motivation

The motivation of our study mainly springs from the recent upsurge in voxelization and various applications involving voxel sets. A significant volume of work on voxelization and related applications have been reported in recent time. The first group of work is on designing efficient voxelization algorithms for 2-manifold surfaces such as triangle mesh. Although some work had initiated on this in 1990s [6,8,13,15,25], its importance has shot up quite lately with the current explosion

© Springer International Publishing AG 2017
V.E. Brimkov and R.P. Barneva (Eds.): IWCIA 2017, LNCS 10256, pp. 212–225, 2017.
DOI: 10.1007/978-3-319-59108-7_17

in computational field and various voxel-based applications. Some recent work related to voxelization algorithms can be seen in [12,19,21,24,26,33,34].

On the application side, a multitude of work have come up over the last few years, which essentially indicates the advantage of voxelized representation of a surface for various purposes. These include reconstruction of surfaces from voxelized data [23,37]; 3D printing using voxel set as input [5,7,9,30,31,36]; ray casting using voxel octrees [17,20]; shadow generation based on voxelized geometry [14,27,32]; texture creation on voxelized surfaces [11,35]; animation with well-formed voxel sets [10]; and physical simulation like fluid flow where particles are modeled as voxels [22].

## 1.2  Our Contribution

Computation of isothetic distance between two simplices can be considered as an optimization problem and hence can be solved by an optimization technique. Since isothetic distance is a continuous but not a differentiable function, derivative-based optimization cannot be used. Moreover, optimization methods are iterative, time consuming, and can provide approximate solutions, in general. For example, computation of isothetic distance between two triangles will be inefficient if an iterative method is used. As an efficient solution, we show how isothetic distance between two simplices, e.g., two triangles, can be computed in constant time, using decomposition to lower-order simplices.

We show that for discretization of a 2-manifold surface, certain constraints need to be satisfied. This constraints can be, for example, based on topological properties of a voxel set in concurrence with those of its preimage in the real space. In short and by intuition, as voxels have their edges or faces parallel to the principal axes or the principal planes, isothetic distance becomes a natural choice for determining the maximum permissible size.

It is worth mentioning here that the discretization problem has been studied by several researchers for continuous analytical surfaces such as the ones that are $r$-regular; see, for example, [18,28,29]. For topological equivalence of an $r$-regular surface with its discrete representation under Gauss digitization, we refer to [28]. Discretization of $n$-dimensional implicit surfaces and related analysis can be seen in [29], and for further details we refer to a recent work in [18]. The major difference of our work with all these work are as follows.

- We consider 2-manifold surface, e.g., triangle mesh, for voxelization. On the contrary, the above-mentioned techniques all deal with $r$-regular implicit surfaces.
- We show how the voxel size can be fixed using inter-simplex isothetic distance for 2-manifold surface. Estimation of voxel size for $r$-regular surface is computationally expensive and not addressed in the related papers.
- A 2-manifold surface such as triangle mesh is not $r$-regular and hence cannot readily be analyzed by the existing techniques. Our technique based on isothetic distance is designed for this and can efficiently voxelize the surface with necessary topological properties.

## 2    Preliminaries

This section contains some basic terminologies to be used in the sequel. The rest are put in the relevant sections.

A 2-*simplex* means, in general, a planar facet (e.g., triangle) of a mesh in the 3D real space. A 2-simplex $M^{(2)}$ consists of a planar interior bounded by three or more 1-*simplices* (line segments), which are denoted by $M_i^{(2:1)}$, where $i = 1, 2, 3, \ldots$, and an equal number of 0-*simplices* (points) denoted by $M_i^{(2:0)}$. Similarly, each 1-simplex $L^{(1)}$, has a linear interior and is bounded by two 0-simplices, denoted as $L_1^{(1:0)}, L_2^{(1:0)}$. For a persistent topology, we consider 2-manifold orientable surface, and hence the constituent 2-simplices intersect each other only at 0- and 1-simplices.

A *pixel* is a 2-cell, or equivalently, an axis-parallel unit square centered at a point in $\mathbb{Z}^2$ [16]. A pixel is this made of four 1-cells (edges) and four 0-cells (vertices); and hence, a 2D digital straight line (2DSL) or segment (2DSS) can be 0- or 1-connected. In a 0-connected 2DSL or 2DSS for example, every two consecutive pixels are 0-adjacent. A *voxel* is a 3-dimensional extension of a pixel, and defined as a 3-cell or an axis-parallel unit cube centered at a point in $\mathbb{Z}^3$. It contains eight 0-cells, twelve 1-cells, and six 2-cells (faces). Two voxels are said to be *k-adjacent* ($k = 0, 1, 2$) if they share a *k*-cell. Note that this notion of 0-, 1-, and 2-adjacencies is equivalent to the notion of 26-, 18-, and 6-neighborhoods used in [8].

Let $p(x_p, y_p, z_p)$ and $q(x_q, y_q, z_q)$ be two 0-simplices or ordinary points in $\mathbb{R}^3$. The *x*-distance, *y*-distance, *z*-distance between them are denoted by $f_x(p,q) := |x_p - x_q|$, $f_y(p,q) := |y_p - y_q|$, and $f_z(p,q) := |z_p - z_q|$, respectively, and are used to define the isothetic distance between them as follows.

**Definition 1.** *The 0-0 or* **inter-point isothetic distance** *between $p$ and $q$ is defined as*

$$f_\perp(p, q) = \max\{f_x(p,q), \ f_y(p,q), \ f_z(p,q)\}. \tag{1}$$

*The axis-parallel box with $p$ and $q$ as two endpoints of its principal diagonal is called the* **isoBox** *of $p$ and $q$, and denoted by $\mathbb{B}(p,q)$.*

Note that $f_\perp(p,q)$ is basically the $L_\infty$ or Chebyshev norm between two points $p$ and $q$ [16]. Geometrically, it is the maximum length over the three sides of their isoBox. We extend Eq. 1 to define and characterize the isothetic distance between two higher-order geometric primitives like points/0-simplices, line segments/1-simplices, triangles/2-simplices, and between two set of simplices. As a generalization, we now introduce the following definition.

**Definition 2.** *For $m, n \in \{0, 1, 2\}$, the* **m-n** *or* **inter-simplex isothetic distance** *between two simplices $M^{(m)}$ and $N^{(n)}$ is defined as*

$$f_\perp\left(M^{(m)}, N^{(n)}\right) = \min\left\{f_\perp(p, q) : \left(p \in M^{(m)}\right) \wedge \left(q \in N^{(n)}\right)\right\}. \tag{2}$$

Note that for an orientable 2-manifold surface, the isothetic distance between two 2-simplices is zero if and only if they are adjacent, i.e., they share a 0- or a 1-simplex. The possible cases arising during the computation of triangle-to-triangle and other types of inter-simplex distance are discussed in Sect. 3.

# 3  Characterization of Isothetic Distance

We start with the following observation.

**Observation 1.** *All three axis-parallel distances and hence the isothetic distance of a fixed point $p$ from a variable point on a line in $\mathbb{R}^3$ are convex continuous functions having their minimums at finite points.*

*Proof.* Let the line be $L = a + (b - a)t$, where $t \in \mathbb{R}$, and let $a$ and $b$ be two fixed and distinct points on $L$. For brevity, let $x_{ab} = x_a - x_b$, $x_{ba} = x_b - x_a$, etc. Then the $x$-distance function $f_x(p, q) := |x_{pq}|$ between $p$ and $q(t) \in L$ can explicitly be written as

$$f_x(p, q) = \begin{cases} x_{pa} - t \cdot x_{ba} \text{ if } t \leqslant \frac{x_{pa}}{x_{ba}} \\ x_{ap} + t \cdot x_{ba} \text{ otherwise.} \end{cases} \tag{3}$$

Hence, $f_x(p, q)$ is a piecewise linear function with a global minimum at $t = \frac{x_{pa}}{x_{ba}}$. The other two functions $f_y(p, q) := |y_{pq}|$ and $f_z(p, q) := |z_{pq}|$ can be shown to be of the same nature in a similar way. Figure 1 shows an example. Since the maximum of two (or more) continuous functions is continuous, and the maximum of two convex functions is convex, the proof follows.    □

Observation 1 can be extended for the isothetic distance between a fixed point and a variable point on a 3D plane or on a 2-simplex. This is in fact true when both the points are allowed to vary on the simplices. We have the following observation for this.

**Observation 2.** *The axis-parallel and the isothetic distance functions defined for two variable points lying on two $k$-simplices, $k \in \{0, 1, 2\}$, are continuous, convex, and piecewise linear.*

We now do a characterization of isoBox and corresponding isothetic distance between 0- and 1-simplices. We use the following lemma for this.

**Lemma 1.** *Let $L$ be a straight line and $p$ be a point in $\mathbb{R}^3$. Let $q$ be a point in $L$ such that $f_\perp(p, q) = f_\perp(p, L)$. Then at least two among $f_x(p, q)$, $f_y(p, q)$, and $f_z(p, q)$ are equal in value.*

*Proof.* Let, w.l.o.g., $f_\perp(p, q) = f_x(p, q)$. From the definition of isothetic distance, $f_x(p, q) \geqslant f_y(p, q)$ and $f_x(p, q) \geqslant f_z(p, q)$. If $f_x(p, q)$ is strictly greater than both $f_y(p, q)$ and $f_z(p, q)$, then we can still move $q$ on $L$ so as to reduce it further until it becomes equal with $f_y(p, q)$ or $f_z(p, q)$. This happens as each of $f_x(p, q)$, $f_y(p, q)$, and $f_z(p, q)$ is a convex continuous function (Observation 1).    □

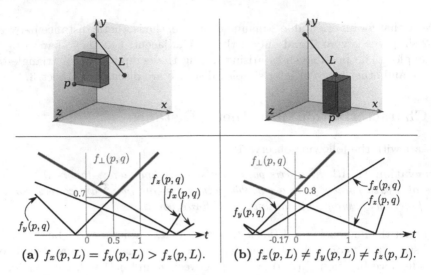

**Fig. 1.** Two possible cases of 0-1 ($p$-to-$L$) isothetic distance. Here, the endpoints of $L$ are $a = (5.2, 5.0, 4.8)$ and $b = (4.6, 6.0, 5.2)$ with respective parameter values $t = 0$ and 1. The variable point $q$ lies on $L$. **(a)** Case 1: $p = (4.2, 4.8, 5.4)$. The isoBox has two of its sides equal and larger in length than the third, which implies it touches an interior point $t = 0.5 := (4.9, 5.5, 5.0)$ of the line $L$. **(b)** Case 2: $p = (5.6, 4.2, 5.4)$. All three axis-parallel distances from $p$ to $L$ are different, which means the isoBox touches $L$ at one endpoint. As shown in the plots, $f_\perp(p, L) = 0.7$ in (a) and 0.8 in (b).

When $L$ is a finite line segment, Lemma 1 implies that there may not exist an interior point $q$ in $L$ at which two axis-parallel distances are of the same value; the isothetic distance in such case occurs at an endpoint of $L$. Figure 1 illustrates these two cases. In particular, we have the following theorem.

**Theorem 1.** *Let $a$ and $b$ be the endpoints of a line segment $L$, and $p$ be a point in $\mathbb{R}^3$. If $q$ is an interior point in $L$ such that $f_\perp(p, q) = f_\perp(p, L)$, then at least two of the axis-parallel distances between $p$ and $q$ are equal in value; otherwise, the isothetic distance occurs when $q$ coincides with either $a$ or $b$.*

*Proof.* Let $L'$ be the straight line containing the segment $L$. Let $q$ be a point in $L'$ such that $f_\perp(p, q) = f_\perp(p, L)$. If $q$ is a point in $L$, then by Lemma 1, the theorem holds. If $q$ lies in $L' \setminus L$, then we can move $q$ on $L'$ to the effect that $f_\perp(p, q)$ increases continuously (Observation 1) until $q$ coincides with an endpoint of $L$. This completes the proof.                                          □

Theorem 1 can be extended to higher-order simplices as well, and is stated in the following theorem.

**Theorem 2.** *If $p$ and $q$ are two points in $M^{(m)}$ and $N^{(n)}$ respectively, such that either $p$ or $q$ is an interior point and $f_\perp(p, q) = f_\perp(M^{(m)}, N^{(n)})$, then at least two of the axis-parallel distances between $p$ and $q$ are equal in value.*

*Proof.* We prove by contradiction. If possible, let $p$ be an interior point in $M^{(m)}$ such that, w.l.o.g., $f_\perp(M^{(m)}, N^{(n)}) = f_x(p, q) > f_y(p, q) > f_z(p, q)$. As $p$ lies in the interior of $M^{(m)}$, there exists a sufficiently small neighborhood $X_\epsilon(p)$ of $p$ in $M^{(m)}$. As the distance functions are all linear and continuous (Observation 1), we always get a point $p'$ in $X_\epsilon(p)$ such that $f_x(p', q) < f_x(p, q)$, which implies $f_x(p, q) > f_\perp(M^{(m)}, N^{(n)})$, whence the contradiction. $\qquad\square$

# 4 Decomposition of Distance Functions

In this section we discuss the algebraic techniques for computing the isothetic distance between two simplices having same or different order. We denote by $\mathbb{B}$ the isoBox of two simplices under consideration.

## 4.1 0-1 Distance

Let $L^{(1)}$ be a 1-simplex and $p$ a 0-simplex or point in $\mathbb{R}^3$. Let $q(t) = tL_1^{(1:0)} + (1 - t)L_2^{(1:0)}$, where $t$ is a real number. Then, by Theorem 1, the function $f_\perp(p, L^{(1)})$ is decomposed into two cases as follows (see Fig. 1).

*Case 1.* $\mathbb{B}(p, L^{(1)})$ touches a point in $L^{(1)} \smallsetminus \{L_1^{(1:0)}, L_2^{(1:0)}\}$, which is true if and only if the following equation has a solution $t \in (0, 1)$.

$$\left(\left|x_p - x_{q(t)}\right| - \left|y_p - y_{q(t)}\right|\right)\left(\left|x_p - x_{q(t)}\right| - \left|z_p - z_{q(t)}\right|\right)\left(\left|y_p - y_{q(t)}\right| - \left|z_p - z_{q(t)}\right|\right) = 0 \quad (4)$$

*Case 2.* Equation 4 produces $t \notin [0, 1]$, which means $\mathbb{B}(p, L^{(1)})$ touches $L^{(1)}$ at $L_1^{(1:0)}$ or at $L_2^{(1:0)}$, and so it reduces to 0-0 distance function, i.e., $f_\perp(p, L^{(1)}) = \min\left\{f_\perp(p, L_1^{(1:0)}), f_\perp(p, L_2^{(1:0)})\right\}$.

## 4.2 0-2 Distance

Let $M^{(2)}$ be a 2-simplex defined by the 1-simplex set $\mathcal{M}^{(1)} := \left\{L_i^{(1)} : 1 \leqslant i \leqslant k\right\}$ and the 0-simplex set $\mathcal{M}^{(0)} := \left\{q_i : 1 \leqslant i \leqslant k\right\}$. Clearly, a point $q(\mathbf{t}) := \sum_{i=1}^{k} t_i q_i$ belongs to the interior of $M^{(2)}$ if and only if $\{t_i\}_{i=1}^k \in (0, 1)^k$ and $\sum_{i=1}^{k} t_i = 1$. The distance function $f_\perp(p, M^{(2)})$ between a point $p$ and the simplex $M^{(2)}$ can be decomposed as follows.

*Case 1.* The isoBox $\mathbb{B}(p, M^{(2)})$ touches an interior point of $M^{(2)}$, which is true if and only if at least two of the axis-parallel distances are equal in value (Theorem 2), or equivalently, the following equation yields $\{t_i\}_{i=1}^k \in (0, 1)^k$ with $\sum_{i=1}^{k} t_i = 1$.

$$\left(\left|x_p - x_{q(t)}\right| - \left|y_p - y_{q(t)}\right|\right)\left(\left|x_p - x_{q(t)}\right| - \left|z_p - z_{q(t)}\right|\right)\left(\left|y_p - y_{q(t)}\right| - \left|z_p - z_{q(t)}\right|\right) = 0 \quad (5)$$

*Case 2.* $\mathbb{B}(p, M^{(2)})$ touches $M^{(2)}$ at one of its 1-simplices, and so reduces to 0-1 function (Sect. 4.1), i.e., $f_\perp(p, M^{(2)}) = \min_{L_i^{(1)} \in \mathcal{M}^{(1)}} \left\{f_\perp(p, L_i^{(1)})\right\}$.

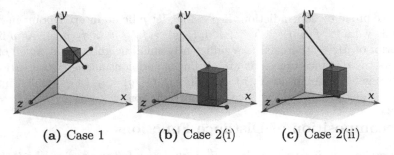

(a) Case 1          (b) Case 2(i)          (c) Case 2(ii)

**Fig. 2.** Decomposition of 1-1 distance into possible cases and sub-cases: isoBox touches (a) both the line segments at their interiors; (b) one at the interior; (c) both at their endpoints. Case 1 is solved in the parametric space, while Case 2(i) as 0-1 and Case 2(ii) as 0-0 distance functions.

### 4.3   1-1 Distance

Let $K^{(1)}$ and $L^{(1)}$ be two 1-simplices.. Let $q(s) = sK_1^{(1:0)} + (1-s)K_2^{(1:0)}$ and $r(t) = tL_1^{(1:0)} + (1-t)L_2^{(1:0)}$, where $s, t$ are two real numbers. The distance function $f_\perp(p, M^{(2)})$ between a point $p$ and the simplex $M^{(2)}$ can be decomposed as follows (see Fig. 2).

*Case 1.*   The isoBox $\mathbb{B}(K^{(1)}, L^{(1)})$ touches the interiors of both the 1-simplices if and only if, by Theorem 2, the following equation yields $(s, t) \in (0, 1)^2$.

$$\left( \left| x_{q(s)} - x_{r(t)} \right| - \left| y_{q(s)} - y_{r(t)} \right| \right) \left( \left| x_{q(s)} - x_{r(t)} \right| - \left| z_{q(s)} - z_{r(t)} \right| \right)$$
$$\left( \left| y_{q(s)} - y_{r(t)} \right| - \left| z_{q(s)} - z_{r(t)} \right| \right) = 0 \quad (6)$$

*Case 2.*   $\mathbb{B}(K^{(1)}, L^{(1)})$ touches one of the 0-simplices, and so reduces to 0-1 function (Sect. 4.1), i.e., $f_\perp(K^{(1)}, L^{(1)}) = \min\limits_{i=1,2} \left\{ f_\perp(K_i^{(1:0)}, L^{(1)}), f_\perp(L_i^{(1:0)}, K^{(1)}) \right\}$.

### 4.4   2-2 Distance

Using the result presented in Sects. 4.2 and 4.3, we get the following theorem on the inter-simplex distance between two 2-simplices.

**Theorem 3.** *For two 2-simplices, there exists an isoBox that does not simultaneously touch their interiors.*

*Proof.* We assume that $M^{(2)}$ and $N^{(2)}$ are mutually in general orientation, that is non-parallel to each other, since otherwise there exist infinitely many positions of the isoBox $\mathbb{B}(M^{(2)}, N^{(2)})$, and the theorem is true for some of them. We prove by contradiction for the general case. Let, if possible, $\mathbb{B}(M^{(2)}, N^{(2)})$ touch $M^{(2)}$ and $N^{(2)}$ at their respective interior points $p$ and $q$ such that $f_\perp(M^{(2)}, N^{(2)}) = f_\perp(p, q)$. By Theorem 2, at least two among $f_x(p, q), f_y(p, q), f_z(p, q)$ are equal in value. So, let, w.l.o.g., $f_\perp(p,q) := f_x(p,q) = f_y(p,q) \geqslant f_z(p,q)$. Let $X_\epsilon(p)$ be a sufficiently small neighborhood of $p$ in $M^{(2)}$. Since the axis-parallel distance functions are linear and continuous (Observation 1), and $M^{(2)}$ and $N^{(2)}$ are mutually non-parallel, we always get another point $p' \in X_\epsilon(p)$ such that $f_x(p',q) < f_x(p,q)$ and $f_y(p',q) < f_y(p,q)$, or $f_\perp(p',q) < f_\perp(p,q)$, whence the contradiction.                                                                    $\square$

Clearly, by Theorem 3, the isothetic distance between $M^{(2)}$ and $N^{(2)}$ is given by

$$f_\perp(M^{(2)}, N^{(2)}) = \min \left\{ \begin{array}{l} \displaystyle\min_{\substack{1 \leqslant i \leqslant m \\ 1 \leqslant j \leqslant n}} \left\{ f_\perp(M_i^{(2:1)}, N_j^{(2:1)}) \right\}, \\[2mm] \displaystyle\min_{1 \leqslant i \leqslant m} \left\{ f_\perp(M_i^{(2:0)}, N^{(2)}) \right\}, \\[2mm] \displaystyle\min_{1 \leqslant j \leqslant n} \left\{ f_\perp(N_j^{(2:0)}, M^{(2)}) \right\} \end{array} \right\} \tag{7}$$

where, $m$ and $n$ denote the respective number of 1-simplices (and the same of 0-simplices thereof) comprising $M^{(2)}$ and $N^{(2)}$.

Equation 7 shows how the isothetic distance between two 2-simplices (triangles) is computed from the isothetic distances among lower-order simplices. An illustration of the concept is given in Fig. 3. Notice in particular that the

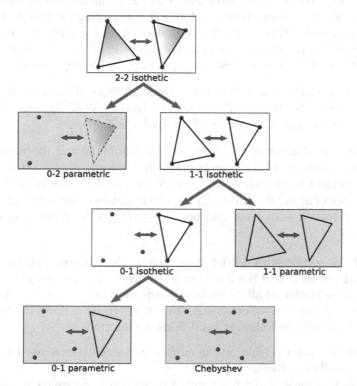

**Fig. 3.** Hierarchy of computation of isothetic distance.

computation of 2-2 isothetic distance is quite simplified, since the isoBox never touches the interior regions of both the triangles in simultaneity, as shown by us; hence, it has only two possibilities: (i) solving in the parametric space for a point and the interior of a triangle and (ii) 1-1 isothetic distance. Clearly, the distance computation between any two simplices down the hierarchy takes $O(1)$ time, wherefore the 2-2 isothetic distance is done in $O(1)$ time.

### 4.5   Voxel Size

The inter-simplex isothetic distance determines the voxel unit for voxelization of a 2-manifold surface. We show some result in Sect. 5. For the underlying theory, we recall here a few concepts from [3, 4, 16].

Let $A$ be a set of 3-cells (i.e., voxels), which may be of finite or of infinite cardinality. For $k = 0, 1, 2$, a $k$-*path* in $A$ is a sequence of voxels in $A$ where every two consecutive voxels are $k$-adjacent. If there is a $k$-path in $A$ between any two voxels of $A$, then $A$ is said to be $k$-*connected*. A $k$-*component* in $A$ is a maximal $k$-connected subset of $A$. If $B$ is a subset of $A$ such that $A \setminus B$ is not $k$-connected, then $B$ is said to be $k$-*separating* in $A$; in addition, $B$ is $k$-*minimal* if it contains no $k$-simple voxel. A $k$-*simple voxel* of $B$ is a voxel $v$ such that $B \setminus \{v\}$ is still $k$-separating.

The *clearance* between two voxel sets $A$ and $B$ is the number of voxels comprising a/the shortest 2-path between $A$ and $B$, discounting the first and the last voxels. We denote by $\mathbb{V}(M^{(2)})$ the 2-minimal voxel set obtained by voxelization of $M^{(2)}$, wherefore the distance of each voxel $u \in \mathbb{V}(M^{(2)})$ is at most half the voxel unit from $M^{(2)}$. We use these concepts in the following theorem.

**Theorem 4.** *Let $M^{(2)}$ and $N^{(2)}$ be two 2-simplices in an orientable surface such that $\delta := f_\perp(M^{(2)}, N^{(2)}) > 0$. The clearance between $\mathbb{V}(M^{(2)})$ and $\mathbb{V}(N^{(2)})$ is at least $\kappa$ voxel units if the voxel length is fixed to $s \leqslant \frac{\delta}{\kappa+2}$.*

*Proof.* As the voxelization of the 2-simplices is 2-minimal and the voxel unit is $s$, the isothetic distance of (the center of) each voxel $u \in \mathbb{V}(M^{(2)})$ from $M^{(2)}$ is at most $\frac{s}{2}$, and so also for each voxel $v \in \mathbb{V}(N^{(2)})$ from $N^{(2)}$ (see [2]: Theorem 1). For a clearance of at least $\kappa$ voxels, the isothetic distance between $\mathbb{V}(M^{(2)})$ and $\mathbb{V}(N^{(2)})$ is at least $\kappa + 1$ voxels, and hence $f_\perp(M^{(2)}, N^{(2)}) \geqslant s(\kappa + 1) + 2 \times \frac{s}{2}$, or, $s \leqslant \frac{\delta}{\kappa+2}$.   $\square$

We end this section with a brief discussion on the homeomorphism between a 2-manifold surface $\mathcal{S}$ and the voxel set $\mathbb{V}(\mathcal{S})$ given by the union of naive (i.e., 2-minimal) voxelization of all 2-simplices comprising the surface $\mathcal{S}$. We refer to a recent work [1] for further details. As shown in [1], by making the voxel size appropriately small, the following conditions are ensured:

(i) for each real point $p$ lying on the surface $\mathcal{S}$, there exists a voxel $v \in \mathbb{V}(\mathcal{S})$, which is sufficiently close to $p$;

(ii) for each voxel $v \in \mathbb{V}(\mathcal{S})$, there exists a point $p \in \mathcal{S}$, which is sufficiently close to $v$.

Theorem 4 can be used to fix the voxel size while setting up the homeomorphism. Its practical usefulness is discussed further in Sect. 5.

## 5  Test Result

Based on the isothetic distance metric, we have designed an algorithm for voxelization that takes as input a 2-manifold surface along with the value of the parameter $\kappa$. The algorithm first computes the inter-simplex isothetic distance for every pair of non-adjacent 2-simplices, using the decomposition technique discussed in Sect. 4. It uses the minimum ($\delta_{min}$) of these inter-simplex distances, and based on the specified value of $\kappa$, finds the appropriate voxel size $s$ using Theorem 4. Each 2-simplex (triangle) is then voxelized to its 2-minimal set as discussed in [4, Sect. 7.3.1].

The algorithm runs in the integer space. This is achieved by scaling up the input surface by a factor of $\frac{1}{s}$. Owing to this, the effective voxel size becomes unity in the transformed voxel space and the voxelization also retains the required property of 2-minimality for the voxel set corresponding to each 2-simplex. This is as per the practice commonly followed in the existing algorithms [19,21,24,26].

We have shown here our test result on two models, namely cogwheel and bunny, in Figs. 4 and 5. A summary of result for these two models and a couple of other models is presented in Table 1. In the figures, we have shown cutout images for seeing the interiors of the voxelized surfaces. It is clear from these result that the number of voxels is minimum when $\kappa$ is set to 1, and with an increasing value of $\kappa$, the voxelized surface improves in size and quality, as the voxel size $s$ becomes smaller and the number of voxels $n_s$ becomes larger, thereby approximating the surface with a higher resolution and precision. Nevertheless, the resultant voxelization is always homeomorphic as long as $\kappa \geqslant 1$. For a low value of $\kappa$, say $\kappa = 1$ in cogwheel, there appear some object voxels around the 'corners' of the triangle mesh, which may be removed without creating any tunnels in

**Table 1.** Summary of test result by our algorithm on 3D models.

| Cogwheel | TorusKnot | Bunny | Dragon |

| | | $\kappa = 1$ | | $\kappa = 2$ | | $\kappa = 5$ | |
|---|---|---|---|---|---|---|---|
| Object, #tris | $\delta_{min}$ | $s$ | $n_s$ | $s$ | $n_s$ | $s$ | $n_s$ |
| Cogwheel, 1k | 5.827 | 1.942 | 2566 | 1.457 | 4615 | 0.832 | 13830 |
| TorusKnot, 2k | 3.841 | 1.280 | 2155 | 0.960 | 3669 | 0.549 | 10868 |
| Bunny, 5k | 19.006 | 6.335 | 1624 | 4.751 | 2588 | 2.715 | 7080 |
| Dragon, 20k | 18.739 | 6.246 | 2226 | 4.685 | 3691 | 2.677 | 9740 |

**Fig. 4.** Result on `cogwheel`. Top, Middle: $\kappa = 1$. Bottom: $\kappa = 5$.

the voxel set. This operation is, however, not permissible under homeomorphism (Sect. 4.5). For, on removal of any such apparently redundant voxel, there arises points on the corresponding real triangle which would not have any object voxel in a sufficiently close neighborhood.

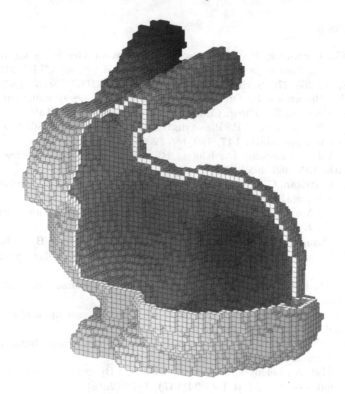

**Fig. 5.** Result on bunny ($\kappa = 5$).

# 6    Conclusion

We have shown how isothetic distance can be decomposed into lower-order distances, which eventually aids in efficient computation of the distance function for simplices of different orders. The significance of isothetic distance in voxelization has also been theoretically explained and experimentally shown through different test result. As this technique produces 2-minimal voxelization of a 2-manifold surface, it can readily be used for construction of voxel sets for different applications mentioned in Sect. 1.1.

We plan to explore further with the usefulness of our technique for solving other digital-geometric problems like curvature and normal estimation on a voxelized surface. These problems are found to have strong connections with different application-oriented problems defined on voxel sets, such as ray casting, shadow generation, and texture creation, which can have efficient solutions thereof, as we foresee.

**Acknowledgments.** We are thankful to the reviewers for their critical comments and suggestions, which helped us in revising the paper up to its merit.

# References

1. Bhalla, G., Bhowmick, P.: DIG: Discrete Iso-contour Geodesics for topological analysis of voxelized objects. In: Bac, A., Mari, J.-L. (eds.) CTIC 2016. LNCS, vol. 9667, pp. 265–276. Springer, Cham (2016). doi:10.1007/978-3-319-39441-1_24
2. Biswas, R., Bhowmick, P.: On different topological classes of spherical geodesics paths and circles in $\mathbb{Z}^3$. Theor. Comput. Sci. **605**, 146–163 (2015)
3. Brimkov, V.E., Barneva, R.P.: Plane digitization and related combinatorial problems. Discrete Appl. Math. **147**, 169–186 (2005)
4. Brimkov, V.E., Coeurjolly, D., Klette, R.: Digital planarity–a review. Discrete Appl. Math. **155**, 468–495 (2007)
5. Brunton, A., Arikan, C.A., Urban, P.: Pushing the limits of 3D color printing: error diffusion with translucent materials. ACM ToG **35**(4), 1–13 (2015)
6. Chandru, V., Monohar, S., Prakash, C.: Voxel-based modeling for layered manifacturing. IEEE Comput. Graph. Appl. **15**, 42–47 (1995)
7. Chen, X., Zhang, H., Lin, J., Hu, R., Lu, L., Huang, Q., Benes, B., Cohen-Or, D., Chen, B.: Dapper: decompose-and-pack for 3D printing. ACM ToG **34**(213), 1–12 (2015)
8. Cohen-Or, D., Kaufman, A.: Fundamentals of surface voxelization. Graph. Models Image Process. **57**(6), 453–461 (1995)
9. Desimone, J., Ermoshkin, A., Samulski, E.: Method and apparatus for three-dimensional fabrication, US Patent 20140361463 (2014)
10. Dionne, O., de Lasa, M.: Geodesic voxel binding for production character meshes. In: Proceedings of SCA 2013, pp. 173–180 (2013)
11. Dumas, J., Lu, A., Lefebvre, S., Wu, J., Dick, C.: By-example synthesis of structurally sound patterns. ACM ToG **34**(137), 1–12 (2015)
12. Fei, Y., Wang, B., Chen, J.: Point-tessellated voxelization. In: Proceedings of Graphics Interface, GI 2012, pp. 9–18 (2012)
13. Huang, J., Yagel, R., Filippov, V., Kurzion, Y.: An accurate method for voxelizing polygon meshes. In: Proceedings of 1998 IEEE Symposium, VVS 1998, pp. 119–126 (1998)
14. Kämpe, V., Sintorn, E., Assarsson, U.: High resolution sparse voxel DAGs. ACM ToG **32**(101), 1–13 (2013)
15. Karabassi, E.A., Papaioannou, G., Theoharis, T.: A fast depth-buffer-based voxelization algorithm. J. Graph. Tools **4**, 5–10 (1999)
16. Klette, R., Rosenfeld, A.: Digital Geometry: Geometric Methods for Digital Picture Analysis. Morgan Kaufmann, San Francisco (2004)
17. Koa, M.D., Johan, H.: ESLPV: enhanced subsurface light propagation volumes. Vis. Comput. **30**, 821–831 (2014)
18. Lachaud, J.O., Thibert, B.: Properties of Gauss digitized shapes and digital surface integration. JMIV **54**(2), 162–180 (2016)
19. Laine, S.: A topological approach to voxelization. Comput. Graph. Forum **32**, 77–86 (2013)
20. Laine, S., Karras, T.: Efficient sparse voxel octrees. In: Proceedings of ACM SIGGRAPH Symposium, I3D 2010, pp. 55–63 (2010)
21. Laine, S.: System, method, and computer program product implementing an algorithm for performing thin voxelization of a three-dimensional model, US Patent 9,245,363 (2016)
22. Lozano-Durán, A., Borrell, G.: Algorithm 964: an efficient algorithm to compute the genus of discrete surfaces and applications to turbulent flows. ACM Trans. Math. Softw. **42**(34), 1–19 (2016)

23. Niebner, M., Zollhöfer, M., Izadi, S., Stamminger, M.: Real-time 3d reconstruction at scale using voxel hashing. ACM ToG **32**(169), 1–11 (2013)
24. Pantaleoni, J.: VoxelPipe: a programmable pipeline for 3D voxelization. In: Proceedings of ACM SIGGRAPH Symposium, HPG 2011, pp. 99–106 (2011)
25. Prakash, C., Manohar, S.: Volume rendering of unstructured grids–a voxelization approach. Comput. Graph. **19**, 711–726 (1995)
26. Schwarz, M., Seidel, H.P.: Fast parallel surface and solid voxelization on GPUs. ACM ToG **29**(179), 1–10 (2010)
27. Sintorn, E., Kämpe, V., Olsson, O., Assarsson, U.: Compact precomputed voxelized shadows. ACM ToG **33**(150), 1–8 (2014)
28. Stelldinger, P., Latecki, L.J., Siqueira, M.: Topological equivalence between a 3D object and the reconstruction of its digital image. IEEE TPAMI **29**(1), 126–140 (2007)
29. Toutant, J.-L., Andres, E., Largeteau-Skapin, G., Zrour, R.: Implicit digital surfaces in arbitrary dimensions. In: Barucci, E., Frosini, A., Rinaldi, S. (eds.) DGCI 2014. LNCS, vol. 8668, pp. 332–343. Springer, Cham (2014). doi:10.1007/978-3-319-09955-2_28
30. Vidimče, K., Wang, S.P., Ragan-Kelley, J., Matusik, W.: OpenFab: a programmable pipeline for multi-material fabrication. ACM ToG **32**(136), 1–12 (2013)
31. Wu, J., Dick, C., Westermann, R.: A system for high-resolution topology optimization. IEEE TVCG **22**, 1195–1208 (2016)
32. Wyman, C.: Voxelized shadow volumes. In: Proceedings of ACM SIGGRAPH Symposium, HPG 2011, pp. 33–40 (2011)
33. Zhang, J.: Speeding up large-scale geospatial polygon rasterization on GPGPUs. In: Proceedings ACM SIGSPATIAL, HPDGIS 2011, pp. 10–17 (2011)
34. Zhang, L., Chen, W., Ebert, D.S., Peng, Q.: Conservative voxelization. Vis. Comput. **23**, 783–792 (2007)
35. Zhao, S., Hašan, M., Ramamoorthi, R., Bala, K.: Modular flux transfer: efficient rendering of high-resolution volumes with repeated structures. ACM ToG **32** (2013). Article No. 131
36. Zhou, Y., Sueda, S., Matusik, W., Shamir, A.: Boxelization: folding 3D objects into boxes. ACM ToG **33**(71), 1–8 (2014)
37. Zollhofer, M., Dai, A., Innmann, M., Wu, C., Stamminger, M., Theobalt, C., Niebner, M.: Shading-based refinement on volumetric signed distance functions. ACM ToG **34**(96), 1–14 (2015)

# Theory and Applications: Image Segmentation, Classification, Reconstruction, Compression, Texture Analysis, and Bioimaging

# Topological Data Analysis for Self-organization of Biological Tissues

M.J. Jimenez[1]([✉]), M. Rucco[2], P. Vicente-Munuera[3,4], P. Gómez-Gálvez[3,4], and L.M. Escudero[3,4]

[1] Departamento Matematica Aplicada I, Universidad de Sevilla,
Campus Reina Mercedes, 41012 Sevilla, Spain
majiro@us.es

[2] School of Science and Technology, Computer Science Division,
University of Camerino, Camerino, Italy
matteo.rucco@unicam.it

[3] Departamento de Biología Celular, Universidad de Sevilla, Sevilla, Spain
{pvicente1,pgomez-ibis,lmescudero-ibis}@us.es

[4] Instituto de Biomedicina de Sevilla (IBiS), Hospital Universitario Virgen del Rocío,
CSIC, Universidad de Sevilla, 41013 Sevilla, Spain

**Abstract.** In this paper we propose a method to topologically analyze segmented images of cells in a biological tissue. This is a mainly experimental paper in which we present initial results of applying persistent homology computation to characterize cell organization. For that aim, a graph is constructed to model the cell organization and a simplicial complex is derived from such a graph. Then a filter function is designed, on the simplicial complex, that reflects neighbouring relations on cells as well as their size. Finally, persistent homology and persistent entropy are computed and the results are analyzed.

**Keywords:** Cell organization · Persistent homology · Bottleneck distance · Persistent entropy

## 1 Introduction

Topological Data Analysis (TDA) has got its main motivation in the study of shape of data. The core concept of computational topology that is used to analyse shapes is *homology*. Homology is an algebraic concept which is a topological invariant of the space. Roughly speaking, homology characterizes "holes" in any dimension (in the case of a 3D space, connected components, tunnels and cavities). The main tool used in TDA, however, is *persistent homology* [7,22],

M.J. Jimenez—Partially supported by Ministerio de Economía y Competitividad de España under grant MTM2015-67072-P.

P. Vicente-Munuera, P. Gómez-Gálvez and L.M. Escudero—Part of this work was supported by Ramon y Cajal program (PI13/01347); two grants from "Fundacion Asociacion Española contra el Cancer" and "Universidad de Sevilla"; BFU2016-74975 grant from the Spanish government.

© Springer International Publishing AG 2017
V.E. Brimkov and R.P. Barneva (Eds.): IWCIA 2017, LNCS 10256, pp. 229–242, 2017.
DOI: 10.1007/978-3-319-59108-7_18

which studies the evolution of homology classes and their life-times (persistence) in an increasing nested sequence of spaces (that is called a filtration) and which is more informative that the homology class of the whole space.

Persistent homology has proved to be a useful tool in the study of shape analysis (in [12] some trends are described). For example, in the paper [4] the authors deal with the application of persistent homology for comparing 3D shapes represented by triangle meshes, or in [15], an algorithm for applying persistent homology to activity recognition is designed. However, as far as we know, there are no applications to organizational study of biological tissues using persistent homology, so far.

We are concerned with a specific application of persistent homology to the analysis of organizational aspects of biological cells in a tissue. As a first step, we deal, in this paper, with the topological analysis of some epithelia samples as well as a group of synthetic images that are usually considered (in the biological context) as a reference scale to study self-organization.

Our main contribution is the introduction of persistent homology as a tool for the analysis of cell organization in biological tissue images.

In the following Section, we describe the biological problem that motivated this work. Section 3 recalls main concepts from TDA that will be used in the sequel. Section 4 describes the particular way in which we make use of persistent homology concepts to topologically analyze the input data. Reports on the computations performed as well as some conclusions are collected in Sect. 5. We draw some ideas for future work in the last section.

## 2    Motivation

The embryo of any animal is constituted by epithelial cells which will undergo lots of transformations through development to acquire the specific functions and shapes of the 4 types of adult tissues (epithelia, connective tissue, nervous tissue and muscle tissue). Epithelia are packed tissues formed by tightly assembled cells. Their apical surfaces are shaped as convex polygons forming a natural tessellation of cells that fill the exterior of the epithelium without leaving any empty space. Epithelial organization has been analyzed in various systems from a topological and biophysical perspective [11,13,19,20]. These studies have been mainly based in the analysis of the polygon distribution of the tissues.

Recently, some of the authors have used Voronoi tessellations as a geometrical tool to understand the physical constraints that drive the organization of biological tissues [21]. They described a "Centroidal Voronoi Tessellation" (CVT) path formed by successive Lloyd iterations of a random Voronoi diagram, and showed that the CVT became predictive of the polygon distribution of the natural packed tissues [19]. They found that packed tissues are under a physical constraint that drives its self-organization.

We consider that understanding how tissues are organized is key to uncover the causes and mechanisms of developmental and pathological variations. For this reason we want to go beyond of a mere quantification of polygon distributions and capture the organization of a tessellation in a more complete manner.

We propose the use of persistent homology as a tool to topologically analyse data coming from tessellations with the goal of characterising epithelial organization under different biological conditions.

# 3  Background

We can always consider that the input data is a point cloud on the euclidean plane that corresponds to centroids of regions or polygons that partition the plane. Hence, those points are endowed with information of "neighbour" points, corresponding to neighbour regions or polygons. By considering the graph from such a set of vertices and edges, one can, in fact, construct a *simplicial complex*.

Intuitively, a simplicial complex is a representation of a topological space by decomposing it into *simple pieces* such that their common intersections are lower-dimensional pieces of the same kind, what allows the use of a data structure that is easily treated computationally.

**Simplicial complex.** A simplicial complex $K$ is given by:

- a set $V$ of vertices or 0–simplices;
- for each $d \geq 1$ a set of $d$–simplices $\{\sigma = \{v_0, v_1, \ldots, v_d\}$, where $v_i \in V\}$;
- each $d$–simplex has $d + 1$ *faces* obtained by removing one of the vertices;
- if $\sigma$ belongs to $K$, then all the faces of $\sigma$ must belong to $K$.

**Homology.** Homology groups are defined from the algebraic structure endowed to the set of $d$-simplices: chain complexes. The set of groups $\{C_d(K)\}_d$ where each $C_d(K)$ is the group (with $\mathbb{Z}_2$ coefficients) of $d$-chains generated by all the $d$-simplices of $K$, and a set of homomorphisms $\{\partial_d : C_d(K) \rightarrow C_{d-1}(K)\}_d$, describing algebraically the boundary of $d$-chains (boundary operators). A $d$-chain $c$ such that $\partial(c) = 0$ is a $d$-cycle and it is a $d$-boundary if there exists a $(d + 1)$-chain $c'$ such that $\partial(c') = c$. This way, the $d$–dimensional homology group $H_d(K)$ is the group of $d$-cycles modulo the group of $d$-boundaries. The $d$–th *Betti number* $\beta_d(K)$ is the rank of the $d$–dimensional homology group $H_d(K)$. Informally, for a fixed $d$, the $d$–th Betti number, $\beta_d$ counts the number of $d$-dimensional holes characterizing the space: $\beta_0$ is the number of connected components, $\beta_1$ counts the number of holes in 2D or tunnels in 3D[1], $\beta_2$ can be thought as the number of voids in geometric solids. See [14] or [16] for an introduction to algebraic topology.

**Persistent homology.** Persistent homology is a method that was created originally for computing $d$–dimensional holes ($d$–holes) at different spatial resolutions. For a more formal description we refer the reader to [6]. In order to compute persistent homology, we need *a filtration* on the simplicial complex, that is a nested sequence of increasing subcomplexes. More formally, a filtered simplicial complex $K$ is a collection of subcomplexes $\{K(t) : t \in \mathbb{R}\}$ of $K$ such that $K(t) \subset K(s)$ for $t < s$ and there exists $t_{max} \in \mathbb{R}$ such that $K_{t_{max}} = K$. The

---

[1] nD refers here to the $n$–dimensional space $\mathbb{R}^n$.

filtration time (or filter value) of a simplex $\sigma \in K$ is the smallest $t$ such that $\sigma \in K(t)$.

Persistent homology describes how the homology of $K$ changes along the filtration. A $d$–dimensional Betti interval, with endpoints $[t_{start}, t_{end})$, corresponds to a $d$–dimensional hole that appears at filtration time $t_{start}$ and remains until time $t_{end}$. We refer to the holes that are still present at $t = t_{max}$ as *persistent topological features*, otherwise they are considered *topological noise* [1]. The set of intervals representing birth and death times of homology classes is called the *persistence barcode* associated to the corresponding filtration. In the case of an interval with no death time, we consider the interval $[t_{start} , m)$ in the persistence barcode, where $m = t_{max}+1$. We call $d$–barcode to the persistence barcode encoding the persistence of $d$-holes (corresponding to $d$–dimensional homology classes).

Instead of bars, we sometimes draw points in the plane such that a point $(x,y) \in \mathbb{R}^2$ (with $x < y$) corresponds to a bar $[x,y)$ in the barcode. This set of points is called *persistence diagram*.

**Bottleneck distance.** The Bottleneck distance is used as a metric on the space of persistence diagrams. The bottleneck distance between the persistence diagrams $D_1$ and $D_2$ is: $d_B(D_1, D_2) = \inf_\gamma \sup_a \{\|a - \gamma(a)\|_\infty\}$, where, for points $a = (x,y)$ and $\gamma(a) = (x',y')$, $\|a - \gamma(a)\|_\infty = \max\{|x - x'|, |y - y'|\}$ and the bijection $\gamma : D_1 \to D_2$ can associate a point off the diagonal with another on or off the diagonal[2]. See [6, p. 229].

**Persistent entropy.** Persistent entropy is defined as the Shannon entropy of the persistence barcode of a given filtration [5,17]. Given a filtered simplicial complex $\{K(t) : t \in \mathbb{R}\}$, and the corresponding persistence diagram $D = \{a_i = (x_i, y_i) : i \in I\}$ (being $x_i < y_i$ for all $i \in I$), the *persistent entropy* $H$ of the filtered simplicial complex is calculated as follows:

$$H = -\sum_{i \in I} p_i \log(p_i)$$

where $p_i = \frac{\ell_i}{L}$, $\ell_i = y_i - x_i$, and $L = \sum_{i \in I} \ell_i$. In [18] the authors made use of persistent entropy to successfully classify the signal of DC motors and, what is more important, they proved the stability of persistent entropy for piece-wise linear functions.

## 4   Methodology

In this section we describe the methodology applied to topologically study the organization of cells. Roughly speaking, we mean to capture organizational information of the cells in a tissue by looking at the evolution of connected components and holes as we add cells of increasingly higher number of neighbours (see Fig. 1). For such aim, we need to model the data by a simplicial complex for defining later a filter function on the simplices.

---

[2] *Diagonal* is the set of points $\{(x,x)\} \subset \mathbb{R}^2$.

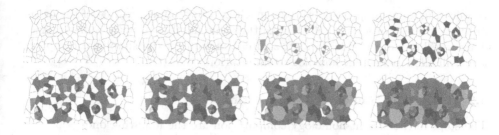

**Fig. 1.** The cells are added increasingly (left to right, top to bottom) according to their number of neighbours.

## 4.1 Modelling Data by Weighted Graphs

The input data could be seen as a map of regions (cells) in the plane. We can consider the centroid of each cell, together with the information of the neighbours of each cell. Then, naturally we have got the graph whose vertices are the centroids and whose edges are determined by each pair of neighbouring centroids (that is, those of neighbouring cells). See Fig. 2, left and center pictures.

If we assume that there are no four mutually adjacent neighbours, the resulting graph would be a triangulation of the convex hull of the set of centroids in the plane. In such a case, a *Delaunay complex* could be considered, induced by such graph. However, in practice, there are cases in which four regions are considered to be neighbours, so those four vertices would form a 4-clique and hence the corresponding simplicial complex (induced by the initial graph) would have, eventually, a 3-simplex.

Since it would be interesting to get information of the organization of different polygonal cells (triangles, squares, pentagons,...) within the tissue, we endow each vertex of the graph with a weight that corresponds to the number of neighbouring cells. Such amount will coincide with the degree of the vertex in the graph in the cases of vertices representing no-boundary cells (see Fig. 2, right).

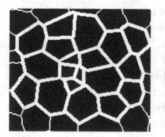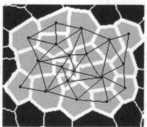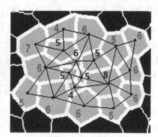

**Fig. 2.** Left: Segmented cells from a tissue image. Center: Graph constructed out of the centroids of the segmented cells. Right: Weights assigned to each vertex, encoding the number of neighbouring cells. This value coincides with the degree of the vertex, except for those points lying on the boundary (in red). (Color figure online)

However, in an attempt to capture as much information as possible of the organization, we also order the cells having the same number of neighbours by size. This ordering will provide, in fact, a filter function on the simplicial complex, that will allow persistent homology computation.

### 4.2 Persistent Homology Computation

The steps for persistent homology computation are the following ones:

1. Consider the simplicial complex $K$ induced by the graph representing neighbouring relations of cells, with set of vertices $V = \{v_1, \ldots, v_m\}$.
2. Define a filter function $f : K \to \mathbb{R}$. First, define the filter value on each vertex, $f(v_i)$. Then, extend the filter value to the rest of simplices $\sigma$ in $K$, such that

$$f(\sigma) = \max\{f(v)|v \in \sigma\}.$$

   This filter function has been designed so that information of both aspects: the polygon distribution as well as the role of the size of the cells within their self-organization, take part in the computation.
3. Now, consider the *lower-star filtration* of $f$ (see [6, p. 135]). Recall that the *lower star* of $v_i$ is the subset of simplices for which $v_i$ is the vertex with maximum function value,

$$St_- v_i = \{\sigma \in St\, v_i : x \in \sigma \Rightarrow f(x) \leq f(v_i)\},$$

   where $St\, v_i$, is the star of a vertex $v_i$, which is the set of simplices in $K$ for which $v_i$ is a face. The considered filtration is the lower-star filtration of $f$: $\emptyset = K_0 \subset K_1 \subset \cdots \subset K_r = K$, in which $K_i$ is the union of the first $i$ lower stars.
4. Persistent barcodes and persistent entropy are computed.

## 5    Experiments

Two different natural images have been taken to be analyzed: chick neuroepithelium (cNT1-cNT16) from chicken embryos and wing imaginal disc in the prepupal stage (dWP1-dWP16) from *Drosophila*. A detailed description of the protocol for obtaining the images can be found in [9,10,19]. The area of 1 pixel is $3.78 \times 10 - 3\,\mu m^2$. A region of interest was selected for each image, and the cells that border the ROI were excluded. Likewise, each cell included in ROI not lying on the border of the picture was considered as a valid cell. Images were processed modifying the bright contrast, and they were segmented using *segment.sci* software [10]. A post-curation process was carried out by visual inspection correcting boundaries with *Photoshop CS2* and processed images were exported to BMP files with 2 pixel wide cell outline. See Fig. 3 as one examples of each type of tissue.

Besides, a group of synthetic images have also been used for experiments. They are the so-called Voronoi diagrams in CVT path [19]. In fact, we have

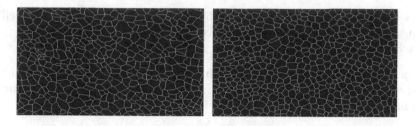

**Fig. 3.** Samples of cNT (left) and dWP (right) segmented images.

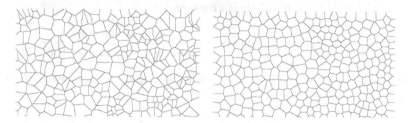

**Fig. 4.** Samples of two Voronoi images: a Voronoi1 on the left; a Voronoi5 on the right.

taken as input a variation of CVT standard path called CVT noise path. This CVT noise path is similar to CVT standard path in proportion of sides in first 20 diagrams. However, as the number of iterations increases, the proportion of hexagons in CVT path is stabilized at while CVT noise path hexagons percentage breaks the 90% wall. The development process of CVT noise path is the same than CVT path modifying the fact in which seeds taken for next iteration diagram are not strictly the centroid regions. So we chose a region of 5 pixels of ratio from centroid region position where seeds could be located randomly. This noise is only included in even iterations, stabilizing the system in odd iterations applying the original Lloyd algorithm. We will call this set of images, Voronoi images (see Fig. 4). The iterations considered (from 20 different initial random sets of points) are: iterations 1 to 10, 20, 30, 50, 100, 300 and 700. So we will denote ImageXXX-VoronoiYYY to the image number XXX within the group of YYY-th iteration of Lloid algorithm.

## 5.1   Implementation

In order to perform the computational experiment we have coded the algorithm that implements the procedure represented in Fig. 2. Generally speaking, the algorithm aims to transform a biological picture already segmented in cells into a weighted graph. Finally, the graph is completed to a filtered simplicial complex that is studied by persistent homology and persistent entropy.

Our input Matlab file contains the following variables: a list of the identifiers of valid cells, a list of the identifiers of all the cells, i.e. valid cells and not valid cells, a list of the identifiers of the neighbors of each cell, and for each cell its area measured in squared pixels computed from the original image space. For each input image:

- Map a valid cell to a vertex. For each vertex $v_i$, compute a weight function $f$ described in Subsect. 4.2 as a filter function. This way, in our experiments, we have used two different filter functions:

$$f_1(v_i) = \text{Area of the cell represented by } v_i,$$

to order the cells in terms of their areas, and

$$f_2(v_i) = \text{Number of neighbors} + \frac{\text{area of the cell} - 1}{\text{maximum area among the cells}},$$

to order the cells by the number of neighbors and the cells with the same number of neighbors, by their area.
- For each filter function, $f = f_1, f_2$:
  - Fill the adjacency matrix between vertices. For each edge $e_{i,j}$ between vertex $v_i$ and vertex $v_j$ compute the following weight function:

$$f(e_{i,j}) = \max\{f(v_i), f(v_j)\}.$$

  - List all the maximal cliques of dimension $k \geq 3$ from the weighted graph by using the Eppstein algorithm [8]. We remark that degeneracy-based Eppstein algorithm has a computational complexity of $\Theta(d(n - d)3^{\frac{d}{3}})$, where $d$ is the degeneracy of the graph and $n$ is the number of maximal cliques.
    * For each maximal clique do a tessellation with triangles, now the clique is a set of triangles $\{t_1, t_2, \ldots, t_n\}$.
    * For each triangle $t_l \in \{t_1, t_2, \ldots, t_n\}$ that is composed by the set of edges $\{e_{i,j}, e_{i,k}, e_{j,k}\}$ compute the following weight function:

$$f(t_l) = \max\{f(e_{i,j}), f(e_{i,k}), f(e_{j,k})\}$$

  - For each clique $c_i$ compute the following weight function:

$$f(c_i) = max\{f(t_1), f(t_2), \ldots, f(t_n)\}$$

  - Build a filtered simplicial complex that contains the weighted vertices of the graph as filtered 0-simplices, the weighted edges as filtered 1-simplices, the weighted triangles as filtered 2-simplices, and the weighted $k$-cliques as filtered $k - 1$-simplices.
  - Compute $i$-barcodes for $i = 0, 1$, of the resulting filtered complex and save birth and death times for each persistent homology class.
  - Export the intervals forming the $i$-barcodes to separate text files.
  - For each $i$-barcode compute persistent entropy (PE): $PE_i(f)$.

Given a set of images, we are initially interested in comparing both the entropy and the barcodes themselves. In order to compare the persistent entropies we normalized them for each image as follows

$$PE_i(f) = \frac{PE_i(f)}{\log(\text{number of bars in the barcode})},$$

as it is suggested in [2]. That way, all the entropy values lie in $[0, 1]$.

Moreover, from the text file, we compute the pair-wise Bottleneck distance matrix among the barcodes, see for example Fig. 5.

The computational experiment has been coded in Matlab R2016a and the java package *JavaPlex* has been used for computing persistent homology [1]. The code for computing the Bottleneck distance has been written by Miro Kramar and it can be freely downloaded[3]. The code for listing all the maximal cliques has been written by Darren Strash and it can be freely downloaded[4]. The experiment is executed on the following laptop: Asus G752VY, equipped with a CPU Intel $i7@2.60\,GHz$ and 32 GB of RAM, the average time for the analysis of a single image is of the order of 30 s.

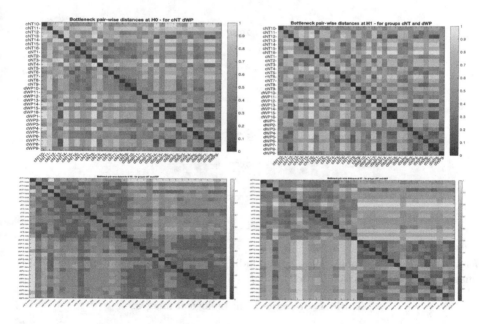

**Fig. 5.** First row: heat maps representing Bottleneck distance between persistence diagrams (0-dimensional diagram on the left, 1-dimensional diagram on the right) of all the samples of cNT and dWP, using filter function $f_1$. Second row: anologous heat maps corresponding to filter function $f_2$.

---

[3] http://www.wpi-aimr.tohoku.ac.jp/hiraoka_labo/miroslav/software/distances-between-the-persi.html.

[4] https://github.com/darrenstrash/quick-cliques.

## 5.2   Results

We have computed persistence barcodes from 32 biological images as well as 320 Voronoi images.

First we have computed Bottleneck distances of pairs of persistence diagrams within each of the two groups using both filter functions $f_1$ and $f_2$. First row of Fig. 5 shows a heat map to visualize such distances for filter function $f_1$, which have been normalized with respect to the maximum distance in the group of 32 biological images. Second row shows analogous heat maps for $f_2$. In the latter, we can appreciate that, for both, 0–dimensional and 1–dimensional persistent homology, two clusters have been formed precisely corresponding to cNT and dWP images. However, no such a clear clustering can be visualized in the case of the filter function that order the cells only by area ($f_1$), so filter function $f_2$ is capturing more characterising properties than $f_1$. Regarding Voronoi images, neither for dimension 0 or 1 and neither for filter function $f_1$ or $f_2$, there is any clear information derived from the corresponding heat maps. This means that persistent homology (via the second filter function) is capturing topological properties of biological images that are not present in the synthetic ones.

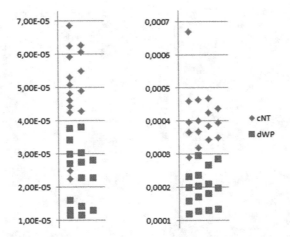

**Fig. 6.** Persistent entropies $PE_0$ corresponding to the 16 samples of cNT and dWP images, using filter function $f_1$ (left) and $f_2$ (right).

**Table 1.** Statistics from biological images.

| Group | Mean $PE_0$ | STD $PE_0$ | Mean $PE_1$ | STD $PE_1$ |
|---|---|---|---|---|
| cNT ($f_1$) | 0.0000496 | 0.0000128 | 0.3885337 | 0.0096051 |
| dWP ($f_1$) | 0.0000235 | 0.0000093 | 0.3920771 | 0.0086374 |
| cNT ($f_2$) | 0.0004082 | 0.0000866 | 0.3936774 | 0.0078098 |
| dWP ($f_2$) | 0.0001950 | 0.0000562 | 0.3880369 | 0.0186688 |

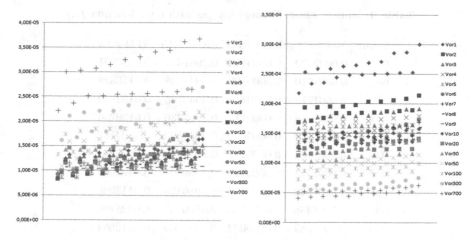

**Fig. 7.** $PE_0$ values of the 20 Voronoi images of each iteration $(1\text{-}10, 20, 30, 50, 100, 300, 700)$ computed over filter function $f_1$ (left) and $f_2$ (right).

**Table 2.** Statistics from Voronoi images with filter function $f_1$.

| Group | Mean $PE_0$ | STD $PE_0$ | Mean $PE_1$ | STD $PE_1$ |
|---|---|---|---|---|
| 001 | 0.0000290 | 0.0000044 | 0.3765092 | 0.0174484 |
| 002 | 0.0000210 | 0.0000030 | 0.3741915 | 0.0199586 |
| 003 | 0.0000173 | 0.0000024 | 0.3673860 | 0.0264625 |
| 004 | 0.0000159 | 0.0000026 | 0.3647423 | 0.0261707 |
| 005 | 0.0000137 | 0.0000021 | 0.3671967 | 0.0219497 |
| 006 | 0.0000140 | 0.0000022 | 0.3708472 | 0.0202959 |
| 007 | 0.0000122 | 0.0000019 | 0.3699749 | 0.0181497 |
| 008 | 0.0000129 | 0.0000019 | 0.3795762 | 0.0144121 |
| 009 | 0.0000113 | 0.0000015 | 0.3788369 | 0.0160855 |
| 010 | 0.0000124 | 0.0000015 | 0.3842456 | 0.0164408 |
| 020 | 0.0000116 | 0.0000012 | 0.3923878 | 0.0125203 |
| 030 | 0.0000111 | 0.0000015 | 0.3964554 | 0.0101165 |
| 050 | 0.0000114 | 0.0000014 | 0.3988734 | 0.0076556 |
| 100 | 0.0000113 | 0.0000011 | 0.4012543 | 0.0062276 |
| 300 | 0.0000108 | 0.0000010 | 0.3964468 | 0.0066500 |
| 700 | 0.0000099 | 0.0000008 | 0.3972066 | 0.0071246 |

Second, we have computed persistent entropy of the persistence 0–barcodes and 1–barcodes, $PE_0$ and $PE_1$, also for the two groups. Figure 6 shows graphic representations of values of $PE_0$ obtained with both filter functions $f_1$ (left) and $f_2$ (right) and we can observe that in the second case, both classes are better separated by more differentiated values of $PE_0$. This fact does not occur

**Table 3.** Statistics from Voronoi images with filter function $f_2$.

| Group | Mean $PE_0$ | STD $PE_0$ | Mean $PE_1$ | STD $PE_1$ |
|-------|-------------|------------|-------------|------------|
| 001 | 0.0002572 | 0.0000194 | 0.3981657 | 0.0090901 |
| 002 | 0.0001910 | 0.0000122 | 0.3947638 | 0.0110328 |
| 003 | 0.0001719 | 0.0000114 | 0.3943745 | 0.0128170 |
| 004 | 0.0001618 | 0.0000114 | 0.3928436 | 0.0129756 |
| 005 | 0.0001533 | 0.0000099 | 0.3940693 | 0.0124182 |
| 006 | 0.0001465 | 0.0000117 | 0.3954044 | 0.0108254 |
| 007 | 0.0001448 | 0.0000119 | 0.3969867 | 0.0122664 |
| 008 | 0.0001430 | 0.0000105 | 0.3986129 | 0.0113020 |
| 009 | 0.0001388 | 0.0000083 | 0.4030867 | 0.0128106 |
| 010 | 0.0001389 | 0.0000084 | 0.4039906 | 0.0112654 |
| 020 | 0.0001254 | 0.0000078 | 0.4109133 | 0.0078082 |
| 030 | 0.0001173 | 0.0000113 | 0.4142859 | 0.0075656 |
| 050 | 0.0001015 | 0.0000060 | 0.4139179 | 0.0054006 |
| 100 | 0.0000852 | 0.0000072 | 0.4080267 | 0.0083150 |
| 300 | 0.0000615 | 0.0000060 | 0.3824104 | 0.0087491 |
| 700 | 0.0000506 | 0.0000056 | 0.3692338 | 0.0100102 |

with values $PE_1$ for any of the two filter functions. Numeric values of mean and standard deviation collected in Table 1 clarify this idea. Regarding Voronoi images, a clearer behaviour can be described in the case of filter function $f_2$ (see Fig. 7), where persistent entropy of first iteration is quite higher than the others and we can observe that, in general, the values decrease as the Voronoi iteration increases (see also Table 3). So $PE_0$, computed over $f_2$, reveals to be of topological significance. In the case of $f_1$ one cannot distinguish such a clear pace, as shown in Table 2. Again, $PE_1$ does not seem to throw any classifying information (neither for $f_1$ or $f_2$) as it can be deduced from Tables 2 and 3.

## 6  Conclusions and Future Work

The work developed in this paper is a first step in a persistent-homology-based topological approach for studying epithelial organization. We have found that the design of the filter function for persistent homology computation is of great importance for getting an informative description of the input data. Our method is able to capture new traits of tissue organization beyond the traditional comparison of polygon distributions used in biology. We have found differences between cNT and Voronoi1 images at the level of persistent entropy that cannot be appreciated comparing the cell sides frequencies (which are similar). Something similar happened when examining dWP and Voronoi5 images. Therefore, we plan to work on this line to improve the initial results presented here and to set persistent entropy as a reference parameter able to quantify topological aspects of cell organization. We will also focus our further efforts on implementing a new version of *jHoles* algorithm that is a tool for computing persistent homology from weighted undirected graphs [3]. The current version of *jHoles* is based on the Bron-Kerbosch algorithm that is characterized by a computational complexity in the worst case bounded by $O(3^{\frac{n}{3}})$, where $n$ is the number of maximal cliques. The new version of *jHoles* will use the Eppstein algorithm instead of Bron-Kerbosch and it will benefit from a reduced computational complexity.

# References

1. Adams, H., Tausz, A.: Javaplex tutorial. Stanford University (2011)
2. Atienza, N., Gonzalez-Diaz, R., Rucco, M.: Persistent Entropy for Separating Topological Features from Noise in Vietoris-Rips Complexes. arXiv:1701.07857
3. Binchi, J., Merelli, E., Rucco, M., Petri, G., Vaccarino, F.: jHoles: a tool for understanding biological complex networks via clique weight rank persistent homology. Electron. Notes Theor. Comput. Sci. **306**, 5–18 (2014)
4. Cerri, A., Di Fabio, B., Jablonski, J., Medri, F.: Comparing shapes through multiscale approximations of the matching distance. Comput. Vis. Image Underst. **121**, 43–56 (2014)
5. Chintakunta, H., Gentimis, T., Gonzalez-Diaz, R., Jimenez, M.J., Krim, H.: An entropy-based persistence barcod. Pattern Recogn. **48**(2), 391–401 (2015)
6. Edelsbrunner, H., Harer, J.: Computational Topology: An Introduction. American Mathematical Society, Providence (2010)
7. Edelsbrunner, H., Letscher, D., Zomorodian, A.: Topological persistence and simplification. In: FOCS 2000, IEEE Computer Society, pp. 454–463 (2000)
8. Eppstein, D., Löffler, M., Strash, D.: Listing all maximal cliques in sparse graphs in near-optimal time. In: Cheong, O., Chwa, K.-Y., Park, K. (eds.) ISAAC 2010. LNCS, vol. 6506, pp. 403–414. Springer, Heidelberg (2010). doi:10.1007/978-3-642-17517-6_36
9. Escudero, L.M., Freeman, M.: Mechanism of G1 arrest in the Drosophila eye imaginal disc. BMC Dev. Biol. **7**, 13 (2007)
10. Escudero, L.M., Costa Lda, F., Kicheva, A., Briscoe, J., Freeman, M., Babu, M.M.: Epithelial organisation revealed by a network of cellular contacts. Nat. Commun. **2**, 526 (2011)
11. Farhadifar, R., Roper, J.C., Aigouy, B., Eaton, S., Julicher, F.: The influence of cell mechanics, cell-cell interactions, and proliferation on epithelial packing. Curr. Biol. **17**(24), 2095–2104 (2007)
12. Ferri, M.: Progress in persistence for shape analysis (extended abstract). In: Bac, A., Mari, J.-L. (eds.) CTIC 2016. LNCS, vol. 9667, pp. 3–6. Springer, Cham (2016). doi:10.1007/978-3-319-39441-1_1
13. Gibson, M.C., Patel, A.B., Nagpal, R., Perrimon, N.: The emergence of geometric order in proliferating metazoan epithelia. Nature **442**(7106), 1038–1041 (2006)
14. Hatcher, A.: Algebraic Topology. Cambridge University Press, Cambridge (2002)
15. Jimenez, M.-J., Medrano, B., Monaghan, D., O'Connor, N.E.: Designing a topological algorithm for 3D activity recognition. In: Bac, A., Mari, J.-L. (eds.) CTIC 2016. LNCS, vol. 9667, pp. 193–203. Springer, Cham (2016). doi:10.1007/978-3-319-39441-1_18

16. Munkres, J.: Elements of Algebraic Topology. Addison-Wesley Co., Reading (1984)
17. Rucco, M., Castiglione, F., Merelli, E., Pettini, M.: Characterisation of the idiotypic immune network through persistent entropy. Proc. ECCS **2014**, 117–128 (2016)
18. Rucco, M., Gonzalez-Diaz, R., Jimenez, M.J., Atienza, N., Cristalli, C., Concettoni, E., Ferrante, A., Merelli, E.: A new topological entropy-based approach for measuring similarities among piecewise linear functions. Sig. Process. **134**, 130–138 (2017)
19. Sánchez-Gutiérrez, D., Tozluoglu, M., Barry, J.D., Pascual, A., Mao, Y., Escudero, L.M.: Fundamental physical cellular constraints drive self-organization of tissues. EMBO J. **35**, 77–88 (2016)
20. Shraiman, B.I.: Mechanical feedback as a possible regulator of tissue growth. Proc. Natl. Acad. Sci. U.S.A. **102**(9), 3318–3323 (2005)
21. Voronoi, G.F.: Nouvelles applications des paramètres continus à la théorie de formes quadratiques. J. Für Die Reine Und Angewandte Mathematik **134**, 198–287 (1908)
22. Zomorodian, A., Carlsson, G.: Computing persistent homology. Discrete Comput. Geometry **33**(2), 249–274 (2005)

# Distance Between Vector-Valued Representations of Objects in Images with Application in Object Detection and Classification

Nataša Sladoje[1,2(✉)] and Joakim Lindblad[1,2]

[1] Centre for Image Analysis, Department of IT, Uppsala University, Uppsala, Sweden
natasa.sladoje@it.uu.se, joakim@cb.uu.se
[2] Mathematical Institute of Serbian Academy of Sciences and Arts, Belgrade, Serbia

**Abstract.** We present a novel approach to measuring distances between objects in images, suitable for information-rich object representations which simultaneously capture several properties in each image pixel. Multiple spatial fuzzy sets on the image domain, unified in a vector-valued fuzzy set, are used to model such representations. Distance between such sets is based on a novel point-to-set distance suitable for vector-valued fuzzy representations. The proposed set distance may be applied in, e.g., template matching and object classification, with an advantage that a number of object features are simultaneously considered. The distance measure is of linear time complexity w.r.t. the number of pixels in the image. We evaluate the performance of the proposed measure in template matching in presence of noise, as well as in object detection and classification in low resolution Transmission Electron Microscopy images.

## 1 Introduction

Fuzzy object representations and related fuzzy image analysis tools show, in general, very good performance when used to model imprecision of images, [1]. Image objects can be well represented by spatial fuzzy sets; that has shown to reduce loss of information caused by discretization and hard decisions made about belongingness of image elements (pixels) to one object exclusively. A variety of image processing tools applicable to fuzzy representations have been proposed, [2,8–11]. Studies confirm increased precision of a number of shape descriptors if fuzzy object representations are used instead of crisp ones, [4,12–14].

Distance measures, being among the most useful image processing tools, have also been proposed for fuzzy objects; a detailed overview can be found in [1]. We have previously suggested a family of fuzzy point-to-fuzzy set distances, which we utilize to develop state-of-the-art performing distance measures between fuzzy sets, [7]. They capture both shape and intensity variations of objects into one distance measure which, as confirmed by performed evaluation, make them well suited for applications in template matching and classification. They can be applied to fuzzy segmented objects, but also directly on gray scale image data.

© Springer International Publishing AG 2017
V.E. Brimkov and R.P. Barneva (Eds.): IWCIA 2017, LNCS 10256, pp. 243–255, 2017.
DOI: 10.1007/978-3-319-59108-7_19

We are aiming at responding to the needs of modern image analysis applications where single channel object representations are not sufficient and novel methods that can handle heterogeneous information-rich representations are needed. Such representations can incorporate information about a variety of different object properties, and may result from fusion of information coming from different sources (e.g., simultaneous imaging by different modalities). We suggest to model each property by a fuzzy set over the image domain, and to create a vector valued fuzzy set to store the membership values assigned to a pixel by each observed fuzzy membership function. Dimension of the representation is equal to the number of observed features/fuzzy sets.

In this paper we extend the path-based point-to-set distance measure and related set-to-set distances proposed in [7]. We introduce the concept of a set distance measure applicable to object representation by vector-valued fuzzy sets and we evaluate the performance of some examples of such measures in template matching and object classification. We observe fuzzy sets representing (1) original image intensities, (2) smoothed image intensities, and (3) gradient magnitude map. The combined use of these fuzzy sets enables simultaneous multi-resolution representation of image data, noise suppression and fine structural analysis.

## 2   Preliminaries

### 2.1   Fuzzy Sets

A **fuzzy set** [16] $S$ on a reference set $X$ is a set of ordered pairs $S = \{(x, \mu_S(x)) : x \in X\}$, where $\mu_S : X \to [0,1]$ is the membership function of the set $S$.

A crisp set $C \subset X$ is a special case of a fuzzy set, defined by its characteristic function as its membership function: $\mu_C(x) = \begin{cases} 1, \text{ for } & x \in C \\ 0, \text{ for } & x \in \overline{C}. \end{cases}$

The *height* of a fuzzy set $S$ is $h(S) = \max_{x \in X} \mu_S(x)$.

The *support* of $S$ is $\text{supp}(S) = \{x \in X : \mu_S(x) > 0\}$.

The *complement* $\overline{S}$ of a fuzzy set $S$, is $\overline{S} = \{(x, 1 - \mu_S(x)) : x \in X\}$.

A *fuzzy point* $p$ defined at $p \in X$, also called a fuzzy *singleton*, with height $h(p)$ is defined by a membership function

$$\mu_p(x) = \begin{cases} h(p), \text{ for } & x = p \\ 0, \quad \text{ for } & x \neq p. \end{cases}$$

### 2.2   Distance Transforms and Path-Based Distances

The (internal) **Distance transform** (DT) of a crisp set $A \subset X$ is

$$\text{DT}[A](x) = \min\{d(x, y) : y \in \overline{A}\}.$$

Most often used point-to-point distance $d$ is the Euclidean distance. In this paper we use its squared version.

The **Fuzzy distance transform** [11] (FDT) of a fuzzy set $\mathcal{A}$ is

$$\text{FDT}[\mathcal{A}](x) = \min\{d_{\mathcal{A}}(x,y) : y \in \overline{\text{supp}(\mathcal{A})}\}. \tag{1}$$

The point-to-point distance $d_{\mathcal{A}}$ in (1) is commonly a path-based distance. The **path-based distance** (a.k.a. shortest path distance) between two points in X is the length $l$ of a shortest path connecting them:

$$d(p,q) = \min_{\pi(p,q)} l(\pi(p,q)).$$

In the discrete case, a path $\pi$ between grid points $p$ and $q$, is a sequence of adjacent grid points, $\pi : p = r_1, r_2, \ldots r_{n-1}, r_n = q$, with respect to a given adjacency relation. The length of a path is commonly computed as a sum of weighted local steps,

$$l(\pi(p,q)) = \sum_{i=1}^{n-1} w(r_i, r_{i+1}) d(r_i, r_{i+1}), \tag{2}$$

where $d$ is a distance between adjacent grid points [3] and $w(r_i, r_{i+1})$ is some non-negative cost function. The approach used in this paper is defined in [11]. We use

$$w(r_i, r_{i+1}) = \frac{1}{2}\big(\mu_{\mathcal{A}}(r_i) + \mu_{\mathcal{A}}(r_{i+1})\big)$$

in (2) to define a path-based-distance $d_{\mathcal{A}}(p,q)$ between points $p$ and $q$ in a fuzzy set $\mathcal{A}$. As shown in [11], $d_{\mathcal{A}}(p,q)$ is a metric on the support of a fuzzy set $\mathcal{A}$.

## 2.3  Path-Based Point-to-Set Distances

Two path-based point-to-set distances, exhibiting several good properties, are introduced in [7]. The *inwards path-based distance* is zero as soon as the height of the point is not larger than the membership function of the set at the same position. The *bi-directional path-based point-to-set distance* is zero if and only if the height of the point is the same as the membership function value of the set at that position. Illustrative "definitions" are given in Fig. 1, where the cost of a path between the point and the set is indicated (red/striped area) and its suitable measure represents the desired distance.

Formal definitions of the distances follow.

For a fuzzy point $p$ with height $h(p)$ and a fuzzy set $S$ with membership function $\mu_S$, let us denote with $\mathcal{D}_{p,S}$ the fuzzy set with membership function

$$\mu_{\mathcal{D}_{p,S}} = \max\{0, h(p) - \mu_S\}. \tag{3}$$

**Definition 1 ([7]).**  *The inwards path-based point-to-set distance between a fuzzy point $p$ and a fuzzy set $S$, where $h(p) \leq h(S)$, is*

$$d^{\pi}(p,S) = \text{FDT}[\mathcal{D}_{p,S}](p). \tag{4}$$

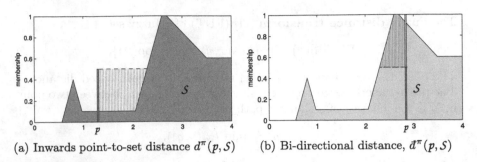

(a) Inwards point-to-set distance $d^\pi(p, S)$     (b) Bi-directional distance, $\bar{d}^\pi(p, S)$

**Fig. 1.** Illustration of distance between a fuzzy point $p$ with height $h(p) = 0.5$ (shown as a vertical stick reaching membership 0.5 at spatial positions $p = 1.3$ (a) resp. $p = 2.8$ (b)) and a one-dimensional fuzzy set $S$, with a piecewise linear membership function (shown in blue). The resp. distance corresponds to the area of the red striped region. (Color figure online)

**Definition 2 ([7]).** *The bi-directional path-based point-to-set distance between a fuzzy point $p$ and a fuzzy set $S$, s.t. $\min_{x \in X} \mu_S(x) \leq h(p) \leq h(S)$, is*

$$\bar{d}^\pi(p, S) = d^\pi(p, S) + \bar{d}^\pi(p, S) = \text{FDT}[\mathcal{D}_{p,S}](p) + \text{FDT}[\mathcal{D}_{\bar{p},\bar{S}}](p). \tag{5}$$

FDT is defined according to (1) and $\text{FDT}[\mathcal{D}_{\bar{p},\bar{S}}]$ is computed on a fuzzy set with membership function

$$\mu_{\mathcal{D}_{\bar{p},\bar{S}}} = \max\{0, h(\bar{p}) - \mu_{\bar{S}}\} = \max\{0, \mu_S - h(p)\}.$$

### 2.4 Distance Between Fuzzy Sets

Distances between sets of points are often based on a point-to-set distance. Examples include the widely used Hausdorff distance, and the Sum of minimal distances (SMD) [5,6]. In [7] a family of distances between fuzzy sets is defined based on different fuzzy point-to-fuzzy set distances. For this study, the most relevant set distance measure is the one based on SMD with path-based point-to-set distances (4) and (5). We take the definition from [7].

For a given fuzzy set $\mathcal{A}$, let $\mathcal{A}(x)$ indicate the fuzzy point at $x \in X$ with height equal to $\mu_{\mathcal{A}}(x)$.

**Definition 3 ([7]).** *The sum of minimal distances (SMD) between fuzzy sets $\mathcal{A}$ and $\mathcal{B}$, based on a point-to-set distance $d^* \in \{d^\pi, \bar{d}^\pi\}$, is*

$$d^*_{\text{SMD}}(\mathcal{A}, \mathcal{B}) = \frac{1}{2}\left( \sum_{x \in X} d^*(\mathcal{A}(x), \mathcal{B}) + \sum_{x \in X} d^*(\mathcal{B}(x), \mathcal{A}) \right).$$

## 3   Novel Point-to-Set Distances for Vector-Valued Fuzzy Sets

When representing an image with a spatial fuzzy set, the membership function reflects spatial distribution of some observed property over the image domain.

Each pixel in the underlying grid is assigned the degree to which it exhibits the particular property. Belongingness to the object can be given by the original gray level map, and/or different fuzzy segmentation techniques can be used to obtain suitable membership functions (coverage of a pixel by the image object, connectedness w.r.t. a seed point, edgeness measured as e.g. gradient intensity). It may be beneficial to observe several of these features simultaneously. In that case each pixel in the image is assigned a sequence of values – its memberships to each of the observed fuzzy sets (properties). A convenient way to handle these values is to use vector-valued membership functions.

Our aim is to develop image analysis tools for such information-rich representations. In this study we propose *a path-based distance between a vector-valued fuzzy point and a vector-valued fuzzy set*. To define such a distance several approaches can be followed. A vector-valued fuzzy set can be aggregated to a single-valued fuzzy set, utilizing some appropriate aggregation principle, which then allows application of any existing distance measures between fuzzy objects. An alternative approach is to apply existing distance measures for single-valued fuzzy objects component-wise, and then aggregate the obtained distances.

We suggest to take an approach in between the two mentioned: instead of aggregating early (vector-valued sets), or late (component-wise distance measures), we aggregate the generated distance transform landscapes internally in the distance measure. We believe that this approach better utilizes the joint information included in the vector-valued fuzzy representations.

## 3.1   Vector-Valued Fuzzy Sets

Let us observe $n$ fuzzy sets, $S_1, S_2, \ldots, S_n$, defined on a reference set $X$. Let $\mu_{S_i} : X \to [0,1]$ be the membership function of the fuzzy set $S_i$, for $i = 1, 2, \ldots, n$.

A *vector-valued fuzzy set* $S$ on $X$, is a set of ordered $(n+1)$-tuples

$$S = \{(x, \mu_{S_1}(x), \mu_{S_2}(x), \ldots, \mu_{S_n}(x)) : x \in X\}.$$

We denote $\mu_S = (\mu_{S_1}, \mu_{S_2}, \ldots, \mu_{S_n})$. The sequence of values assigned to $x \in X$ by the observed membership functions is denoted by $\mu_S(x) = (\mu_{S_1}(x), \mu_{S_2}(x), \ldots, \mu_{S_n}(x))$.

A *vector-valued fuzzy point* $p$ at a point $p \in X$, with the (vector-valued) height $h(p) = (h_1(p), h_2(p), \ldots, h_n(p))$, w.r.t. the components of a vector-valued fuzzy set, is defined by a membership function

$$\mu_p(x) = \begin{cases} h(p) = (h_1(p), \ldots, h_n(p)), & \text{for} \quad x = p \\ 0, & \text{for} \quad x \neq p. \end{cases}$$

The complement of a vector-valued fuzzy set is given by the complement of its components: $\bar{S} = \{(x, 1 - \mu_{S_1}(x), \ldots, 1 - \mu_{S_n}(x)) : x \in X\}$.

The definitions of the proposed distance measures rely on the following aggregation of the fuzzy distance transform landscape, as a generalization of (3):

For a vector-valued fuzzy point $p$ with height $h(p)$ and a vector-valued fuzzy set $S$ with membership function $\mu_S$, let us denote with $\mathcal{D}_{p,S}$ the (scalar valued) fuzzy set with membership function

$$\mu_{\mathcal{D}_{p,S}} = \max\{0, h_1(p) - \mu_{S_1}, \ldots, h_n(p) - \mu_{S_n}\}.$$

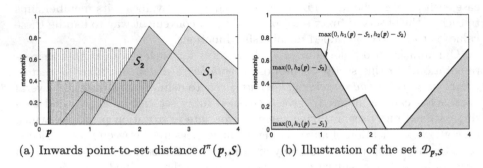

(a) Inwards point-to-set distance $d^\pi(p, S)$     (b) Illustration of the set $\mathcal{D}_{p,S}$

**Fig. 2.** Illustration of distance between a vector valued fuzzy point $p$ with height $h(p) = (0.4, 0.7)$ (shown as a green and blue vertical stick reaching memberships 0.4 and 0.7 at spatial position $p = 0.2$) and a two-component one-dimensional fuzzy set $S$, with piecewise linear membership functions $S_1$ and $S_2$ (shown in green and blue). Cf. Fig. 1(a). (Color figure online)

Using this extension, the *inwards* and the *bi-directional path-based point-to-set distances* follow directly, as generalizations of Definitions 1 and 2.

**Definition 4.** *The inwards path-based point-to-set distance between a vector-valued fuzzy point $p$ and a vector-valued fuzzy set $S$, where $h_i(p) \leq h(S_i)$, for $i = 1, 2, \ldots, n$, is*

$$d^\pi(p, S) = \text{FDT}[\mathcal{D}_{p,S}](p). \tag{6}$$

The inwards path based point-to-set distance $d^\pi(p, S)$ is illustrated in Fig. 2, for a two component fuzzy point $p$ and set $S$ on a 1D reference set on $\mathbb{R}$.

**Definition 5.** *The bi-directional path-based point-to-set distance between a vector-valued fuzzy point $p$ and a vector-valued fuzzy set $S$, s.t. $\min\limits_{x \in X} \mu_{S_i}(x) \leq h_i(p) \leq h(S_i)$, for $i = 1, 2, \ldots, n$, is*

$$\bar{d}^\pi(p, S) = d^\pi(p, S) + \vec{d}^\pi(p, S) = \text{FDT}[\mathcal{D}_{p,S}](p) + \text{FDT}[\mathcal{D}_{\bar{p},\bar{S}}](p). \tag{7}$$

## 4   Distances Between Vector-Valued Fuzzy Sets

The point-to-set distances defined in the previous section can be inserted into different expressions for set-to-set distances. In this study we insert (7) into the weighted SMD to define a distance measure between vector-valued fuzzy sets.

For a vector-valued fuzzy set $\mathcal{A}$, let $\mathcal{A}(x)$ indicate the vector-valued fuzzy point at $x \in X$, such that $h_i(\mathcal{A}(x)) = \mu_{\mathcal{A}_i}(x)$.

**Definition 6.** *Given two weight functions* $w_\mathcal{A}, w_\mathcal{B} : X \to \mathbb{R}$, *the Weighted sum of minimal distances (wSMD) between vector-valued fuzzy sets $\mathcal{A}$ and $\mathcal{B}$, based on the point-to-set distance* $d^* \in \{d^\pi, \bar{d}^\pi\}$, *is*

$$d^*_{\text{wSMD}}(\mathcal{A}, \mathcal{B}, w_\mathcal{A}, w_\mathcal{B}) = \frac{1}{2}\left( \sum_{x \in X} w_\mathcal{A}(x) d^*(\mathcal{A}(x), \mathcal{B}) + \sum_{x \in X} w_\mathcal{B}(x) d^*(\mathcal{B}(x), \mathcal{A}) \right).$$

As commented in [7], the symmetric treatment of the two observed sets, provided by Definition 6, is desired when the involved sets are similar (in the sense of noise level, size, etc.) This is often the case for, e.g., image registration. However, in applications such as, e.g., template matching, an asymmetric treatment of the two sets may be more appropriate.

**Definition 7.** *Given a weight functions* $w_\mathcal{A} : X \to \mathbb{R}^+$, *the asymmetric Weighted sum of minimal distances from a fuzzy sets $\mathcal{A}$ to a fuzzy set $\mathcal{B}$, based on the point-to-set distance* $d^* \in \{d^\pi, \bar{d}^\pi\}$, *is*

$$\underrightarrow{d}^*_{\text{wSMD}}(\mathcal{A}, \mathcal{B}, w_\mathcal{A}) = \sum_{x \in X} w_\mathcal{A}(x) d^*(\mathcal{A}(x), \mathcal{B}). \tag{8}$$

# 5  Implementation and Complexity Analysis

The Fuzzy Distance Transform is of linear complexity w.r.t. number of pixels, the same holds for the proposed set distance. A separate distance transform is required for each combination of membership values present. From a computational point of view, it is often beneficial to compute all required distance transforms in advance. Remaining computation required for the proposed set distances then reduces to one or two lookup-table accesses per pixel.

The selection of the number of gray/membership levels to use in object representation affects both computational time and performance of the distance measures. A good balance between gain in information preservation from gray-levels and loss in speed should be made. It is often beneficial to use a coarser quantization than what is provided from the input image; intensity variations which are mainly attributed to image noise do not need to be well preserved. In Sect. 6.2 we evaluate performance vs. number of membership levels. To further limit the impact of noise, it may be beneficial to set an upper limit to the point-to-set-distance by restricting the values of the fuzzy distance transform to the range $[0, d_{\max}]$. This circumvents the height requirements in Definitions 4 and 5.

# 6  Performance Analysis

The proposed distance measures use not only spatial and intensity information, but allow to incorporate a variety of image features. We illustrate the performance of the proposed distance measures between vector-valued fuzzy sets on examples related to template matching and object classification. We compare them with the state-of-the-art distance measure proposed in [7].

## 6.1  Template Matching

We evaluate performance of the proposed distances in template matching, observing a well known data set. From the *Lena* image, Fig. 3(a), we cut out the region corresponding to Lena's right eye, Fig. 3(b), and use that as a template which is to be appropriately positioned in the original image. We are particularly interested in the performance in presence of noise. The distance measures proposed in [7] exhibit relatively high noise-sensitivity; many pixels appear close to noise (offering all possible membership values). Utilizing more features, even if extracted from a noisy image, decreases the probability that noise will simultaneously provide a suitable match for several features of a pixel. We evaluate the noise sensitivity of the distance measure in template matching, searching for a noise free template in an observed image which is heavily corrupted (signal-to-noise ratio $-7.2\,\mathrm{dB}$) by a sum of Gaussian white noise and blurred ($\sigma = 5$) Gaussian noise.

We consider the following components in scalar and vector valued representations: (i) The intensity values of the observed noisy *Lena* image, Fig. 3(f); (ii) The observed image convolved by a Gaussian smoothing filter with $\sigma = 2$, Fig. 3(g); (iii) Gradient magnitude map, Fig. 3(h), (computed from convolutions with derivative of Gaussian with $\sigma = 1$) of the observed image. In Fig. 3(d)–(e) the corresponding blur and gradient magnitude membership functions of the template, obtained by blurring and gradient map computation on the noise-free template (b), are shown. The values in all images are scaled to the interval $[0,1]$ and considered as membership values of fuzzy sets.

We evaluate separately all combinations of the three observed membership functions in vector-valued representations (of both template and image), incorporating one, two, or all three, values assigned to each pixel. Appropriately for template matching, we use the asymmetric set distance formulation. Figure 3(c) shows the weight mask used in asymmetric wSMD, according to Definition 7. We use $d_{\max} = 30$. For the special case of a single valued fuzzy representation, Definition 7 is consistent with the one in [7]. This evaluation therefore enables direct performance comparison with the corresponding measure presented in [7].

For each observed vector-valued fuzzy representation, the template is translated over the image and the asymmetric wSMD between the template and the image is evaluated at each position. This exhaustive evaluation of the search space, computed using fast convolutions, enables us to draw conclusions on how gradient based optimization would work on the same task. We observe the number of regional minima (NoM) in the search space, and the size of the catchment basin (CB) (region of attraction) of the global minimum relative to the total size of the search space. A large CB of the global optimum increases probability that gradient based optimization converges to the correct global optimum.

**Results.** The first row of Fig. 4 illustrates the performance of the asymmetric wSMD in finding the template when relying on individual (scalar-valued) fuzzy sets (intensity, $I$, blurred, $B$, and gradient magnitude, $G$). The second row shows performance when combining two sets ($IB$, $BG$, and $IG$), and the last row

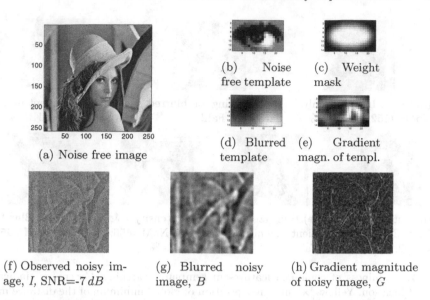

(a) Noise free image

(b) Noise free template

(c) Weight mask

(d) Blurred template

(e) Gradient magn. of templ.

(f) Observed noisy image, $I$, SNR=-7 $dB$

(g) Blurred noisy image, $B$

(h) Gradient magnitude of noisy image, $G$

**Fig. 3.** Template matching: (a) Reference *Lena* image, (b) selected template and (c) the weight mask used. (f) Observed noisy image in which we search for (b). (d, e) and (g, h) additional component representations utilized in the matching.

shows the result when all three sets ($IBG$, raw intensity, blurred, and gradient magnitude) are utilized together.

The green "+" sign indicates the correct position of the template. Yellow "×" indicates position of the global minimum of the distance map computed for a particular representation (in a noise-free case this distance is equal to zero, but in presence of noise that is not granted). In the case of a successful detection, the signs overlap, forming a "✳"; otherwise the matching failed. The red region around the yellow "×" corresponds to the CB of the global minimum; the larger the CB, the better (however, this is meaningful only for successful matches). The NoM (smaller is better) and the size of the CB relative to the image size are indicated for successful detections.

The best performance is observed when all the three fuzzy membership functions are used together, combining the good properties of the three; a large CB (from the blurred image) with the higher specificity of the other two. Following in performance is the $IB$ vector-valued fuzzy representation. The $IG$ representation produces smaller CB than $IB$, but larger than $I$, which is, on the other hand, the only single valued representation that provides successful matching (and corresponds to the result presented in [7]).

## 6.2 Cilia Detection and Classification

We explore the applicability of the proposed distance measure in a template matching based method for automated detection of cilia objects in low

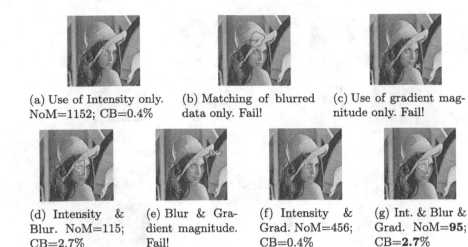

(a) Use of Intensity only. NoM=1152; CB=0.4%

(b) Matching of blurred data only. Fail!

(c) Use of gradient magnitude only. Fail!

(d) Intensity & Blur. NoM=115; CB=2.7%

(e) Blur & Gradient magnitude. Fail!

(f) Intensity & Grad. NoM=456; CB=0.4%

(g) Int. & Blur & Grad. NoM=**95**; CB=**2.7%**

**Fig. 4.** Usefulness of combined features in template matching. Green "+" indicates correct location. Yellow "×" indicates position of global minimum of the distance map. Successful detection is indicated by "*"; otherwise the matching failed. The red region corresponds to the CB of the global minimum; larger is better (for successful matching). The NoM (smaller is better) and the relative size of the CB are indicated for successful detections. (a) $I$; (b) $B$; (c) $G$; (d) $IB$; (e) $BG$; (f) $IG$; (g) $IBG$. We observe that the combined use of several membership functions in the proposed distance measure provides most reliable detection. (Color figure online)

magnification Transmission Electron Microscopy (TEM) images. Cilia are hair-like cell organelles protruding from cells; their dysfunctionality (often due to genetic disorders) causes a number of serious health problems. For setting a pathological diagnosis, it is required to efficiently detect regions highly populated by cilia, at very low image resolution, where a cilium instance does not have more than 20 pixels in diameter. At such a resolution, the characteristic cilia structure is barely resolved and can hardly be used as a reliable discriminative feature. However, the shape and size can be used to detect relevant areas, as shown in [15].

The task is so far addressed by template matching utilizing Normalized Cross Correlation and a highly optimized synthetic template. We believe that the excellent discriminative power of our proposed distance measure may allow to directly utilize a few cilia cut-outs as templates, avoiding the tedious synthetic template optimization which requires a large number of annotated cilia observations and has to be repeated whenever image resolution, or other imaging conditions, are changed. To compensate for the very low presence of characteristic structural details in cilia instances at so low resolutions as well as to take into account within class variations, we utilize multiple templates, i.e., a number of object cut-outs, to detect all resembling objects in the image. At each position, distance from each of the templates and the image is computed, and the lowest value is assigned to the position.

We start from 9 cilia cut-outs (three are shown in Fig. 5(a)–(c)). The cut-outs are rotated in steps of $90°$ and mirrored to produce $9 \times 4 \times 2 = 72$ templates $T_i$. For each template $T_i$, the corresponding distance to the observed image $I$, $\underrightarrow{d}_{\text{wSMD}}(T_i, I, w)$, is computed for every translation of $T_i$ over $I$ (using FFT convolutions). We limit the point-to-set distance values to $d_{\text{max}} = 300$. Similarly as in Sect. 6.1, we evaluate 7 combinations of one, two, or three, membership functions to create vector valued representations.

For each image position, the smallest value (the best fitting template) over the 72 distance maps is selected. Every local minimum in this final distance map which has a distance value below a detection threshold is considered a detected cilium. We set the detection threshold at the level which maximizes $F_1$ score, $F_1 = 2 \cdot \frac{precision \cdot recall}{precision + recall}$. This approach, which provides easy comparison between methods, is possible since ground truth is provided by a pathologist. We compare the performance of the different combinations of membership functions, and we analyze the performance w.r.t. the number of membership levels in the observed fuzzy representations.

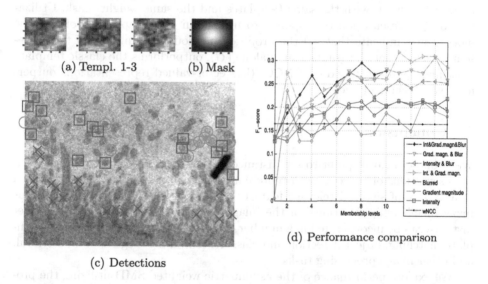

(a) Templ. 1-3        (b) Mask

(d) Performance comparison

(c) Detections

**Fig. 5.** Detection of cilia in a TEM image by template matching. The proposed distance (Definition 7) and different vector-valued fuzzy representations are utilized. (a) Cut-outs from the original image used as templates; (b) Weight mask used for wSMD computation. (c) Detected objects in the original image when using $IG$ and 11 membership levels: green circles are correctly detected, red crosses indicate false positives (wrongly detected objects), and red squares indicate false negatives (missed cilia); (d) Performance comparison: $F_1$-score w.r.t number of quantization levels, for all observed representations. (Color figure online)

**Results.** A low magnification (LM) TEM image of cilia, of size $4096 \times 4096$ pixels, with a Field of View (FOV) $60.6\,\mu m$ is used in the study. A part of the image is shown in Fig. 5(c). Nine templates are cut out from the LM image, each of a size $23 \times 23$ pixels, and containing one cilium; three examples are shown in Fig. 5(a), reflecting the difficulty of the task. The same types of vector-valued fuzzy representations as in Sect. 6.1 are used, however in this case no noise-free image is available and both the reference and the template, as acquired, contain high level of noise. Asymmetric wSMD is used, with a weight mask $w$ shown in Fig. 5(b).

The detection results are presented in Fig. 5(c)–(d). The plot shows $F_1$ score as a function of the number of membership levels used, for the proposed distance measure applied to 7 different vector-valued fuzzy representations: three are single-channeled ($I,G,B$) and four are their different combinations ($IG$, $IB$, $GB$, and $IGB$). The $IGB$ combination is, however, rather computationally demanding; our naive MATLAB implementation ran out of memory past 10 membership levels, indicating the need for a more efficient implementation. As a base-line performance, the weighted Normalized Cross Correlation (wNCC) on the intensity image ($I$) is used, with the same templates and the same weight mask. Utilization of $I$ representation corresponds to utilization of the approach of [7]. The plots clearly indicate that: (i) the proposed method, utilizing any of the representations, except the gradient map alone ($G$), outperforms the classic template matching approach based on wNCC; (ii) vector-valued representations outperform single-channel ones.

## 7   Conclusion

We propose a novel point-to-set distance measure applicable to vector-valued representations of objects. Such representations enable simultaneous utilization of a variety of selected relevant features of the observed images, presented as fuzzy membership functions on the image domain. The proposed point-to-set distance can be incorporated in a number of set-to-set distances and used in template matching, object detection and classification, image registration, retrieval, and other image processing tasks.

We explore performance of the asymmetric weighted SMD utilizing the proposed point-to-set distance in template matching and object detection and classification. We observe evident improvement compared to the previously suggested approach utilizing one fuzzy set as an object representation. Considering comparative analysis conducted in [7], where the distance measures applicable to (single) fuzzy representations outperformed several other widely used approaches, we can conclude that the improvements proposed in this paper are highly relevant for the field.

We note that an alternative and closely related approach is recently presented in [17]. Due to simultaneous submission and processing/reviewing times, performance comparison of the two methods remains as future work.

**Acknowledgment.** The authors acknowledge Amit Suveer, Anca Dragomir, and Ida-Maria Sintorn for acquired and annotated TEM images of Cilia. Ministry of Science of the Republic of Serbia is acknowledged for support through the Projects ON 174008 and III 44006 of MI-SANU. N. Sladoje is also supported by Swedish Governmental Agency for Innovation Systems (VINNOVA).

# References

1. Bloch, I.: On fuzzy distances and their use in image processing under imprecision. Pattern Recogn. **32**, 1873–1895 (1999)
2. Bloch, I., Maître, H.: Fuzzy mathematical morphologies: a comparative study. Pattern Recogn. **28**(9), 1341–1387 (1995)
3. Borgefors, G.: Distance transformations in digital images. Comput. Vis. Graph. Image Process. **34**, 344–371 (1986)
4. Dražić, S., Sladoje, N., Lindblad, J.: Accurate estimation of feret's diameter of a shape from pixel coverage digitization. Pattern Recogn. Lett. **80**, 37–45 (2016)
5. Eiter, T., Mannila, H.: Distance measures for point sets and their computation. Acta Informatica **34**(2), 103–133 (1997)
6. Lindblad, J., Ćurić, V., Sladoje, N.: On set distances and their application to image registration. In: Proceedings of the IEEE International Symposium Image Signal Processing and Analysis (ISPA), pp. 449–454 (2009)
7. Lindblad, J., Sladoje, N.: Linear time distances between fuzzy sets with applications to pattern matching and classification. IEEE Trans. Image Process. **23**(1), 126–136 (2014)
8. Rosenfeld, A.: Fuzzy digital topology. Inf. Control **40**, 76–87 (1979)
9. Rosenfeld, A.: The fuzzy geometry of image subsets. Pattern Recogn. Lett. **2**, 311–317 (1984)
10. Saha, P.K., Udupa, J.K.: Relative fuzzy connectedness among multiple objects: theory, algorithms, and applications in image segmentation. Comput. Vis. Image Underst. **82**(1), 42–56 (2001)
11. Saha, P.K., Wehrli, F.W., Gomberg, B.R.: Fuzzy distance transform: theory, algorithms, and applications. Comput. Vis. Image Underst. **86**, 171–190 (2002)
12. Sladoje, N., Lindblad, J.: Estimation of moments of digitized objects with fuzzy borders. In: Roli, F., Vitulano, S. (eds.) ICIAP 2005. LNCS, vol. 3617, pp. 188–195. Springer, Heidelberg (2005). doi:10.1007/11553595_23
13. Sladoje, N., Lindblad, J.: High precision boundary length estimation by utilizing gray-level information. IEEE Trans. Pattern Anal. Mach. Intell. **31**(2), 357–363 (2009)
14. Sladoje, N., Nyström, I., Saha, P.K.: Measurements of digitized objects with fuzzy borders in 2D and 3D. Image Vis. Comput. **23**, 123–132 (2005)
15. Suveer, A., Sladoje, N., Lindblad, J., Dragomir, A., Sintorn, I.-M.: Automated detection of cilia in low magnification transmission electron microscopy images using template matching. In: Proceedings of IEEE International Symposium on Biomedical Imaging (ISBI), pp. 386–390 (2016)
16. Zadeh, L.: Fuzzy sets. Inf. Control **8**, 338–353 (1965)
17. Öfverstedt, J., Sladoje, N., Lindblad, J.: Distance between vector-valued fuzzy sets based on intersection decomposition with applications in object detection. In: Angulo, J., et al. (eds.) ISMM 2017. LNCS, vol. 10225, pp. 395–407. Springer, Cham (2017). doi:10.1007/978-3-319-57240-6_32

# A Statistical-Topological Feature Combination for Recognition of Isolated Hand Gestures from Kinect Based Depth Images

Soumi Paul[1], Hayat Nasser[2,3(✉)], Mita Nasipuri[1], Phuc Ngo[2,3],
Subhadip Basu[1], and Isabelle Debled-Rennesson[2,3]

[1] Department of Computer Science and Engineering,
Jadavpur University, Kolkata 700032, India
soumip@research.jdvu.ac.in, {mnasipuri,subhadip}@cse.jdvu.ac.in
[2] Université de Lorraine, LORIA, UMR 7503, 54506 Vandoeuvre-lès-nancy, France
[3] CNRS, LORIA, UMR 7503, 54506 Vandoeuvre-lès-nancy, France
{hayat.nasser,hoai-diem-phuc.ngo,isabelle.debled-rennesson}@loria.fr

**Abstract.** Reliable hand gesture recognition is an important problem for automatic sign language recognition for the people with hearing and speech disabilities. In this paper, we create a new benchmark database of multi-oriented, isolated ASL numeric images using recently launched Kinect V2. Further, we design an effective statistical-topological feature combinations for recognition of the hand gestures using the available V1 sensor dataset and also over the new V2 dataset. For V1, our best accuracy is 98.4% which is comparable with the best one reported so far and for V2 we achieve an accuracy of 92.2% which is first of its kind.

**Keywords:** Hand gesture recognition · Sign language · Kinect · Depth data · Statistical-topological features · Discrete curve · Polygonal simplification

## 1 Introduction

Hand gesture recognition is of great importance due to its potential applications in contactless human-computer interaction (HCI). In particular, reliable hand gesture recognition is crucial for many applications, including automatic sign language recognition for the HCI of hearing and speech impaired persons. Some of these techniques require wearing of an electronic glove [5] so that the key features of hand can be accurately measured, but the device is somewhat costly and inconvenient for domestic applications. Another class of methods uses optical markers [8] instead of electronic gloves but it requires rather complex configuration. On the other hand, affordable depth-based systems are coming up with promising results in the field of depth-based hand gesture recognition. Microsoft Kinect [15] is one such RGB-D sensor providing synchronized color and depth images. It was launched as a gaming device, but computer vision research community has taken interest into it and extended it for a lower cost replacement for

V.E. Brimkov and R.P. Barneva (Eds.): IWCIA 2017, LNCS 10256, pp. 256–267, 2017.
DOI: 10.1007/978-3-319-59108-7_20

traditional 3D cameras, such as stereo cameras and time-of-flight (TOF) cameras. In just two years after Kinect V1 was released, a large number of scientific papers with technical demonstrations have started appearing in diverse publication venues. Recently, the new version of the Kinect sensor V2 has been launched with more accurate depth sensing technology. The work presented in this paper involves both V1 and V2 sensors for recognition of isolated depth images of ASL.

**Literature Survey and Our Contributions.** Many vision based hand gesture recognition algorithms have been proposed in the past years and comprehensive reviews can be found in [9]. Methods based on skin color model [21] and hand shape model [20] have also been proposed. However, they are not robust in the dynamic environment and rely significantly on the models. The recent development of depth cameras, such as Microsoft Kinect [15], Creative Senz3D or Mesa Swiss-Ranger etc., opens up new avenues for hand gesture recognition. Therefore, how the depth information can be efficiently utilized and how the depth camera can be incorporated in the hand gesture recognition system is an active topic of research [17]. In early studies, hand detection mainly relies on vision-based features which was sensitive to variations of skin colors and lighting. On the other hand, depth camera offers a much simpler way of isolating hands by depth thresholding. After the hand localization and segmentation, various hand features can be extracted from either the depth maps, e.g. Histogram of 3D Facets (H3DF) [22], or the corresponding color images such as Histogram of Oriented Gradients (HOG) [3], which will then be used for hand gesture recognition.

We have already explored that depth based image recognition has more advantage over vision based systems. Within depth based systems, recent trend is to use low cost device like Kinect V1 to input images, which gives a color image of $640 \times 480$ resolution and depth map of $320 \times 240$ at 30 FPS. Whereas recently launched V2 has better RGB resolution $1920 \times 1080$ and depth resolution $512 \times 424$ at 30 FPS. Not only that, the field of view (i.e., the solid angle through which the detector is sensitive to electromagnetic radiation) has also been expanded, skeleton joint point has also been upgraded from 20 to 26 and most importantly, with USB 3.0, the speed has been increased to get more support for real time applications. A detailed comparison of Kinect V1 and V2 can be found in [16].

With the use of depth data we can detect hand gestures robustly in the cluttered background independent of lighting conditions. So the objective of our proposed work is to, (1) create a new challenging benchmark of multi-oriented, isolated ASL numeric image dataset using recently launched Kinect V2, (2) design an effective statistical-topological feature combinations for recognition of the hand gestures using the available V1 sensor dataset and also the new V2 dataset.

## 2    Tools to Study Discrete Contours

In this section, we recall a method of contour simplification based on selected dominant points. They are computed using a discrete structure, named adaptive tangential cover (ATC) [11], well adapted to analyse irregular noisy contours.

### 2.1    Adaptive Tangential Cover [11]

An **adaptive tangential cover (ATC)** is composed of a sequence of maximal straight segments, called maximal blurred segments, of the studied contour. The notion of maximal blurred segment has been introduced in [4] as an extension of arithmetical discrete line [14] with a width parameter for noisy or disconnected digital contours.

**Definition 1.** *An **arithmetical discrete line** $\mathcal{D}(a, b, \mu, \omega)$, with a main vector $(b, a)$, a lower bound $\mu$ and an arithmetic thickness $\omega$ (with $a, b, \mu, \omega \in \mathbb{Z}$ and $gcd(a, b) = 1$) is the set of integer points $(x, y)$ verifying $\mu \leq ax - by < \mu + \omega$.*

**Definition 2.** *A set $S_f$ is a **blurred segment of width** $\nu$ if the discrete line $\mathcal{D}(a, b, \mu, \omega)$ containing $S_f$ has the vertical (or horizontal) distance $d = \frac{\omega - 1}{\max(|a|, |b|)}$ equal to the vertical (or horizontal) thickness of the convex hull of $S_f$, and $d \leq \nu$ (see Fig. 1).*

Let $C$ be a discrete curve and $C_{i,j}$ a sequence of points of $C$ indexed from $i$ to $j$. Let denote the predicate "$C_{i,j}$ is a blurred segment of width $\nu$" as $BS(i, j, \nu)$.

**Definition 3.** *$C_{i,j}$ is called a **maximal blurred segment (MBS) of width** $\nu$ and denoted $MBS(i, j, \nu)$ iff $BS(i, j, \nu)$, $\neg BS(i, j + 1, \nu)$ and $\neg BS(i - 1, j, \nu)$ (see Fig. 1).*

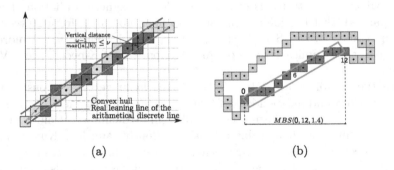

(a)                                    (b)

**Fig. 1.** (a) Example of arithmetical discrete line $\mathcal{D}(2, -3, -5, 5)$ (grey and blue points) and a blurred segment of width $\nu = 1.4$ (grey points) bounded by $D$. (b) Maximal blurred segment of width $\nu = 1.4$ (green points). (Color figure online)

An *ATC* consists of MBS of different widths, which are a function of the noise perturbations of the studied contour. In particular, we use the local noise estimator, namely **meaningful thickness** [6,7], to determine the significant width locally at each point of the contour. This meaningful thickness is used as an input parameter to compute the $ATC$ with appropriate widths w.r.t. noise. A non-parametric algorithm is developed in [11] to compute the $ATC$ of a given discrete curve. In the $ATC$, the obtained $MBS$ decomposition of various widths transmits the noise levels and the geometrical structure of the given discrete curve (see Fig. 2(a, c)).

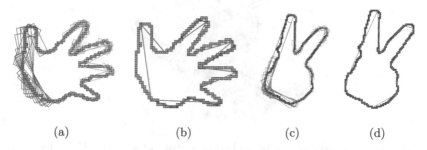

|  (a)  |  (b)  |  (c)  |  (d)  |

**Fig. 2.** (a, c): Adaptive tangential cover, (b, d) polygonal representation (in red) using the dominant points and polygonal simplification results (in green). (Color figure online)

## 2.2 Polygonal Simplification [10–12]

**Dominant points** are significant points on a curve with local maximum curvature. Such points contain a rich information which allows to characterize and describe the curve. Issued from the dominant point detection proposed in [10, 12] and the notion of $ATC$, an algorithm is developed in [11] to determine the dominant points of a given noisy curve $C$. The main idea is that the candidate dominant points are localized in the common zones of successive $MBS$ of the $ATC$ of $C$. An angle measure $m$ is used to determine the dominant points with local extreme curvature in the common zones. More precisely, this measure $m$ is the angle between the considered point and the two left and right endpoints of the left and right MBS involved in the studied common zone. When the considered point varies, $m$ becomes a function of it. A dominant point is defined as a local minimum of $m$. Dominant points are illustrated in Fig. 2(b, d) in red points. Red lines represent the polygonal representation of the shape.

First goal of finding the dominant points is to have an approximate description of the input curve, called **polygonal simplification**. Dominant points are sometimes redundant or stay very near, which is presumably undesirable in particular for polygonal simplification. So, we associate to each detected dominant point a **weight**, i.e., the ratio of integral sum of square errors and the angle with the two dominant point neighbours, indicating its importance with respect to the approximating polygon of the curve. Polygonal simplification is illustrated in Fig. 2(b, d) with green lines.

# 3    Feature Descriptors

In this work, we are using a combination of topological, statistical and geometric features. In the following, we describe each feature descriptor in details.

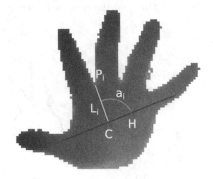

**Fig. 3.** Histogram of contour angles and contour distances: $C$ and $H$ are the center of Contour points and center of Convex Hull points; $P_i$, $L_i$, $a_i$ are the $i$-th Contour point, Contour Distance and Contour Angle respectively.

## 3.1    Histogram of Contour Angles (HoCa) and Contour Distances (HoCd)

Suppose there are $m$ contour points $(X_1, Y_1), \ldots, (X_m, Y_m)$ and $n$ convex hull points $(X'_1, Y'_1), \ldots, (X'_n, Y'_n)$.

The image moments with pixel intensity $f(x, y)$ at location $(x, y)$ are given by

$$M_{ij} = \sum_{x,y} x^i y^j f(x, y). \tag{1}$$

For a contour or a set of points without any associated intensity, $f(x, y)$ is taken to be 1.

In our case, let $M_{ij}^c$ denote the moments of the contours and let $M_{ij}^h$ be the moments of the convex hull points. Then the center $C = (X_c, Y_c)$ of the contour points is given by

$$X_c = \frac{M_{10}^c}{M_{00}^c}, Y_c = \frac{M_{01}^c}{M_{00}^c}, \tag{2}$$

and the center $H = (X'_c, Y'_c)$ of the convex hull is given by

$$X'_c = \frac{M_{10}^h}{M_{00}^h}, Y'_c = \frac{M_{01}^h}{M_{00}^h}. \tag{3}$$

Then we fix the line segment $L_{CH}$ between the points $C$ and $H$. Let $L_i$ be the line segment joining $C$ to the contour point $P_i = (X_i, Y_i)$ and let $a_i$ be the angle between $L_{CH}$ and $L_i$ formed at the point $C$, for $i = 1, \ldots, m$.

Now we create a histogram with 10 bins from these $m$ angles $a_1, \ldots, a_m$ and get 10 descriptors say $h_1, \ldots, h_{10}$.

Let $|L_i|$ be the length of the line segment $L_i$, $i = 1, \ldots, m$. We create a histogram with 10 bins from these $m$ lengths and thus get 10 descriptors $l_1, \ldots, l_{10}$.

In Fig. 3 we explain the above set of descriptors.

## 3.2   Moments

We use different kinds of moments as follows.

**Raw or Spatial Moments.** We use Eq. (1) on the entire image, where $(x, y)$ denotes a pixel location and $f(x, y)$ denotes the corresponding greyscale intensity value. From this, we generate 10 descriptors $M_{ij}$, with $0 \leq i + j \leq 3$. More explicitly, we use $M_{00}, M_{01}, M_{10}, \ldots, M_{03}, M_{30}$.

**Central Moments.** First we calculate the spatial moments as above. Then we define

$$\bar{x} = \frac{M_{10}}{M_{00}} \text{ and } \bar{y} = \frac{M_{01}}{M_{00}}. \tag{4}$$

Now, the central moments are given by

$$\mu_{ij} = \sum_{x,y} (x - \bar{x})^i (y - \bar{y})^j f(x, y), \tag{5}$$

where $(x, y)$ denotes a pixel location and $f(x, y)$ denotes the corresponding greyscale intensity value. From this, we generate 7 descriptors $\mu_{ij}$, with $2 \leq i + j \leq 3$. More explicitly, we use $\mu_{11}, \ldots, \mu_{03}, \mu_{30}$.

**Central Standardized or Normalized or Scale Invariant Moments.** These moments are normalized versions of the central moments, defined as follows:

$$\nu_{ij} = \frac{\mu_{ij}}{(M_{00})^{\frac{i+j}{2}+1}}. \tag{6}$$

From this, we generate 7 descriptors $\nu_{11}, \nu_{12}, \nu_{02}, \nu_{20}, \nu_{21}, \nu_{03}, \nu_{30}$.

## 3.3   Geometric Descriptors from Polygonal Simplification of Shape Contours

We propose different descriptors from the selected dominant points (DP), obtained with the method presented in Sect. 3.3, applied on a shape $S$. The polygon obtained with the DP is called $DP(S)$. Let $per(DP(S))$ and $area(DP(S))$ be respectively the perimeter and the area of $DP(S)$. Let $ch(DP(S))$ be the convex hull of $DP(S)$. The following descriptors give indications about compacity and convexity of the shape $S$:

$$\frac{per(DP(S))^2}{area(DP(S))} \qquad (7)$$

$$\frac{per(ch(DP(S)))^2}{area(ch(DP(S)))} \qquad (8)$$

$$\frac{area(ch(DP(S)))}{area(DP(S))} \qquad (9)$$

We compute descriptors that indicate if the contour of the shape $S$ is regular or contains big irregularities:

- mean value of angles between two successive segments formed by three successive dominant points of $DP(S)$
- variance value of angles between two successive segments formed by three successive dominant points of $DP(S)$
- minimum distance from the centroid of $S$ to the dominant points of $S$
- maximum distance from the centroid of $S$ to the dominant points of $S$
- the difference between minimum and maximum distances
- variance of segment lengths in $DP(S)$
- number of peaks detected in $S$

We should notice that the centroid of $S$ is the centroid of the polygonal simplification of the shape $S$. All our geometric descriptors are based on the dominant points selected by the polygonal simplification process. We do not work with all points of the contour $S$.

Moreover we detect the number of "peaks" in each shape contour.

A peak is a dominant point located in a convex part of the shape contour with an angle greater than a given threshold. We fix this threshold angle to 1,38. Figure 4 shows that in this convex part, point $P_{i+1}$ is considered as a peak point because its angle is greater than the threshold and angle at point $P_{i+2}$ is less than the threshold.

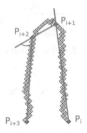

**Fig. 4.** Convex part of a shape contour shows the angle at each dominant points ($P_i$, $P_{i+1}$, $P_{i+2}$, $P_{i+3}$) to determine peak points.

# 4  Dataset and Pre-processing for Experimental Evaluation

This section describes the collection of dataset and pre-processing that we performed to extract stong features.

## 4.1  Dataset from Kinect V1

As a benchmark dataset for Kinect V1, we are using the NTU dataset [13]. This dataset is collected from 10 subjects and it contains 10 gestures for numbers 0 to 9. Each subject performs 10 different poses for the same gesture. Thus this database has 1000 cases in total and it consists of a color image and the corresponding depth map.

## 4.2  Development of New Benchmark Dataset from Kinect V2

We are proposing a new hand gesture dataset using Kinect V2. In this dataset we have collected standard ASL hand gestures from 10 subjects for numbers 0 to 9. Each subject performs 10 different poses for the same gesture. Thus in total the dataset has 10 people × 10 gestures/people × 10 cases/gesture = 1000 cases, each of which consists of a contour map and the corresponding depth map. This dataset is a real-life dataset, which is collected in cluttered backgrounds. Besides, for each gesture, the subject poses with variations in hand orientation (almost 180 degree variation), scale, articulation, etc.

## 4.3  Data Collection

For NTU dataset collected from Kinect V1, we have annotated the ROI region in RGB images. Then after depth thresholding, we have cropped the exact ROI regions from depth images by comparing it with the RGB images. For V1 dataset, it was not a problem because depth resolution is $320 \times 240$ and RGB resolution is $640 \times 480$ which is just the double. Whereas, in our proposed database collected from Kinect V2, RGB resolution is $1920 \times 1080$ and Depth resolution is $512 \times 424$. So in our database, we collected a joint color-depth hand gesture along with wrist joint point. This gives us the flexibility of locating the ROI in the whole frame. Gesture samples are shown in Fig. 5 which are labeled from 0 to 9. It should be noted that this dataset is a real life dataset collected in uncontrolled environment with different illumination and different orientations.

## 4.4  Hand Localization and Segmentation

In previous depth camera-based approaches [17], the hand is required to be the front-most object from the depth camera. Moreover, a black belt on the gesturing hand's wrist is also required in some cases [13], which is rather inconvenient for real world applications. In our system, we relax these restrictions by utilizing the rather stable joints from Kinect's skeleton tracking. The Kinect joints are directly used to locate the hands, wrists and elbows. By assuming that the hand is visible to the camera without any occlusion, it allows us to quickly separate the hands from background objects using depth information alone. Using the hand joint point as the center, a pair of color texture and depth map blocks is extracted first and then the hand shape is segmented quickly using a depth threshold value.

### 4.5   Noise Removal

In practical applications, the extracted hand gestures usually have different scales due to various distances from the camera to hand, or different rotations caused by the body postures. Moreover, different people's hands always have distinct characteristics even for the same gesture. Hence it is necessary to perform some pre-processing to normalize and align the shape representation before recognition. However, the palm size varies from one person to another, which affects the recognition between different subjects. So instead of depth thresholding in our dataset, we have done histogram thresholding calculating mean and standard deviation of the depth values and extracted gray scale cropped hand shapes, shown in Fig. 5. In Fig. 5, it also can be seen that the hand shapes from different persons are correctly segmented, even when the hands are cluttered by the face or background.

## 5   Performance Evaluation

Now we evaluate the performance of the proposed system from mean accuracy, time efficiency and comparisons with other methods. We are using Random Forest classifier on extracted features using the Weka version 3.6.12 (c) 1999–2014 [19] machine learning tool. We performed all experiments on an Intel i7-6500U CPU @2.50 GHz processor with 8 GB RAM and 64-bit Windows 8.1 Pro OS.

In our experiments, leave-$p$-out (LpO) cross-validation (CV) is conducted to evaluate the recognition performance, where with $M$ instances, $p < M$ subjects are used for testing and the remaining for training. This process is repeated for every combination of subjects so that the average accuracy can be computed. For NTU dataset, we only calculated LOO CV to compare the results. In our dataset, two values of (1 and 100) are considered, which are respectively referred to as leave-one-out CV (LOO CV) and 10-fold CV. Experiments based on these two CVs are presented in next section.

### 5.1   Results and Comparison

Here we produce comparative results (see Table 1) between our proposed work and other different algorithms on standard V1 NTU hand digit dataset. Our mean accuracy is 98.4% which outperforms all other works mentioned in the

**Table 1.** Comparison between Mean Accuracy of Shape Contexts, Skeleton Matching, HOG, H3DF, FEMD and our method on the NTU HAND DIGIT DATASET.

| Algorithms | Mean Accuracy |
| --- | --- |
| Skeleton Matching [1] | 78.6% |
| Shape Context with bending cost [2] | 79.1% |
| Shape Context without bending cost [2] | 83.2% |
| HOG [3] | 93.1% |
| Thresholding Decomposition+FEMD [13] | 93.2% |
| Near-convex Decomposition+FEMD [13] | 93.9% |
| H3DF [22] | 95.5% |
| Current Work | **98.4%** |

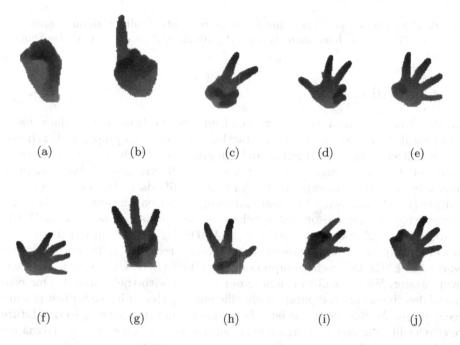

(a)   (b)   (c)   (d)   (e)

(f)   (g)   (h)   (i)   (j)

**Fig. 5.** Gesture samples (0–9) of ASL captured in Kinect V2.

table. Figure 5 shows our proposed benchmark V2 dataset containing ASL gestures 0 to 9 captured using Kinect V2. On this dataset, we get an accuracy of 92.2% with the same feature set as V1.

Figure 6 illustrates the confusion matrix of hand gesture recognition on standard V1 dataset and our benchmark V2 dataset (Fig. 5) using our feature set presented in Sect. 4.

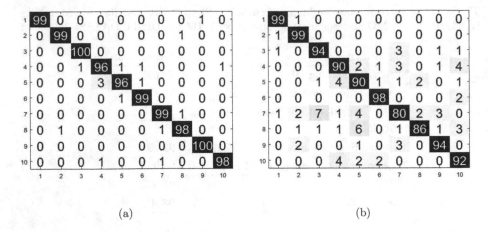

(a)                                   (b)

**Fig. 6.** The confusion matrices of hand gesture recognition using with our feature set: (a) LOO CV on NTU hand digit dataset, (b) 10-fold CV on our own V2 benchmark dataset.

# 6    Conclusion

In this work, a hand gesture recognition system from depth data using topological-statistical features for contactless HCI has been proposed. It is based on an efficient feature extraction and classification which recognizes gestures based on the shape and depth features. The effectiveness of the system is illustrated by extensive experiments on two real-life datasets, NTU hand digit dataset [13] and our own V2 dataset (achieving high mean accuracies 98.4%, 92% respectively). In comparison with previous distance measures such as FEMD [13], shape context [2], Skeleton Matching [1], H3DF [22], HOG [3], our feature set achieves comparable performance for hand gesture recognition. Recently, another work of SPEMD [18] has been proposed with 99.1% (LOO CV) accuracy on their own dataset. We are still to compare our features with their dataset. Our proposed descriptors are computationally efficient and thus suitable for fast gesture recognition. As this work has only been done using depth map, so our future research will focus on exploring robust color features and extending it to dynamic hand gesture, body posture and generic object recognition.

# References

1. Bai, X., Latecki, L.J.: Path similarity skeleton graph matching. IEEE Trans. Pattern Anal. Mach. Intell. **30**(7), 1282–1292 (2008)
2. Belongie, S., Malik, J., Puzicha, J.: Shape matching and object recognition using shape contexts. IEEE Trans. Pattern Anal. Mach. Intell. **24**(4), 509–522 (2002)
3. Dalal, N., Triggs, B., Schmid, C.: Human detection using oriented histograms of flow and appearance. In: Leonardis, A., Bischof, H., Pinz, A. (eds.) ECCV 2006. LNCS, vol. 3952, pp. 428–441. Springer, Heidelberg (2006). doi:10.1007/11744047_33

4. Debled-Rennesson, I., Feschet, F., Rouyer-Degli, J.: Optimal blurred segments decomposition of noisy shapes in linear time. Comput. Graph. **30**(1), 30–36 (2006)
5. Dewaele, G., Devernay, F., Horaud, R.: Hand motion from 3D point trajectories and a smooth surface model. In: Pajdla, T., Matas, J. (eds.) ECCV 2004. LNCS, vol. 3021, pp. 495–507. Springer, Heidelberg (2004). doi:10.1007/978-3-540-24670-1_38
6. Kerautret, B., Lachaud, J.O.: Meaningful scales detection: an unsupervised noise detection algorithm for digital contours. Image Process. On Line **4**, 98–115 (2014)
7. Kerautret, B., Lachaud, J.O., Said, M.: Meaningful thickness detection on polygonal curve. In: Proceedings of the 1st International Conference on Pattern Recognition Applications and Methods, pp. 372–379. SciTePress (2012)
8. Kry, P.G., Pai, D.K.: Interaction capture and synthesis. ACM Trans. Graph. (TOG) **25**, 872–880 (2006). ACM
9. Mitra, S., Acharya, T.: Gesture recognition: a survey. IEEE Trans. Syst. Man Cybern. Part C (Appl. Rev.) **37**(3), 311–324 (2007)
10. Ngo, P., Nasser, H., Debled-Rennesson, I.: Efficient dominant point detection based on discrete curve structure. In: Barneva, R.P., Bhattacharya, B.B., Brimkov, V.E. (eds.) IWCIA 2015. LNCS, vol. 9448, pp. 143–156. Springer, Cham (2015). doi:10. 1007/978-3-319-26145-4_11
11. Ngo, P., Nasser, H., Debled-Rennesson, I., Kerautret, B.: Adaptive tangential cover for noisy digital contours. In: Normand, N., Guédon, J., Autrusseau, F. (eds.) DGCI 2016. LNCS, vol. 9647, pp. 439–451. Springer, Cham (2016). doi:10.1007/ 978-3-319-32360-2_34
12. Nguyen, T.P., Debled-Rennesson, I.: A discrete geometry approach for dominant point detection. Pattern Recogn. **44**(1), 32–44 (2011)
13. Ren, Z., Yuan, J., Meng, J., Zhang, Z.: Robust part-based hand gesture recognition using kinect sensor. IEEE Trans. Multimedia **15**(5), 1110–1120 (2013)
14. Reveillès, J.P.: Gèométrie discrète, calculs en nombre entiersgorithmique, et al.: thèse d'état. Université Louis Pasteur, Strasbourg (1991)
15. Shotton, J., Sharp, T., Kipman, A., Fitzgibbon, A., Finocchio, M., Blake, A., Cook, M., Moore, R.: Real-time human pose recognition in parts from single depth images. Commun. ACM **56**(1), 116–124 (2013)
16. Paul, S., Basu, S.: Microsoft kinect in gesture recognition: a short review. Int. J. Control Theor. Appl. **8**(5), 2071–2076 (2015)
17. Suarez, J., Murphy, R.R.: Hand gesture recognition with depth images: a review. In: RO-MAN, 2012 IEEE, pp. 411–417. IEEE (2012)
18. Wang, C., Liu, Z., Chan, S.C.: Superpixel-based hand gesture recognition with kinect depth camera. IEEE Trans. Multimedia **17**(1), 29–39 (2015)
19. WEKA: Fibonacci notes (1996). http://www.cs.waikato.ac.nz/ml/weka/ downloading.html
20. Wu, Y., Lin, J., Huang, T.S.: Analyzing and capturing articulated hand motion in image sequences. IEEE Trans. Pattern Anal. Mach. Intell. **27**(12), 1910–1922 (2005)
21. Yang, M.H., Ahuja, N., Tabb, M.: Extraction of 2D motion trajectories and its application to hand gesture recognition. IEEE Trans. Pattern Anal. Mach. Intell. **24**(8), 1061–1074 (2002)
22. Zhang, C., Yang, X., Tian, Y.: Histogram of 3D facets: a characteristic descriptor for hand gesture recognition. In: 10th IEEE International Conference and Workshops on Automatic Face and Gesture Recognition (FG), pp. 1–8. IEEE (2013)

# Image Segmentation via Weighted Carving Decompositions

Derek Mikesell$^{(\boxtimes)}$ and Illya V. Hicks

Computational and Applied Mathematics, Rice University,
Houston, TX 77005, USA
{djm13,ivhicks}@rice.edu

**Abstract.** In this paper we propose a graph-theoretic method of image segmentation, borrowing ideas from finding community structure in social networks using edge betweenness. This method constructs a weighted carving decomposition, or a partial carving decomposition to some resolution, of the image. From this structure image segments can be obtained in a hierarchical manner. We apply this method to multiple generated images with well defined image segments of varying complexity. Additionally, we apply this method to the *Mona Lisa*, an image without such well defined partitions. Results suggest that the method provides a hierarchical segmentation framework that is well suited for finding features in images.

## 1 Introduction

Numerous studies have explored the subject of image segmentation. The main task in image segmentation is to decompose an image into multiple segments or features. Depending on the goal, this can take place as dividing the image in two distinct regions up to any number less than the size of the pixel set. Within this framework an *image* is not to be confused with the output of a function, rather it is a collection of pixels assembled into an array. A *pixel*, coming from "picture" and "element", is the minimal element of an image, generally showing some color [4].

A *graph* $G$ is an ordered pair $(V, E)$, where $V$ is the *vertex set* and $E$ is the *edge set*. Elements of $E$, or *edges*, are unordered pairs of elements of $V$ that defines a relation between the two vertices. A *planar graph* is a graph that can be drawn onto a plane or sphere without having any edges cross. A *subgraph* $F$ of $G$ is a graph $F = (\bar{V}, \bar{E})$ with $\bar{V} \subseteq V, \bar{E} \subseteq E$. Given a graph $G = (V, E)$ and $\bar{V} \subseteq V$, the *induced subgraph* $G[\bar{V}]$ is the graph with vertex set $\bar{V}$ in which the edge set is composed of all edges that have both endpoints in $\bar{V}$. In a graph $G$ a *walk* is a sequence $v_0, e_1, v_1, e_1, ..., e_k, v_k$, where $v_i \in V$ and $e_i \in E$ for $0 \leq i \leq k$. We will only consider walks in which all elements of the sequence are distinct, known as *paths*. If $v_0 = v_k$ then the walk is said to be *closed* or a *cycle*.

Let $G = (V, E)$ be a graph. Then, let $T$ be a tree such that $T$ has $|V|$ leaves, with every non-leaf vertex having degree 3. Additionally, let $\mu$ be a bijection between the nodes of $G$ and the leaves of $T$. The pair $(T, \mu)$ is said to be a *carving decomposition* or *minimum-congestion routing tree* of $G$.

© Springer International Publishing AG 2017
V.E. Brimkov and R.P. Barneva (Eds.): IWCIA 2017, LNCS 10256, pp. 268–279, 2017.
DOI: 10.1007/978-3-319-59108-7_21

Later, we will need a centrality measure known as *edge betweenness (EB)*. Let $G = (V, E)$ be a graph with $u, v \in V$. Additionally, let $\sigma_{u,v}$ be the number of shortest paths between $u$ and $v$ and $\sigma_{u,v}(e)$ be the number of shortest paths between $u$ and $v$ containing edge $e$. Then, the betweenness of edge $e$ can be expressed as:

$$EB(e) = \sum_{u \in V} \sum_{v \in V} \frac{\sigma_{u,v}(e)}{\sigma_{u,v}} \tag{1}$$

The edge betweenness represents the extent to which an edge is between groups of nodes. Thus, the removal of an edge with high edge betweenness is more likely to disconnect a graph, than one of low edge betweenness.

In this paper we present a graph-theoretic image segmentation method borrowing techniques from biological and social network community structure. In Sect. 1.1 we discuss various methods of image segmentation. Additionally, some context into the graph objects and community structures is explored. In Sect. 2 a full exploration of the method and algorithm is given. Section 3 introduces the results of the method, with Sects. 3.1 through Sect. 3.4 exploring different test cases. We end with some concluding remarks in Sect. 4.

## 1.1   Literature Review

A number of image segmentation techniques have been created with a varying levels of complexity. Thresholding techniques may be the most simple; they seek to set some threshold and choose pixels on either side of that threshold. An early method of thresholding is Otsu's method (1975) [21]. This method operates on a greyscale image and acts to construct a binary image by placing pixels into two groups such that their interclass variance is maximized. Another method used as a means of thresholding an image is the k-means algorithm. Originally used in scalar quantization, Lloyd's k-means algorithm (1982) [17] was shown to be an adequate thresholding method for computer vision by Barghout and Sheynin [1]. More current models of thresholding have been developed for computed tomography images using local and adaptive thresholding to improve on global techniques, and providing higher quality segmentation in the presence of artifacts [2,3].

Region growing techniques are another set of methods that rely on one simple rule that adjacent pixels share similar values. An early method, building on the work in [5], uses a split-and-merge and a tree traversal algorithm to recursively build image segments [15,16]. Pixel intensities and neighborhood-linking paths are used by Chen et al. to form a partial connectedness measure, $\lambda$-connectedness. This connectedness is applied to "fuzzy subfibers" to construct a higher dimensional region growing segmentation method [6]. Statistical region merging uses the same idea of neighboring pixel intensity as a means of determining likeness. This method improves on previous techniques by introducing a priority queue and merging via a statistical basis [19].

Another set of methods, borrowing techniques from encoding, is image segmentation via compression. The main idea in compression based segmentation is that the best segmentation comes from finding the minimal coding length of the data. Mobahi et al. constructed a method that uses the minimum description length for image segmentation after noticing that homogenous textures can be modeled via Gaussian distributions and their boundaries can be encoded efficiently [18, 22].

Graph based techniques are a natural method of segmentation as they encode pixel connectivity, while segments can be found by looking for clusters of nodes. It will be seen that the model proposed in this paper lies within this category. In 1971, Zahn proposed a family of minimum-spanning tree based algorithms capable of detecting cluster structure and applicable to image segmentation [28]. Wu and Leary proposed a method that removes edges in the network to "form mutually exclusive subgraphs such that the largest inter-subgraph maximum flow is minimized" [27]. In 2000, Shi and Malik proposed a method using normalized cuts, which attempts to find a global impression of an image [26]. Grady proposed a method utilizing random walks. In this technique a small number of user given labels exist within a graph and the pixels with high probability of reaching these pixels via a random walk form segments in an image [11].

The aforementioned methods are only a small subset of the field and more can be found in the *Survey on Image Segmentation Techniques* by Zaitoun [29] and *A Study of Digital Image Segmentation Techniques* by Oak [20].

Inspiration for the present paper came from Diestel and Whittle [7], who stated that an image can be represented by the tangles of its pixels, but left the development and implementation of this technique as an open problem. Tangles are introduced and studied extensively in Robertson and Seymour's *Graph Minors X* [23]. As tangles relate to separations of nodes in graphs there seems to exist a potential relation to image segmentation. Further objects of interest, carvings and carving decompositions, are explored in [14, 24]. Efficient computation of carving width and branch decompositions is given by Hicks in [12, 13].

In the forthcoming method, the edge betweenness centrality is used to find edges that lie between segments. This centrality is originally defined by Girvan and Newman [10], who provide an algorithm for clustering that, while different, parallels the method in this paper in its use of edge betweenness as a metric for node selection.

## 2   Method

### 2.1   Color Expression

Rather than using the standard Euclidean norm in the familiar RGB color space, the CIEDE2000 color space [25] is used instead. Numerous weaknesses of the RGB color space have been fixed in CIEDE2000, such as a hue rotation that fixes difficulties in the blue range.

While feature differentiation is an easy task for most humans, it can be a rather difficult task in computer vision. The CIEDE2000 color space is said to better represent human vision, and therefore is a natural candidate for completing computer vision tasks. Computing the color CIEDE2000 color difference requires the evaluation of multiple expressions and thus increases the cost of computing the difference compared to simpler color difference formulas.

Rather than representing color in the RGB space, CIEDE2000 represents color in $\mathbf{L}$ for lightness, $\mathbf{a}$ a red-green opponent axis, and $\mathbf{b}$ a blue-yellow opponent axis. Thus, this space gives a different three dimensional representation of color with neutral grey at $\mathbf{a} = 0$ and $\mathbf{b} = 0$, green for $\mathbf{a} < 0$, red for $\mathbf{a} > 0$, and similarly blue for $\mathbf{b} < 0$ and yellow for $\mathbf{b} > 0$.

The formula for color difference in the CIEDE2000 color space is as follows:

$$\Delta E_{00}^{i,j} = \sqrt{\left(\frac{\Delta L'}{k_L S_L}\right)^2 + \left(\frac{\Delta C'}{k_C S_C}\right)^2 + \left(\frac{\Delta H'}{k_H S_H}\right)^2 + R_T \frac{\Delta C'}{k_C S_C} \frac{\Delta H'}{k_H S_H}} \quad (2)$$

where $\Delta L'$ is the difference in lightness, $\Delta C'$ is the difference in chroma, is a difference in hue, and the cross term is a hue rotation factor. The values $k_L, k_C$, and $k_H$ are constants associated with each distance and $S_L, S_C$, and $S_H$ are factors computed in the algorithm. A full explanation and algorithm can be found in [25].

## 2.2   Image Representation

Construction of a network from an image is a simple task as graphs are naturally suited for handling connectivity. Within this framework we will consider a five-point stencil for interior pixels and adjust accordingly for pixels along the boundary. It should be noted that a higher order stencil can be used within this method at a cost of increased runtime and a loss of planarity. Additionally, this method is not restricted to square images or two dimensional images and is not affected by the range of colors.

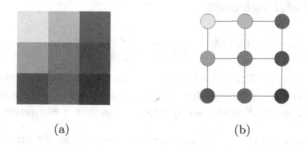

(a)                                    (b)

**Fig. 1.** (a) shows an image and its corresponding graph in (b). The weights along each edge are given as given by (2).

We will assume the image is a $m \times n$ rectangular array. Therefore, we arrive at a $m \times n$ grid graph as seen in Fig. 1. Weights along each edge in the graph are given by:

$$w_G(i,j) = exp(-\Delta E_{00}^{i,j}) \tag{3}$$

where $\Delta E_{00}^{i,j}$ is the CIEDE2000 color difference as described in (2).

### 2.3   Segmentation Method

Consider the scenario in which a person is tasked with selecting a feature within an image. Equipped with only a pen and the image, it is natural for the individual to circle the feature on the image. This action of circling features can be modeled as follows.

Let $\Sigma$ be a sphere, and let $G$ be a graph drawn in $\Sigma$, constructed from some image $I$. The set of regions, $r_i$, enclosed by the edges in $E(G)$ in addition to the region surrounding $G$, will be denoted $R(G)$. Each region $r \in R(G)$ is an open set in $\Sigma$, with the edges that enclose a given region, $r$, labeled as *incident* to $r$.

Now, drawn within $\Sigma$, form the geometric dual, $G^*$, to the graph $G$. This construction occurs in the following way: for each $r \in R(G)$ there exists a $v^* \in V(G^*)$ and for each $r_i, r_j \in R(G)$ that share an incident edge $e \in E(G)$ there exists an edge $f^* \in E(G^*)$ connecting the vertices corresponding to $r_i, r_j$. Hence, $G^*$ contains a unique node for each region of $G$. Similarly, for each edge $e \in E(G)$ there exists a dual edge $f^* \in E(G^*)$ such that the endpoints of $f^*$ are contained within the regions incident to $e$; $f^*$ is said to *cross* $e$.

In the hypothetical situation above, a feature, $F$, is a set of nodes in $G$, while the act of circling the image is equivalent to finding a cycle, $C$, in the geometric dual such that $E(C)$ cross all of the edges connecting $G \backslash F$ to $F$. Further, the concept of finding a "best" segment could be posed as finding a minimal cycle in $G^*$. It should be noted that using a higher order stencil for connectivity eliminates planarity and thus limits or previously stated intuition. Thus, it will be assumed that graphs are constructed using the five-point stencil.

**Proposition 1.** *Let $G$ be a graph constructed from an image $I$. Let $G^*$ be the geometric dual of $G$ such that*

$$w_{G^*}(f) = \frac{1}{BE(e)} \quad \forall f \in E(G^*)$$

*such that $e$ crosses $f$. Then the first separation formed by removing edges of $G$ in ascending inverse order of edge betweenness is equivalent to the separation formed by $G \backslash S$, where $S$ is the set of edges crossed by a minimum cycle in $G^*$.*

*Proof.* Let $F$ be a priority queue of edges of $G$, with priority proportional to its edge betweenness centrality. Remove edges of $G$ according to $F$ until $G$ is split into two components. Label the vertices of each component $V_1, V_2$.

Consider the subgraphs $G_1, G_2$ induced by $V_1, V_2$. As $\Sigma$ is continuous, there exists a continuous closed curve, $\ell$, in $\Sigma$ such that $\ell$ encloses $G_1, G_2$ (note that

$\ell$ encloses both as $G$ is embedded in a sphere and is minimal with respect to the regions in $R(G)$ it traverses as $G_1, G_2$ are induced by the partitioned vertex set). Therefore, we can construct a cycle, $C$, in $G^*$ by taking the vertex that corresponds to each region that $\ell$ crosses. As $G_1, G_2$ are constructed via $F$, $C$ is minimal in $G^*$. □

Given the above proposition, we can begin to derive an algorithm that mimics a natural process in finding segments in images. Starting with an image $I$, we construct the graph, $G$, of that image as is stated in Sect. 2.2. From $G$ construct a priority queue $F$, containing edges of $G$ with priority set to be the edge betweenness. Iteratively remove edges from $G$, from the queue $F$ until $G$ separates into two components. This marks one iteration of the algorithm. By recomputing the edge betweenness after each separation, and recursively applying this algorithm on the components until we reach individual nodes, we arrive at a full hierarchy of image segments. The algorithm can be seen in detail in Algorithm 1

**Data:** Graph $G$ constructed from image, Graph $H$ with single node
        containing list of $V(G)$
**Result:** Tree $H$ of segments
**while** $|E(G)| > 0$ **do**
    numComp0 = number of components of $G$;
    numCompNew = numComp0;
    Construct priority queue $F$;
    **while** $numCompNew \leq numComp0$ **do**
        $e$ = highest priority edge in $F$;
        $F = F \setminus e$ ;
        $G = (V, E \setminus e)$;
        newComp = connected components of $G$;
        numCompNew = |newComp|;
    **end**
    Append new component nodes to parent node in $H$;
**end**

**Algorithm 1.** Betweenness Segmentation

*Remark 1.* It is advised to add an edge selection tie-breaker to Algorithm 1. This can be done by constructing the priority queue $F$ from the edge betweenness and then sorting ties by the largest $\Delta E_{00}^{i,j}$

Upon completion of the algorithm we are left with a tree $H$ containing all of the segmentation information. It should be noted that the leaves of the tree correspond to individual pixels of the original image. Additionally, each additional node in the tree has degree 3 (aside from the original root node). Therefore, by removing the root node and adding an edge in between its two children we arrive at a carving decomposition. Research into branch decompositions of graphs have shown significant complexity benefits. Having constructed a weighted carving decomposition of our image, it is left to future works to take advantage of this structure in an attempt to produce lower complexity image algorithms.

Edge betweenness can be computed via finding the shortest paths between all pairs of vertices in the graph. Given the relative sparsity of the grid structure assumed in this model, this can be computed via repeated Dijkstra's algorithm [8]. Using Fibonacci heaps, Fredman and Tarjan give a method that can be implemented in $\mathcal{O}(|E||V| + |V|^2 \log |V|)$ [9]. Additionally, the task of finding the number of connected components in the graph can be found linearly in $|V|$ and $|E|$ via a depth first search. The complexity of these algorithms are given only as a starting point for the complexity analysis as it is not the focus of this work to provide an efficient implementation of the algorithm, but will be explored in future works.

## 3  Results

The following sections contain various test cases used in checking the efficacy of the aforementioned model. When two images are presented side by side, as in Fig. 4(d) and (e), the figures do not display all pixels, but rather a separation of those in the parent node of the tree constructed by the model. This method of display is used to illustrate the separation at a given layer. Note that a cut in the hierarchy tree, such as in Fig. 3(b), will yield a complete and disjoint separation of pixels.

### 3.1  Times New Roman "L"

The first test image used to test the method is a greyscale text image; see Fig. 2. The file is a white on black, $16 \times 16$ pixel image of the letter "L" in Times New Roman font. While the image is not binary, it provides a test case with two primary regions and some intensity variations (in the form of grey pixels). Inspiration for this test case comes from [7]. Figure 2(a) shows the original test image. After the first separation in the algorithm we are left with Fig. 2(b) and (c). In Fig. 2(c) we see that the majority of the "L" has been separated away from the black background, while six gray pixels remain in Fig. 2(b). Figure 2(d) and (e) show a separation of the right serif from the baseline bar, while Fig. 2(f) and (g) show a separation of the baseline bar from the stem. Looking at further segments (figures not shown) one would hope to find the remaining serifs, which are indeed found, but one cannot expect to find the reflected serifs as one segment given that the stem of the "L" is much wider than the serifs.

### 3.2  Embedded Full Color

Now consider the test image as seen in Fig. 3. This image is a full color, $4 \times 4$ pixel image. A blue square is embedded inside a red and orange hemisphere. Given the visual similarity of red and orange compared to that of blue, it is natural to assume that the primary separation will occur via removing the blue segment. The segmentation of the blue pixels would expected to be followed by a separation of the orange and red segments. Figure 3(b) gives a full hierarchical

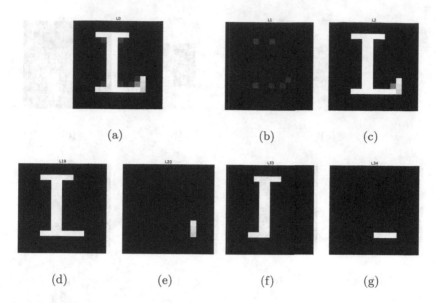

(a)                    (b)                    (c)

(d)            (e)            (f)            (g)

**Fig. 2.** The original image can be seen in (a). A 16pt Times New Roman "L" in white on black background. (b) through (g) show the steps of segmentation using the method outlined in this paper. It should be noted that the black portions of (c) are not black pixels, rather the lack of pixels.

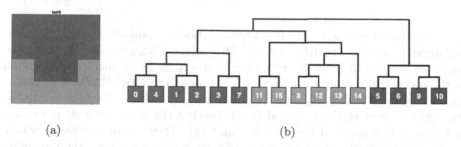

(a)                                    (b)

**Fig. 3.** The original image can be seen in (b). (b) shows the hierarchical breakdown after running the segmentation model; each numbered box represents a pixel in (b) from top left to bottom right. (Color figure online)

breakdown of Fig. 3(a), in which the values at the leaves correspond to pixel indices (labeled from top left to bottom right). One can see that by cutting the topmost edge in the tree we obtain our desired result. Additionally, cutting the next topmost edge separates the red and orange sections.

### 3.3   Embedded Full Color 2

The test image in Fig. 4(a) is a more complex $8 \times 8$ pixel version of Fig. 3(a). A green and blue vertically aligned rectangle are embedded in the middle of orange

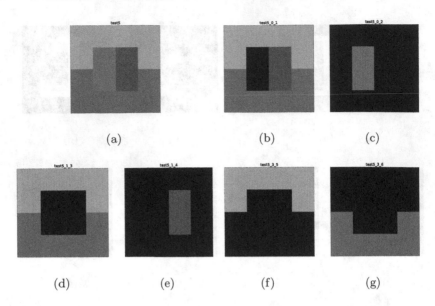

**Fig. 4.** The original image can be seen in (a). An $8 \times 8$ image of imbedded shapes of various colors. (b) and (c) show the first step of segmentation using the method outlined in this paper. (d) and (e) show the splitting of (b). Further, (f) and (g) show the splitting of (d). (Color figure online)

and red hemispheres. After the first segmentation, the algorithm finds the green rectangle as seen in Fig. 4(b) and (c). This aligns with intuition, as the green region is visually different from the red and orange hemisphere (the selection of the green region over the blue is a function of the CIEDE2000 color difference formula). The next segment found via the algorithm is the blue embedded rectangle as seen in Fig. 4(d) and (e). Following the blue segment, the image of Fig. 4(d) is segmented into Fig. 4(f) and (g). These segmentations continue to align with intuition as the blue segment varies the most from the remaining pixels, while the splitting of the two hemispheres is a natural step.

### 3.4   Mona Lisa

The last test image is a low resolution ($48 \times 32$) picture of the *Mona Lisa* inspired by [7], see Fig. 5(a). While the previous methods were able to immediately find the most prominent features of the image, the *Mona Lisa* proved to be a bit more difficult. For instance, there were a few small sections of pixels that were chosen early in the computation. After choosing these small groups of pixels, later separations in the tree provide high quality features. In Fig. 5(b) and (c) one can see a separation of the chest from the rest of the image. This feature selection makes intuitive sense as it is a large section of nearly uniform color. Next, in Fig. 5(d) and (e) the algorithm separated out the hands and arms from

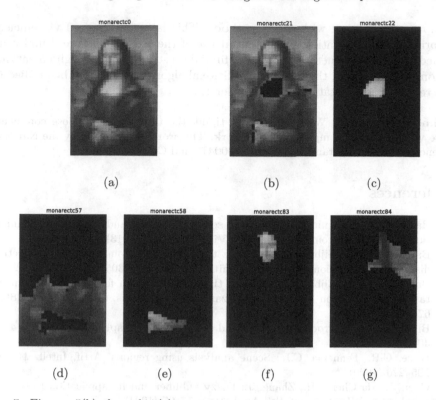

(a)  (b)  (c)

(d)  (e)  (f)  (g)

**Fig. 5.** Figures 5(b) through 5(g) contain interesting segmentations of Figure 5(a). Figures 5(b) and 5(c) separate the chest from the background and body. Figures 5(d) and 5(e) separate the hands and arms from the body; Figures 5(f) and 5(g) separates the face from the surrounding features.

the body. Finally, in Fig. 5(f) and (g) we see the a selection of the face, containing the famous smile, from the hair and background.

## 4   Conclusion

In this paper we explored an image segmentation method that looks for edges of high betweenness in "communities" of pixels. This method attempts to approximate the human process of image segmentation. We have tested this technique on multiple generated images and find results in accordance with our expectations. Further, the method was tested on a low resolution image of the *Mona Lisa* and we found it was able to capture natural features in the image in line with intuition. There are numerous potential avenues for improvement, such as implementing a faster, parallel edge betweenness algorithm. Additionally, the construction of a method that focuses more on the boundaries of segments as opposed to their centers may provide for a more scalable algorithm. The primary focus of this paper is to develop the aforementioned method and an efficient implementation will be the focus of future work, allowing us to explore and compare the runtime with other methods. This is to be achieved via efficient algorithm implementation and potential use of the graph structure. Finally, an added bonus to this method is that the final data structure is a weighted carving decomposition. Thus, there may be additional algorithms that can be applied to the resulting segmentation in lower order time.

**Acknowledgements.** We would like to thank the three referees whose comments were valuable to the completion of this work. This work is supported by the National Science Foundation, Grants No. CMMI-1300477 and CMMI-1404864.

# References

1. Barghout, L., Sheynin, J.: Real-world scene perception and perceptual organization: lessons from computer vision. J. Vis. **13**(9), 709 (2013)
2. Batenburg, K.J., Sijbers, J.: Adaptive thresholding of tomograms by projection distance minimization. Pattern Recognit. **42**(10), 2297–2305 (2009)
3. Batenburg, K.J., Sijbers, J.: Optimal threshold selection for tomogram segmentation by projection distance minimization. IEEE Trans. Med. Imaging **28**(5), 676–686 (2009)
4. Billingsley, F.C.: Processing ranger and mariner photography. Optical Eng. **4**(4), 404147 (1966)
5. Brice, C.R., Fennema, C.L.: Scene analysis using regions. Artif. Intell. **1**(3–4), 205–226 (1970)
6. Chen, L., da Cheng, H., Zhang, J.: Fuzzy subfiber and its application to seismic lithology classification. Inf. Sci. Appl. **1**(2), 77–95 (1994)
7. Diestel, R., Whittle, G.: Tangles and the Mona Lisa. arXiv preprint arXiv:1603.06652 (2016)
8. Dijkstra, E.W.: A note on two problems in connexion with graphs. Numerische mathematik **1**(1), 269–271 (1959)
9. Fredman, M.L., Tarjan, R.E.: Fibonacci heaps and their uses in improved network optimization algorithms. J. ACM (JACM) **34**(3), 596–615 (1987)
10. Girvan, M., Newman, M.E.: Community structure in social and biological networks. Proc. Natl. Acad. Sci. **99**(12), 7821–7826 (2002)
11. Grady, L.: Random walks for image segmentation. IEEE Trans. Pattern Anal. Mach. Intell. **28**(11), 1768–1783 (2006)
12. Hicks, I.V.: Planar branch decompositions i: the ratcatcher. INFORMS J. Comput. **17**(4), 402–412 (2005)
13. Hicks, I.V.: Planar branch decompositions ii: The cycle method. INFORMS J. Comput. **17**(4), 413–421 (2005)
14. Hicks, I.V., Koster, A.M., Kolotoğlu, E.: Branch and tree decomposition techniques for discrete optimization. In: Emerging Theory, Methods, and Applications, pp. 1–29 (2005)
15. Horowitz, S.L., Pavlidis, T.: Picture segmentation by a tree traversal algorithm. J. ACM (JACM) **23**(2), 368–388 (1976)
16. Horowitz, S.L., Pavlidis, T.: Picture segmentation by a directed split and merge procedure. In: CMetImAly77 (1977)
17. Lloyd, S.: Least squares quantization in PCM. IEEE Trans. Inf. Theor. **28**(2), 129–137 (1982)

18. Mobahi, H., Rao, S.R., Yang, A.Y., Sastry, S.S., Ma, Y.: Segmentation of natural images by texture and boundary compression. Int. J. Comput. Vis. **95**(1), 86–98 (2011)
19. Nock, R., Nielsen, F.: Statistical region merging. IEEE Trans. Pattern Anal. Mach. Intell. **26**(11), 1452–1458 (2004)
20. Oak, R.: A study of digital image segmentation techniques. Int. J. Eng. Comput. Sci. **5**(12), 19779–19783 (2016)
21. Otsu, N.: A threshold selection method from gray-level histograms. Automatica **11**(285–296), 23–27 (1975)
22. Rao, S.R., Mobahi, H., Yang, A.Y., Sastry, S.S., Ma, Y.: Natural image segmentation with adaptive texture and boundary encoding. In: Zha, H., Taniguchi, R., Maybank, S. (eds.) ACCV 2009. LNCS, vol. 5994, pp. 135–146. Springer, Heidelberg (2010). doi:10.1007/978-3-642-12307-8_13
23. Robertson, N., Seymour, P.D.: Graph minors. X. Obstructions to tree-decomposition. J. Comb. Theor. Ser. B **52**(2), 153–190 (1991)
24. Seymour, P.D., Thomas, R.: Call routing and the ratcatcher. Combinatorica **14**(2), 217–241 (1994)
25. Sharma, G., Wu, W., Dalal, E.N.: The CIEDE2000 color-difference formula: implementation notes, supplementary test data, and mathematical observations. Color Res. Appl. **30**(1), 21–30 (2005)
26. Shi, J., Malik, J.: Normalized cuts and image segmentation. IEEE Trans. Pattern Anal. Mach. Intell. **22**(8), 888–905 (2000)
27. Wu, Z., Leahy, R.: An optimal graph theoretic approach to data clustering: theory and its application to image segmentation. IEEE Trans. Pattern Anal. Mach. Intell. **15**(11), 1101–1113 (1993)
28. Zahn, C.T.: Graph-theoretical methods for detecting and describing gestalt clusters. IEEE Trans. Comput. **100**(1), 68–86 (1971)
29. Zaitoun, N.M., Aqel, M.J.: Survey on image segmentation techniques. Procedia Comput. Sci. **65**, 797–806 (2015)

# An Image Texture Analysis Method for Minority Language Identification

Darko Brodić[1(✉)], Alessia Amelio[2], and Zoran N. Milivojević[3]

[1] Technical Faculty in Bor, University of Belgrade, V.J. 12, 19210 Bor, Serbia
dbrodic@tfbor.bg.ac.rs
[2] DIMES, University of Calabria, Via Pietro Bucci Cube 44, 87036 Rende, CS, Italy
aamelio@dimes.unical.it
[3] College of Applied Technical Sciences, Aleksandra Medvedeva 20, 18000 Niš, Serbia
zoran.milivojevic@vtsnis.edu.rs

**Abstract.** This paper introduces an image texture analysis method for minority language identification. In the first stage, each letter is associated with a given script type according to its energy status in the text-line area. Mapping is carried out by extracting unicode text and transforming it into coded text. There are four different script types, which correspond to four grey levels of an image. Then, the obtained image is subjected to a feature extraction process performed by the texture analysis. This way, the grey level co-occurrence matrix and its derivative features are calculated. Extracted features are compared and classified using the K-Nearest Neighbors and Naive Bayes methods to establish a difference that can identify a minority language such as Serbian language among other world languages in the text. Very good accuracy results prove the efficiency of the proposed approach, when compared to other state-of-the-art methods.

**Keywords:** Image processing · Natural language processing · Classification · Statistical analysis · Feature extraction

## 1 Introduction

As of July 2013, the United Nations website was available in six languages, while the official website of the European Union could be read in 24 languages. Furthermore, Google supported 90 languages, while Wikipedia supported 295 [25]. However, just a few of them have an above-average dispersion. As a consequence, the Web mainly employs texts written in the so-called world languages, such as English, French, German and Spanish (over 70% of all web content). In contrast, the vast majority, i.e., over 95% of the languages have already lost the capacity to ascend digitally [16]. A number of minority languages in Europe still exists despite the strong pressure from the majority languages such as English, French, German and Spanish. The later languages are clearly widespread and dominant languages in Europe and on the Web. As a consequence, most speakers of minority languages are bilingual or multilingual. Hence, minority languages

© Springer International Publishing AG 2017
V.E. Brimkov and R.P. Barneva (Eds.): IWCIA 2017, LNCS 10256, pp. 280–293, 2017.
DOI: 10.1007/978-3-319-59108-7_22

are changing because the websites in these languages also have to include some of the widespread world languages, creating the concept of multilingual websites. Hence, the extraction of some text fragments in minority languages and their classification is a real challenge and worths investigation. If we take as an example of minority language the Serbian one, then it is used in around 0.1% of the websites [25].

Language identification is the process of language recognition in a certain text. Many methods have been proposed for language identification. They are classified into the following groups [13]: (i) Letter based approach, (ii) Word based approach, (iii) N-gram approach, and (iv) Language identification using a Markov model. Previous research has included the statistical analysis of the text content. In the letter based approach, the frequency distribution of certain letters or common letter combination is analyzed for making easier the language identification [23]. The problem is that the obtained results would be reliable if the number of words in the sentences was high (above 21). Word based approach uses only words up to a specific length or the most frequent word's appearance for establishing the language model [12,22]. The main limitation is the training phase needing a high number of documents to create the language model. Another technique generates a language $n$-gram model for each of the languages, extracting substrings of length $n$ and computing their frequency in the text [4,8]. A problem can occur when the pieces of input text are composed of several languages. Unfortunately, this approach cannot solve such a case. The methods in the last group use Markov models in combination with Bayesian decision rules to produce models for each language [9]. It is worth noting that the training process is computationally intensive, needing at least 50-100 K words for successfully testing small parts of text of length above 100 characters [18].

In this paper, an image-based method for language identification is proposed to overcome the limitations of the previous approaches. In fact, it has the following advantages: (i) it does not require a large piece of text for training, (ii) it is not computer time intensive, (iii) it is able to identify a minority language among the most widespread languages on the Internet, (iv) it needs unicode or prior to Optical Character Recognition (OCR) input, and (v) it is not dependent on the certain alphabet extension with specific letters. The first stage of the method is based on coding. It maps the initial text into a coded text established according to the energy characteristic of each letter based on its position in the text line. A similar approach with six established elements, which is not converted to image elements, was proposed in [21]. Still, this method uses a typical $n$-gram method for language discrimination. Unlike that, in our method the coded text, which includes four different script types, corresponds to an image with four grey levels. Then, it is subjected to co-occurrence statistical analysis in order to extract texture features. This way, the feature extraction and further text classification are performed in the image processing area. To test the proposed approach, an experiment is conducted on a custom-oriented dataset containing text mainly from Web documents (unicode format) given in different world languages: German, Spanish, English and French and minority language such as Serbian. The classification is performed by employing K-Nearest Neighbors and Naive Bayes algorithms. The differences in the feature values establish an important aspect for classification that can identify a minority language such as Serbian among the world languages (in unicode, i.e. web, PDF, or in a scanned text document). This presents a new application of the method which has not been investigated in the previous literature [1,2]. Classification results are compared with the results obtained by the $n$-gram language model for identification of the minority language. This comparison confirms the superiority of the proposed method in language identification. It was also confirmed in a complex problem such as discrimination of evolving languages [3].

The remainder of the paper is organized as follows. Section 2 addresses all aspects concerning the proposed algorithm, including script coding, four grey level image definition, texture features extraction and classification. Section 3 discusses the experiment. Section 4 presents the classification results. Section 4 draws the conclusions and outlines future work directions.

## 2    Proposed Algorithm

The proposed algorithm is a multi-stage method that consists of the Following steps: (i) unicode text is mapped into the coded text according to the energy characteristics, (ii) creation of four grey level image that fully corresponds to the coded text, (iii) feature extraction by co-occurrence analysis, (iv) feature classification, and (v) identification of the language. Figure 1 shows the flow of the proposed algorithm.

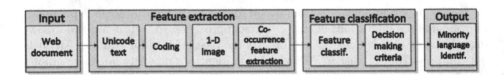

**Fig. 1.** The flow of the algorithm.

### 2.1    Unicode Text Mapping

Typically, the text in the documents is divided into text lines. Each character can be separated according to its energy characteristic, i.e. horizontal projections profile. Figure 2 shows the different characters and their corresponding energy characteristics. We can realize that the height of different characters corresponds to their energy. Obviously, all characters have different energy characteristic. The most diffused energy is given by the characters that outspread over the full text line height such as character $lj$ in Serbian or Croatian language. Taking into account all aforementioned, we can draw virtual lines in each text line. They can be represented as: (i) top-line, (ii) upper-line, (iii) base-line, and (iv) bottom-line. Furthermore, they establish three vertical zones [26]: (i) upper zone, (ii) middle

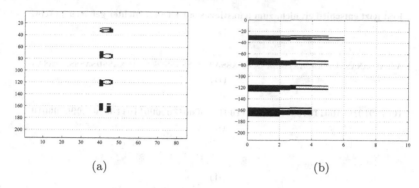

(a)    (b)

**Fig. 2.** Energy characteristic of different characters: (a) different characters in different text lines, (b) their corresponding energy.

**Fig. 3.** Definition of the script characteristics according to baseline status of the text.

zone, and (iii) lower zone. All letters can be classified according to these vertical zones. The short letters (S) take the middle zone. The capital letters and letters with ascenders (A) take the middle and upper zones. The descendent ones (D) occupy the middle and lower zones. The full letters (F) outspread over the upper, middle and lower zones. Figure 3 shows the script characteristics according to the letter baseline position.

According to the vertical zone classification, all letters from the alphabet are substituted with the script types. Each letter is positioned into a given zone(s) in the text line. Consequently, it is mapped to a unique element of the set {S, A, D, F}.

## 2.2 Image Creation

In order to easily apply a statistical analysis, the script type should be injectively coded in the following way:

$$S \to 0, A \to 1, D \to 2, F \to 3. \tag{1}$$

These codes can be transformed into the grey level pixels of an image. Hence, it corresponds to an image with four grey levels. Figure 4 shows an example of the coding procedure for a text sample given in German language.

## 2.3 Feature Extraction

Currently, the initial text is transformed into an image through a variable reduction process. The obtained image is then subjected to the texture analysis. It

**Erst jetzt entschied er, nicht zum Künstler, sondern zum Dichter geboren zu sein.**

(a)

ASSAFSASASSASSAASASSSSASAASSSAASSAASSSSSASSSSSSAASAASSDSASSSSSSSSAS

(b)

100130101001001101000101100011001100000100000011011002010000000010

(c)

(d)

**Fig. 4.** An example of coding procedure for a German text: (a) initial text, (b) extraction of characters, (c) coding according to script types, (d) image creation.

includes the extraction of the co-occurrence probabilities, which provide second-order texture features [14]. These features are extracted from the image in two steps. At the first step, the pairwise spatial co-occurrences of pixels separated by a particular angle $\theta$ and distance $d$ are tabulated using a Grey Level Co-occurrence Matrix (GLCM). At the second step, a set of texture measures is calculated from GLCM. The GLCM shows how often different combinations of grey level co-occur in a part or in the whole image [14].

Let's suppose that our grey scale image is given as $\mathbf{I}$, featuring $M$ rows, $N$ columns, and $T$ number of grey levels. GLCM represents the spatial relationship of grey levels in the image $\mathbf{I}$. It is a $T \times T$ square matrix. To compute GLCM $\mathbf{C}$, a central pixel $I(x, y)$ with a neighborhood defined by the Window Of Interest (WOI) is taken. WOI is defined by inter-pixel distance $d$ and orientation $\theta$. Hence, for the given image $\mathbf{I}$, GLCM $\mathbf{C}$ is defined as [10]:

$$C(i,j) = \sum_{x=1}^{T} \sum_{y=1}^{T} \begin{cases} 1 \text{ if } I(x,y) = i, \text{and} \\ \quad I(x + \Delta x, y + \Delta y) = j, \\ 0 \qquad \text{otherwise} \end{cases} \tag{2}$$

where $i$ and $j$ are the intensity values of the image $\mathbf{I}$, $x$ and $y$ are the spatial positions in the image $\mathbf{I}$, the offset $(\Delta x, \Delta y)$ is the distance between the pixel-of-interest and its neighbor. It should be noted that the offset depends on the direction $\theta$ that is used and the distance $d$ at which the matrix is computed.

In our case, the neighborhood is given as 2-connected only, due to the nature of the text. Accordingly, $\theta$ is $0°$, while $d$ is typically used as first neighborhood, i.e. $d = 1$. Then, the normalized matrix $\mathbf{P}$ of GLCM $\mathbf{C}$ is calculated as [5]:

$$P(i,j) = C(i,j) / \sum_{i}^{T} \sum_{j}^{T} C(i,j). \tag{3}$$

**Table 1.** Twelve co-occurrence elements.

| | |
|---|---|
| $\mu_x$ | $\sum_{i=1}^{T} i \sum_{j=1}^{T} P(i,j),$ |
| $\mu_y$ | $\sum_{j=1}^{T} j \sum_{i=1}^{T} P(i,j),$ |
| $\sigma_x$ | $\sqrt{\sum_{i=1}^{T} (i - \mu_x)^2 \sum_{j=1}^{T} P(i,j)},$ |
| $\sigma_y$ | $\sqrt{\sum_{j=1}^{T} (j - \mu_y)^2 \sum_{i=1}^{T} P(i,j)},$ |
| Correlation | $\sum_{i=1}^{T} \sum_{j=1}^{T} \frac{(i \cdot j) \cdot P(i,j) - (\mu_x \cdot \mu_y)}{\sigma_x \cdot \sigma_y},$ |
| Energy | $\sum_{i=1}^{T} \sum_{j=1}^{T} P(i,j)^2,$ |
| Entropy | $-\sum_{i=1}^{T} \sum_{j=1}^{T} P(i,j) \cdot log P(i,j),$ |
| Maximum | $\max\{P(i,j)\},$ |
| Dissimilarity | $\sum_{i=1}^{T} \sum_{j=1}^{T} P(i,j) \cdot |i - j|,$ |
| Contrast | $\sum_{i=1}^{T} \sum_{j=1}^{T} P(i,j) \cdot (i - j)^2,$ |
| Invdmoment | $\sum_{i=1}^{T} \sum_{j=1}^{T} \frac{1}{1 + (i-j)^2} P(i,j),$ |
| Homogeneity | $\sum_{i=1}^{T} \sum_{j=1}^{T} \frac{P(i,j)}{1 + |i - j|}$ |

Still, GLCM provides only a quantitative description of the spatial patterns. Hence, it is not used for practical image analysis. Ref. [14] proposed a set of texture measures, which summarize the information from GLCM. Although a total of 14 quantities, i.e. features was originally proposed, only subsets of them are used [17]. These are the following twelve GLCM texture measures: (i) mean value $\mu_x$, (ii) mean value $\mu_y$, (iii) standard deviation $\sigma_x$, (iv) standard deviation $\sigma_y$, (v) correlation, (vi) energy, (vii) entropy, (viii) maximum, (ix) dissimilarity, (x) contrast, (xi) inverse difference moment and (xii) homogeneity. The twelve co-occurrence elements are shown in Table 1. Hence, after this phase, the four grey level image is represented by a 12-dimensional feature vector.

## 2.4 Feature Classification

In order to classify the obtained feature vector, the K-Nearest Neighbors and Naive Bayes methods have been used, which are well-known algorithms for data classification.

**K-Nearest Neighbors.** K-Nearest Neighbors (K-NN) is a very easy approach to classify feature vectors [7,11]. Let $Tr$ be the training set composed of $n$ feature vectors with associated class labels and $x_t$ be a test feature vector to classify. Classification of $x_t$ is performed by computing the distance between $x_t$ and each training vector in $Tr$ from 1 to $n$. The $K$ training vectors $Tr'$ which are the nearest to $x_t$ are finally considered. $K$ is a fixed parameter of the algorithm, determining the amplitude of the neighborhood. The predicted class label for $x_t$ is the one occurring most frequently in $Tr'$. Because even values of $K$ can determine class labels with the same frequency in $Tr'$ [19], the value of the $K$ parameter is usually fixed to a small odd integer. When the instances are fixed-length vectors of real-value features, the distance function often adopted for

K-NN is the Euclidean one. However, different distance functions can determine variations in the similarity evaluation [19]. In fact, based on the chosen distance function, two vectors $x_i$ and $x_j$ can be considered more or less similar to each other. Consequently, other possible functions are selected for K-NN, such as the Manhattan distance or the Chebyshev distance.

**Naive Bayes.** Naive Bayes (NB) classifier is a probabilistic learning method based on the assumption that all variables are mutually independent, given the class variable [20]. The classifier is defined as: $f_{nb}(x_i) = \frac{p(Y=1)}{p(Y=0)} \prod_{k=1}^{h} \frac{p(x_i^k|Y=1)}{p(x_i^k|Y=0)}$, where $x_i = \{x_i^1, ..., x_i^h\}$ represents a vector of $h$ features and class variable $Y$. In order to classify the test feature vector $x_i$ in class 1 or class 0, the probability of each of its features conditioned to class 1 or 0 and the probability of occurrence of class 1 and 0 in the training set are computed. $x_i$ is predicted to be in class 1 if and only if $f_{nb}(x_i) \geq 1$. Otherwise, it is predicted to be in class 0.

For numerical features, the normal distribution is considered for computing the probability values:

$$f(w, \mu, \sigma) = \frac{1}{\sqrt{2\pi}\sigma} e^{-\frac{(w-\mu)^2}{\sigma^2}}. \tag{4}$$

Accordingly, $p(x_i^k|Y = 1) = f(x_i^k, \mu_{y_i}, \sigma_{y_i})$, where $\mu_{Y=1}$ and $\sigma_{Y=1}$ are respectively the mean and standard deviation of the values of $k$-th feature with class 1.

## 3  Experiment

A test is conducted to evaluate the quality of the proposed method in correctly identifying a minority language among a set of world languages. Accordingly, a custom-oriented dataset extracted from the web text given in unicode format of different languages is employed. It represents a set of text excerpts in German, Spanish, English, French and Serbian languages. It consists of a total of 150 texts, divided into two classes respectively of 25 Serbian and 125 world language texts (German, Spanish, English and French). The class label for each text in the dataset corresponds to Serbian or world language. All texts have different contents and size. The size of the text excerpts is between 378 and 2822 characters. The length of the texts is chosen according to the size standard in factor analysis, which means that the total number of analyzed elements would be higher than 300 [24]. It means that our text samples contain approx. more than 300 characters. Figure 5 illustrates a web page in Serbian language from which the unicode text is extracted.

The proposed features are extracted from each text in the dataset, in order to create 150 feature vectors. Then, K-NN and NB algorithms are adopted on the feature representation of the dataset for identification of the minority language. Finally, classification results are compared and discussed.

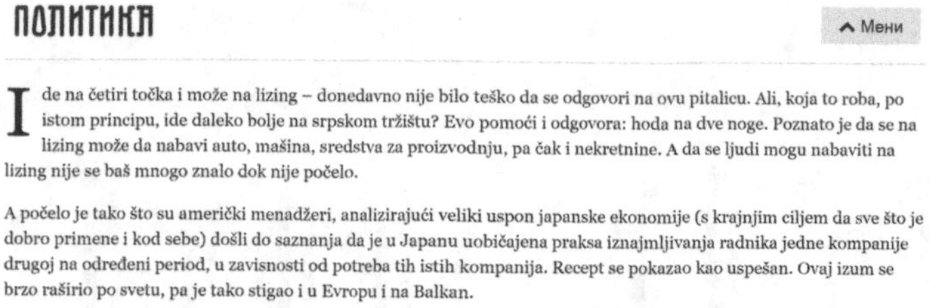

ПОЛИТИКА                                    ∧ Мени

I de na četiri točka i može na lizing – donedavno nije bilo teško da se odgovori na ovu pitalicu. Ali, koja to roba, po istom principu, ide daleko bolje na srpskom tržištu? Evo pomoći i odgovora: hoda na dve noge. Poznato je da se na lizing može da nabavi auto, mašina, sredstva za proizvodnju, pa čak i nekretnine. A da se ljudi mogu nabaviti na lizing nije se baš mnogo znalo dok nije počelo.

A počelo je tako što su američki menadžeri, analizirajući veliki uspon japanske ekonomije (s krajnjim ciljem da sve što je dobro primene i kod sebe) došli do saznanja da je u Japanu uobičajena praksa iznajmljivanja radnika jedne kompanije drugoj na određeni period, u zavisnosti od potreba tih istih kompanija. Recept se pokazao kao uspešan. Ovaj izum se brzo raširio po svetu, pa je tako stigao i u Evropu i na Balkan.

**Fig. 5.** Web page sample in Serbian language.

# 4 Results and Discussion

Figure 6 and Table 2 show the GLCM features obtained by the dataset for the world and Serbian languages in a min-max manner. $\mu_x$ and $\mu_y$ are the same as well as $\sigma_x$ and $\sigma_y$ count on two decimal places. Hence, only one graph representing both is enough. It is worth noting that Serbian language can mostly be discriminated from the world languages by the GLCM dissimilarity and contrast. In fact, the overall characteristics of Serbian text have much smaller values of dissimilarity and contrast.

Because the classification accuracy depends on the choice of training and test sets, the dataset is processed by a $k$-fold cross-validation strategy [15], for obtaining different results on multiple training and test sets. The dataset is randomly divided into $k$ folds. Then, each fold is considered as the test set and the remaining $k - 1$ folds are considered as the training set. Each fold has roughly equal dimension and roughly the same language class proportions as in the dataset. Consequently, the test set is composed of a small number of texts in Serbian and a number of texts in world languages. Model learning by K-NN and NB is performed each time by using the current training set, then classification evaluation is established on the current test set. The $K$ value of the K-Nearest Neighbor has been fixed to small odd values (see Sect. 2.4), i.e. 1, 3 and 5. Furthermore, the K-NN algorithm has been executed with the traditional Euclidean distance, which revealed to be particularly reliable in this context with respect to other distance measures. Because the feature vectors have numerical values, probabilities of the NB have been computed by using (4).

Our task is in the domain of binary classification, because we need to classify a model and correctly predict the classes that represent texts written in the world languages (German, Spanish, English and French) and in minority Serbian language. The problem of language classification and identification is similar to the information retrieval one. Hence, precision, recall and f-measure are preferred metrics for the evaluation of the proposed algorithm [6]. They are calculated from the confusion matrix between the classification results obtained by the test set and the ground truth partitioning of the test set in Serbian and world languages. Performance measures have been computed for each selection of the test and training folds and the average values together with the standard deviation have been reported for each of the measures. Furthermore, $k$-fold cross-validation has

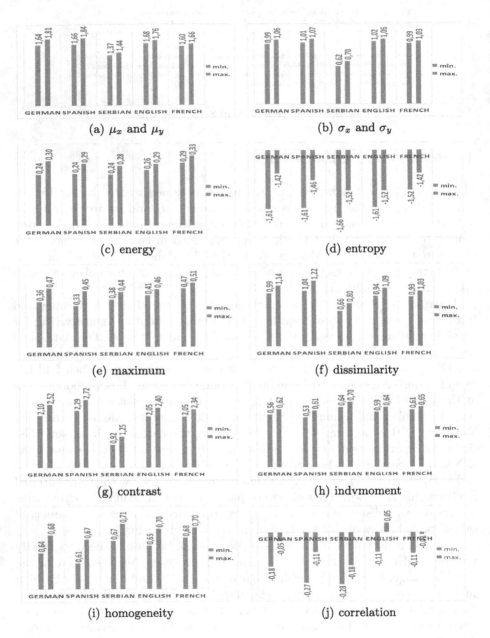

**Fig. 6.** The twelve GLCM features obtained by the dataset for the world and Serbian languages in a min-max manner

**Table 2.** GLCM feature values from Fig. 6 in the min-max manner.

|  |  | German | Spanish | Serbian | English | French |
|---|---|---|---|---|---|---|
| $\mu_x$ | min. | 1.6418 | 1.6557 | 1.3700 | 1.6809 | 1.6007 |
|  | max. | 1.8090 | 1.8353 | 1.4382 | 1.7610 | 1.6634 |
| $\mu_y$ | min. | 1.6399 | 1.6526 | 1.3691 | 1.6813 | 1.5999 |
|  | max | 1.8081 | 1.8329 | 1.4375 | 1.7612 | 1.6618 |
| $\sigma_x$ | min. | 0.9865 | 1.0088 | 0.6153 | 1.0209 | 0.9904 |
|  | max. | 1.0586 | 1.0665 | 0.7003 | 1.0611 | 1.0303 |
| $\sigma_y$ | min. | 0.9857 | 1.0077 | 0.6151 | 1.0204 | 0.9901 |
|  | max. | 1.0585 | 1.0655 | 0.7002 | 1.0606 | 1.0298 |
| Energy | min. | 0.2388 | 0.2431 | 0.2395 | 0.2602 | 0.2928 |
|  | max. | 0.3041 | 0.2911 | 0.2811 | 0.2890 | 0.3270 |
| Entropy | min. | −1.6149 | −1.6106 | −1.6641 | −1.6062 | −1.5151 |
|  | max. | −1.4249 | −1.4624 | −1.5171 | −1.5180 | −1.4245 |
| Maximum | min. | 0.3639 | 0.3283 | 0.3805 | 0.4136 | 0.4684 |
|  | max. | 0.4695 | 0.4496 | 0.4401 | 0.4636 | 0.5127 |
| Dissimilarity | min. | 0.9913 | 1.0364 | 0.6593 | 0.9385 | 0.9300 |
|  | max. | 1.1361 | 1.2244 | 0.8037 | 1.0867 | 1.0324 |
| Contrast | min. | 2.0975 | 2.2870 | 0.9201 | 2.0532 | 2.0479 |
|  | max | 2.5232 | 2.7229 | 1.2532 | 2.4019 | 2.3350 |
| Invdmoment | min. | 0.5645 | 0.5309 | 0.6431 | 0.5882 | 0.6141 |
|  | max. | 0.6211 | 0.6068 | 0.6965 | 0.6422 | 0.6468 |
| Homogeneity | min. | 0.6356 | 0.6081 | 0.6685 | 0.6549 | 0.6768 |
|  | max. | 0.6832 | 0.6717 | 0.7134 | 0.7005 | 0.7038 |
| Correlation | min. | −0.1833 | −0.2696 | −0.2779 | −0.1055 | −0.1130 |
|  | max. | −0.0488 | −0.1068 | −0.1758 | 0.0498 | −0.0122 |

been repeated separately for three different values of $k$ equal to 2, 5 and 10 [15]. Finally, for avoiding the dependence of the classification results from the particular division in folds, $k$-fold cross-validation has been executed 50 times for each value of $k$.

The classification results obtained by the proposed method using K-NN or NB classifier are very positive. Evaluation reveals a perfect identification of the Serbian texts in the test set. In fact, precision, recall and f-measure obtain a value of 1 in all the cases, when the $k$ value of fold cross-validation is equal to 2, 5 and 10 for all the 50 runs and when the $K$ value of the Nearest Neighbor classifier is fixed to 1, 3 and 5. The classification results of the proposed method are compared with those obtained by the $n$-gram language model. In particular, each text in the dataset is represented by the normalized frequency values of the extracted bi-grams. The same classification experiment is performed with the bi-gram feature vectors by adopting K-NN and NB algorithms and $k$-fold cross-validation. Also, the same parameter values for the classifiers are selected in the experiment with bi-grams.

Table 3 reports the classification results of bi-grams with the K-NN classifier and Euclidean distance at the different values of $k = 2, 5, 10$ folds and with $K = 1, 3, 5$.

**Table 3.** Average results in terms of precision, recall and f-measure, together with the standard deviation (in parenthesis), obtained by bi-grams and K-NN classifier using $k$-fold cross-validation.

| | | 2-fold | | 5-fold | | 10-fold | |
|---|---|---|---|---|---|---|---|
| | | World lang | Serbian | World lang | Serbian | World lang | Serbian |
| $K = 1$ | Precision | 0.9994 | 1.0000 | 0.9997 | 1.0000 | 1.0000 | 1.0000 |
| | | (0.0001) | (0.0000) | (0.0001) | (0.0000) | (0.0000) | (0.0000) |
| | Recall | 1.0000 | 0.9962 | 1.0000 | 0.9984 | 1.0000 | 1.0000 |
| | | (0.0000) | (0.0054) | (0.0000) | (0.0036) | (0.0000) | (0.0000) |
| | F-Measure | 0.9997 | 0.9979 | 0.9998 | 0.9990 | 1.0000 | 1.0000 |
| | | (0.0004) | (0.0030) | (0.0003) | (0.0022) | (0.0000) | (0.0000) |
| $K = 3$ | Precision | 0.9935 | 1.0000 | 0.9988 | 1.0000 | 1.0000 | 1.0000 |
| | | (0.0091) | (0.0000) | (0.0026) | (0.0000) | (0.0000) | (0.0000) |
| | Recall | 1.0000 | 0.9587 | 1.0000 | 0.9924 | 1.0000 | 1.0000 |
| | | (0.0000) | (0.0584) | (0.0000) | (0.0170) | (0.0000) | (0.0000) |
| | F-Measure | 0.9967 | 0.9771 | 0.9994 | 0.9950 | 1.0000 | 1.0000 |
| | | (0.0046) | (0.0324) | (0.0013) | (0.0112) | (0.0000) | (0.0000) |
| $K = 5$ | Precision | 0.9859 | 1.0000 | 0.9954 | 1.0000 | 0.9976 | 1.0000 |
| | | (0.0134) | (0.0000) | (0.0103) | (0.0000) | (0.0077) | (0.0000) |
| | Recall | 1.0000 | 0.9120 | 1.0000 | 0.9710 | 1.0000 | 0.9843 |
| | | (0.0000) | (0.0854) | (0.0000) | (0.0648) | (0.0000) | (0.0495) |
| | F-Measure | 0.9929 | 0.9519 | 0.9976 | 0.9831 | 0.9987 | 0.9893 |
| | | (0.0068) | (0.0475) | (0.0053) | (0.0379) | (0.0040) | (0.0337) |

Although precision, recall, and f-measure are in the range of 0.91–1.00, it is worth noting that bi-grams are not able to obtain the perfect identification of the minority language among the world languages in all the cases. In fact, in 10-fold the f-measure value for Serbian class is in the range 0.98–1.00. However, in 5-fold it varies in the range 0.98–0.99. Finally, in 2-fold the f-measure value is in the range 0.95–0.99, depending on the $K$ value.

Table 4 shows the classification results of bi-grams with the NB classifier at the different values of $k = 2, 5, 10$ folds. We may observe a poor classification result w.r.t that obtained by our method. In particular, the method is mostly poor in recognition of the minority Serbian language, with the highest f-measure

**Table 4.** Average results in terms of precision, recall and f-measure, together with the standard deviation (in parenthesis), obtained by bi-grams and NB classifier using *k*-fold cross-validation.

| | 2-fold | | 5-fold | | 10-fold | |
|---|---|---|---|---|---|---|
| | World lang | Serbian | World lang | Serbian | World lang | Serbian |
| *Precision* | 0.9141 | 0.6409 | 0.8837 | 0.2000 | 0.8701 | 0.0867 |
| | (0.0754) | (0.4821) | (0.0495) | (0.3332) | (0.0433) | (0.2574) |
| *Recall* | 0.9984 | 0.4045 | 1.0000 | 0.1690 | 0.9992 | 0.0850 |
| | (0.0022) | (0.5310) | (0.0000) | (0.3187) | (0.0024) | (0.2477) |
| *F-Measure* | 0.9532 | 0.4356 | 0.9375 | 0.1747 | 0.9296 | 0.0847 |
| | (0.0397) | (0.5410) | (0.0269) | (0.3180) | (0.0234) | (0.2491) |

value of 0.44 in the 2-fold. Also, classification of the world languages is not perfect, reaching a peak of 0.95 in the 2-fold.

Figure 7 illustrates the f-measure values obtained by our method and by bi-grams using K-NN and NB classifiers for the different *k*-folds. Bars represent the f-measure values obtained by bi-grams. The black dashed lines represent the value of f-measure equal to 1 obtained by our method in all cases. This graphical comparison confirms that our method outperforms the competitor in most cases.

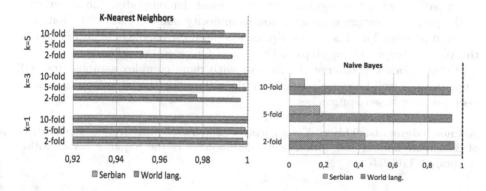

**Fig. 7.** F-measure values obtained by our method and by the bi-grams feature representation. Bars represent the f-measure values obtained by bi-grams. The black dashed lines represent the value of f-measure equal to 1 obtained by our method in all cases

These results indicate that the bi-grams are not totally robust and reliable in the identification of a minority language such as Serbian among the world languages. Also, the tri-grams did not obtain a meaningful improvement w.r.t. bi-grams, similarly as in [3]. Consequently, their results will be omitted. On the contrary, the proposed feature representation demonstrated to be robust and effective in solving the same task on the same dataset and in the same operating conditions when two different machine learning algorithms are employed. Hence, it is a very promising approach in language identification.

The experiment has been performed in MATLAB R2015a, on a Desktop computer quad-core 2.3 GHz CPU with 8 GB RAM and operating system Windows 7. Based on these specifications, the CPU time for the feature extraction procedure is below 0.1 s. Again, the CPU time for feature classification by K-NN and NB is around 0.003 s. Obviously, the method has proven to be computationally non-intensive.

## 5    Conclusion

This paper proposed an image-based method for minority language identification by considering statistical analysis of the text based on the text-line status of each script element. The statistical analysis was performed by the grey level co-occurrence matrix. Due to the difference in the language characteristics, the results of the statistical analysis in terms of features showed significant dissimilarity. Classification of the introduced features was performed by the well-known supervised learning algorithms K-Nearest Neighbors and Naive Bayes. The proposed method was tested on text excerpts from a custom oriented dataset. It incorporated texts given in German, Spanish, English, French (world languages) and Serbian (minority language). The experiments showed encouraging results: minority language was perfectly classified by the proposed method. Furthermore, a comparison with the $n$-gram language model demonstrated the superiority of the proposed feature representation in minority language identification. The research presented in this manuscript can be used for language identification on the Web, in preprocessing steps of OCR, and for video text identification.

Future work will enlarge the dataset with more complex samples and will employ other well-known classification algorithms for the experiment, such as deep learning based approaches.

**Acknowledgments.** This work was partially supported by the Grant of the Ministry of Education, Science and Technological development of the Republic Serbia within the project TR33037.

## References

1. Brodić, D., Amelio, A., Milivojević, Z.N.: An approach to the language discrimination in different scripts using adjacent local binary pattern. J. Exp. Theor. Artif. Intell., 1–19 (2016, in press). doi:10.1080/0952813X.2016.1264090
2. Brodić, D., Amelio, A., Milivojević, Z.N.: Language discrimination by texture analysis of the image corresponding to the text. Neural Comput. Appl., 1–22 (2016, in press). doi:10.1007/s00521-016-2527-x
3. Brodić, D., Amelio, A., Milivojević, Z.N.: Clustering documents in evolving languages by image texture analysis. Appl. Intell. **46**(4), 916–933 (2017)
4. Cavnar, W.B., Trenkle, J.M.: N-gram-based text categorization. In: Document Analysis and Information Retrieval, Las Vegas, USA, pp. 161–175 (1994)

5. Clausi, D.A.: An analysis of co-occurrence texture statistics as a function of grey level quantization. Can. J. Remote Sens. **28**(1), 45–62 (2002)
6. Confusion    Matrix.    http://www2.cs.uregina.ca/~dbd/cs831/notes/confusion_matrix/confusion_matrix.html
7. Dasarathy, B.V.: Nearest Neighbor: Pattern Classification Techniques (Nn Norms: Nn Pattern Classification Techniques). IEEE Computer Society Press, Los Alamitos (1990)
8. Dunning, T.: Statistical Identification of Language. Technical report MCCS 94–273, New Mexico State University (1994)
9. Dunning, T.: Statistical Identification of Language. Technical report CRLMCCS-94-273, Computing Research Lab, New Mexico State University (1994)
10. Eleyan, A., Demirel, H.: Co-occurrence matrix and its statistical features as a new approach for face recognition. Turkish J. Electr. Eng. Comput. Sci. **19**(1), 97–107 (2011)
11. Elkan, C.: Nearest Neighbor Classification (2011). http://cseweb.ucsd.edu/~elkan/250Bwinter2010/nearestn.pdf
12. Grefenstette, G.: Comparing two language identification schemes. In: Statistical Analysis of Textual Data, Rome, Italy, pp. 1–6 (1995)
13. Grothe, L., De Luca, E.W., Nurnberger, A.: A comparative study on language identification methods. In: Language Resources and Evaluation, Marrakech, Morocco, pp. 980–985 (2008)
14. Haralick, R.M., Shanmugan, K., Dinstein, I.: Textural features for image classification. IEEE Trans. Syst. Man Cybern. **3**(6), 610–621 (1978)
15. Hastie, T., Tibshirani, R., Friedman, J.: The Elements of Statistical Learning. Springer Series in Statistics. Springer, New York (2009)
16. Kornai, A.: Digital language death. PLoS ONE **8**(10), 1–11 (2013)
17. Newsam, S., Kamath, C.: Comparing shape and texture features for pattern recognition in simulation data. In: Image Processing: Algorithms and Systems IV, San Jose, USA, pp. 1–14 (2005)
18. Padro, M., Padro, L.: Comparing methods for language identification. In: XXCongreso de la Sociedad Espanola para el Procesamiento del Lenguage Natural, Barcelona, Spain, pp. 155–161 (2004)
19. Proietti, A., Panella, M., Leccese, F., Svezia, E.: Dust detection and analysis in museum environment based on pattern recognition. Measurement **66**, 62–72 (2015)
20. Russell, S., Norvig, P.: Artificial Intelligence: A Modern Approach, 2nd ed. Prentice Hall (2003). [1995]
21. Sibun, P., Spitz, A.L.: Language determination: natural language processing from scanned document images. In: 4th Conference on Applied Natural Language Processing, Stuttgart, Germany, pp. 15–21 (1994)
22. Souter, C., Churcher, G., Hayes, J., Hughes, J., Johnson, S.: Natural language identification using corpus-based models. Hermes J. Linguist. **13**, 183–203 (1994)
23. Takcı, H., Soğukpınar, İ.: Letter based text scoring method for language identification. In: Yakhno, T. (ed.) ADVIS 2004. LNCS, vol. 3261, pp. 283–290. Springer, Heidelberg (2004). doi:10.1007/978-3-540-30198-1_29
24. Wackerly, D.D., Mendenhall, W., Scheaffer, R.L.: Mathematical Statistics with Applications. Duxbury Press, Belmont (1996)
25. Web 2014. http://w3techs.com/technologies/overview/content_language/all
26. Zramdini, A.W., Ingold, R.: Optical font recognition using typographical features. IEEE Trans. Pattern Anal. Mach. Intell. **20**(8), 877–882 (1998)

# JPEG Quantization Table Optimization by Guided Fireworks Algorithm

Eva Tuba[1], Milan Tuba[2](✉), Dana Simian[3], and Raka Jovanovic[4]

[1] Faculty of Mathematics, University of Belgrade, Belgrade, Serbia
etuba@ieee.org
[2] Graduate School of Computer Science, John Naisbitt University, Belgrade, Serbia
tuba@ieee.org
[3] Faculty of Science, Lucian Blaga University, Sibiu, Romania
dana.simian@ulbsibiu.ro
[4] Qatar Environment and Energy Research Institute,
Hamad Bin Khalifa University, Doha, Qatar
rjovanovic@qf.org.qa

**Abstract.** Digital images are very useful and ubiquitous, however there is a problem with their storage because of their large size and memory requirement. JPEG lossy compression algorithm is prevailing standard that solves that problem. It facilitates different levels of compression (and the corresponding quality) by using recommended quantization tables. It is possible to optimize these tables for better image quality at the same level of compression. This presents a hard combinatorial optimization problem for which stochastic metaheuristics proved to be efficient. In this paper we propose an adjustment of the recent guided fireworks algorithm from the class of swarm intelligence algorithms for quantization table optimization. We tested the proposed approach on standard benchmark images and compared results with other approaches from literature. By using various image similarity metrics our approach proved to be more successful.

**Keywords:** Image processing · JPEG algorithm · Quantization tables · Fireworks algorithm · Swarm intelligence

## 1 Introduction

Widespread use of digital images facilitated advances in numerous scientific areas. Biology, astronomy, medicine and many other fields were significantly improved by digital images introduction [3,10]. Common requirement in all these areas that use digital images is some kind of image processing which reduces to

M. Tuba was supported by the Ministry of Education, Science and Technological Development of Republic of Serbia, Grant no. III-44006.

D. Simian was supported by the research grant LBUS-IRG-2015-01, project financed by Lucian Blaga University of Sibiu.

© Springer International Publishing AG 2017
V.E. Brimkov and R.P. Barneva (Eds.): IWCIA 2017, LNCS 10256, pp. 294–307, 2017.
DOI: 10.1007/978-3-319-59108-7_23

application of different algorithms and mathematical formulas to the matrix of numbers that represent digital image. Simplicity and power of digital image processing is one of the major factors that contributes to ubiquity of digital images.

Benefits of using digital images are numerous, however there are also some problems. One of the biggest problems is the space needed for storing digital images. One high quality digital image typically consists of millions of pixels and accordingly needs more that ten megabytes of memory. One solution to that problem is to use compression techniques to record image data in some format that would use less memory. All compression algorithms can be divided into two groups: lossless and lossy algorithms. Lossless algorithms try to find some consistency and redundancy in the data and to rewrite it in more compact but reversible way so that decompressed data are identical to the original [21]. Such algorithms can achieve typical compression ratios of 2:1 or 3:1 which is inadequate for huge digital images. Fortunately, lossy compression algorithms, by discarding some informations which results in minimal quality loss, achieve much high compression rates of 10:1 or even 100:1.

JPEG is one of the most used lossy compression algorithm for digital images. It is well known that by using JPEG algorithm space needed for image information can be reduced twenty or even fifty times [11]. The main compression is done by quantization step where, based on quantization table, less important information is neglected.

In this paper a recent and very successful swarm intelligence algorithm, guided fireworks algorithm, was used to solve combinatorial problem of selecting elements of quantization table for JPEG algorithm in order to optimize decompressed image according to some metrics.

In Sect. 2 basic steps of JPEG algorithm are described while in Sect. 3 the use of quantization table in JPEG algorithm is briefly explained. In Sect. 4 guided fireworks algorithm is presented. The proposed algorithm for quantization table selection is explained in Sect. 5. Experimental results are shown in Sect. 6 and at the end in Sect. 7 conclusion and suggestion for further work are given.

## 2 JPEG Algorithm

JPEG algorithm the most used lossy algorithm for digital image compression. It is a very powerful algorithm that can significantly reduce the size of the image file without any significant visible consequences.

JPEG algorithm consists of several steps and the first preprocessing step is based on the fact that human eye is less sensitive to changes in color than in the light intensity. HSV chromaticity components resolution reduction facilitates stronger compression for color images. JPEG algorithm is less suitable for the simple drawings (drawings with sharp edges or text). For such images even moderate compression will noticeably damage them.

The main step in the JPEG algorithm that results in significant compression is quantization, which is the subject of research and experiments in this paper.

Some of the steps in JPEG algorithm will be just mentioned since they do not have impact on results of our proposed algorithm.

The first step, after preprocessing, is to transform image into the frequency domain by applying two-dimensional discrete cosine transform (DCT) which is performed on non-overlapping blocks of size $8 \times 8$ of light intensities.

Result of this step is that the image is transformed to blocks of size $8 \times 8$ where each block consists of 64 frequency coefficients. The very first coefficient in the block represents average intensity and it is named DC component, while the rest 63 coefficients are AC components. DC coefficient contain the most information about the block of the image. Coefficients close to DC coefficient represent low frequencies and coefficients closer to the lower-right corner represent high frequencies. Higher frequencies describe sudden changes in intensity values, i.e. edges and noise. These features are less important and usually these coefficients are close to zero.

The next step in JPEG algorithm is quantization where the main compression is done, but also some information is lost.

## 3   Quantization Tables

Compression level and the quality of compressed image in JPEG algorithm are mainly determined by the quantization table. The idea is to discard DCT coefficient that are less important and to reduce precision level for others. At quantization step each DCT coefficient from $8 \times 8$ matrix is divided by the corresponding element from the quantization table. By arranging elements in the quantization table different compression levels and qualities of compressed images can be determined. If all elements in the quantization table are equal to one, then there will be no compression and the quality of the image will not be decreased.

JPEG standard provides recommendation for quantization tables that was determined empirically based on human perception. Quantization table that provides high compression and good decompressed image quality is named $Q_{50}$. Based on this quantization table, tables $Q_1$ to $Q_{100}$ can be calculated. The number in index represents the scale for quality of the compressed image. For low quality index higher compression will be achieved, but at the expense of image quality. On the other hand, with high quality index better image quality will be achieved, but the compression level will be low. For higher quality indices over 50 (less compression), quantization tables are obtained by multiplying the table $Q_{50}$ by $(100 - quality\_level)/50$ while for lower quality indices below 50 but higher compression, quantization tables are defined by multiplying standard $Q_{50}$ by $50/quality\_level$. In both cases values are clipped to be between 1 and 255.

As mentioned before, these $Q_i$ matrices are quantization tables constructed based on human subjective opinion about image quality, but for many applications some more objective metrics are necessary. For such kinds of applications compressed images are further processed with some specific goal, and that goal achievement represents the quality metrics [1,9,13,24].

A method for determining customized JPEG quantization table for low bit rate mobile visual search was proposed in [8]. In the proposed method pairwise image matching precision was incorporated into distortion measure and quantization table was optimized to achieve better trade-off between compression level and visual quality.

In [5] an algorithm for finding the optimal quantization table that enables improvement of feature detection performance was proposed. Optimal quantization table was based on the measured impact that scale-space processing has on the DCT.

A comparison between compression by the traditional quantization matrix and by a set of quantization matrices especially optimized for ultrasound images was performed in [33]. Experimental results have shown that images compressed by optimized tables have significantly better quality in the sense of the medical information.

Selecting elements of the quantization table represents a combinatorial problem: each of the 64 elements can be any number from some range. Theoretically, that range should be $[1, 1023]$, but in practice table elements are usually in the range $[1, 255]$. The only certain way to find the best quantization table is exhaustive search. However, that is not possible since the computational time is measured in hours even for 5 coefficients and then increased 255 times for each additional coefficient up to 64. For such hard optimization problems during last decades algorithms that imitate some natural processes were successfully used. Very promising branch of such algorithms are swarm intelligence algorithms that simulate simple individuals that collectively produce significant intelligence.

Many different swarm intelligence algorithms have been proposed so far and were successfully used for various purposes $[2, 4, 6, 14, 22, 26, 28, 30]$. These algorithms were also used for JPEG quantization table optimization. Evolutionary approach for quantization table selection was proposed in [16]. In [15] genetic algorithm was used, while in [19] particle swarm optimization was used for optimizing quantization table. In [27] firefly algorithm was proposed to solve this combinatorial problem. Bacterial foraging optimization algorithm for quantization table selection for color images was proposed in [7]. In [20, 29] brief reviews on swarm intelligence algorithms applied to JPEG algorithm were given. In this paper one of the latest swarm intelligence algorithm, fireworks algorithm, will be used for quantization table selection.

## 4  Guided Fireworks Algorithm

Guided fireworks algorithm (GFWA) is the latest improvement of the fireworks algorithm and it was proposed by Li, Zheng and Tan in 2016 [18]. The original fireworks algorithm (FWA) proposed in 2010 [23] simulates fireworks explosion with two different types of the fireworks. Well manufactured fireworks produce numerous sparks around explosion center which is used to define exploitation, while badly manufactured fireworks produce only a few sparks scattered in the space which represents exploration [23]. During last few years fireworks algorithm was used as part of many different applications for solving hard optimization problems. It was used for SVM parameters tuning in [25] and in [12] it was used for parameter tuning of local-concentration model for spam detection.

Since the initial version of the fireworks algorithm introduction, several improved versions were proposed. The first modification was named enhanced fireworks algorithm where five modification of the initial fireworks algorithm were introduced [31]. After enhanced FWA, cooperative FWA (CoFWA) was proposed in [32]. CoFWA enhanced the exploitation ability by using independent selection operator and increased the exploration capacity by crowdness-avoiding cooperative strategy among the fireworks. In [17] another two methods for improving exploration were proposed. One is the mechanism that allows FWA to dynamically adjust the number of sparks based on the fitness function results and on the search results. Additionally, better diversity of the fireworks population was achieved by sharing the fitness information among the fireworks. This version of the FWA is also known as the FWA with dynamic resource allocation (FWA-DRA). The latest version of the FWA, guided fireworks algorithm, will be briefly described.

Guided fireworks algorithm uses $n$ fireworks and for each of them some number of sparks is generated. Fireworks and sparks represent points in $d$-dimensional space, where $d$ is the dimension of the problem. The number of the sparks for each firework $x_i$ is calculated as:

$$\lambda_i = \hat{\lambda} \ \frac{\max_{j}(f(x_j)) - f(x_i)}{\sum_{j=1}^{n}(\max_{k}(f(x_k)) - f(x_i))}, \tag{1}$$

where $\hat{\lambda}$ represents parameter that controls the overall number of sparks generated by $n$ fireworks, $y_{max} = \max(f(x_i))$, $(i = 1, 2, ..., n)$ represents the worst solution in the population and $\eta$ is a small constant used to avoid division-by-zero error.

For each firework, explosion amplitude is defined by the following equation:

$$A_i = \hat{A} \ \frac{f(x_i) - y_{min} + \eta}{\sum_{i=1}^{n}(f(x_i) - y_{min}) + \eta}, \tag{2}$$

where $\hat{A}$ defines the highest value of the explosion amplitude and $y_{min} = \min(f(x_i))$, $(i = 1, 2, ..., n)$ represents the best solution in the population of $n$ fireworks.

In each generation, the firework with the best fitness is named core firework (CF). For CF, explosion amplitude is adjusted according to the following equation [18]:

$$A_{CF}(t) = \begin{cases} A_{CF}(1) & \text{if } t = 1, \\ C_r A_{CF}(t-1) & \text{if } f(X_{CF}(t)) = f(X_{CF}(t-1)), \\ C_a A_{CF}(t-1) & \text{if } f(X_{CF}(t)) < f(X_{CF}(t-1)) \end{cases} \tag{3}$$

where $t$ represents the number of the current generation, while $C_a > 1$ and $C_r < 1$ are constants.

In each generation, a guiding spark (GS) is generated for each firework. The GS is generated by adding to the firework's position a guiding vector (GV). The position of the GS and $G_i$ for firework $X_i$ is determined by the following algorithm [18]:

---

**Algorithm 1.** Generating the Guiding Spark for $X_i$ [18]

**Require:** $X_i, s_{ij}, \lambda_i$ and $\sigma$

Sort the sparks by their fitness values $f(s_{ij})$ in ascending order.

$\Delta_i \leftarrow \frac{1}{\sigma\lambda_i}(\sum_{j=1}^{\sigma\lambda_i})s_{ij} - \sum_{j=\lambda_i - \sigma\lambda_i + 1}^{\lambda_i} s_{ij}$

$G_i \leftarrow X_i + \Delta_i$

**return** $G_i$

---

The guiding vector $\Delta_i$ is the mean of $\sigma\lambda_i$ vectors which is defined by the following equation:

$$\Delta_i = \frac{1}{\sigma\lambda_i} \sum_{j=1}^{\sigma\lambda_i}(s_{ij} - s_{i,\lambda_i - j + 1}) \tag{4}$$

The complete guided fireworks algorithm overview is shown as Algorithm 2:

---

**Algorithm 2.** Guided fireworks algorithm [18]

Randomly initialize $\mu$ fireworks in the potential space.

Evaluate the fireworks' fitness.

**repeat**

  Calculate $\lambda_i$ according to the Eq. 1

  Calculate $A_i$ according to the Eq. 2 and Eq. 3

  For each firework, generate $\lambda_i$ sparks within the amplitude $A_i$

  For each firework, generate guiding sparks according to previous algorithm.

  Evaluate all the sparks' fitness.

  Keep the best individual as a firework.

  Randomly choose other $\mu - 1$ fireworks among the rest of individuals.

**until** termination criteria is met.

**return** the position and the fitness of the best individual.

---

In this paper the GFWA will be used for selecting coefficients in the JPEG quantization table.

## 5    The Proposed Algorithm

Compression level and the corresponding image quality are mainly determined by the quantization table. In this paper the goal is to find equivalent compression level to some compression level achieved by using the recommended quantization tables $Q_i$, but with the aim that image compressed by new quantization table has better quality. As mentioned before, the quality is often measured by human perception, but in many different applications some objective measurements are necessary. We will use two different metrics that are used in literature to measure similarity of two images. The compressed image has better quality if it is more similar to the original one. For our proposed algorithm the goal is to get the best quality of the compressed image with some constraints, so the objective function for the optimization algorithm, GFWA, will be appropriate image similarity metrics.

The first standard metrics for image similarity is the mean square error (MSE) defined by:

$$MSE = \frac{1}{NM} \sum_{i=1}^{N} \sum_{j=1}^{M} (x_{i,j} - x'_{i,j})^2 \qquad (5)$$

where $x_{i,j}$ represents the intensity value of the pixel $(i, j)$ in the original image, $x'_{i,j}$ represents the intensity value of the corresponding pixel in the compressed image, while $M$ and $N$ are dimensions of the image. For two identical images, MSE is equal to zero since all the differences in the sum are zero which means that in the case when MSE is used as an objective function, the goal is to minimize it. Standard metrics based on MSE also used for image similarity is peak signal to noise ratio (PSNR). This metrics is defined by:

$$PSNR = 10 \log \frac{255^2}{MSE + \epsilon} \qquad (6)$$

where $\epsilon$ is a small constant to prevent dividing by zero. PSNR is larger for more similar images.

The second metrics that will be used as an objective function is normalized cross correlation (NK) that is defined as:

$$NK = \frac{\sum_{i=1}^{N} \sum_{j=1}^{M} x_{i,j} x'_{i,j}}{\sum_{i=1}^{N} \sum_{j=1}^{M} x_{i,j}^2} \qquad (7)$$

For identical images NK is equal to 1 which is also the maximal possible value. Therefore, the objective is to minimize -NK.

When the metrics are established, GFWA should find elements of the quantization table so that the best quality of the image with some predetermined compression level is achieved. Since the elements of the quantization table need to be determined, they will represent input vector for the GFWA. Problem dimension is 64. Even though theoretically elements can be in range $[1, 1023]$, in practice, usually it is enough to set the range to $[1, 255]$ because DCT coefficients are rarely larger then 255 (with minor consequence that in rare situations it will be impossible to completely cancel some DCT frequency coefficients).

Condition that some compression level should be achieved can be described in different ways. One measure of compression level could be the sum of all bits that are needed by all quantized non-zero DCT coefficients i.e. coefficient between $2^{k-1}$ and $2^k - 1$ requires $k$ bits. This measure is not convenient since the number of required bits changes for each block. Element from the quantization table determines the maximum number of bits that is necessary for saving the quantized coefficient, but not the exact number. However, that is rectified by later Huffman coding so the sum of all the elements in the quantization table can be an appropriate measure of the level of compression. Larger sum corresponds to larger elements in the quantization table and consequently higher compression. In this paper this measure was used as a requirement that the sum of elements in the optimized quantization table should be equal to the sum of elements in the corresponding $Q_i$ table.

In order to incorporate this condition the problem of selecting elements in the quantization table becomes a constrained optimization problem. The most difficult constraints are equality constraints where all feasible solutions are in one hyperplane since the search space is extremely reduced and it is difficult to find any feasible solutions. Equality constraints are usually relaxed by allowing some tolerance, larger in the beginning in order to find some feasible solutions and later dynamically reduced to zero.

In our case, equality constraint can be changed to inequality constraint that the sum elements in the optimized quantization table need to be larger or equal than the sum of elements in the corresponding $Q_i$ table. This is possible because the objective function searches for the best possible quality of the image, and the image will have better quality if more information is saved, i.e. DCT coefficients were divided by smaller numbers so the sum will be as small as possible. In this way the objective function will force the solutions towards equality constraint and play the role of dynamic tolerance for equality relaxation.

In constraint optimization problems not all generated solution will be feasible. In this example, solutions where the sum of quantization table elements is less then given number are not acceptable. In order to guide optimization algorithm, Deb's rules are usually used. Between feasible and non-feasible solution, feasible solution is better regardless of the value of the objective function. Between two non-feasible solutions, better is the one that has smaller constraint violation. For two feasible solution, value of the objective function is used to determine the better one. In this paper, GFWA was modified so that non-feasible solutions were discarded immediately after they were generated, without computing objective function value which is computationally very expensive operation (compression and decompression of the whole image). This was possible since the constraint was not on objective function but on the property of the input vector (bound-constrained optimization).

Another adjustment in the proposed algorithm deals with the fact that elements of the quantization table are integers while GFWA works with real numbers. We performed optimization with the standard GFWA that generates real number solutions and rounded them to the nearest integer. That is possible since pixel intensities as well as DCT coefficients are originally real numbers, artificially converted to integers because of our conventions for storing digital images.

## 6    Experimental Results

The proposed algorithm was implemented in Matlab version R2016b. All experiments were performed on Intel ® Core™ i7-3770K CPU @ 4 GHz, 8 GB RAM computer with Windows 10 Professional OS.

Performance of our proposed algorithm was tested on several standard test images. Experimental results are shown for image "Lena" (Fig. 1), gray version, resolution $512 \times 512$.

Results are first shown for the level of compression where the degradation of image quality is easily visible. For that purpose we selected recommended $Q_{10}$ table and by using GFWA we computed optimized quantization table $Q_{10\_opt}$ that achieves the same compression level. JPEG recommended table $Q_{10}$ and our optimized $Q_{10\_opt}$ when MSE was used as objective function are shown in Table 1.

**Table 1.** Quantization table $Q_{10}$ (left) and $Q_{10\_opt}$ optimized by GFWA (right)

| 80 | 55 | 50 | 80 | 120 | 200 | 255 | 255 |
|----|----|----|----|-----|-----|-----|-----|
| 60 | 60 | 70 | 95 | 130 | 255 | 255 | 255 |
| 70 | 65 | 80 | 120 | 200 | 255 | 255 | 255 |
| 70 | 85 | 110 | 145 | 255 | 255 | 255 | 255 |
| 90 | 110 | 185 | 255 | 255 | 255 | 255 | 255 |
| 120 | 175 | 255 | 255 | 255 | 255 | 255 | 255 |
| 245 | 255 | 255 | 255 | 255 | 255 | 255 | 255 |
| 255 | 255 | 255 | 255 | 255 | 255 | 255 | 255 |

| 7 | 5 | 9 | 2 | 18 | 226 | 231 | 255 |
|----|----|----|----|-----|-----|-----|-----|
| 26 | 17 | 35 | 68 | 177 | 254 | 255 | 255 |
| 8 | 35 | 15 | 84 | 252 | 255 | 255 | 255 |
| 118 | 172 | 244 | 247 | 255 | 255 | 255 | 255 |
| 243 | 250 | 252 | 255 | 255 | 255 | 255 | 255 |
| 138 | 201 | 255 | 255 | 255 | 255 | 255 | 255 |
| 133 | 255 | 255 | 255 | 255 | 255 | 255 | 255 |
| 255 | 255 | 255 | 255 | 255 | 255 | 255 | 255 |

It can be seen that these tables are rather different even though their level of compression is very similar. Elements under and on anti-diagonal are all 255 in both cases. The difference is in elements above anti-diagonal. Sum of the elements in $Q_{10}$ is 12,610 while the sum of elements in our $Q_{10\_opt}$ is larger, 12,647. This means that the compression level is slightly larger then by $Q_{10}$ table. This is due to the constraint that the sum has to be larger or equal to the sum of elements in $Q_{10}$ table. Even with larger sum in quantization table, i.e. higher compression level, better quality of decompressed image was achieved because elements of quantization table are more appropriate. Resulting images are shown in Fig. 2. By visual inspection it is easy to see that the quality image in Fig. 2(b) is much better. Image is smoother and block edges are not visible like in the image compressed by JPEG recommended $Q_{10}$ table (Fig. 2(a)).

Next, we used the level of compression 20 where degradation of the image is almost unnoticeable by human eye, but detectable by the metrics described by Eqs. 5, 6 and 7. JPEG standard recommended $Q_{20}$ table and computed optimized

**Table 2.** Quantization table $Q_{20}$ (left) and $Q_{20\_opt}$ optimized by GFWA (right)

| 40 | 28 | 25 | 40 | 60 | 100 | 128 | 153 |
|----|----|----|----|----|-----|-----|-----|
| 30 | 30 | 35 | 48 | 65 | 145 | 150 | 138 |
| 35 | 33 | 40 | 60 | 100 | 143 | 173 | 140 |
| 35 | 43 | 55 | 73 | 128 | 218 | 200 | 155 |
| 45 | 55 | 93 | 140 | 170 | 255 | 255 | 193 |
| 60 | 113 | 138 | 160 | 203 | 255 | 255 | 230 |
| 123 | 160 | 195 | 218 | 255 | 255 | 255 | 253 |
| 180 | 230 | 238 | 245 | 255 | 250 | 255 | 248 |

| 4 | 1 | 22 | 17 | 138 | 96 | 184 | 9 |
|----|----|----|----|-----|----|-----|----|
| 2 | 10 | 8 | 2 | 102 | 220 | 156 | 191 |
| 9 | 6 | 16 | 236 | 8 | 248 | 60 | 187 |
| 32 | 2 | 4 | 178 | 30 | 205 | 235 | 255 |
| 39 | 205 | 213 | 5 | 43 | 107 | 255 | 255 |
| 227 | 160 | 120 | 162 | 238 | 255 | 255 | 255 |
| 170 | 139 | 188 | 159 | 255 | 255 | 255 | 255 |
| 120 | 219 | 226 | 255 | 255 | 255 | 255 | 255 |

by GFWA quantization table $Q_{20\_opt}$ that achieves the same compression level are shown in Table 2.

Again, these tables are rather different even though their level of compression is very similar. Compression is smaller then in the previous case, thus less elements have value 255. Elements in the lower right triangle of $Q_{20}$ are not all 255 but they are close. In quantization table obtained by our proposed method elements are somewhat unexpected where some rather small values appear on the anti-diagonal and even below it. Sum of the elements in $Q_{20}$ is 9070 while the sum of the elements in our quantization table is, as in the previous case, slightly larger, 9183.

Resulting images are shown in Fig. 3. In this case the difference in quality is hardly noticeable by human eye. Skin texture on the shoulder is smoother in the image decompressed by our quantization table in Fig. 3(b) while the blocks edges can be seen when $Q_{20}$ table was used (Fig. 3(a)).

Besides perceptual results which are interesting, numerical results using mentioned metrics are more important and they are also better. In Table 3 metrics obtained for three different test images are shown. As it can be seen, our proposed algorithm successfully selected elements of the quantization tables so that the quality of the images was significantly improved for the same compression level.

Compression by quantization tables obtained by our proposed method achieved better results for all considered metrics compared to images compressed by standard quantization tables. Improvements were higher for compression level $Q_{10}$ than for $Q_{20}$. It could be expected since for higher compression level degradation is larger thus there is more space for improvements. For example, image quality measured by MSE for $Q_{10}$ was improved by 45.72% for "Lena", 26.20% for "Barbara" and 35.21% for "Boat" while for $Q_{20}$ improvements were 22.64%, 15.42% and 20.42% respectively.

In [27] average pixel intensity distance between the original and compressed image was used as similarity measure. For images that were using $Q_{10}$ table average pixel intensity distance was 5.886, while for optimized quantization table it was 5.100. For quantization table obtained by our proposed algorithm, average pixel intensity level was 4.085 which is better.

**Fig. 1.** Original image

**Fig. 2.** Decompressed image by (a) $Q_{10}$ and (b) quantization table $Q_{10\_opt}$ obtained by GFWA with MSE as objective function.

**Fig. 3.** Decompressed image by (a) $Q_{20}$ and (b) quantization table $Q_{20\_opt}$ obtained by GFWA with MSE as objective function.

**Table 3.** Comparison of experimental results obtained by recommended JPEG standard quantization tables and quantization tables optimized by the proposed GFWA.

| Image | MSE | | PSNR | | NK | |
|---|---|---|---|---|---|---|
| | $Q_{10}$ | $Q_{GFWA}$ | $Q_{10}$ | $Q_{GFWA}$ | $Q_{10}$ | $Q_{GFWA}$ |
| Lena | 59.3049 | 32.1913 | 30.3999 | 33.0534 | 0.9999 | 1.0000 |
| Barbara | 175.1430 | 129.2543 | 25.6969 | 27.0164 | 0.9997 | 0.9999 |
| Boat | 99.9588 | 61.7657 | 28.1326 | 30.2233 | 0.9979 | 0.9996 |
| Image | MSE | | PSNR | | NK | |
| | $Q_{20}$ | $Q_{GFWA}$ | $Q_{20}$ | $Q_{GFWA}$ | $Q_{20}$ | $Q_{GFWA}$ |
| Lena | 32.9300 | 25.4735 | 32.9549 | 34.0699 | 0.9987 | 0.9999 |
| Barbara | 97.0153 | 82.0541 | 28.2624 | 28.9898 | 0.9974 | 0.9982 |
| Boat | 57.9879 | 46.1490 | 30.4974 | 31.4892 | 0.9985 | 0.9991 |

# 7  Conclusion

In this paper an algorithm for JPEG quantization table selection was proposed. For selecting elements in quantization table novel swarm intelligence algorithm, guided fireworks algorithm, was used. Quantization table elements were optimized so that desired compression level was achieved while the quality of the image was maximized according to selected metrics. For quality measurement two standard metrics were used, mean square error and normalized cross correlation, while peak signal to noise ratio was also used. Our proposed algorithm significantly improved the quality of the compressed image. In further work more similarity metrics that are adjusted for specific applications can be used.

**Acknowledgement.** M. Tuba was supported for this research by the Ministry of Education, Science and Technological Development of Republic of Serbia, Grant No. III-44006.

# References

1. Alam, L., Dhar, P.K., Hasan, M.A.R., Bhuyan, M.G.S., Daiyan, G.M.: An improved JPEG image compression algorithm by modifying luminance quantization table. Int. J. Comput. Sci. Netw. Secur. (IJCSNS) **17**(1), 200 (2017)
2. Alihodzic, A., Tuba, M.: Improved bat algorithm applied to multilevel image thresholding. Sci. World J. **2014**, 1–17 (2014)
3. Aschwanden, M.J.: Image processing techniques and feature recognition in solar physics. Sol. Phys. **262**(2), 235–275 (2010)
4. Bacanin, N., Tuba, M.: Firefly algorithm for cardinality constrained mean-variance portfolio optimization problem with entropy diversity constraint. Sci. World J. **2014**, 1–16 (2014)
5. Chao, J., Chen, H., Steinbach, E.: On the design of a novel JPEG quantization table for improved feature detection performance. In: IEEE International Conference on Image Processing, pp. 1675–1679 (2013)

6. Dorigo, M., Gambardella, L.M.: Ant colonies for the travelling salesman problem. Biosystems **43**(2), 73–81 (1997)

7. Dua, R.L., Gupta, N.: Fast color image quantization based on bacterial foraging optimization. In: Fourth International Conference on Advances in Recent Technologies in Communication and Computing (ARTCom), pp. 100–102 (2012)

8. Duan, L.Y., Liu, X., Chen, J., Huang, T., Gao, W.: Optimizing JPEG quantization table for low bit rate mobile visual search. In: Visual Communications and Image Processing, pp. 1–6 (2012)

9. Ernawan, F., Nugraini, S.H.: The optimal quantization matrices for JPEG image compression from psychovisual threshold. J. Theor. Appl. Inform. Technol. **70**(3), 566–572 (2014)

10. Gunda, N.S.K., Choi, H.W., Berson, A., Kenney, B., Karan, K., Pharoah, J.G., Mitra, S.K.: Focused ion beam-scanning electron microscopy on solid-oxide fuel-cell electrode: Image analysis and computing effective transport properties. J. Power Sources **196**(7), 3592–3603 (2011)

11. Gupta, M., Garg, A.K.: Analysis of image compression algorithm using DCT. Int. J. Eng. Res. Appl. (IJERA) **2**(1), 515–521 (2012)

12. He, W., Mi, G., Tan, Y.: Parameter optimization of local-concentration model for spam detection by using fireworks algorithm. In: Tan, Y., Shi, Y., Mo, H. (eds.) ICSI 2013. LNCS, vol. 7928, pp. 439–450. Springer, Heidelberg (2013). doi:10.1007/978-3-642-38703-6_52

13. Jiang, C., Pang, Y., Xiong, S.: A high capacity steganographic method based on quantization table modification and F5 algorithm. Circuits Syst. Sig. Process. **33**(5), 1611–1626 (2014)

14. Karaboga, D.: An idea based on honey bee swarm for numerical optimization. Technical report - TR06, pp. 1–10 (2005)

15. Kumar, B.V., Karpagam, M.: Differential evolution versus genetic algorithm in optimising the quantisation table for JPEG baseline algorithm. Int. J. Adv. Intell. Paradigms **7**(2), 111–135 (2015)

16. Lazzerini, B., Marcelloni, F., Vecchio, M.: A multi-objective evolutionary approach to image quality/compression trade-off in JPEG baseline algorithm. Appl. Soft Comput. **10**(2), 548–561 (2010)

17. Li, J., Tan, Y.: Enhancing interaction in the fireworks algorithm by dynamic resource allocation and fitness-based crowdedness-avoiding strategy. In: IEEE Congress on Evolutionary Computation (CEC), pp. 4015–4021 (2016)

18. Li, J., Zheng, S., Tan, Y.: The effect of information utilization: Introducing a novel guiding spark in the fireworks algorithm. IEEE Trans. Evol. Comput. **21**(1), 153–166 (2017)

19. Ma, H., Zhang, Q.: Research on cultural-based multi-objective particle swarm optimization in image compression quality assessment. Optik-Int. J. Light and Electron. Opt. **124**(10), 957–961 (2013)

20. Naresh, S., Kumar, B.V., Karpagam, G.: A literature review on quantization table design for the JPEG baseline algorithm. Int. J. Eng. Comput. Sci. 4(10), 14686–14691 (2015)

21. Starosolski, R.: New simple and efficient color space transformations for lossless image compression. J. Vis. Commun. Image Represent. **25**(5), 1056–1063 (2014)

22. Subotic, M., Tuba, M., Stanarevic, N.: Parallelization of the artificial bee colony (ABC) algorithm. In: Proceedings of the 11th WSEAS International Conference on Evolutionary Computing, vol. 10, pp. 191–196 (2010)

23. Tan, Y., Zhu, Y.: Fireworks algorithm for optimization. In: Tan, Y., Shi, Y., Tan, K.C. (eds.) ICSI 2010. LNCS, vol. 6145, pp. 355–364. Springer, Heidelberg (2010). doi:10.1007/978-3-642-13495-1_44

24. Thai, T.H., Cogranne, R., Retraint, F., et al.: JPEG quantization step estimation and its applications to digital image forensics. IEEE Trans. Inf. Forensics Secur. **12**(1), 123–133 (2017)

25. Tuba, E., Tuba, M., Beko, M.: Support vector machine parameters optimization by enhanced fireworks algorithm. In: Tan, Y., Shi, Y., Niu, B. (eds.) ICSI 2016. LNCS, vol. 9712, pp. 526–534. Springer, Cham (2016). doi:10.1007/978-3-319-41000-5_52

26. Tuba, M., Bacanin, N.: Improved seeker optimization algorithm hybridized with firefly algorithm for constrained optimization problems. Neurocomputing **143**, 197–207 (2014)

27. Tuba, M., Bacanin, N.: JPEG quantization tables selection by the firefly algorithm. In: International Conference on Multimedia Computing and Systems (ICMCS), pp. 153–158. IEEE (2014)

28. Tuba, M., Bacanin, N., Stanarevic, N.: Guided artificial bee colony algorithm. In: Proceedings of the 5th European Conference on European Computing Conference, pp. 398–403 (2011)

29. Viswajaa, S., Kumar, V., Karpagam, G.R.: A survey on nature inspired meta-heuristics algorithm in optimizing the quantization table for JPEG baseline algorithm. Int. Adv. Res. J. Sci. Eng. Technol. **2**(4), 114–123 (2015)

30. Yang, X.-S.: Firefly algorithms for multimodal optimization. In: Watanabe, O., Zeugmann, T. (eds.) SAGA 2009. LNCS, vol. 5792, pp. 169–178. Springer, Heidelberg (2009). doi:10.1007/978-3-642-04944-6_14

31. Zheng, S., Janecek, A., Tan, Y.: Enhanced fireworks algorithm. In: 2013 IEEE Congress on Evolutionary Computation, pp. 2069–2077 (2013)

32. Zheng, S., Li, J., Janecek, A., Tan, Y.: A cooperative framework for fireworks algorithm. IEEE/ACM Trans. Comput. Biol. Bioinform. **PP**(99), 1 (2016)

33. Zimbico, A., Schneider, F., Maia, J.: Comparative study of the performance of the JPEG algorithm using optimized quantization matrices for ultrasound image compression. In: 5th ISSNIP-IEEE Biosignals and Biorobotics Conference: Biosignals and Robotics for Better and Safer Living (BRC), pp. 1–6 (2014)

# Shape Matching for Rigid Objects by Aligning Sequences Based on Boundary Change Points

Abdullah N. Arslan[1](✉) and Nikolay M. Sirakov[1,2]

[1] Department of Computer Science and Information Systems,
Texas A&M University – Commerce, Commerce, TX 75428, USA
{Abdullah.Arslan,Nikolay.Sirakov}@tamuc.edu
[2] Department of Mathematics, Texas A&M University – Commerce,
Commerce, TX 75428, USA

**Abstract.** This paper presents a new boundary (shape) matching algorithm for $2D$ rigid objects without voids. Our new algorithm presents a new shape representation that uses the outcome from an active contour (AC) model. An object's shape is partitioned into a clockwise ordered sequence of *edges*, where every edge is a boundary segment enclosed by *reference points*. These points are convex hull vertices which lie on boundary corners. Further, the reference points are used to generate angles. Hence, a boundary shape maps to a sequence of angles, turning the shape matching problem to alignment of cyclic sequences of angles. The latter makes our method scaling and rotational invariant. Experiments validate the theoretical concept, and provide qualitative comparison with other methods in the field.

**Keywords:** Shape · Active contour · Concavity · Representation · Cyclic alignment · Matching

## 1 Introduction

Shape is an effective discriminator for objects in many domains. Shape matching has been used to classify objects in computer vision [13,16,18,28,29,35], in medical imaging [2,24,37], in molecular pharmacology [11,22,23]. This study is part of an on-going project for assessment of threat posed by firearms [4,5].

There are many different shape matching methods based on various representation and distance definitions [9,15,19,25,26,28] (e.g. Haussdorff distance in [19], Inner-Distance in [25]). Some shape matching methods are developed for rigid objects (e.g. [36]) while some others are for non-rigid objects (e.g. [10]). In the present paper, we consider rigid objects without voids.

The method we propose in this paper combines an AC model, and a cyclic sequence alignment method [3,30,32].

A number of methods are published on shape representation and matching using sequences derived from 2D shapes [12,20,21]. Methods in [12,20] are most relevant to our work in this paper. They generate sequences by collecting local

© Springer International Publishing AG 2017
V.E. Brimkov and R.P. Barneva (Eds.): IWCIA 2017, LNCS 10256, pp. 308–321, 2017.
DOI: 10.1007/978-3-319-59108-7_24

information (based on *turning angle*) from the points on the boundary of a
given shape. Analogously our method generates sequences from boundary. Unlike
[12,20] it considers concavities, convexities, and line segments. The literature
has works on concavity extraction, and their use in shape representation and
matching [14,31,33].

The basics of the new shape matching method can be found in [3]. Similarly
to [3,5] the present paper partitions object boundary into *segments* separated by
*reference points* (convex hull vertices which lie on boundary corners). However,
unlike previous methods it considers each segment and its enclosing reference
points as an *edge*. The shape is described as a sequence of angles constructed from
edges. Thus the shape matching consists of sequence matching. The resulting
shape matching is more accurate compared to the one in [4,5] which considers
reference points and segments separately. The constructed angles do not change
under shape rotation and scaling. Therefore, they are invariant according to
these plane transformations, and they make the method rotation invariant as
well. Furthermore, the method attains scaling invariance in several steps. First,
it defines a neighborhood based on the relative sizes of shapes, and removes
boundary points in close vicinity. By this way, we aim to have similarly sized
boundary representations (in terms of both number of boundary points and
lengths of the sequences of angles) for compared shapes. A similarity score is
calculated for aligned sequences of angles in two shapes. The total score obtained
is divided by the average of the lengths of the compared sequences. This results
in scores in $[0, 1]$. The shape similarity score is calculated by using cyclic sequence
alignment at a higher level to handle rotations.

We organize sections with the following purposes: Sect. 2 for the notation;
Sect. 3 for the new shape representation; Sect. 4 for the new shape matching
method; Sect. 5 for experimental evidence; Sect. 6 for concluding remarks.

## 2   Notations and Basic Definitions

On a given shape without voids, let $B = b_1 b_2 \ldots b_n$ be a clockwise ordered
sequence of boundary points, $b_j = (x_j, y_j)$ such that $b_n = b_1$. Let $R = r_1 r_2 \ldots r_m \subseteq B$ be a clockwise ordered sequence of *reference points*. We call
*reference points* of a shape the convex hull points which lie on boundary

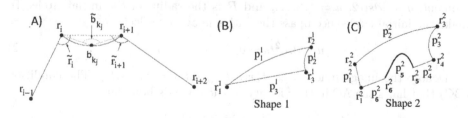

**Fig. 1.** (A) Edge $e_i = r_i p_i r_{i+1}$; (B, C) Two shapes.

corners, where the convex hull means the convex hull of the shape. For example, in Fig. 1(C), the reference points are $r_1^2, r_2^2, r_3^2, r_4^2, r_5^2$ and $r_6^2$.

Denote by

$p_k$: a nonempty boundary *segment* as a sequence $b_{k_1} b_{k_2} \ldots b_{k_{|p_k|}}$, where
$\quad b_{k_j} \in B$ denotes the $j$'th boundary point in segment $k$;

$e_k$: an *edge* $r_k p_k r_{k+1}$ which is a segment enclosed by two reference points, i.e.
$\quad r_k b_{k_1} b_{k_2} \ldots b_{k_{|p_k|}} r_{k+1}$;

$\tilde{b}_j$: angle $\angle r_k b_j r_{k+1}$ with a vertex at $b_j$ and arms through $r_k$ and $r_{k+1}$;

$\tilde{r}_i$: an angle $\angle r_{i-1} r_i r_{i+1}$ at reference point $r_i$;

$\tilde{p}_k$: the sequence of angles $\tilde{b}_{k_1} \tilde{b}_{k_2} \ldots \tilde{b}_{k_{|p_k|}}$ at the points in segment $p_k$;

$\tilde{s}_k$: a sequence of angles $\tilde{r}_k \tilde{p}_k \tilde{r}_{k+1}$.

To summarize, for every segment $p_i$, there is a corresponding sequence of angles $\tilde{p}_i = \tilde{b}_{i_1} \tilde{b}_{i_2} \ldots \tilde{b}_{i_{|p_i|}}$.

Figure 1(A) illustrates an edge $e_i = r_i p_i r_{i+1}$ that includes a (concave) boundary segment. The angles $\tilde{r}_i = \angle r_{i-1} r_i r_{i+1}$, $\tilde{r}_{i+1} = \angle r_i r_{i+1} r_{i+2}$, and $\tilde{b}_{k_j} = \angle r_i b_{k_j} r_{i+1}$ are also illustrated. Shape 1 in (B) has 3 reference points, and 3 edges that include one convex, one concave, and one line segment. Shape 2 in (C) has 6 reference points, and 6 edges that include 3 line segments, one convex, and two concave segments. No angles are shown in (B) and (C) in Fig. 1, but they are calculated in a similar way as shown in (A).

## 3   Shape Representation

We abstract a shape by a clockwise ordered sequence of edges $e_k$ obtained from boundary.

### 3.1   Boundary Extraction

In the present study we apply a shrinking active contour model (S-ACES) to extract the boundary of objects of interest. The model is developed in [32], and uses the following evolution equation to converge S-ACES toward objects' boundaries:

$$r(s,t) = R e^{as - 4a^2(t_0 + u\partial t)} [\cos(cas), \sin(sas)], \tag{1}$$

In Eq. 1, $s \in [0, \frac{2\pi}{ca}]$ is a space parameter, $t$ is a time parameter, $t_0$ is initial time moment, $a = |\partial s|/2$, $u = 1, 2, \ldots$, and $R$ is the radius of the initial circle. To make the initial circle encompass the entire image we select:

$$u = 0, a^2 t_0 = 0.001, c = 1000. \tag{2}$$

Denote the image function as $f(x(s,t), y(s,t)) = f(r(s,t))$. The condition (BC) that halts the AC in the vicinity of the object's boundary is:

$$\begin{aligned} &r(s,t) = r(s, t + \partial t)) \text{ which holds if} \\ &\frac{\partial f(r(s,t))}{\partial t} > \varepsilon \text{ where } t = t_0 + u\partial t \text{ such that} \\ &2.5 \geq ta^2 \geq 0.001. \end{aligned} \tag{3}$$

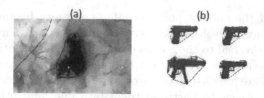

**Fig. 2.** Extracted CH and boundary of (a) a skin lesion; (b) weapons.

To evolve into deep concavities, a curve re-parametrization is conducted [32]. If Ineq. 4 is satisfied the AC point $(x_i, y_i)$ moves to the right of the AC if a clockwise direction is considered.

$$(y_i - y_{i-1})(x_{i+1} - x_i) < (y_{i+1} - y_i)(x_i - x_{i-1}). \tag{4}$$

The AC was validated on the extraction of 162 skin lesion boundaries and 170 weapon and non-weapon images [5]. Sample results are shown in Fig. 2.

### 3.2   Generating the Shape Sequence

Let $(x_{i-1}, y_{i-1}), (x_i, y_i), (x_{i+1}, y_{i+1})$ be any three clockwise ordered points in a 2D Euclidean plane. Consider the clockwise traversal of $(x_{i-1}, y_{i-1}), (x_i, y_i),$ $(x_{i+1}, y_{i+1})$. If Ineq. 4 is satisfied, we say that $(x_i, y_i)$ is a *concavity point* with respect to $(x_{i-1}, y_{i-1})$ and $(x_{i+1}, y_{i+1})$. If the reverse of Ineq. 4 is satisfied, then we say that $(x_i, y_i)$ is a convexity point with respect to $(x_{i-1}, y_{i-1})$ and $(x_{i+1}, y_{i+1})$. If Ineq. 4 becomes an equality, then we say that these points are co-linear.

We define concavity as a sequence of clockwise-ordered boundary points $r_k b_{k_1} b_{k_2} \ldots r_{k+1}$ on $B$ such that all $b_{k_j}$ are concavity points with respect to $r_k$ and $r_{k+1}$. We say that $r_k b_{k_1} b_{k_2} \ldots r_{k+1}$ is a *level-1 concavity* if it is a concavity not included in another (larger) concavity (this definition is in parallel with the definition in [31]). In this paper we consider only level-1 concavities. Figure 1(A) illustrates a concave boundary segment.

Our method collects a sample of boundary points on the boundary via S-ACES. We process the initial boundary obtained by S-ACES to eliminate boundary points which are "too close" to each other. For this purpose, we calculate the minimum-area rectangular bounding box enclosing the object. By dividing the perimeter of this rectangle by a parameter, we obtain a threshold length. If any two neighbors are within this threshold (horizontally and vertically) from each other, only one of them is kept in $B$ and the other one is removed. Our goal is to obtain similarly sized sequence representation regardless of the size of the object. We perform a complete clockwise traversal on $B$ starting at an arbitrary point, and find all level-1 concavities. We do this by considering every visited point as a potential concavity beginning, and all successors as potential concavity ends. Then we apply Ineq. 4 to check if all the points between the two potential concavity beginning and end points are concavity points with respect to these points. If the current point is not a concavity beginning we move to the next point. Once a level-1 concavity is found, we mark the found concavity and advance the traversal to the concavity end point, and continue iterating the same logic described above. By this way, we find all level-1 concave segments. All other points on $B$ are labelled as convexity points initially. Additional clockwise traversals are performed to partition them into line, and other (convex) segments using Ineq. 4. Concave and convex segments which are almost linear are replaced by line segments. We merge short line segments into larger ones. This includes consecutive line segments which are almost linear, too. Similarly, two consecutive convex segments can be merged into a larger convex segment. We test if two consecutive segments can be merged by taking the beginning of the first and the end of the second one, and checking if all the points between them have the same characteristic (convex or concave) with respect to these reference points. We continue iterating until there are no such consecutive segments.

We define a *shape* as a cyclic sequence $\tilde{s}_1\tilde{s}_2\tilde{s}_3\ldots\tilde{s}_{|\tilde{s}|}$, where each $\tilde{s}_i$ is a sequence of angles obtained from edge $e_i$ ordered clockwise. In this sequence, each angle at a reference point is calculated by using two other reference points (the predecessor and the successor reference of this point clockwise), and each angle at boundary point is calculated using the reference points enclosing this point. We assign a sign to segment types as follows: for any segment $p_i$, $sign(p_i) = -1$ indicates that $p_i$ is concave; $sign(p_i) = 0$ indicates that $p_i$ is a line; $sign(p_i) = 1$ indicates that $p_i$ is convex. Figure 1 includes two example shapes. Shape number is shown in the superscript. Shape 1 in (B) is represented by $\tilde{s}_1^1\tilde{s}_2^1\tilde{s}_3^1$, where $\tilde{s}_i^1 = \tilde{r}_i^1 \ \tilde{p}_i^1 \ \tilde{r}_{i+1}^1$, for all $i, 1 \le i \le 3$, and $sign(p_1^1) = 1$, $sign(p_2^1) = -1$, $sign(p_3^1) = 0$, and Shape 2 in (C) is represented by $\tilde{s}_1^2\tilde{s}_2^2\tilde{s}_3^2\tilde{s}_4^2\tilde{s}_5^2\tilde{s}_6^2$, where $\tilde{s}_i^2 = \tilde{r}_i^2 \ \tilde{p}_i^2 \ \tilde{r}_{i+1}^2$, for all $i, 1 \le i \le 6$, and $sign(p_6^2) = sign(p_4^2) = sign(p_1^2) = 0$, $sign(p_2^2) = 1$, and $sign(p_5^2) = sign(p_3^2) = -1$.

## 4   Shape Matching

We convert the differences between angles $\tilde{b}_i^1$ and $\tilde{b}_j^2$, and $\tilde{r}_k^1$ and $\tilde{r}_\ell^2$ to similarity scores in $[0, 1]$. All angles are represented in radian and as a factor of $\pi$. We convert the difference $\Delta = |\tilde{b}_i^1 - \tilde{b}_j^2|$ between angles $\tilde{b}_i^1$ and $\tilde{b}_j^2$, to a similarity score in $[0, 1]$ using

$$f(\Delta) = \begin{cases} 1, & \text{if } (\Delta < \beta_1); \\ 1 - \sqrt{\Delta}, & \text{if } \beta_1 \le \Delta < \beta_2; \\ 0, & \text{otherwise,} \end{cases} \tag{5}$$

where in the current implementation we set $\beta_1 = 0.02$, and $\beta_2 = 0.05$. Via these parameters, the differences are either ignored or amplified. The purpose is to distinguish very close matches from other similarities. When the difference $\Delta$ is within $\beta_1$, the angles are considered perfectly matching, and the similarity score $f(\Delta)$ is maximum (i.e. 1). When $\Delta$ is larger than or equal to $\beta_2$ the angles are considered completely different (not similar at all), and $f(\Delta)$ is minimum (i.e. 0). In between $\beta_1$ and $\beta_2$, as the difference $\Delta$ increases the similarity score $f(\Delta)$

decreases at a faster rate. We note that with $\beta_1 = 0.02$, and $\beta_2 = 0.05$, $f(\Delta) = 0$ if $\Delta > 0.05$; $f(\Delta) = 1$ if $\Delta \leq 0.02$; and $f(\Delta)$ is in $[0.77, 0.85]$ for $\Delta$ in $[0.02, 0.05]$. Analogously, the difference $\Delta = |\tilde{r}_k^1 - \tilde{r}_\ell^2|$ between angles $\tilde{r}_k^1$ and $\tilde{r}_\ell^2$, is converted to a similarity score in $[0, 1]$ using $f(\Delta)$. Based on these, the similarity score for a pair of boundary segments $p_i^1$, and $p_j^2$, with the same sign, is the alignment score $score_s(p_i^1, p_j^2)$ for the sequences $\tilde{b}_{i_1}^1 \tilde{b}_{i_2}^1 \ldots \tilde{b}_{i_{|p_i^1|}}^1$, and $\tilde{b}_{j_1}^2 \tilde{b}_{j_2}^2 \ldots \tilde{b}_{j_{|p_j^2|}}^2$. This can be computed by a special case of the global sequence alignment algorithm [27] in which score of insertions, deletions are zeros, and substitutions (matches) have positive scores. We can also formulate the objective of the optimization as the following:

$$score_s(p_i^1, p_j^2) = \max_{i', j'} \sum_{m=1}^{u} f(|b_{i'_m}^1 - b_{j'_m}^2|) \qquad (6)$$

over all index sequences $i', j'$ such that $i'_1, i'_2, \ldots, i'_u$ is a subsequence of $i_1, i_2, \ldots, i_{|p_i^1|}$, and $j'_1, j'_2, \ldots, j'_u$ is a subsequence of $j_1, j_2, \ldots, j_{|p_j^2|}$, for some $u \in [1, \min\{|p_i^1|, |p_j^2|\}]$. For $p_i^1$ and $p_j^2$, when one is concave and the other one is convex then $score_s(p_i^1, p_j^2) = 0$; when one is concave or convex, and the other one is a line segment then the similarity score calculated by using Eq. 6 is halved.

For two sequences of angles $\tilde{s}_i^1$, $\tilde{s}_j^2$ obtained from edges $e_i^1 = r_i^1 p_i^1 r_{i+1}^1$, $e_j^2 = r_j^2 p_j^2 r_{j+1}^2$ with $sign(p_i^1) = sign(p_j^2)$, $wscore_s(\tilde{s}_i^1, \tilde{s}_j^2)$ is one fourth of the sum of the scores between $r_i^1$ and $r_j^2$, and between $r_{i+1}^1$ and $r_{j+1}^2$ plus half of the alignment score for segments $\tilde{p}_i^1$ and $\tilde{p}_j^2$ divided by average length $(|p_i^1| + |p_j^2|)/2$. The resulting score is in $[0, 1]$ and denoted by $wscore_s(\tilde{s}_i^1, \tilde{s}_j^2)$. More formally,

$$wscore_s(\tilde{s}_i^1, \tilde{s}_j^2) = \frac{1}{4} \left( f(|r_i^1 - r_j^2|) + f(|r_{i+1}^1 - r_{j+1}^2|) \right) + \frac{score_s(p_i^1, p_j^2)}{|p_i^1| + |p_j^2|} \qquad (7)$$

For any two shape sequences $\tilde{s}^1 = \tilde{s}_1^1 \tilde{s}_2^1 \ldots \tilde{s}_{|\tilde{s}^1|}^1$ and $\tilde{s}^2 = \tilde{s}_1^2 \tilde{s}_2^2 \ldots \tilde{s}_{|\tilde{s}^2|}^2$ we consider sequence alignment [27] for calculating their similarity score $score_{seq}(\tilde{s}^1, \tilde{s}^2)$. Let $|\tilde{s}^1|$ and $|\tilde{s}^2|$ denote the number of edges in shapes 1, and 2, respectively.

The objective function can be described as the following:

$$score_{seq}(\tilde{s}^1, \tilde{s}^2) = \max_{i, j} \sum_{m=1}^{r} wscore_s(\tilde{s}_{i_m}^1, \tilde{s}_{j_m}^2) \qquad (8)$$

over all index sequences $i, j$ such that $i_1, i_2, \ldots, i_r$ is a subsequence of $1, 2, \ldots, |\tilde{s}^1|$, and $j_1, j_2, \ldots, j_r$ is a subsequence of $1, 2, \ldots, |\tilde{s}^2|$, for some $r \in [1, \min\{|\tilde{s}^1|, |\tilde{s}^2|\}]$.

The dynamic programming formulation of the sequence alignment in this case is based on deleting $\tilde{s}_i^1$, inserting $\tilde{s}_j^2$, and matching $\tilde{s}_i^1$ to $\tilde{s}_j^2$. We create a model with the following similarity score parameters described by the real score function $\gamma$. We set the insert and delete scores to zero. That is, $\gamma\left(\begin{bmatrix} - \\ \tilde{s}_j^2 \end{bmatrix}\right) = \gamma\left(\begin{bmatrix} \tilde{s}_i^1 \\ - \end{bmatrix}\right) = 0$. The similarity score for matching $\tilde{s}_i^1$ to $\tilde{s}_j^2$ is

$\gamma \left( \begin{bmatrix} \tilde{s}_i^1 \\ \tilde{s}_j^2 \end{bmatrix} \right) = wscore_s(\tilde{s}_i^1, \tilde{s}_j^2)$, where $wscore_s(\tilde{s}_i^1, \tilde{s}_j^2)$ is the alignment score calculated as we described in Eq. 7. Then the alignment score $score_{seq}$ $(\tilde{s}^1, \tilde{s}^2) = E_{|\tilde{s}^1|,|\tilde{s}^2|}$, where $E$ is the matrix calculated by the following dynamic programming formula for sequence alignment [27]: For all $i, j$, $i \in [0, |\tilde{s}^1|]$, $j \in [0, |\tilde{s}^2|]$, $E_{i,-1} = E_{-1,j} = -\infty$, and for all other $i, j$, if $i = j = 0$ then $E_{0,0} = 0$ else $E_{i,j}$ is calculated from $E_{i,j-1}, E_{i-1,j-1}$, and $E_{i-1,j}$ using the following formula

$$E_{i,j} = \max \left\{ \begin{array}{c} E_{i,j-1} + \gamma \left( \begin{bmatrix} \tilde{s}_i^1 \\ - \end{bmatrix} \right), \; E_{i,j-1} + \gamma \left( \begin{bmatrix} - \\ \tilde{s}_j^2 \end{bmatrix} \right), \\ E_{i-1,j-1} + \gamma \left( \begin{bmatrix} \tilde{s}_i^1 \\ \tilde{s}_j^2 \end{bmatrix} \right) \end{array} \right\} \tag{9}$$

The alignment score $score_{seq}(\tilde{s}^1, \tilde{s}^2) = E_{|\tilde{s}^1|,|\tilde{s}^2|}$ is normalized by dividing it by an upper bound for a maximal attainable score $(|s^1|+|s^2|)/2$, where $n_1, n_2$ are respectively the number of edges in aligned shapes $\tilde{s}^1, \tilde{s}^2$. The resulting score is in $[0, 1]$ and denoted by $|score_{seq}(\tilde{s}^1, \tilde{s}^2)|$. That is, $|score_{seq}(\tilde{s}^1, \tilde{s}^2)| = \frac{score_{seq}(\tilde{s}^1, \tilde{s}^2)}{(|s^1|+|s^2|)/2}$. If $\tilde{s}^1 = \tilde{s}^2$, $|score_{seq}(\tilde{s}^1, \tilde{s}^2)| = 1$, otherwise, $|score_{seq}(\tilde{s}^1, \tilde{s}^2)| < 1$.

Given $\tilde{s}^2$, the *cyclic shift* of $\tilde{s}^2$ by $k$ positions is $\tilde{s}^{2,k} = \tilde{s}_{k+1}^2 \tilde{s}_{k+2}^2 \cdots \tilde{s}_{|\tilde{s}^2|}^2 \tilde{s}_1^2 \cdots \tilde{s}_k^2$. Therefore we define the shape similarity score for two shapes (cyclic sequences) $\tilde{s}^1$ and $\tilde{s}^2$ as

$$cscore(\tilde{s}^1, \tilde{s}^2) = \max_{0 \le k < |\tilde{s}^2|} |score_{seq}(\tilde{s}^1, \tilde{s}^{2,k})| \tag{10}$$

If $\tilde{s}^1 = \tilde{s}^{2,k}$, for some $k$, then $cscore(\tilde{s}^1, \tilde{s}^2) = 1$, else, $cscore(\tilde{s}^1, \tilde{s}^2) < 1$.

On the bases of above concepts, we develop the following shape matching algorithm: (1) Extract the boundary; (2) Generate shape sequences $\tilde{s}^1 = \tilde{s}_1^1 \tilde{s}_2^1 \cdots \tilde{s}_{|\tilde{s}^1|}^1$ and $\tilde{s}^2 = \tilde{s}_1^2 \tilde{s}_2^2 \cdots \tilde{s}_{|\tilde{s}^2|}^2$; (3) Shape matching is performed as follows: (3.1) Define $T$ as a two dimensional matrix whose elements are described by the following:

$$T[i, j] = wscore(\tilde{s}_i^1, \tilde{s}_j^2), \; 1 \le i \le |\tilde{s}^1|, \; 1 \le j \le |\tilde{s}^2| \tag{11}$$

The entire matrix $T$ is built by calculating $wscore(\tilde{s}_i^1, \tilde{s}_j^2)$ for all $i, j$. (3.2) Sequences of angles from edges (Shapes) $\tilde{s}^1$ and $\tilde{s}^2$ are cyclically aligned using matrix $T$, the optimum score is found by using the dynamic programming formulation given by Eq. 9, and calculating Eq. 10. That is, for $k = 0$ to $|\tilde{s}^2| - 1$, $|score_{seq}(\tilde{s}^1, \tilde{s}^{2,k})|$ is computed, and an optimal alignment is returned.

Boundary extraction takes $O(MN)$ time, where $M \times N$ is the size of the input images. Boundary sequence generation can be done within the same theoretical time complexity using convex hull. In our implementation, we only used the boundary points as described in Sect. 3.2, and it took less than 16 ms to generate boundary sequences. For different $k$, the scores of the insertions, deletions, and substitutions at given positions in the alignment computation for $\tilde{s}^1$ and $\tilde{s}^{2,k}$ are different because of the cyclic shifting of symbols in $\tilde{s}^2$. Each pair of positions $(i, j + k)$ ($i$ in $\tilde{s}^1$ and $j$ in $\tilde{s}^{2,k}$) in aligning $\tilde{s}^1$ and $\tilde{s}^{2,k}$ corresponds to $(i, j)$ in aligning $\tilde{s}^1$ and $\tilde{s}^2$. Therefore, matrix $T$ is computed only once, and all necessary values are available there at different indices. Each pairwise segment alignment takes time $O(|p_i^1||p_j^2|)$. Let $\ell^1$ and $\ell^2$ be the number of boundary points of the input shapes, Shape 1 and Shape 2, respectively, after reducing close neighbors as described in Sect. 3.2. That is, $\ell^1 = \Sigma_{k=1}^{|\tilde{s}^1|}|p_k^1|$, and $\ell^2 = \Sigma_{k=1}^{|\tilde{s}^2|}|p_k^2|$. The total time required for building the table $T$ is $O(\Sigma_{i,j}|p_i^1||p_j^2|) = O(|p_1^1|\ell^2 + |p_2^1|\ell^2 + \ldots + |p_{|\tilde{s}^1|}^1|\ell^2) = O((\Sigma_{k=1}^{|\tilde{s}^1|}|p_k^1|)\ell^2) = O(\ell^1\ell^2)$. After constructing the table $T$, each pairwise edge alignment takes time $O(|\tilde{s}^1||\tilde{s}^2|)$. The algorithm performs $|\tilde{s}^2|$ such alignments. Therefore, the total time spent in this step is $O(|\tilde{s}^1||\tilde{s}^2|^2)$, where $|\tilde{s}^1|$ and $|\tilde{s}^2|$ are the number of edges (much smaller than the perimeters), respectively. Hence the total time of our shape matching algorithm is $O(\ell^1\ell^2 + |\tilde{s}^1||\tilde{s}^2|^2)$.

## 5 Experimental Results

An earlier version of the shape representation and matching method was used in [5]. The results there validated our theoretical concepts on a visual weapon ontology composed by 153 weapons [5]. A visual hierarchy was designed by creating clusters such as machine guns, pistols, riffles. Figure 3 includes a cluster from this hierarchy. The clustering was done based on the algorithm in [17] and using as the measure of similarity an earlier version of the cyclic shape sequence alignment score described by the present paper. The ontology was queried by objects. The results of identifying queried objects were encouraging [5].

In the present paper we tested our method on the dataset of Aslan and Tari [7], and shown some of our results out of 56 shapes in Fig. 4. In each row we give a query on the leftmost column, and in the next four columns we present the nearest matches to the query in descending order of similarity as computed by our method.

The results in Fig. 4 show the accuracy of 100% of our shape model in finding identical 2D shapes. When segments -in particular concavities- appear similarly in two compared shapes, the similarity score is high. For example, turtle looks similar to human when hands and legs are in similar gesture.

We also want to note an implementation detail. The normalized similarity scores distinguish the nearest neighbors. The scores are numbers in $[0, 1]$. We

**Fig. 3.** A cluster from the visual hierarchy in [5].

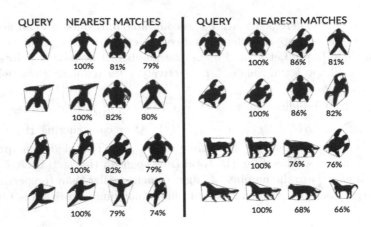

**Fig. 4.** Select queries and nearest matches in a subset of the database from [7].

note that because of the scoring function for angles in Eq. 5, only very closely matching angles in edges (at reference points, and at boundary points) contribute to the total score; others have no contribution. Our observation is that in this model we discriminate real similarities from other random matches. That is, the scores we obtain is an effective measure of similarity. However, the resulting normalized scores are small, even for near matches. Therefore, by taking the fourth power of the normalized score in $[0, 1]$, we maintain the same ordering for matches, yet we obtain numbers corresponding 70% or higher percentages for near matches. These numbers are shown as percentages in our results.

The dataset of Aslan and Tari [7] contains 56 images. The sizes of these images vary from $190 \times 111$ to $222 \times 250$ pixels. In [8,34] the same dataset is used for shape classification. The methods in [8,34] took above 5 min to process all images. In our method on this dataset, on average per image, the AC extraction took 400 ms; the sequence generation time, and the total time for alignments were 16 ms. The total time to process the entire data set with our method and answer queries with all 56 images is 47 s using a PC with 1.6 GHz clock, 512 MB RAM. The comparisons show that our method is faster than those in [8,34].

We remark that our shape representation is based on the boundary features (e.g. concavities). This is different than models based on symmetry axis in [7]. Naturally, a symmetry axis-based model performs very well classifying all human shapes with different arm and leg positions. Our shape representation and comparison method performs very well for objects whose boundaries are rigid such as firearms. The effectiveness of the seed ideas and initial method were proven empirically in [5]. The effectiveness in detecting partial matches was illustrated in identifying partially occluded firearms in [4]. We also note that two dissimilar objects can have similar axis of symmetry such as a broom and a long gun, however, boundary features can be the discriminating features in this case. Our method can differentiate these objects in this example from each other (see [5]).

To validate our shape matching method on weapons on a simple illustrative example, we select six weapons from the weapon ontology presented in [5]. For these weapons, the number of boundary points range from 300 to 998, and the boundary sequence lengths range from 97 to 129. These weapons come from three different clusters such that there are two weapons from each cluster. In Fig. 5, on the very left, enclosed by a dashed rectangle these weapons are shown. We perform a query with each weapon. In every case, the weapon itself is the nearest match, and the next nearest match is the other weapon from the same cluster.

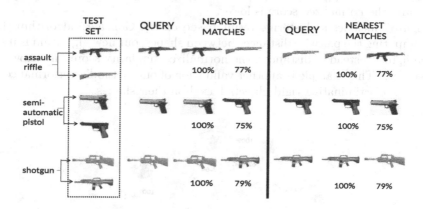

**Fig. 5.** Select queries and nearest matches in a subset of weapon ontology in [5].

To validate and have an experimental evidence of the scale invariance capability of our method, we compared a query human figure with its $2 \times 2$ and $3 \times 3$ enlargements shown in Fig. 6(A). The similarity score remains very high even there is a significant scaling difference.

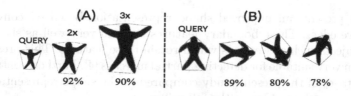

**Fig. 6.** (A) Comparison of a human figure with its $2\times$ and $3\times$ enlargements; (B) Comparison of a human figure with its rotations.

To validate the rotation invariance capability of our method, we compared a query human figure with its rotations by 30, 90, and 150° angles (clockwise.) We show the results in Fig. 6(B). The similarity score between the object and its rotated version remains high. However, the 90° angle rotation yields better scores compared to 30-degree and 150-degree rotations. We believe the reason lies with the rectangular bounding box. We bound an object with its minimum horizontal, and minimum vertical boundary position as the bottom left corner, and maximum horizontal, and maximum vertical boundary position on the top right corner. To see this take a vertical line segment of length $x$ with width 1; rotate it at 45° angle, the rectangular box that encloses the new object has a diagonal of length $\sqrt{2}\ x$, which has larger perimeter. In 90-degree case, the perimeter stays the same. However, in 30 and 150-degree cases, the perimeter increases. As a result, in these cases, shape sequences are longer, the total length of the compared sequences is larger while the similarity score remains nearly the same, and the normalized score is lower.

Figure 7 includes a clustering result based on the Gonzalez' algorithm [17]. For computing the pairwise distances between shapes, our new algorithm is used. In the figure, instead of distances, the normalized similarity scores are shown as percentages. This example is another validation of our method's performance on clustering/discriminating rigid objects based on their shapes.

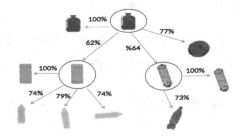

**Fig. 7.** Clusters of some containers.

## 6   Conclusions

The paper presents an improved shape representation based on convex, line, and concave edges. These boundary features perform very well as shown in [5] for rigid objects such as weapons which retain these features. This preservation makes them very suitable for detecting partial matches [4]. Further, contributions and advantages of the present study compared to the shape representation and matching approach in [3,5] are the following:

– Here the shape is represented as a single sequence of angles obtained from edges. The previous methods created separate convex hull (CH) and boundary sequences of angles, and aligned them separately and independently. Therefore, the new method is better applicable for local matches when only parts of the boundary are visible. The new method aligns sequences of edges maintaining the original clockwise ordering with respect to each other (in the order of edges). The AC was reformulated for the tasks.

– The new shape matching method generates similarly sized sequences and similar angles even though the object sizes are different, and even though the objects are rotated. This makes the method not only rotation invariant, but also, scale invariant, which is a missing property in [3,5].

One shortcoming of the new representation and method is that some unrelated objects may look similar in different orientation. For example, with respect to the new shape model, a human and a turtle can be very similar, and two human shapes in different posses (e.g. a jumping man in two different poses) may look very different (see. Fig. 4). The new representation and method apply better to rigid objects. This is because rigid objects retain their shapes better and concavities in them could be identifying features.

One disadvantage of the elaborated method is that it uses a number of user-defined parameters such as a number of thresholds.

An area in which our shape matching method can be applied is the Ribonucleic Acid (RNA) 2D structure analysis. RNA molecule makes interesting 2D formations. Similar functions and evolutionary relatedness can be analysed via structural similarities. New representations and algorithms for RNA 2D structures continue to be popular (e.g. see recent articles [1,6]). RNA 2D structures have distinguishable boundary features such as bulges and hairpin loops in their drawings. A linear sequence representation can be developed based on these boundary features, and partitioning the boundary into segments. This would yield a cyclic sequence alignment RNA structure comparison algorithm similar to the one we use in this paper. Such a representation would also be useful for searching boundary for given segment types. Our method will also be applied to compare malignant and benign skin lesion boundaries.

**Acknowledgements.** We are thankful to reviewers for their insightful and constructive comments. Addressing them yielded valuable additions to the present work. This work is partially supported by USA NSF Award No: IIS-1528027.

# References

1. Anandan, J., Fry, E., Monschke, Arslan, A.N.: A fast algorithm for finding largest common substructures in multiple RNAs. In: Proceeedings of the 9th International Conference on Bioinformatics and Computational Biology, BICOB'07, Honolulu, HI, USA, 20–22 March, pp. 51–57 (2017)
2. Antani, S., Xu, X., Long, L.R., Thoma, G.R.: Partial shape matching for CBIR of spine X-ray images. In: SPIE Proceedings of Storage and Retrieval Methods and Applications for Multimedia, vol. 5307, pp. 1–8 (2004)
3. Arslan, A.N., Sirakov, N.M., Attardo, S.: Weapon ontology annotation using boundary describing sequences. In: Proceedings of IEEE SSIAI 2012, pp. 101–104, Santa Fe, 22–24 April 2012
4. Arslan, A.N., Hempelmann, C.F., Attardo, S., Blount, G.P., Sirakova, N.N., Sirakov, N.M.: Identification of partially occluded firearms through partonomy. In: Sadjadi, M. (ed.) Proceedings of SPIE 2015 (2015). doi:10.1117/12.2184102

5. Arslan, A.N., Hempelmann, C.F., Attardo, S., Blount, G.P., Sirakov, N.M.: Threat assessment using visual hierarchy and conceptual firearms ontology. Opt. Eng. **54**(5), 053109 (2015). doi:10.1117/1.OE.54.5.053109

6. Arslan, A.N., Anandan, J., Fry, E., Pandey, R., Monschke, K.: A new structure representation for RNA and fast RNA substructure search. In: Proceedings of CSCI 2016, Las Vegas, USA, 14–17 December, pp. 1226–1231. IEEE CPS (2016). doi:10. 1109/CSCI.2016.230

7. Aslan, C., Tari, S.: An axis based representation for recognition. In: Proceedings of ICCV, pp. 1339–1346 (2005)

8. Bai, X.Y., Yui, D., Latecki, L.J.: Skeleton-based classification using path similarity. Int. J. Pattern Recogn. Artif. Intell. **22**(4), 733–746 (2008)

9. Belongie, S., Malik, J., Puzicha, J.: Shape matching and object recognition using shape contexts. Trans. PAMI **24**, 209–222 (2002)

10. Bronstein, A.M., Bronstein, M.M., Kimmel, R., Mahmoudi, M., Sapiro, G.: A Gromov-Hausdorff framework with diffusion geometry for topologically-robust non-rigid shape matching. Int. J. Comput. Vis. **89**(2), 266–286 (2010)

11. Chekmarev, D.S., Kholodovych, V., Balakin, K.V., Ivanenkov, Y., Ekins, S., Welsh, W.J.: Shape signatures: new descriptors for predicting cardiotoxicity in silico. Chem. Res. Toxicol. **21**, 1304–1314 (2008)

12. Chen, L., Feris, R.S., Turk, M.: Efficient partial shape matching using the Smith-Waterman algorithm. In: Proceedings of CVPR Workshop on Non-Rigid Shape Analysis and Deformable Image Alignment, Anchorage, Alaska, June 2008. doi:10. 1109/CVPRW.2008.4563078

13. Cour, T., Shi, J.: Recognizing objects by piecing together the segmentation puzzle. In: Proceedings of CVPR (2008). doi:10.1109/CVPR.2007.383051

14. Badawy, O., Kamel, M.: Matching concavity trees. In: Fred, A., Caelli, T.M., Duin, R.P.W., Campilho, A.C., de Ridder, D. (eds.) SSPR/SPR 2004. LNCS, vol. 3138, pp. 556–564. Springer, Heidelberg (2004). doi:10.1007/978-3-540-27868-9_60

15. Felzenszwalb, P., Schwartz, J.: Hierarchical matching of deformable shapes. In: Proceedings of CVPR, pp. 1–8. IEEE (2007)

16. Gavrila, D.M.: Pedestrian detection from a moving vehicle. In: Vernon, D. (ed.) ECCV 2000. LNCS, vol. 1843, pp. 37–49. Springer, Heidelberg (2000). doi:10.1007/3-540-45053-X_3

17. Gonzalez, T.F.: Clustering to minimize the maximum intercluster distance. Theoret. Comput. Sci. **38**(2–3), 293–306 (1985)

18. Grega, M., Matiolański, A., Guzik, P., Leszczuk, M.: Automated detection of firearms and knives in a CCTV image. Sensors **16**(1), 47 (2016). doi:10.3390/s16010047

19. Huttenlocher, D., Klanderman, G., Rucklidge, W.: Comparing images using the Hausdorff distance. Trans. PAMI **15**(9), 850–863 (1993)

20. Huang, R., Pavlovic, V., Metaxas, D.N.: A profile Hidden Markov Model framework for modeling and analysis of shape. In: Proceedings of International Conference on Image Processing, Atlanta, GA. IEEE, October 2006. doi:10.1109/ICIP.2006. 312827

21. Kim, H.-S., Chang, H.-W., Liu, H., Lee, J., Lee, D.: BIM: Image matching using biological gene sequence alignment. In: Proceedings of International Conference on Image Processing, pp. 205–108. IEEE ICIP (2009)

22. Kortagere, S., Chekmarev, D., Welsh, W.J., Ekins, S.: Hybrid scoring and shape based classification approaches for human pregnane X receptor. Pharm. Res. **26**(4), 1001–1011 (2009)

23. Kortagere, S., Krasowski, M.D., Sean Ekins, S.: The importance of discerning shape in molecular pharmacology. Trends Pharmacol. Sci. **30**(3), 138–147 (2009)
24. Korotkov, K.: Automatic change detection in multiple skin lesions. Ph.D. thesis, Universitat de Girona (2014)
25. Ling, H., Jacobs, D.: Using the Inner-Distance for classification of articulated shapes. In: Proceedings of Conference on Computer Vision and Pattern Recognition, vol. II, pp. 719–726. IEEE CVPR (2005)
26. Mori, G., Malik, J.: Recognizing objects in adversarial clutter: breaking a visual CAPTCHA. In: Proccedings of Computer Vision and Pattern Recognition, pp. 134–141. IEEE CVPR (2003)
27. Needleman, S.B., Wunsch, C.D.: A general method applicable to the search for similarities in the amino acid sequence of two proteins. J. Mol. Biol. **48**(3), 443–53 (1970)
28. Opelt, A., Pinz, A., Zisserman, A.: A boundary-fragment-model for object detection. In: Leonardis, A., Bischof, H., Pinz, A. (eds.) ECCV 2006. LNCS, vol. 3952, pp. 575–588. Springer, Heidelberg (2006). doi:10.1007/11744047_44
29. Shotton, J., Blake, A., Cipolla, R.: Contour-based learning for object detection. In: Proceedings of ICCV (2005). doi:10.1109/ICCV.2005.63
30. Sirakov, N.M.: A new active convex hull model for image regions. J. Math. Imaging Vis. **26**(3), 309–325 (2006)
31. Sirakov, N.M., Simonelli, I.: A new automatic concavity extraction model. In: Proceedings of SSIAI, pp. 178–182 (2006)
32. Sirakov, N.M., Ushkala, K.: An integral active contour model for convex hull and boundary extraction. In: Bebis, G., Boyle, R., Parvin, B., Koracin, D., Kuno, Y., Wang, J., Pajarola, R., Lindstrom, P., Hinkenjann, A., Encarnação, M.L., Silva, C.T., Coming, D. (eds.) ISVC 2009. LNCS, vol. 5876, pp. 1031–1040. Springer, Heidelberg (2009). doi:10.1007/978-3-642-10520-3_99
33. Sklansky, J.: Measuring concavity on a rectangular mosaic. IEEE Trans. Comput. **C-21**, 1355–1364 (1972)
34. Sun, K.B., Super, B.J.: Classification of contour shapes using class segment sets. In: Proceedings of CVPR, pp. 727–733 (2005). doi:10.1109/CVPR.2005.98
35. Thayananthan, A., Stenger, B., Torr, P., Cipolla, R.: Shape context and chamfer matching in cluttered scenes. In: Proceedings of CVPR, pp. 127–134 (2003). doi:10.1109/CVPR.2003.1211346
36. Wang, H., Oliensis, J.: Rigid shape matching by segmentation averaging. IEEE Trans. Pattern Anal. Mach. Intell. **32**(4), 619–635 (2010). doi:10.1109/TPAMI.2009.199
37. Xu, X., Lee, D.-J., Antani, S., Long, L.R.: A spine x-ray image retrieval system using partial shape matching. IEEE Trans. Inf Technol. Biomed. **12**(1), 100–108 (2008)

# Gradient and Graph Cuts Based Method for Multi-level Discrete Tomography

Tibor Lukić and Marina Marčeta[✉]

Faculty of Technical Sciences, University of Novi Sad, Novi Sad, Serbia
{tibor,marina.marceta}@uns.ac.rs

**Abstract.** In this paper, we are proposing a new energy-minimization reconstruction method for the multi gray level discrete tomography. The proposed reconstruction approach combines a gradient based algorithm with the graph cuts optimization. This new technique is able to reconstruct images that consist of an arbitrary number of gray levels. We present the experimental evaluation of the new method, where we compare its performance with performance of the already suggested methods for multi-level discrete tomography. The obtained experimental results give an advantage to the proposed approach, especially regarding the quality of the reconstructed test images.

## 1  Introduction

*Tomography* [14] reconstructs images of non accessible or non visible objects. It deals with recovering images from a number of projections. Tomography will be our focus in this paper. From a mathematical point of view, the object corresponds to a function. The problem posed, is to reconstruct this function from its integrals, or its sums over subsets of its domain. In general, the tomographic reconstruction problem may be continuous or discrete. In *Discrete Tomography* (DT) [15,16] the range of the function is a finite set. In practice, DT often deals with reconstructions of digital images that consist of a number of gray levels. DT has a wide range of applications in areas where the materials of the object under investigation are known before, such as industrial non-destructive testing or electron tomography [15,16].

To the best of our knowledge, there are only a few reconstruction algorithms suggested for this DT problem, that deal with multi gray level tomography image reconstruction. These are the Discrete Algebraic Reconstruction Technique (DART) [2], the Multi-Well Potential based method (MWPDT) [22], method which combines non-local projection constraints, continuous convex relaxation of the Multilabeling problem and DC programming (MDC) [25], and the Non-Linear Discretization function based reconstruction algorithm (NLD) [30]. The DART method uses a fixed threshold function for the discretization process (without any regularization), which can lead to radical solutions and less accurate reconstructions, especially in the case of reduced projection data. The MDC is a powerful method, but less flexible related to adding new regularization terms, because the energy function has to be expressed as a difference of convex functions. The MWPDT and NLD methods applies a non-convex energy function in

ⓒ Springer International Publishing AG 2017
V.E. Brimkov and R.P. Barneva (Eds.): IWCIA 2017, LNCS 10256, pp. 322–333, 2017.
DOI: 10.1007/978-3-319-59108-7_25

the reconstruction process, which can stuck in local minimum, i.e., in a semi-continuous solution. The proposed method in this paper is developed in such a way to avoids the above listed disadvantages.

One of the approaches used for solving problems in image processing and computer vision has been developed based on graph cuts. The core of this approach is to construct a specialized graph for the energy function to be minimized such that the minimum cut on the graph also minimizes the energy (either globally or locally). The minimum cut, in turn, can be computed very efficiently by max-flow algorithms. The output of these algorithms is generally a solution with some interesting theoretical quality guarantees. In [20] is given, which conditions the energy function needs to satisfy in order to be minimized via graph cuts.

In this paper, we propose a new deterministic reconstruction method for the DT problem, which combines a gradient based method, with a graph cuts type optimization method. The proposed method uses a smooth regularization prior and allows reconstruction of images that contain an arbitrary number of different gray levels.

The structure of the paper is the following. In Sect. 2, the basic reconstruction problem is described. In Sect. 3, we present the new reconstruction method based on the graph cuts approach. Our experimental results are provided in Sect. 4 and finally, Sect. 5 is the conclusion.

## 2   Reconstruction Problem

In this paper we consider the DT reconstruction problem, represented by a linear system of equations

$$A\,u = b, \text{ where}$$
$$A \in \mathbb{R}^{M \times N}, \quad u \in \Lambda^N, \quad b \in \mathbb{R}^M, \quad \Lambda = \{\lambda_1, \lambda_2, ..., \lambda_k\}, \quad \lambda_i \in [0, 1], \quad k \geq 2. \tag{1}$$

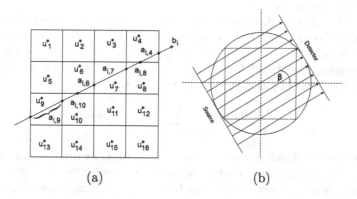

(a)                                              (b)

**Fig. 1.** (a) Example of a projection value calculation on an image $u^*$ of size $N = 4 \times 4 = 16$. A projection ray penetrates through the image pixels. The projection value $b_i$ is calculated by $b_i = a_{i,4}u_4^* + a_{i,6}u_6^* + a_{i,7}u_7^* + a_{i,8}u_8^* + a_{i,9}u_9^* + a_{i,10}u_{10}^*$. (b) Parallel beam projection. The source-detector system can rotate around a center point. The projection direction is determined by the angle $\beta$.

The value of $k$ represents the number of different gray level values. The set $\Lambda$ is given by the user. The matrix $A$ is a so-called *projection matrix*, whose each row corresponds to one projection ray. The corresponding components of the vector $b$ contain the detected projection values, while the vector $u$ represents the unknown binary image to be reconstructed. The $i$-th row entries $a_{i,.}$ of $A$ represent the length of the intersection of the pixels and the $i$-th projection ray passing through them, see Fig. 1(a). The projection value measured by a projection ray is calculated as a sum of products of the pixel's intensity and the corresponding length of the projection ray through that pixel. The side length of each pixel is one. Therefore, vertical and horizontal projection rays represent the sum of the gray intensity values of pixels in corresponding columns and rows, respectively. Projections are taken from different directions. For each projection direction, a number of parallel projection rays are taken (parallel beam projection), as shown in Fig. 1(b). The distance between two adjacent parallel projection rays can vary depending on the reconstruction problem. We set this distance to be equal to the side length of pixels.

The reconstruction problem means finding the solution image $u$ of the linear system of Eq. (1), where the projection matrix $A$ and the projection vector $b$ are given. This system is often undetermined ($N > M$), and therefore additional regularization (based on a priori information) is needed for the determination of quality and acceptable solutions.

## 3   Reconstruction Method Based on the Graph Cuts Method

A directed, weighted graph $G = (X, \rho)$, consists of a set of nodes $X$ and a set of directed edges $\rho$ that connect them. The nodes, in image processing interpretations, mostly correspond to pixels or voxels in 3D. All edges of graph are assigned some weight or cost.

Let $G = (X, \rho)$ be a directed graph with non-negative edge weights that has two special nodes or terminals, the source $A$ and the sink $B$. An $a - b$-cut (which is referred informally as a cut) $C = A, B$ is a partition of the terminals in $X$ into two disjoint sets $A$ and $B$ so that $a \in A$ and $b \in B$. The cost of the cut is the sum of the costs of all edges that go from $A$ to $B$:

$$c(A, B) = \sum_{x \in A, y \in B, (x,y) \in \rho} c(x, y).$$

The minimum $a - b$-cut problem is to find a cut $C$, with the minimum cost among all cuts. Algorithms to solve this problem can be found in [8].

The approach that uses graph cuts for energy minimization has, as a basic technique, the construction of a specialized graph for the energy function to be minimized, so that the minimum cut on the graph also minimizes the energy. The form of the graph depends on the exact form of $X$ and on the number of labels. The minimum cut, in turn, can be computed very efficiently by max flow algorithms.

These methods have been successfully used in the last 20 years for a wide variety of problems, naming image restoration [9,10], stereo and motion [3,19], image synthesis [21], image segmentation [7] and medical imaging [6,18].

## 3.1   Potts Model

The Potts model in graph cuts theory is based on the minimization of the following energy

$$E(d) = \sum_{p \in \mathcal{P}} D(p, d_p) + \sum_{(p,q) \in \mathcal{N}} K_{(p,q)} \cdot T(d_p \neq d_q), \tag{2}$$

where $d = \{d_p | p \in \mathcal{P}\}$ represents the labelling of the image pixels $p \in \mathcal{P}$. By $D(p, d_p)$ we denote the data cost term, where $D(p, d_p)$ is a penalty or cost for assigning a label $d_p$ to a pixel $p$. $K_{(p,q)}$ is an interaction potential between neighboring pairs $p$ and $q$, $\mathcal{N}$ is a set of neighboring pairs. Function $T(\cdot)$ is 1 if the condition inside parenthesis is true and 0 otherwise.

## 3.2   Proposed Reconstruction Method

Our tomography reconstruction approach is a combination of the graph cuts method and the quadratic iterative minimization method. In the first step, we determine the data cost values for each image pixels. The data cost values are determined as intensity values of the continuous/smooth approximation of the final reconstruction image, obtained as a solution of the following energy-minimization problem

$$\min_{u \in [0,1]^N} E_Q(u) := \|Au - b\|^2. \tag{3}$$

Function $E_Q$ is quadratic type and $\Omega = [0, 1]^N$ is a feasible set. Therefore, the problem (3) is a constrained and quadratic type energy-minimization problem. This minimization problem can be solved by several optimization methods. According to our earlier experiences in similar problems [23,24,26] we chose the Spectral Projected Gradient (SPG) optimization algorithm [4] for this task.

For THE application of this algorithm two conditions must be satisfied [4]: (i) The objective function has continuous partial derivatives on an open set that contains $\Omega$; (ii) The projection function $P_\Omega$ of an arbitrary vector onto the set $\Omega$ is provided. The objective function in (3) is a multiple differentiable function in $\mathbb{R}^N$, therefore requirement (i) is satisfied. The projection $P_\Omega$ of an arbitrary vector $u \in \mathbb{R}^n$ onto the set $\Omega$ we define as

$$[P_\Omega(u)]_i = \begin{cases} 0, & u_i \leq 0 \\ 1, & u_i \geq 1 \\ u_i, \text{ elsewhere} \end{cases} \quad \text{where } i = 1, \ldots, N.$$

$P_\Omega$ is a projection with respect to the Euclidean distance, i.e. $P_\Omega(x) = \arg\{\min_{y \in \Omega} d_2(x, y)\}$. Hence, requirement (ii) is also satisfied.

The pseudo-code of the SPG is presented in Algorithm 1. The reconstruction process, starts with the initial solution $u^0$, where each pixel intensity is set as 0.5, as the middle of the interval $[0, 1]$. The SPG algorithm combines the non-monotone line search algorithm [13] and the spectral gradient step-length selection [1, 5, 27].

---

**Algorithm 1.** SPG optimization algorithm.

$u^0 = [0.5, 0.5, ..., 0.5]^T$; $d^0 = P_\Omega(u^0 - \nabla E_Q(u^0)) - u^0$; $k = 0$;
**repeat**
  Determine the step-length $\lambda^k > 0$ by a line search approach, see [4];
  $u^{k+1} = u^k + \lambda^k d^k$;
  Calculate the gradient spectral step-length $\theta_{k+1} > 0$, see [4];
  $d^{k+1} = P_\Omega(u^{k+1} - \theta_{k+1}\nabla E_Q(u^{k+1})) - u^{k+1}$; $k = k + 1$;
**until** $\|u^k - u^{k-1}\|_\infty < 10^{-2}$;
$u^{new} = u^k$;

---

In the next step we have to discretize the smooth solution of the problem 3 $u$, obtained by the SPG algorithm. For this task we apply the graph cuts method based on the Potts model, described in Sect. 3.1. The energy model in (2) is successfully used in many energy minimization problems with similar energy structure: sum of a data and a regularization/neighboring interaction terms. We mention discrete tomography reconstruction algorithms proposed by Schüle et al. [28,31] and Lukić et al. [23,24]. The Potts interaction model (second term in (2)) showed good ability to enhance compactness of the solution (see [8,12,29]), in a similar way as the compactness saving regularization terms do in already suggested reconstruction methods [24,28], which also motivate our choice for application of this model. We note that other interaction models, for example the linear model [8], can also be taken into consideration, but this issue is out of focus of this paper. The data cost term $D$ in (2) is determined using information provided by the smooth solution. More precisely, we define it in the following way

$$D(p, 0) = |u(p) - \lambda_1|,$$
$$D(p, 1) = |u(p) - \lambda_2|,$$
$$D(p, 2) = |u(p) - \lambda_3|,$$
$$\vdots$$
$$D(p, k-1) = |u_p - \lambda_k|,$$

where $u(p)$ represents the intensity of a pixel $p$. The idea is to make data cost small/cheap in the vicinity of the given gray values. The neighbor pairs are defined based on 1-neighboring system, i.e., $(p, q) \in \mathcal{N}$ if the image coordinates of $p$ and $q$ differs for one value only. The interaction potential $K_{(p,q)}$ (see (2)) in our experiments is set as a constant and its value is 1. Now, the energy function in (2) is determined and ready to be minimized. For this task we use the GCO graph cuts based optimization algorithm, introduced in [10] and further analyzed in [8,11,20]. The GCO algorithm determines the label values $d_p$ for each pixel $p$. Each label value is assigned to one predefined gray level in the following way: $d_p = 0 \rightarrow \lambda_1$, $d_p = 1 \rightarrow \lambda_2$, ..., $d_p = (k-1) \rightarrow \lambda_k$. Therefore, the obtained label values also determine intensities of pixels (from the given set of gray levels) in the final (discrete) solution, therefore the reconstruction process is terminated. We denote this method by Graph Cuts Discrete Tomography (GCDT) reconstruction method.

Naturally arises the simplest, but less powerful, way for discretization of the smooth solution $u$ provided as a result of the minimization problem (3). This approach is based on the application of the thresholding function, defined by

$$t(v) = \begin{cases} \lambda_1 & v < \tau_1 \\ \lambda_2 & \tau_1 \le v < \tau_2 \\ ... \\ \lambda_k & \tau_{l-1} \le v \end{cases},$$

where $v \in \mathbb{R}$ and $\tau_l = \frac{\lambda_i + \lambda_{i+1}}{2}$, $l = 1, 2, ..., k - 1$. The final solution $u^r$ is obtained by application of the thresholding function to the smooth solution $u$, i.e., $u^r = [t(u_1), t(u_2), t(u_3), ..., t(u_N)]$. We denote this method by TRDT, and use it in experimental work as a control method.

## 4  Experimental Results

In this section we experimentally evaluate the proposed graph cuts based reconstruction method, denoted by GCDT. In the experiments we use 4 test images (phantoms), as originals in reconstructions, presented in Fig. 2. Phantoms PH1, PH2 and PH3 contain 3 gray levels, while the well-known Shepp-Logan phantom [17] contains 6 gray levels. We consider reconstructions of these images obtained from different projection directions. A total of 128 parallel rays are taken for each projection direction. In all cases, the projection directions are uniformly selected between 0 and 180°. The obtained results are compared with the results provided by the Multi Well Potential based method (MWPDT) [22], already

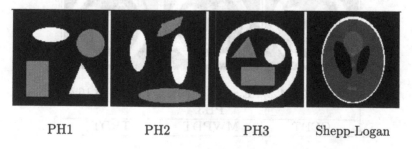

PH1          PH2          PH3          Shepp-Logan

**Fig. 2.** Original test images (128 × 128). Phantoms PH1, PH2 and PH3 contain 3 different gray levels (0, 0.5, 1), while Shepp-Logan contains 6 different gray levels (0, 0.1, 0.2, 0.3, 0.4, 1).

suggested for multi-level discrete tomography reconstruction, and with the simple method based on the classical thresholding, denoted by TRDT. Related to the Shepp-Logan test image, the DART method, proposed in [2], is also included into the evaluation process.

In the evaluation process, we analyze the quality of the reconstructions and required running times. The quality of the reconstructions are expressed by the pixel error $(PE)$, i.e., the absolute number of the misclassified pixels, and by the misclassification rate $(m.r.)$, i.e., the pixel error measure relative to the total number of image pixels. Also, as a qualitative error measure, we consider the projection error, defined by $PRE = \|Au^r - b\|$, where $u^r$ represents the reconstructed image. This error expresses the accordance of the reconstruction with the given projection data.

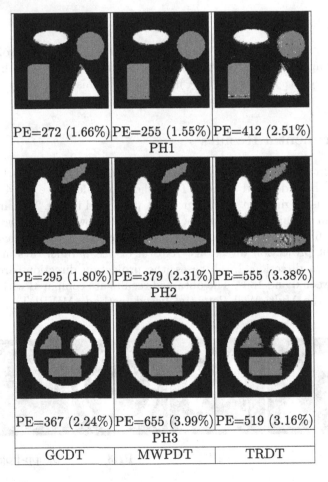

**Fig. 3.** Reconstructions of the test images using data from 6 projection directions.

**Table 1.** Experimental results for Shepp-Logan image, using three different reconstruction methods. The abbreviation m.r. indicates misclassification rate and $d$ indicates the number of projections.

|  | $d$ | TRDT (m.r.%) | DART (m.r.%) | GCDT (m.r.%) |
|---|---|---|---|---|
| Shepp-Logan | 12 | 12.74 | 14.21 | 5.72 |
|  | 15 | 10.44 | 8.44 | 3.17 |
|  | 18 | 10.03 | 2.56 | 2.14 |

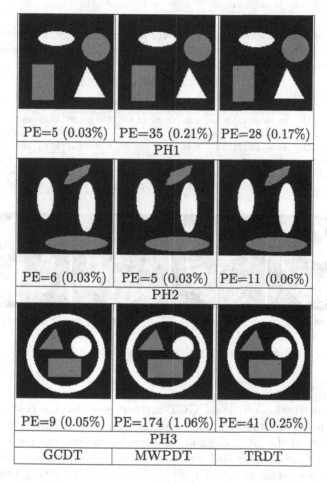

**Fig. 4.** Reconstructions of the test images using data from 15 projection directions.

In Table 2 we present pixel errors for reconstructions of three phantom images (PH1, PH2 and PH3) obtained from a different number of projections by three different methods (MWPDT, GCDT and TRDT). In Figs. 3 and 4 reconstructions from 6 and 15 projection directions are presented. From a total of 12

**Table 2.** Experimental results for PH1, PH2 and PH3 images, using three different reconstruction methods. The abbreviation $d$ indicates the number of projections.

| $d$ | | PH1 | | | | PH2 | | | | PH3 | | | |
|---|---|---|---|---|---|---|---|---|---|---|---|---|---|
| | | 6 | 9 | 12 | 15 | 6 | 9 | 12 | 15 | 6 | 9 | 12 | 15 |
| MWP | (PE) | 255 | 159 | 59 | 35 | 379 | 242 | 56 | 5 | 655 | 456 | 275 | 174 |
| | (m.r.%) | 1.55 | 0.97 | 0.36 | 0.21 | 2.31 | 1.47 | 0.34 | 0.03 | 3.99 | 2.78 | 1.67 | 1.06 |
| TRDT | (PE) | 412 | 175 | 48 | 28 | 555 | 290 | 37 | 11 | 412 | 301 | 101 | 41 |
| | (m.r.%) | 2.51 | 1.06 | 0.29 | 0.17 | 3.38 | 1.77 | 0.22 | 0.06 | 2.51 | 1.83 | 0.61 | 0.25 |
| GCDT | (PE) | 272 | 69 | 8 | 5 | 295 | 121 | 15 | 6 | 272 | 116 | 20 | 9 |
| | (m.r.%) | 1.66 | 0.42 | 0.04 | 0.03 | 1.80 | 0.73 | 0.09 | 0.03 | 1.66 | 0.70 | 0.12 | 0.05 |

**Table 3.** Experimental results for PH1, PH2 and PH3 images, using three different reconstruction methods. The abbreviation e.t. means elapsed time in minutes and $d$ indicates the number of projections.

| $d$ | | PH1 | | | | PH2 | | | | PH3 | | | |
|---|---|---|---|---|---|---|---|---|---|---|---|---|---|
| | | 6 | 9 | 12 | 15 | 6 | 9 | 12 | 15 | 6 | 9 | 12 | 15 |
| MWPDT | (PRE) | 14.70 | 12.19 | 9.96 | 9.08 | 15.32 | 15.50 | 9.68 | 3.09 | 19.83 | 18.77 | 18.80 | 16.43 |
| | (e.t.) | 1.76 | 2.63 | 3.17 | 4.06 | 1.78 | 2.82 | 3.21 | 3.06 | 2.19 | 2.87 | 4.30 | 4.66 |
| TRDT | (PRE) | 18.66 | 14.72 | 10.61 | 8.87 | 19.01 | 19.09 | 8.89 | 6.23 | 23.64 | 17.87 | 13.66 | 10.61 |
| | (e.t.) | 7.73 | 12.58 | 14.55 | 17.77 | 5.44 | 10.90 | 12.67 | 15.74 | 7.28 | 11.07 | 13.39 | 16.00 |
| GCDT | (PRE) | 23.24 | 11.12 | 6.52 | 4.39 | 18.31 | 13.94 | 7.10 | 4.57 | 25.87 | 14.96 | 7.59 | 5.60 |
| | (e.t.) | 7.73 | 12.58 | 14.55 | 17.77 | 5.45 | 10.91 | 12.67 | 15.74 | 7.29 | 11.07 | 13.40 | 16.01 |

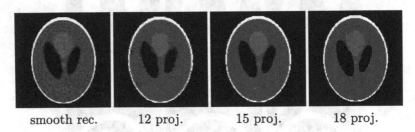

smooth rec.　　　12 proj.　　　15 proj.　　　18 proj.

**Fig. 5.** Reconstructions of the Shepp-Logan test images by the proposed GCDT method.

different reconstruction problems, GCT method provided the best results in 10 cases, while in 2 cases the dominant was the MWPDT method. We emphasize that the results of the GCT method, in cases when they are the best, are significantly better, at least by 50%, compared with other results. Table 3 presents the obtained projection errors ($PRE$) and the needed running times in these experiments. Regrading the $PRE$ values, the proposed GCT method dominated in 8 cases, while MWPDT in 4 cases.

All reconstruction methods (MWPDT, GCDT and TRDT) are implemented completely in Matlab. The best running times in all of the experiments was achieved by the MWPDT method (see Table 3). GCDT and TRDT methods uses the smooth solution/reconstruction as a first step, before the "binarization process" starts by GCO graph cuts optimization [10]. This smooth solution is achieved as a final termination, with high precision. This process, because of the high precision, requires significantly higher number of iterations than is needed for MWPDT method in total, resulting in a greater consumption of time.

Reconstruction results of the well-known Shepp-Logan [17] phantom image is presented in Table 1 and Fig. 5. This phantom is considered to be one of the most complex, containing 6 different gray levels. We compare the results obtained by the three different reconstruction methods: TRDT, DART and GCDT. The results for DART are taken from [2]. The projection data is acquired from 12, 15 and 18 projection directions. The GCDT method provides the best results in all cases (smallest m.r. values).

Summarizing the results obtained by the total of the 15 analyzed reconstruction tasks, see Tables 1 and 2, the quality of the reconstruction, indicated by m.r., for the proposed GCDT method was the best in 13 cases, i.e., in 87% of the cases. According to these results, we conclude that the experiments confirm the capability of the proposed method to provide high quality reconstructions.

## 5 Conclusions

In this paper, a new energy-minimization based reconstruction method for multi-level tomography is proposed. It combines a gradient based method, with the graph cuts optimization method. Experiments show advantages of the proposed method in comparison with three formerly published reconstruction methods. Based on the obtained experimental results and analysis presented in this paper, we conclude that the combination of a gradient based method with graph cuts optimization method is suitable for providing high quality reconstructions.

**Acknowledgement.** Tibor Lukić acknowledges the Ministry of Education and Sciences of the R. of Serbia for support via projects OI-174008 and III-44006. Marina Marčeta acknowledges the Ministry of Education and Sciences of the R. of Serbia for support via project OI-174008.

## References

1. Barzilai, J., Borwein, J.M.: Two point step size gradient methods. IMA J. Num. Anal. **8**, 141–148 (1988)
2. Batenburg, K.J., Sijbers, J.: DART: a fast heuristic algebraic reconstruction algorithm for discrete tomography. In: Proceedings of International Conference on Image Processing (ICIP), pp. 133–136 (2007)
3. Birchfield, S., Tomasi, C.: Multiway cut for stereo and motion with slanted surfaces. In: Proceedings of International Conference on Computer Vision, pp. 489–495 (1999)
4. Birgin, E.G., Martínez, J.M., Raydan, M.: Algorithm: 813: SPG - software for convex-constrained optimization. ACM Trans. Math. Softw. **27**, 340–349 (2001)
5. Birgin, E., Martínez, J.: Spectral conjugate gradient method for unconstrained optimization. Appl. Math. Optim. **43**, 117–128 (2001)

6. Boykov, Y., Jolly, M.P.: Interactive graph cuts for optimal boundary and region segmentation of objects in n-d images. In: Proceedings of International Conference on Computer Vision, pp. 105–112 (2001)

7. Boykov, Y., Kolmogorov, V.: Computing geodesics and minimal surfaces via graph cuts. In: Proceedings of International Conference on Computer Vision, pp. 26–33 (2003)

8. Boykov, Y., Kolmogorov, V.: An experimental comparison of min-cut/max-flow algorithms for energy minimization in vision. IEEE Trans. PAMI **26**(9), 1124–1137 (2004)

9. Boykov, Y., Veksler, O., Zabih, R.: Markov random fields with efficient approximations. In: Proceedings of IEEE Conference on Computer Vision and Pattern Recognition, pp. 648–655 (1998)

10. Boykov, Y., Veksler, O., Zabih, R.: Fast approximate energy minimization via graph cuts. IEEE Trans. PAMI **23**(11), 1222–1239 (2001)

11. Delong, A., Osokin, A., Isack, H.N., Boykov, Y.: Fast approximate energy minimization with label costs. In: Proceedings of IEEE Conference on Computer Vision and Pattern Recognition vol. 96, no. 1, pp. 1–27 (2010)

12. Greig, D., Porteous, B., Seheult, A.: Exact maximum a posteriori estimation for binary images. J. R. Stat. Soc. **51**(2), 271–279 (1989)

13. Grippo, L., Lampariello, F., Lucidi, S.: A nonmonotone line search technique for Newton's method. SIAM J. Numer. Anal. **23**, 707–716 (1986)

14. Herman, G.T.: Fundamentals of Computerized Tomography: Image Reconstruction from Projection. Advances in Computer Vision and Pattern Recognition, 2nd edn. Springer, London (2009)

15. Herman, G.T., Kuba, A.: Discrete Tomography: Foundations, Algorithms and Applications. Applied and Numerical Harmonic Analysis. Birkhäuser, Boston (1999)

16. Herman, G.T., Kuba, A.: Advances in Discrete Tomography and Its Applications. Birkhäuser, Boston (2006)

17. Kak, A.C., Slaney, M.: Principles of Computerized Tomographic Imaging. SIAM, Philadelphia (2001)

18. Kim, J., Zabih, R.: Automatic segmentation of contrast-enhanced image sequences. In: Proceedings of International Conference on Computer Vision, pp. 502–509 (2003)

19. Kolmogorov, V., Zabih, R.: Visual correspondence with occlusions using graph cuts. In: Proceedings of International Conference on Computer Vision, pp. 508–515 (2001)

20. Kolmogorov, V., Zabih, R.: What energy functions can be minimized via graph cuts? IEEE Trans. PAMI **26**(2), 147–159 (2004)

21. Kwatra, V., Schoedl, A., Essa, I., Turk, G., Bobick, A.: Graphcut textures: image and video synthesis using graph cuts. In: Proceedings of SIGGRAPH 2003, pp. 277–286 (2003). ACM Trans. Graphics

22. Lukić, T.: Discrete tomography reconstruction based on the multi-well potential. In: Aggarwal, J.K., Barneva, R.P., Brimkov, V.E., Koroutchev, K.N., Korutcheva, E.R. (eds.) IWCIA 2011. LNCS, vol. 6636, pp. 335–345. Springer, Heidelberg (2011). doi:10.1007/978-3-642-21073-0_30

23. Lukić, T., Balázs, P.: Binary tomography reconstruction based on shape orientation. Pattern Recognit. Lett. **79**, 18–24 (2016)

24. Lukić, T., Nagy, B.: Deterministic discrete tomography reconstruction method for images on triangular grid. Pattern Recognit. Lett. **49**, 11–16 (2014)

25. Zisler, M., Petra, S., Schnörr, C., Schnörr, C.: Discrete tomography by continuous multilabeling subject to projection constraints. In: Rosenhahn, B., Andres, B. (eds.) GCPR 2016. LNCS, vol. 9796, pp. 261–272. Springer, Cham (2016). doi:10.1007/978-3-319-45886-1_21

26. Nagy, B., Lukić, T.: Dense projection tomography on the triangular tiling. Fundamenta Informaticae **145**, 125–141 (2016)

27. Raydan, M.: The Barzilai and Browein gradient method for the large scale unconstrained minimization problem. SIAM J. Optim. **7**, 26–33 (1997)

28. Schüle, T., Schnörr, C., Weber, S., Hornegger, J.: Discrete tomography by convex-concave regularization and D.C. programming. Discrete Appl. Math. **151**, 229–243 (2005)

29. Snow, D., Viola, P., Zabih, R.: Exact voxel occupacy with graph cuts. In: Proceedings of IEEE Conference on Computer Vision and Pattern Recognition, vol. 1, pp. 345–352 (2000)

30. Varga, L., Balázs, P., Nagy, A.: An energy minimization reconstruction algorithm for multivalued discrete tomography. In: Proceedings of 3rd International Symposium on Computational Modeling of Objects Represented in Images, pp. 179–185. Taylor & Francis, Rome (2012)

31. Weber, S., Nagy, A., Schüle, T., Schnörr, C., Kuba, A.: A benchmark evaluation of large-scale optimization approaches to binary tomography. In: Kuba, A., Nyúl, L.G., Palágyi, K. (eds.) DGCI 2006. LNCS, vol. 4245, pp. 146–156. Springer, Heidelberg (2006). doi:10.1007/11907350_13

# Reconstruction of Nearly Convex Colored Images

Fethi Jarray[1,2] and Ghassen Tlig[1,3(✉)]

[1] Cedric-CNAM, 292 rue St-Martin, Paris, France
fethi_jarray@yahoo.fr, ghassen.tlik@gmail.com
[2] Higher Institute of Computer Science of Medenine, Medenine, Tunisia
[3] LIMTIC, Higher Institute of Computer Science of Tunis, Tunis, Tunisia

**Abstract.** This paper studies the problem of reconstructing hv-convex images with a small number of discrete colors from two projections for each color in horizontal and vertical directions. A new integer programming based method is proposed to reconstruct nearly hv-convex colored images. Firstly, we model the reconstruction problem by a quadratic binary program. Secondly, we linearize the program into two linear binary programs. Thirdly, we solve the continuous relaxation of these programs by using IBM ILOG Cplex. Finally, we use a min-cost/max-flow model to transform the continuous solution into an approximate binary solution which may contains overlapping between colors.

**Keywords:** Discrete tomography · Image reconstruction · Integer programming

## 1 Introduction

Discrete tomography (DT) deals with the reconstruction of discrete objects such as matrices and images from a small number of projections (see [14,15]). The reconstruction techniques are used in many real-life applications such as workforce scheduling [18,20], data compression, and data security and networks [19]. Colored image reconstruction problem from horizontal and vertical projections is considered to be one of the most important problems in DT (see Fig. 1). The projections count the number of cells of each color in each row and column.

Ryser [25] and Gale [12] establish necessary and sufficient conditions for the existence of monocolored images satisfying horizontal and vertical projections. The definition of the reconstruction problem raises the question of uniqueness of the reconstruction. To make the reconstruction unique, a priori information on the nature of the object to be reconstructed needs to be integrated. The convexity is one of such useful assumptions for the reconstruction of binary matrices [6,10,21,22] and bicolored images [5,7,17,23].

A colored image is called a polyomino if there exists a path between all pairs of cells with the same color. There are other definitions of connexity in discrete images. A path is a sequence of cells with the same color that are horizontally

© Springer International Publishing AG 2017
V.E. Brimkov and R.P. Barneva (Eds.): IWCIA 2017, LNCS 10256, pp. 334–346, 2017.
DOI: 10.1007/978-3-319-59108-7_26

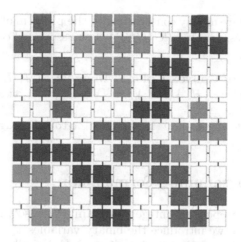

**Fig. 1.** A 3-colored convex image.

or vertically adjacent. A colored image is said to be h-convex if all the cells with the same color in each row are contiguous. Similarly an image is said to be v-convex if all the cells with the same color in each column are contiguous. An image is hv-convex if it is both h-convex and v-convex. The reconstruction problems of polyominoes [4], h-convex [4], v-convex [4] and hv-convex images are NP-complete [26]. However, the reconstruction of hv-convex polyominoes can be solved in polynomial time [4,9].

The general problem of reconstructing colored images was proved to be NP-complete for two or more colors [8,11,13]. Recently, the reconstruction of hv-convex colored disjoint polyominoes was proved to be NP-complete for an unbounded number of colors [1].

In this paper, we examine a new method to reconstruct nearly hv-convex colored images based on an integer programming approach. In Sect. 2, we introduce some definitions and notation. In Sect. 3, we provide an integer programming formulation. In Sect. 4, we propose two linearization approaches. In Sect. 5, we present and discuss the numerical results. We conclude in Sect. 6.

## 2   Definitions and Notations

**Definition 1.** *Given a set of colors $C$ and an $m \times n$ image $M$ whose items are elements of $C$, the horizontal projection of color $c$ is the vector $H^c = (h_1^c, \ldots, h_m^c) \in \mathbb{N}^m$ where $h_i^c = |\{j : M_{ij} = c\}|$, i.e., the number of cells colored with $c$ in row $i$, the vertical projection of color $c$ is the vector $V^c = (v_1^c, \ldots, v_n^c) \in \mathbb{N}^n$, for $c \in C$ where $v_j^c = |\{i : M_{ij} = c\}|$, i.e., the number of cells colored with $c$ in column $j$. We note $H = \{H^1, \ldots, H^C\}$ and $V = \{V^1, \ldots, V^C\}$ the orthogonal projections of image $M$.*

In this paper, we are mainly concerned with the problem of reconstructing hv-convex colored images, denoted $RCI(H,V)$. The associated consistency problem can be defined as:

**Reconstruction of Convex Colored Images:** $RCI(H,V)$
**Given:** $H = \{H^1,\ldots,H^C\}$ and $V = \{V^1,\ldots,V^C\}$ two collections of integer vectors.
**Goal:** Construct an $m \times n$ hv-convex colored image satisfying $H$ and $V$.

## 3    Integer Programming Formulation

In this section, we propose an integer program $P$ for solving $RCI(H,V)$. To simplify the notation, we introduce the binary variables $x_{i,j}^c$, where $1 \leq i \leq m$, $1 \leq j \leq n$ and $c \in \{1,\ldots,C\}$ such that $x_{i,j}^c = 1$ if the cell $(i,j)$ is colored with color $c$. If $x_{i,j}^c = 0$ for all $c$, then the cell is said to be white, or uncolored. The following function $F_{adj}(X)$ counts the number of adjacent pairs of cells having the same color in $X$ where $X$ is the vector of all $(x_{i,j}^c)$:

$$F_{adj}(X) = \sum_{c=1}^{C} \left( \sum_{i=1}^{m-1} \sum_{j=1}^{n} x_{i,j}^c x_{i+1,j}^c + \sum_{i=1}^{m} \sum_{j=1}^{n-1} x_{i,j}^c x_{i,j+1}^c \right)$$

Our optimization formulation is

$$P \begin{cases} max\ F_{adj}(X) \\ s.t. \\ \sum_{j=1}^{n} x_{i,j}^c = h_i^c & i = 1,\ldots,m; c = 1,\ldots,C & (i) \\ \sum_{i=1}^{m} x_{i,j}^c = v_j^c & j = 1,\ldots,n; c = 1,\ldots,C & (ii) \\ \sum_{c=1}^{C} x_{i,j}^c \leq 1 & i = 1,\ldots,m; j = 1,\ldots,n & (iii) \\ x_{i,j}^c \in \{0,1\} & i = 1,\ldots,m; j = 1,\ldots,n; c = 1,\ldots,C \end{cases}$$

Constraints (i) and (ii) ensure the satisfaction of the orthogonal projections of each color. Constraint (iii) represents the exclusiveness condition, i.e., no more than one color per cell.

**Definition 2.** *We define the conflict between the colors of a $C$-colored image as:*

$$conf(X) = \sum_{c_1=1}^{C} \sum_{c_2>c_1}^{C} \sum_{i=1}^{m} \sum_{j=1}^{n} x_{ij}^{c_1} x_{ij}^{c_2}$$

We derive the following result:

**Proposition 1.** *Let $M$ be a $m \times n$ $C$-colored image whose orthogonal projections $H = \{H^1,\ldots,H^C\}$ and $V = \{V^1,\ldots,V^C\}$, are given. Then $M$ is hv-convex image if and only if there exists a vector $X \in [0,1]^{m \times n \times C}$ that satisfies constraints (i), (ii) and (iii) with $F_{adj}(X) = 2\sum_{c=1}^{C} \sum_{i=1}^{m} h_i^c - Cm - Cn$.*

*Proof.* Firstly suppose that $M$ is hv-convex and set $X = M$. $X$ is hv-convex if and only if the cells in each row and column for every color $c$ form a contiguous interval and there is no conflict between colors for every cell $(i, j)$.

So $\sum_{j=1}^{n-1} x_{i,j}^c x_{i,j+1}^c = h_i^c - 1$, for $i = 1, \ldots, m; c = 1, \ldots, C$ and $\sum_{i=1}^{m-1} x_{i,j}^c x_{i+1,j}^c = v_j^c - 1$, for $j = 1, \ldots, n; c = 1, \ldots, C$.

By summing over the rows and the columns, we get $\sum_{i=1}^{m} \sum_{j=1}^{n-1} x_{i,j}^c x_{i,j+1}^c = \sum_{i=1}^{m} h_i^c - m$ and $\sum_{j=1}^{n} \sum_{i=1}^{m-1} x_{i,j}^c x_{i+1,j}^c = \sum_{j=1}^{n} v_j^c - n$. Since $\sum_{i=1}^{m} h_i^c = \sum_{j=1}^{n} v_j^c$ for every color $c$, therefore $\sum_{c=1}^{C} \left( \sum_{i=1}^{m} \sum_{j=1}^{n-1} x_{i,j}^c x_{i,j+1}^c + \sum_{j=1}^{n} \sum_{i=1}^{m-1} x_{i,j}^c x_{i+1,j}^c \right) = 2 \sum_{c=1}^{C} \sum_{i=1}^{m} h_i^c - Cm - Cn = F_{adj}(X)$.

Conversely suppose there exists a vector $X$ satisfying $(i), (ii), (iii)$ and $F_{adj}(X) = 0$.

In one hand, we can not have simultaneously $x_{i+1,j}^{c_p} x_{i+1,j}^{c_q} = 1$ with ensures the exclusiveness constraint.

On the other hand, for each color $c$: $\sum_{j=1}^{n} x_{i,j}^c = h_i^c$ so $\sum_{j=1}^{n-1} x_{i,j}^c x_{i,j+1}^c \leq h_i^c - 1$.

Therefore $\sum_{i=1}^{m} \sum_{j=1}^{n-1} x_{i,j}^c x_{i,j+1}^c \leq \sum_{i=1}^{m} h_i^c - m$ for every color $c$.

So $\sum_{c=1}^{C} \sum_{i=1}^{m} \sum_{j=1}^{n-1} x_{i,j}^c x_{i,j+1}^c \leq \sum_{c=1}^{C} \sum_{i=1}^{m} h_i^c - C m$.

Similarly $\sum_{c=1}^{C} \sum_{i=1}^{m-1} \sum_{j=1}^{n} x_{i,j}^c x_{i+1,j}^c \leq \sum_{c=1}^{C} \sum_{j=1}^{n} v_j^c - C n$.

$\implies F_{adj}(X) \leq 2 \sum_{c=1}^{C} \sum_{i=1}^{m} h_i^c - C m - C n$

Observe that $F_{adj}(X)$ is the sum of non negative terms $\sum_{i=1}^{m-1} x_{i,j}^c x_{i+1,j}^c$ and $\sum_{j=1}^{n-1} x_{i,j}^c x_{i,j+1}^c$. So if $F_{adj}$ reaches its upper bound then each term reaches its upper bound too.

So finally $\sum_{i=1}^{m-1} x_{i,j}^c x_{i+1,j}^c = h_i - 1$ and $\sum_{j=1}^{n-1} x_{i,j}^c x_{i,j+1}^c = v_j - 1$ and $X$ is hv-convex.

## 4   Linearization

It is difficult to directly solve program $P$ because the objective function is quadratic. So, we consider the classical linearization of program $P$ obtained by replacing the quadratic terms $x_{ij}^c x_{i,j+1}^c$ by the 0–1 variables $hx_{ij}^c$ ($h$ stands for horizontally adjacent) and $x_{ij}^c x_{i+1,j}^c$ by $vx_{ij}^c$. We get the following equivalent integer linear program with additional constraints $(1.a; 1.b)$ and $(2.a; 2.b)$ to ensure the equivalence between $P$ and $MIP$ [24]

$$
MIP \quad
\begin{cases}
max \left\{ F_{adj}(X) = \sum_{c=1}^{C} (\sum_{i=1}^{m} \sum_{j=1}^{n-1} hx_{ij}^c + \sum_{i=1}^{m-1} \sum_{j=1}^{n} vx_{ij}^c) \right\} \\
s.t. \\
(i), (ii), (iii) \\
hx_{ij}^c \leq x_{i,j}^c \quad i = 1, \ldots, m; \ j = 1, \ldots, n-1; c = 1, \ldots, C \quad (1.a) \\
hx_{ij}^c \leq x_{i,j+1}^c \quad i = 1, \ldots, m; \ j = 1, \ldots, n-1; c = 1, \ldots, C \quad (1.b) \\
vx_{ij}^c \leq x_{ij}^c \quad i = 1, \ldots, m-1; \ j = 1, \ldots, n; c = 1, \ldots, C \quad (2.a) \\
vx_{ij}^c \leq x_{i+1,j}^c \quad i = 1, \ldots, m-1; \ j = 1, \ldots, n; c = 1, \ldots, C \quad (2.b) \\
x_{ij}^c \in \{0,1\} \quad i = 1, \ldots, m-1; \ j = 1, \ldots, n; c = 1, \ldots, C \\
hx_{ij}^c, vx_{ij}^c \in [0,1] \quad i = 1, \ldots, m-1; \ j = 1, \ldots, n; c = 1, \ldots, C
\end{cases}
$$

It is also possible to linearize $P$ by replacing the constraints $(1.a)$ and $(1.b)$ by $2hx_{ij}^c \le x_{ij}^c + x_{i,j+1}^c$ and the constraints $(2.a)$ and $(2.b)$ by $2vx_{ij}^c \le x_{ij}^c + x_{i+1,j}^c$. We get the following program $IP$

$$IP \begin{cases} max\ \{F_{adj}(X) = \sum_{c=1}^{C}(\sum_{i=1}^{m}\sum_{j=1}^{n-1} hx_{ij}^c + \sum_{i=1}^{m-1}\sum_{j=1}^{n} vx_{ij}^c)\} \\ s.t. \\ (i),(ii),(iii) \\ 2hx_{ij}^c \le x_{ij}^c + x_{i,j+1}^c \quad i = 1,\dots,m;\ j = 1,\dots,n-1; c = 1,\dots,C \\ 2vx_{ij}^c \le x_{ij}^c + x_{i+1,j}^c \quad i = 1,\dots,m-1;\ j = 1,\dots,n; c = 1,\dots,C \\ x_{ij}^c, hx_{ij}^c, vx_{ij}^c \in \{0,1\} \quad i = 1,\dots,m-1;\ j = 1,\dots,n; c = 1,\dots,C \end{cases}$$

We note $\overline{IP}$ the continuous relaxation of $IP$ obtained by relaxing the integrality constraints $(x_{ij}^c, hx_{ij}^c, vx_{ij}^c \in [0,1])$. It's worth to note that $IP$ is a feasibly program. Since $hx_{ij}^c$ and $vx_{ij}^c$ can be chosen freely in $[0,1]$, constraints $2hx_{ij}^c \le x_{ij}^c + x_{i,j+1}^c$ and $2vx_{ij}^c \le x_{ij}^c + x_{i+1,j}^c$ will become equalities:

$$\begin{cases} 2hx_{ij}^c = x_{ij}^c + x_{i,j+1}^c \quad i = 1,\dots,m;\ j = 1,\dots,n-1; c = 1,\dots,C \\ 2vx_{ij}^c = x_{ij}^c + x_{i+1,j}^c \quad i = 1,\dots,m-1;\ j = 1,\dots,n; c = 1,\dots,C \end{cases}$$

If we extend this into the objective function of $IP$, we get:

$$F_{adj}(X) = \sum_{c=1}^{C}\left(\sum_{i=1}^{m}\sum_{j=1}^{n-1} hx_{ij}^c + \sum_{i=1}^{m-1}\sum_{j=1}^{n} vx_{ij}^c\right)$$

$$= \frac{1}{2}\sum_{c=1}^{C}\left(\sum_{i=1}^{m}\sum_{j=1}^{n-1} x_{ij}^c + x_{i,j+1}^c + \sum_{i=1}^{m-1}\sum_{j=1}^{n} x_{ij}^c + x_{i+1,j}^c\right)$$

Applying constraints $(i)$ and $(ii)$ of horizontal and vertical projections, we see that the objective function of $IP$ is equal to a constant:

$$F_{adj}(X) = \frac{1}{2}\sum_{c=1}^{C}\left[\sum_{j=1}^{n-1}(v_j^c + v_{j+1}^c) + \sum_{i=1}^{m-1}(h_i^c + h_{i+1}^c)\right] = \tau$$

Note every feasible solution of the relaxed linear program $IP$ has the same value, and is therefore an optimal solution to the program.

**Proposition 2.** *Let $M$ be $C$-colored image whose orthogonal projections $H = \{H^1,\dots,H^C\}$ and $V = \{V^1,\dots,V^C\}$ are given. Any solution of the continuous relaxation of $MIP$ verifies $F_{adj}(X) \le \tau$*

*Proof.* We prove that the objective value of any solution of $P$ is less than $\tau$. Constraints (a) and (b) imply that $\sum_{i=1}^{m}\sum_{j=1}^{n-1} hx_{i,j}^c \le \frac{1}{2}\sum_{j=1}^{n-1}(v_j^c + v_{j+1}^c)$. Similarly, we obtain $\sum_{i=1}^{m-1}\sum_{j=1}^{n} vx_{i,j}^c \le \frac{1}{2}\sum_{i=1}^{m-1}(h_i^c + h_{i+1}^c)$.
Thus

$$\sum_{c=1}^{C}\left[\sum_{i=1}^{m}\sum_{j=1}^{n-1} hx_{ij}^c\right] \le \frac{1}{2}\sum_{c=1}^{C}\left[\sum_{j=1}^{n-1}(v_j^c + v_{j+1}^c)\right] \tag{1}$$

and

$$\sum_{c=1}^{C}\left[\sum_{i=1}^{m-1}\sum_{j=1}^{n} vx_{ij}^c\right] \le \frac{1}{2}\sum_{c=1}^{C}\left[\sum_{i=1}^{m-1}(h_i^c + h_{i+1}^c)\right] \tag{2}$$

Then $(1) + (2) \Longrightarrow F_{adj}(X) \le \tau$.

## 4.1   Min-cost Max-flow Associated Problem

Both programs $MIP$ and $IP$ remain hard to solve since they deal with binary variables. Thus we consider the continuous relaxations $\overline{MIP}$ and $\overline{IP}$ of $MIP$ and $IP$, respectively, and we investigate the solutions of $\overline{MIP}$ and $\overline{IP}$.

Let $S = (x, hx, vx)$ be an optimal solution provided by the program $\overline{MIP}$ or $\overline{IP}$. If $S$ is integer then $S$ is an optimal solution to $P$ and $RCI(H, V)$. Otherwise, we solve the following integer program to get an approximate solution $z$ satisfying $(H, V)$ and as close as possible to $x$, i.e., the variables $(x_{i,j}^c)$ and $(z_{i,j}^c)$ have the same value.

The program $Q$ seeks to find a matrix $Z$ satisfying both $H$ and $V$ and being close to $x$

$$Q = \begin{cases} max \ \sum_{c=1}^{C} \sum_{i=1}^{m} \sum_{j=1}^{n} x_{ij}^c z_{ij}^c \\ \sum_{j=1}^{n} z_{ij}^c = h_i^c \quad i = 1, \ldots, m; c = 1, \ldots, C \\ \sum_{i=1}^{m} z_{ij}^c = v_j^c \quad j = 1, \ldots n; c = 1, \ldots, C \\ z_{ij}^c \in \{0, 1\} \quad i = 1, \ldots, m; \ j = 1, \ldots, n; \ c = 1, \ldots, C \end{cases}$$

We note that if we consider the exclusiveness constraint, i.e., at most one color per cell. The program $Q$ may be equivalent to the original program $P$. The matrix $z$ satisfies the orthogonal projections and is as close as possible to $x$ since the objective function expresses the number of cells sharing the same color on $x$ and $z$. The main remark is that program $Q$ can be separated into $C$ programs $Q_c$, $c = 1, \ldots, C$, one program for each color. The program $Q_c$ can be represented as:

$$Q_c = \begin{cases} max \ \sum_{i=1}^{m} \sum_{j=1}^{n} x_{ij}^c z_{ij}^c \\ \sum_{j=1}^{n} z_{ij}^c = h_i^c \quad i = 1, \ldots, m; \\ \sum_{i=1}^{m} z_{ij}^c = v_j^c \quad j = 1, \ldots n; \\ z_{ij}^c \in \{0, 1\} \quad i = 1, \ldots, m; \ j = 1, \ldots, n \end{cases}$$

The program $Q_c$ is equivalent to a min-cost/max-flow in a complete bipartite graph $G(R^c, S^c, E)$ [10,21]. $R^c = \{r_i^c, \ i = 1, \ldots, m\}$ represents the rows and

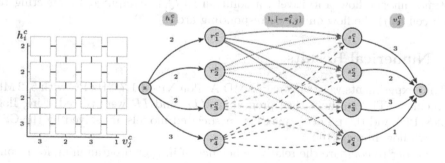

**Fig. 2.** Program $Q_c$ and the associated min-cost max-flow in $G(R^c, S^c, E)$; $H^c = (2, 2, 2, 3)$ and $V^c = (3, 2, 3, 1)$.

| | Image 1 | Image 2 | Image 3 |
|---|---|---|---|
| Test Image | | | |
| Best reconstruction by $MIP$ | | | |

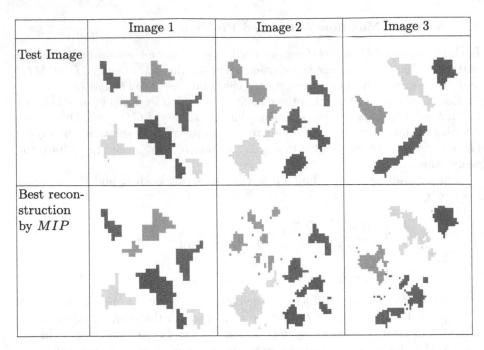

**Fig. 3.** Test images and their approximate reconstructions.

$S^c = \{s_j^c, \ j = 1, \ldots, n\}$ represents the columns (see Fig. 2). We add a source $s$ and a directed edge $(s, r_i^c)$ with capacity equals $h_i^c$ which is the horizontal projection of row $i$. Similarly, We add a sink $t$ and a directed edge $(s_j^c, t)$ with capacity $v_j^c$ which is the vertical projection of column $j$. There is an arc from every pair of row node $r_i^c$ and column node $s_j^c$ with a unit capacity and a cost $-x_{ij}^c$. These arcs correspond to the cells of the image to reconstruct. The problem $Q$ is equivalent to the min-cost/max-flow problem in $G$. $Q$ admits a solution if and only if the maximum flow from the source to the sink is of value $\sum_{i=1}^m h_i^c = \sum_{j=1}^n v_j^c$ for $c \in \{1, \ldots, C\}$. Since the capacities are integer, there exists an optimal integer flow. Intuitively, a solution to $Q$ is computed by affecting to each cell $(i, j)$ the flow on the corresponding arc $(r_i^c, s_j^c)$.

## 5 Numerical Results

All our experiments were run on an AMD Athlon XP-M 1.7 GHz PC with 512 Mb of memory. The mathematical programs $MIP$ and $IP$ were solved using Ilog Cplex 12.0 and the min-cost max-flow associated models are solved by the CS2 network flow library [16].

In order to compare the relative efficiency of integer programming for reconstructing hv-convex images, we have used a systematic approach to generate hv-convex colored images from a set of square hv-convex binary matrices of various sizes as described and generated in [2,3]. We generate a large set of colored

**Table 1.** Numerical comparisons of the heuristics for two colored images, (average on 10 instances for each pair of size and number of components. (∗) indicates that the instance was not solved with one hour.

| Size | $MIP$ based approach | | | | $IP$ based approach | | | $\tau$ | CPLEX Solver | |
|------|-----------------------|------|--------|-------|-----------------------|--------|-------|--------|--------------|------|
|      | $\overline{F}_{adj}$ | $F_{adj}$ | $\%Conf$ | Time | $F_{adj}$ | $\%Conf$ | Time | | $F_{adj}$ | Time |
| (10, 1) | 56 | 48 | 0.00 | **0.013** | 50 | 0.00 | **0.016** | 75 | 54 | **0.06** |
| (10, 2) | 85 | 44 | 0.00 | **0.022** | 41 | 0.00 | **0.028** | 109.5 | 78 | **4.05** |
| (10, 3) | 57 | 16 | 4.12 | **0.031** | 18 | 2.00 | **0.041** | 77 | 46 | **11.68** |
| (10, 4) | 57 | 18 | 3.26 | **0.041** | 12 | 0.00 | **0.053** | 74.5 | 40 | **27.79** |
| (20, 1) | 410.89 | 332 | 3.25 | **0.059** | 304 | 2.75 | **0.073** | 465 | 401 | **26.72** |
| (20, 2) | 307 | 193 | 4.50 | **0.079** | 121 | 1.75 | **0.094** | 367 | 294 | ∗ |
| (20, 3) | 189 | 89 | 3.00 | **0.101** | 62 | 2.00 | **0.116** | 245.5 | 176 | **2307.49** |
| (20, 4) | 199 | 80 | 3.00 | **0.121** | 69 | 2.25 | **0.135** | 251.5 | 177 | ∗ |
| (30, 1) | 437 | 272 | 0.88 | **0.153** | 249 | 0.55 | **0.163** | 501 | 413 | ∗ |
| (30, 2) | 655 | 455 | 6.33 | **0.192** | 297 | 2.55 | **0.198** | 744 | 626 | ∗ |
| (30, 3) | 541 | 315 | 6.11 | **0.233** | 204 | 2.88 | **0.234** | 631 | 506 | ∗ |
| (30, 4) | 373 | 209 | 4.66 | **0.273** | 128 | 1.11 | **0.268** | 458 | 334 | ∗ |
| (40, 1) | 949 | 758 | 1.93 | **0.329** | 567 | 1.75 | **0.315** | 1040.5 | 926 | ∗ |
| (40, 2) | 1076 | 831 | 5.18 | **0.401** | 479 | 2.68 | **0.375** | 1209 | 1017 | ∗ |
| (40, 3) | 737 | 405 | 5.25 | **0.471** | 237 | 1.68 | **0.433** | 850.5 | 502 | ∗ |
| (40, 4) | 637 | 369 | 2.50 | **0.541** | 237 | 1.43 | **0.492** | 755 | 518 | ∗ |
| (50, 1) | 1035 | 807 | 0.00 | **0.623** | 771 | 0.00 | **0.555** | 1127 | 1024 | ∗ |
| (50, 2) | 1283 | 812 | 6.16 | **0.734** | 453 | 2.16 | **0.646** | 1420.5 | 921 | ∗ |
| (50, 3) | 1293 | 883 | 3.84 | **0.846** | 530 | 2.16 | **0.738** | 1451 | 1119 | ∗ |
| (50, 4) | 1029 | 695 | 4.64 | **0.967** | 370 | 1.20 | **0.827** | 1181 | 259 | ∗ |
| (60, 1) | 1253 | 1075 | 0.13 | **1.088** | 670 | 0.00 | **0.937** | 1352 | 1182 | ∗ |
| (60, 2) | 2520 | 2007 | 7.66 | **1.265** | 1093 | 3.38 | **1.068** | 2717.5 | 861 | ∗ |
| (60, 3) | 2003 | 1450 | 5.38 | **1.438** | 898 | 2.55 | **1.208** | 2203.5 | 0 | ∗ |
| (60, 4) | 1671 | 1094 | 3.91 | **1.631** | 637 | 1.83 | **1.355** | 1861 | 472 | ∗ |
| (70, 1) | 1398 | 1159 | 0.04 | **1.816** | 575 | 0.00 | **1.494** | 1502 | 1252 | ∗ |
| (70, 2) | 2211 | 1523 | 3.12 | **2.054** | 825 | 1.28 | **1.686** | 2398.5 | 0 | ∗ |
| (70, 3) | 2115 | 1449 | 3.51 | **2.326** | 747 | 1.16 | **1.889** | 2329 | 0 | ∗ |
| (70, 4) | 2171 | 1589 | 2.63 | **2.579** | 927 | 1.04 | **2.099** | 2400.5 | 702 | ∗ |
| (80, 1) | 2203 | 1931 | 0.03 | **2.845** | 1536 | 0.01 | **2.308** | 2346.5 | 2064 | ∗ |
| (80, 2) | 4038 | 3380 | 6.32 | **3.195** | 2001 | 3.23 | **2.601** | 4288 | 0 | ∗ |
| (80, 3) | 3240 | 2320 | 1.76 | **3.554** | 1587 | 1.23 | **2.897** | 3490 | 0 | ∗ |
| (80, 4) | 2414 | 1606 | 2.04 | **3.916** | 1029 | 0.98 | **3.201** | 2659 | 0 | ∗ |

**Table 2.** Numerical comparisons of the heuristics for three colored images, (average on 10 instances for each pair of size and number of components). (*) indicates that the instance was not solved with one hour.

| Size | $MIP$ based approach | | | | $IP$ based approach | | | $\tau$ | CPLEX Solver | |
|---|---|---|---|---|---|---|---|---|---|---|
| | $\overline{F}_{adj}$ | $F_{adj}$ | %Conf | Time | $F_{adj}$ | %Conf | Time | | $F_{adj}$ | Time |
| (10, 1) | 71 | 56 | 1.00 | **0.013** | 55 | 0.00 | **0.011** | 94.5 | 65 | **0.33** |
| (10, 2) | 76 | 26 | 5.15 | **0.025** | 25 | 2.00 | **0.021** | 106 | 63 | **4.92** |
| (10, 3) | 81 | 27 | 5.17 | **0.036** | 25 | 2.13 | **0.033** | 107.5 | 62 | **18.51** |
| (10, 4) | 69 | 15 | 6.27 | **0.048** | 17 | 2.16 | **0.044** | 95.5 | 46 | **227.98** |
| (20, 1) | 238 | 199 | 0.00 | **0.069** | 174 | 0.29 | **0.061** | 285.5 | 231 | **4.96** |
| (20, 2) | 185 | 79 | 2.00 | **0.093** | 88 | 2.05 | **0.081** | 242.5 | 175 | **139.94** |
| (20, 3) | 275 | 125 | 7.75 | **0.121** | 89 | 5.05 | **0.104** | 352 | 244 | * |
| (20, 4) | 258 | 116 | 8.13 | **0.149** | 68 | 5.25 | **0.128** | 330 | 210 | * |
| (30, 1) | 544 | 451 | 0.00 | **0.191** | 444 | 0.00 | **0.164** | 618.5 | 536 | **18.89** |
| (30, 2) | 236 | 241 | 1.55 | **0.239** | 174 | 1.88 | **0.206** | 459.5 | 345 | * |
| (30, 3) | 519 | 246 | 7.11 | **0.293** | 148 | 3.55 | **0.255** | 626 | 444 | * |
| (30, 4) | 597 | 306 | 8.21 | **0.351** | 207 | 3.77 | **0.307** | 710 | 419 | * |
| (40, 1) | 814 | 687 | 0.00 | **0.424** | 605 | 0.00 | **0.368** | 921.5 | 803 | **1717.53** |
| (40, 2) | 852 | 589 | 2.37 | **0.517** | 391 | 1.68 | **0.434** | 977 | 820 | * |
| (40, 3) | 877 | 550 | 5.06 | **0.613** | 328 | 2.87 | **0.519** | 1031.5 | 666 | * |
| (40, 4) | 585 | 309 | 2.81 | **0.709** | 190 | 1.68 | **0.597** | 725 | 434 | * |
| (50, 1) | 1102 | 873 | 2.48 | **0.837** | 591 | 1.28 | **0.704** | 1226 | 961 | * |
| (50, 2) | 1440 | 940 | 8.12 | **0.998** | 548 | 3.56 | **0.836** | 1621.5 | 424 | * |
| (50, 3) | 1609 | 981 | 8.48 | **1.172** | 549 | 4.88 | **0.975** | 1827 | 392 | * |
| (50, 4) | 1672 | 1046 | 8.58 | **1.348** | 628 | 5.36 | **1.115** | 1899.5 | 0 | * |
| (60, 1) | 2725 | 2390 | 0.44 | **1.543** | 1041 | 0.05 | **1.258** | 1766 | 1493 | * |
| (60, 2) | 2892 | 2402 | 8.63 | **1.792** | 1329 | 3.91 | **1.451** | 3130.5 | 0 | * |
| (60, 3) | 2671 | 1789 | 8.75 | **2.065** | 1056 | 5.77 | **1.663** | 2944.5 | 0 | * |
| (60, 4) | 2132 | 1454 | 5.25 | **2.348** | 803 | 3.00 | **1.874** | 2400.5 | 699 | * |
| (70, 1) | 3042 | 2646 | 0.00 | **2.632** | 2491 | 0.00 | **2.091** | 3236.5 | 2967 | * |
| (70, 2) | 3128 | 2640 | 7.06 | **3.015** | 1173 | 4.02 | **2.387** | 3397 | 0 | * |
| (70, 3) | 3505 | 2640 | 8.36 | **3.406** | 1343 | 4.83 | **2.697** | 3822.5 | 0 | * |
| (70, 4) | 2773 | 1963 | 6.12 | **3.808** | 1067 | 3.12 | **3.016** | 3094 | 2540 | * |
| (80, 1) | 3090 | 2613 | 2.23 | **4.226** | 2028 | 1.67 | **3.329** | 3293.5 | 0 | * |
| (80, 2) | 4970 | 3933 | 6.73 | **4.732** | 2926 | 5.39 | **3.732** | 5296.5 | 0 | * |
| (80, 3) | 4220 | 3304 | 7.17 | **5.264** | 1684 | 4.41 | **4.149** | 4565.5 | 0 | * |
| (80, 4) | 3783 | 2761 | 8.89 | **5.825** | 1402 | 3.87 | **4.615** | 4180 | 0 | * |

**Table 3.** Numerical comparisons of the heuristics for four colored images, (average on 10 instances for each pair of size and number of components). (∗) indicates that the instance was not solved with one hour.

| Size | $MIP$ based approach | | | | $IP$ based approach | | | $\tau$ | CPLEX Solver | |
|------|------------|-----------|--------|-------|-----------|--------|-------|------|-----------|--------|
|      | $\overline{F}_{adj}$ | $F_{adj}$ | %Conf | Time | $F_{adj}$ | %Conf | Time |      | $F_{adj}$ | Time |
| (10, 1) | 57.13 | 51 | 2.00 | **0.032** | 50 | 2.00 | **0.011** | 91 | 56 | **0.26** |
| (10, 2) | 43 | 17 | 2.00 | **0.044** | 15 | 1.00 | **0.021** | 65.5 | 32 | **0.54** |
| (10, 3) | 87 | 24 | 8.32 | **0.059** | 26 | 4.00 | **0.032** | 123 | 67 | **44.11** |
| (10, 4) | 80.65 | 19 | 7.27 | **0.074** | 12 | 5.16 | **0.043** | 108.5 | 49 | **55.91** |
| (20, 1) | 373.09 | 291 | 3.25 | **0.105** | 268 | 3.25 | **0.065** | 447.5 | 360 | **23.38** |
| (20, 2) | 242 | 130 | 8.75 | **0.141** | 75 | 5.25 | **0.088** | 311 | 208 | ∗ |
| (20, 3) | 325 | 147 | 8.50 | **0.174** | 105 | 4.10 | **0.115** | 423.5 | 289 | ∗ |
| (20, 4) | 277 | 121 | 9.75 | **0.213** | 84 | 6.25 | **0.144** | 372 | 227 | ∗ |
| (30, 1) | 691 | 541 | 1.77 | **0.272** | 450 | 0.77 | **0.187** | 797.5 | 659 | ∗ |
| (30, 2) | 367 | 239 | 0.00 | **0.334** | 190 | 0.00 | **0.228** | 464 | 352 | **488.99** |
| (30, 3) | 367 | 403 | 7.44 | **0.408** | 284 | 4.88 | **0.283** | 825.5 | 588 | ∗ |
| (30, 4) | 603 | 262 | 9.66 | **0.487** | 153 | 6.44 | **0.342** | 753 | 118 | ∗ |
| (40, 1) | 1113 | 947 | 0.00 | **0.585** | 815 | 0.00 | **0.415** | 1243 | 1093 | ∗ |
| (40, 2) | 1047 | 747 | 6.37 | **0.705** | 455 | 3.31 | **0.507** | 1227.5 | 931 | ∗ |
| (40, 3) | 1256 | 675 | 9.06 | **0.844** | 440 | 6.62 | **0.614** | 1486 | 335 | ∗ |
| (40, 4) | 1064 | 544 | 8.50 | **0.988** | 359 | 5.37 | **0.723** | 1293 | 229 | ∗ |
| (50, 1) | 1833 | 1596 | 0.20 | **1.155** | 1373 | 0.12 | **0.846** | 1982.5 | 1799 | ∗ |
| (50, 2) | 1003 | 763 | 0.12 | **1.342** | 524 | 0.08 | **0.983** | 1154 | 959 | ∗ |
| (50, 3) | 923 | 620 | 3.68 | **1.545** | 298 | 1.60 | **1.137** | 1125.5 | 728 | ∗ |
| (50, 4) | 1095 | 672 | 4.88 | **1.764** | 365 | 2.72 | **1.301** | 1329.5 | 289 | ∗ |
| (60, 1) | 1284 | 1017 | 0.00 | **2.027** | 776 | 0.11 | **1.491** | 1421.5 | 1212 | ∗ |
| (60, 2) | 2059 | 1628 | 5.63 | **2.331** | 932 | 3.36 | **1.721** | 2287.5 | 1649 | ∗ |
| (60, 3) | 2451 | 1725 | 7.13 | **2.681** | 985 | 4.97 | **1.989** | 2762 | 0 | ∗ |
| (60, 4) | 1277 | 841 | 5.05 | **3.015** | 436 | 2.16 | **2.244** | 1513 | 1028 | ∗ |
| (70, 1) | 2826 | 2466 | 0.00 | **3.402** | 2194 | 0.00 | **2.531** | 3029 | 2709 | ∗ |
| (70, 2) | 2830 | 1732 | 3.46 | **3.866** | 1631 | 2.51 | **2.893** | 3120 | 2237 | ∗ |
| (70, 3) | 2471 | 1732 | 3.36 | **4.344** | 1090 | 2.51 | **3.287** | 2777 | 826 | ∗ |
| (70, 4) | 2301 | 1531 | 3.12 | **4.853** | 891 | 2.97 | **3.739** | 2635.5 | 0 | ∗ |
| (80, 1) | 2774 | 1446 | 0.03 | **5.393** | 2256 | 0.01 | **4.131** | 2967 | 2722 | ∗ |
| (80, 2) | 3353 | 2696 | 3.39 | **6.034** | 1505 | 1.93 | **4.674** | 3663.5 | 1254 | ∗ |
| (80, 3) | 3769 | 2753 | 9.03 | **6.728** | 1372 | 2.96 | **5.232** | 4134 | 0 | ∗ |
| (80, 4) | 3642 | 2731 | 7.21 | **7.431** | 1541 | 3.87 | **5.792** | 4036 | 0 | ∗ |

images with size varying from $10 \times 10$ to $80 \times 80$ and four possible colors for each size. For each size and for each color we generate four classes of images. For each class we generate instances with 1, 2, 3 and 4 components. A component is a maximal hv-convex connected set colored by one color. However, each image can be decomposed into a minimum number of components per color (see Fig. 3). For a given number of colors (2, 3, 4), we take the average of 10 instances for each pair of size and number of components.

The results of computational experiments are summarized in Table 1 for two colored images, Table 2 for three colored images and Table 3 for four colored images. In these tables, the size column gives the size and the number of hv-convex components for each color. The subcolumn labelled $F_{adj}$ contains the number of adjacent pairs cells having the same color. The subcolumn $Conf$ contains the ratio of the number of cells with conflict over the total number of cells $(m\,n)$ provided by each method. The subcolumn $\overline{F}_{adj}$ gives the objective value of the relaxed program. The sub-column labelled Time contains the running CPU time (in seconds) required by each method. The column labelled $\tau$ is bound to $F_{adj}$. We note that for any method if $F_{adj} = \tau$ the image is hv-convex.

The performance measures used in the numerical study are the running time and the final solution quality. For each method, the quality of a solution is expressed by the pair $(F_{adj})$ and $(Conf)$. We have also solved directly the program $MIP$ by CPLEX Solver. We note that for this solver $(Conf = 0)$.

The results showed that for almost all the sizes and the colors, the $MIP$ method gives a higher number of adjacent pair that share the same color and a greater number of conflicts than the $IP$ method. However, we note that the ratio of conflicts is less than 10% even for the large sized images which proves the efficiency of the proposed methods. We mention also that the $MIP$ based approach cannot perform the CPLEX Solver. In fact the CPLEX Solver can only solve small image sizes with one hour.

## 6    Conclusion

Since the reconstruction of hv-convex colored images is NP-complete, it is computationally too expensive to attack the problem directly. We have proposed a method based on mixed integer programming and linear relaxation to reconstruct hv-convex colored images that satisfy horizontal and vertical projections. As a future research in discrete tomography problems, the integer programming formulation can be enriched with other definitions of connectivity and convexity, with more linearization and convexification techniques.

## References

1. Bains, A., Biedl, T.: Reconstructing hv-convex multi-coloured polyominoes. Theor. Comput. Sci. **411**, 3123–3128 (2010)
2. Balázs, P.: A benchmark set for the reconstruction of hv-convex discrete sets. Discrete Appl. Math. **157**, 3447–3456 (2009)

3. Balázs, P.: A framework for generating some discrete sets with disjoint components by using uniform distributions. Theor. Comput. Sci. **406**, 15–23 (2008)
4. Barcucci, E., Del Lungo, A., Nival, M., Pinzani, R.: The reconstruction of polyominoes from their orthogonal projections. Theor. Comput. Sci. **155**, 321–347 (1996)
5. Barcucci, E., Brocchi, S., Frosini, A.: Solving the two color problem: an heuristic algorithm. In: Aggarwal, J.K., Barneva, R.P., Brimkov, V.E., Koroutchev, K.N., Korutcheva, E.R. (eds.) IWCIA 2011. LNCS, vol. 6636, pp. 298–310. Springer, Heidelberg (2011). doi:10.1007/978-3-642-21073-0_27
6. Billionnet, A., Jarray, F., Tlig, G., Zagrouba, E.: Reconstructing convex matrices by integer programming approaches. J. Math. Model. Algorithms OR **12**(4), 329–343 (2013)
7. Billionnet, A., Jarray, F., Tlig, G., Zagrouba, E.: Reconstruction of bicolored images. In: Barneva, R.P., Bhattacharya, B.B., Brimkov, V.E. (eds.) IWCIA 2015. LNCS, vol. 9448, pp. 276–283. Springer, Cham (2015). doi:10.1007/978-3-319-26145-4_20
8. Chrobak, M., Dürr, C.: Reconstructing hv-convex polyominoes from orthogonal projection. Inf. Process. Lett. **69**, 283–289 (1999)
9. Chrobak, M., Dürr, C.: Reconstructing polyatomic structures from discrete x-ray. Theor. Comput. Sci. **259**(3), 81–98 (2001)
10. Dahl, G., Flatberg, T.: Optimization and reconstruction of hv-convex (0, 1)-matrices. Discrete Appl. Math. **151**, 93–105 (2005)
11. Dürr, C., Guiñez, F., Matamala, M.: Reconstructing 3-colored grids from horizontal and vertical projections Is NP-hard. In: Fiat, A., Sanders, P. (eds.) ESA 2009. LNCS, vol. 5757, pp. 776–787. Springer, Heidelberg (2009). doi:10.1007/978-3-642-04128-0_69
12. Gale, D.: A theorem on flows in networks. Discrete Math. **187**, 1073–1082 (1957)
13. Gardner, R.J., Gritzmann, P., Prangenberg, D.: On the computational complexity of determining polyatomic structures by x-rays. Theor. Comput. Sci. **233**, 91–106 (2000)
14. Herman, G.T., Kuba, A.: Discrete Tomography: Foundations, Algorithms and Applications. Birkhäuser, Boston (1999)
15. Herman, G.T., Kuba, A.: Advances in Discrete Tomography and Its Applications. Birkhäuser, Boston (2007)
16. Goldberg, A.V.: An efficient implementation of a scaling minimum-cost flow algorithm. J. Algorithms **22**, 1–29 (1997)
17. Costa, M.-C., Jarray, F., Picouleau, C.: Reconstructing an alternate periodical binary matrix from its orthogonal projections. In: Coppo, M., Lodi, E., Pinna, G.M. (eds.) ICTCS 2005. LNCS, vol. 3701, pp. 173–181. Springer, Heidelberg (2005). doi:10.1007/11560586_14
18. Jarray, F.: A 4-day or a 3-day workweeks scheduling problem with a given workforce size. Asia Pac. J. Oper. Res. **26**(5), 685–696 (2009)
19. Jarray, F., Wynter, L.: An optimal smart market for the pricing of telecommunication services. Technical report 4310, INRIA, Rocquencourt, France (2001)
20. Jarray, F.: Solving problems of discrete tomography: applications in workforce scheduling. Ph.D. thesis, University of CNAM, Paris (2004)
21. Jarray, F., Costa, M.-C., Picouleau, C.: Approximating hv-convex binary matrices and images from discrete projections. In: Coeurjolly, D., Sivignon, I., Tougne, L., Dupont, F. (eds.) DGCI 2008. LNCS, vol. 4992, pp. 413–422. Springer, Heidelberg (2008). doi:10.1007/978-3-540-79126-3_37
22. Jarray, F., Tlig, G.: A simulated annealing for reconstructing hv-convex binary matrices. Electron. Notes Discrete Math. **36**, 447–454 (2010)

23. Jarray, F., Tlig, G.: Approximating bicolored images from discrete projections. In: Aggarwal, J.K., Barneva, R.P., Brimkov, V.E., Koroutchev, K.N., Korutcheva, E.R. (eds.) IWCIA 2011. LNCS, vol. 6636, pp. 311–320. Springer, Heidelberg (2011). doi:10.1007/978-3-642-21073-0_28
24. McCormick, G.P.: Computability of global solutions to factorable nonconvex programs. Math. Program. **10**, 147–175 (1976)
25. Ryser, H.R.: Combinatorial properties of matrices of zeros and ones. Can. J. Math. **9**, 371–377 (1957)
26. Woeginger, G.J.: The reconstruction of polyominoes from their orthogonal projections. Inf. Process. Lett. **77**, 225–229 (2001)

# A Greedy Algorithm for Reconstructing Binary Matrices with Adjacent 1s

Fethi Jarray[1,2] and Ghassen Tlig[1,3(✉)]

[1] Cedric-CNAM, 292 rue St-Martin, Paris, France
fethi_jarray@yahoo.fr, ghassen.tlik@gmail.com
[2] Higher Institute of Computer Science of Medenine, Medenine, Tunisia
[3] LIMTIC, Higher Institute of Computer Science of Tunis, Tunis, Tunisia

**Abstract.** This paper deals with the reconstruction of special cases of binary matrices with adjacent 1s. Each element is horizontally adjacent to at least another element. The projections are the number of elements on each row and column. We give a greedy polynomial time algorithm to reconstruct such matrices when satisfying only the vertical projection. We show also that the reconstruction is NP-complete when fixing the number of sequence of length two and three per row and column.

**Keywords:** Discrete objects · Polynomial time algorithm · Binary matrices reconstruction · NP-completeness

## 1 Introduction

Discrete Tomography (DT) deals with the reconstruction of discrete objects such as matrices and images that are usually represented as integer matrices from finite projections [15,16]. Often, the projections of a structure are the number of elements on each discrete line. One of the main issue is to find a structure that corresponds to the given projections. The reconstruction of discrete objects arises on a number of applications, including Scheduling [8,18,19], Data compression and Data security and networks [17,20], Estimation of parameters in Internet [25], Industrial nondestructive testing [2], Medical imaging [14,23] and Timetabling [4].

The basic problem of reconstructing binary matrices consists in finding an $m \times n$ binary matrix satisfying a given couple of axes projections. The computational complexity of the basic problem is $O(n(m + logn))$ [15,24]. However the problem is usually highly undetermined and a large number of solutions may exist [26]. In many cases, prior knowledge on size, shape, smouthless, etc. are incorporated to uniquely reconstruct the original structure from the projections.

In literature several additional constraints and prior knowledge have been considered to reduce the set of the feasible solutions of the basic problem. These constraints include periodicity [9,11], convexity [5,6,10,12,22], adjacency [4], connectedness [1,7,27] and timetabling [4]. In this paper we are concerned with the adjacency constraint.

© Springer International Publishing AG 2017
V.E. Brimkov and R.P. Barneva (Eds.): IWCIA 2017, LNCS 10256, pp. 347–355, 2017.
DOI: 10.1007/978-3-319-59108-7_27

The remainder of this paper is organized as follows. In Sect. 2, we introduce some definitions and notation. In Sect. 3, we consider the problem of reconstructing binary matrices satisfying only the vertical projection. In Sect. 4, we address the reconstruction under the number of each sequence type per line. We conclude in the last section.

## 2  Definitions and Notations

Given an $m \times n$ binary matrix $A$ we denote by $H = (h_1, ..., h_m)$ the horizontal projection of $A$, $h_i$ being the number of 1's on row $i$, and by $V = (v_1, ..., v_n)$ the vertical projection of $A$, $v_j$ being the number of 1's on column $j$ (see Fig. 1). The condition $\sum_{i=1}^{m} h_i = \sum_{j=1}^{n} v_j$ is obviously necessary for the existence of a solution respecting both projections in the reconstruction problem for the binary matrix from $H$ and $V$.

A *sequence of 1s* (*sequence* for short) is a set of adjacent 1s on a row. By convention, we say that a sequence begins on the column of its leftmost cell. The distance between two consecutive sequences is the number of 0s between them. A *2-sequence* is a sequence of length 2 and a *3-sequence* is a sequence of length 3.

We denote by $HV - adjacent(m, n)$ the problem of reconstructing an $m \times n$ binary matrix from its horizontal projection $H = (h_1, \ldots, h_m)$ and its vertical projection $V = (v_1, \ldots, v_n)$ such that each 1 is adjacent to at least another 1 per row or column.

Brocchi et al. [3] proposed a general framework that solves in polynomial time the problem $HV - adjacent(m, n)$ whenever $m$ or $n$ is fixed. Its time complexity is $O(m^n 2^{2n})$. Jarray [21] proved that if the number of columns is not greater than six the problem could be solved in a quadratic time.

| $h_i$ | | | | | | |
|---|---|---|---|---|---|---|
| 4 | 0 | 1 | 0 | 1 | 1 | 1 |
| 4 | 1 | 0 | 0 | 1 | 1 | 1 |
| 4 | 1 | 1 | **1** | (0) | 1 | 0 |
| 2 | 0 | 1 | (0) | **1** | 0 | 0 |
| 4 | 1 | 1 | 0 | 1 | 1 | 0 |
| | 3 | 4 | 1 | 4 | 4 | 2 | $v_j$ |

**Fig. 1.** A binary matrix with horizontal projection $H = (4, 4, 4, 2, 4)$ and vertical projection $V = (3, 4, 1, 4, 4, 2)$.

Since each row contains at least a sequence of length 2, we have now the obvious necessary condition for the feasibility of the problem.

**Proposition 1.** $HV - adjacent(m, n)$ *admits a solution only if*

(i) $m \geq max_j\, v_j$
(ii) $\sum_{j=1}^{n} v_j \geq 2m$
(iii) $v_1 \leq v_2,\ v_n \leq v_{n-1}$ and $v_j \leq v_{j+1} + v_{j-1},\ j = 2, \ldots, n-1$.

## 3    Reconstruction Under the Vertical Projection $V - adjacent(m, n)$

In this section, we provide the polynomial time algorithm to reconstruct a binary matrix respecting the vertical projection $V$ without isolated cell of value 1 per row and column. The related consistency problem is defined as follows:

**Instance:** $V = (v_1, \ldots, v_n)$ integer positive vector and an integer $m$.
**Question:** Is there an $m \times n$ binary matrix respecting the vertical projection $V$ without isolated cell of value 1 per row and column?

We propose the polynomial time algorithm $A - V - adjacent(m, n)$ that gives a solution $\hat{S}$ or says no such solution exists. At each step $j$, the rows are divided into three classes (as depicted in Fig. 2):

$C_1^j$: the rows, $i$, such that $\hat{S}(i, j-1) = 1$ and $\hat{S}(i, j-2) = 0$.
$C_2^j$: the rows, $i$, such that $\hat{S}(i, j-1) = 1$ and $\hat{S}(i, j-2) = 1$.
$C_3^j$: the rows, $i$, such that $\hat{S}(i, j-1) = 0$.

The algorithm $A - V - adjacent(m, n)$ assigns at each iteration $j$, the value 1 to $d_j$ cells of column $j$ and starts by the rows of the class $C_1^j$, then the highest priority available rows of $C_3^j$ and finally the rows of class $C_2^j$. The highest priority row of $C_3^j$ is the row that admits the leftmost cell of value 1.

We note that for each row $i$ of class $C_1^j$, we should have $\hat{S}(i, j) = 1$ to satisfy the constraint of at least two adjacent consecutive ones.

**Algorithm** $A - V - adjacent(m, n)$
Set $d_j = v_j,\ j = 1, \ldots, n$;
**For** $j = 1$ **to** $n$ **do**
    $\hat{S}(i, j) = 1$ for any row $i$ of $C_1^j$ and $d_j = v_j - |C_1^j|$;
    $\hat{S}(i, j) = 1$ for $\min(d_j, v_{j+1})$ highest priority available rows of $C_3^j$;
    **If** $d_j - \min(d_j, v_{j+1}) > |C_2^j|$ **then** exit with error;
    $\hat{S}(i, j) = 1$ for $d_j - \min(d_j, v_{j+1})$ rows of $C_2^j$;

**Proposition 2.** *The algorithm $A - V - adjacent(m, n)$ solves the problem $V - adjacent(m, n)$ in polynomial time.*

*Proof.* Suppose that the problem $V - adjacent(m, n)$ admits a solution $S$. Denote by $S_{(1..j)}$ the sub-matrix of $S$ corresponding to the 1s placed on columns $1, \ldots, j$. Similarly, denote by $\hat{S}_{(1..j)}$ the partial solution obtained by the algorithm $A - V - adjacent(m, n)$ at step $j$. We will demonstrate by induction on $j$, that there is a solution $S$ such that $S_{(1..j)} = \hat{S}_{(1..j)}$ for $j = 1, \ldots, n$.

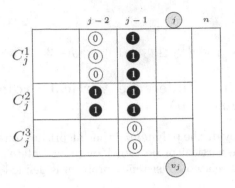

**Fig. 2.** Classes of rows at step $j$.

## 1. Initialization of recurrence

Suppose that $S_{(1)} \neq \hat{S}_{(1)}$. There exist two rows $k$ and $l$ such that $\hat{S}(l,1) = 1, S(l,1) = 0$ and $\hat{S}(k,1) = 0, S(k,1) = 1$. By swapping the rows $k$ and $l$ in

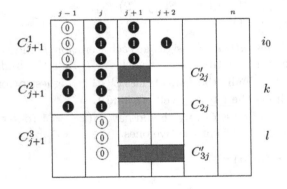

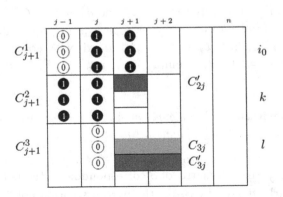

**Fig. 3.** Fifth case: Swapping of lines k, $i_0$ and l at step $j + 1$.

$S$, we get $S(k, 1) = 0$ and $S(l, 1) = 1$. By reiterating this type of swapping, we obtain a solution $S$ such that $S_{(1)} = \hat{S}_{(1)}$.

2. **Suppose that there exists a solution $S$ such that $S_{(1..j)} = \hat{S}_{(1..j)}$ and $S_{(1..j+1)} \neq \hat{S}_{(1..j+1)}$.**

Let us provide an equivalent solution $S$ such that $S_{(1..j+1)} = \hat{S}_{(1..j+1)}$. At step $j + 1$, there exist at least $d_{j+1}$ available rows because $S$ is a solution. There exist two rows $k$ and $l$ such that $\hat{S}(l, j + 1) = 1, S(l, j + 1) = 0$ and $\hat{S}(k, j + 1) = 0, S(k, j + 1) = 1$. We distinguish all the cases according to the classes of rows $k$ and $l$ and we propose an appropriate transformation.

**First case**  if $l \in C_3^{j+1}$ and $k \in C_3^{j+1}$ or $l \in C_2^{j+1}$ and $k \in C_2^{j+1}$ then by swapping rows $k$ and $l$ in $S$ from column $j + 1$, we get $S(k, j + 1) = 0$ and $S(l, j + 1) = 1$.

**Second case** if $k \in C_3^{j+1}$ and $l \in C_2^{j+1}$ then swap rows lines $k$ and $l$ from column $j + 1$.

**Third case** if $k \in C_2^{j+1}$, $l \in C_3^{j+1}$ and $S(k, j + 2) = 1$ then swap $k$ and $l$ from column $j + 1$.

**Fourth case** if $k \in C_2^{j+1}$, $l \in C_3^{j+1}$, $S(k, j + 2) = 0$ and $S(l, j + 2) = 1$ then set $S(k, j + 1) = 0$ and $S(l, j + 1) = 1$.

**Fifth case** if $k \in C_2^{j+1}$, $l \in C_3^{j+1}$, $S(k, j + 2) = 0$ and $S(l, j + 2) = 0$ then search another pair $(k, l)$ satisfying one of the above cases and make the associated transformation in $S$. Suppose that if $S_{(1...j+1)} \neq \hat{S}_{(1...j+1)}$ then the rows $k$ and $l$ belong to the fifth case, i.e. $\hat{S}(l, j + 1) = 1, S(l, j + 1) = 0$, $\hat{S}(k, j + 1) = 0$ and $S(k, j + 1) = 1$. The transformation is done through an auxiliary row $i_0$. To determine $i_0$, we introduce the following subclasses:

$C'2j = \{i \in C_2^{j+1} / \hat{S}(i, j + 1) = 1, \ S(i, j + 1) = 1\}$.

$C2j = \{i \in C_2^{j+1} / \hat{S}(i, j + 1) = 0, \ S(i, j + 1) = 1\}$.

$C'3j = \{i \in C_3^{j+1} / \hat{S}(i, j + 1) = 1, \ S(i, j + 1) = 1\}$.

$C3j = \{i \in C_3^{j+1} / \hat{S}(i, j + 1) = 1, \ S(i, j + 1) = 0\}$.

We deduce the following properties:
(a) $|C3j| = |C2j|$.
(b) $v_{j+2} \geq |C'3j| + |C3j|$ because each sequence starting in row $j + 1$ covers also row $j + 2$.
(c) In column $j + 1$, $|C3j| + |C'3j|$ sequences start in $\hat{S}$ and $|C'3j|$ sequences start in $S$. Even more, in $S$, there are $|C3j|$ rows other than $C'3j$ having a cell of value 1 in column $j + 2$ because of (b).

(d) according to (a) and (c), for each row $k \in C2j$, there is a row $l \in C3j$ and a row $i_0 \notin C'3j$ with $S(i_0, j+2) = 1$ (see Fig. 3).

The transformation of the fifth case is the following: We firstly swap the rows $k$ and $i_0$ from column $j+2$, then, we swap the rows $k$ and $l$ from column $j+1$ (third case).                                                                □

We easily develop the following properties of $A - V - adjacent(m, n)$:

**Property 1.**

(i) The algorithm $A - V - adjacent(m, n)$ constructs a solution with $max_j v_j$ rows since in each step all the rows are available, if $m > max_j v_j$ then the remaining rows are empty.

(ii) The sequences of 0s have a minimum length Because the 1s are cyclically placed on class $C_3^j$.

## 4 Reconstruction Under the Projections of Each Sequence Type $23adjacent(m, n)$

Since each sequence of 1s can be decomposed into a set of 2-sequences and 3-sequences, we suppose that the projections are the number of 2-sequences and 3-sequences per row and column, i.e. we are given the projection of each sequence type. The related consistency problem is the following:

**Instance:** $H^2 = (h_1^2, \ldots, h_m^2)$, $H^3 = (h_1^3, \ldots, h_m^3)$, $V^2 = (v_1^2, \ldots, v_n^2)$ and $V^3 = (v_1^3, \ldots, v_n^3)$ four integer positive vectors.

**Question:** Is there a binary matrix without isolated 1s per row satisfying $(H^2, V^2)$ for the 2-sequences and $(H^3, V^3)$ for the 3-sequences?

We recall that in a 2-colored image, each cell is either uncolored, red or blue. The projections are the number of red cells and blue cells per row and column [13, 19]. The related consistency problem is the following:

**Instance:** $H^r = (h_1^r, \ldots, h_m^r)$, $H^b = (h_1^b, \ldots, h_m^b)$, $V^r = (v_1^r, \ldots, v_n^r)$ and $V^b = (v_1^b, \ldots, v_n^b)$ four integer positive vectors.

**Question:** Is there a 2-colored image respecting $(H^r, V^r)$ for color red and $(H^b, V^b)$ for color blue?

We establish the following result:

**Proposition 3.** *The problem 23adjacent(m, n) is NP-complete.*

*Proof.* We will show that an instance of the 2-colored image can be reduced in polynomial time to an instance of $23adjacent(m, n)$. Consider an instance $I'$ of the 2-colored image consistent with $(H'^r, V'^r)$ and $(H'^b, V'^b)$. We define an instance $I$ of size $m \times 3n$ to $23adjacent(m, n)$ satisfying the projections $(H^2, V^2)$ and $(H^3, V^3)$ by associating the colors red and blue to the 2-sequences and 3-sequences respectively:

- $v^2_{3j-2} = v^2_{3j-1} = v'^r_j, v^2_{3j} = 0, j = 1, \ldots, n,$
- $v^3_{3j-2} = v^3_{3j-1} = v^3_{3j} = v'^b_j, j = 1, \ldots, n,$
- $h^2_i = h'^r_i, i = 1, \ldots, m,$
- $h^3_i = h'^b_i, i = 1, \ldots, m.$

$I'$ and $I$ are equivalent since from the projections $(H^2, V^2)$ and $(H^3, V^3)$, the sequences begin only on columns $3j - 2$. The numbers of 2-sequences and 3-sequences beginning on column $3j - 2$ in $I$ are equal to the number of red cells and blue color respectively on column $j$ in $I'$ (see Fig. 4). Hence the problem $23adjacent(m, n)$ is NP-complete as the problem of reconstructing 2-colored image [13]. □

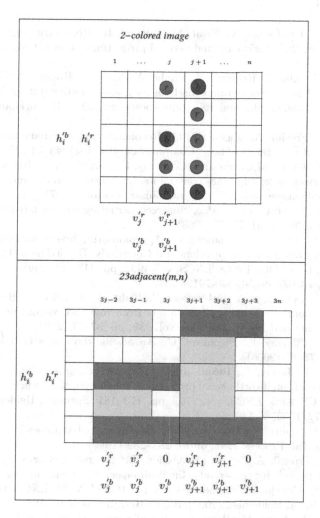

**Fig. 4.** Reduction from *2-colored image* to *23adjacent(m, n)*. (Color figure online)

# 5   Conclusion

In this paper, we have studied the complexity of reconstructing binary matrices from their orthogonal projections without horizontally isolated 1s. We have proved that when the horizontal projection is not considered the problem could be solved in a polynomial time. We have established also that when the projections give the number of 2-sequences and 3-sequences per line, the consistency becomes NP-complete. Nevertheless, the problem $HV$-$adjacent(m,n)$ remains an open problem.

# References

1. Barcucci, E., Del Lungo, A., Nivat, M., Pinzani, R.: Reconstructing convex polyominoes from their horizontal and vertical projections. Theor. Comput. Sci. **155**, 321–347 (1996)
2. Baumann, J., Kiss, Z., Krimmel, S., Kauba, A., Nagy, A., Rodek, L., Schillinger, S., Stephan, J.: Discrete tomography methods for non-destrictive testing, In: Advances in Discrete Tomography and Its Applications, pp. 303–331. Birkhäuser, Boston (2007)
3. Brocchi, S., Frosini, A., Picouleau, C.: Reconstruction of binary matrices under neighborhood constraints. Theor. Comput. Sci. **406**(1–2), 43–54 (2008)
4. Brunetti, S., Costa, M.C., Frosini, A., Jarray, F., Picouleau, C.: Reconstruction of binary matrices under adjacency constraints. In: Advances in Discrete Tomography and Its Applications, pp. 125–150. Birkhäuser, Boston (2007)
5. Brunetti, S., Daurat, A.: An algorithm reconstructing convex lattice sets. Theor. Comput. Sci. **304**(1–3), 35–57 (2003)
6. Jarray, F., Costa, M.-C., Picouleau, C.: Approximating hv-convex binary matrices and images from discrete projections. In: Coeurjolly, D., Sivignon, I., Tougne, L., Dupont, F. (eds.) DGCI 2008. LNCS, vol. 4992, pp. 413–422. Springer, Heidelberg (2008). doi:10.1007/978-3-540-79126-3_37
7. Brunetti, S., Del Lungo, A., Del Ristoro, F., Kuba, A., Maurice, N.: Reconstruction of 4- and 8-connected convex discrete sets from row and column projections. In: Linear Algebra and Its Applications, vol. 339, pp. 37–57 (2001)
8. Costa, M.C., Jarray, F., Picouleau, C.: An acyclic days-off scheduling problem. 4OR **4**(1), 73–85 (2006)
9. Costa, M.-C., Jarray, F., Picouleau, C.: Reconstructing an alternate periodical binary matrix from its orthogonal projections. In: Coppo, M., Lodi, E., Pinna, G.M. (eds.) ICTCS 2005. LNCS, vol. 3701, pp. 173–181. Springer, Heidelberg (2005). doi:10.1007/11560586_14
10. Chrobak, M., Dürr, C.: Reconstructing hv-convex polyominoes from orthogonal projections. Inf. Process. Lett. **69**(6), 283–289 (1999)
11. Lungo, A., Frosini, A., Nivat, M., Vuillon, L.: Discrete tomography: reconstruction under periodicity constraints. In: Widmayer, P., Eidenbenz, S., Triguero, F., Morales, R., Conejo, R., Hennessy, M. (eds.) ICALP 2002. LNCS, vol. 2380, pp. 38–56. Springer, Heidelberg (2002). doi:10.1007/3-540-45465-9_5
12. Del Lungo, A., Nivat, M.: Reconstruction of connected sets from two projections. In: Discrete Tomography: Foundations, Algorithms, and Applications, pp. 163–188. Birkhäuser, Boston (1999)

13. Dürr, C., Guiñez, F., Matamala, M.: Reconstructing 3-colored grids from horizontal and vertical projections Is NP-hard. In: Fiat, A., Sanders, P. (eds.) ESA 2009. LNCS, vol. 5757, pp. 776–787. Springer, Heidelberg (2009). doi:10.1007/978-3-642-04128-0_69

14. Hall, P.: A model for learning human vascular anatomy. In: DIMACS Serie in Discrete Mathematical Problems with Medical Applications, vol. 55, pp. 11–27 (2000)

15. Herman, G.T., Kuba, A. (eds.): Discrete Tomography: Foundations, Algorithms and Applications. Birkhäuser, Boston (1999)

16. Herman, G.T., Kuba, A. (eds.): Advances in Discrete Tomography and its Applications. Birkhäuser, Boston (2007)

17. Ivring, R.W., Jerrum, M.R.: Three-dimensional data security problems. SIAM J. Comput. **23**, 170–184 (1994)

18. Jarray, F.: Solving problems of discrete tomography: applications in workforce scheduling. Ph.D. thesis, University of CNAM, Paris (2004)

19. Jarray, F.: A 4-day or a 3-day workweeks scheduling problem with a given workforce size. Asia Pac. J. Oper. Res. **26**(5), 685–696 (2009)

20. Jarray, F., Wynter, L.: An Optimal Smart Market for the Pricing of Telecommunication Services. Technical report 4310, INRIA, Rocquencourt, France (2001)

21. Jarray, F., Tlig, G.: A simulated annealing for reconstructing hv-convex binary matrices. Electron. Notes Discrete Math. **36**, 447–454 (2010)

22. Kuba, A., Balogh, E.: Reconstruction of convex 2D discrete sets in polynomial time. Theor. Comput. Sci. **283**, 223–242 (2000)

23. Onnasch, D.G.W., Prause, G.P.M.: Heart chamber reconstruction from biplane angiography. In: Discrete Tomography: Foundations, Algorithms and Applications, pp. 385–403. Birkhäuser, Boston (1999)

24. Ryser, H.J.: Combinatorial properties of matrices of zeros and ones. Can. J. Math. **9**, 371–377 (1957)

25. Vardi, Y.: Network tomography: estimating source-destination traffic intensities from link data. J. Am. Stat. Assoc. **91**(433), 65–377 (1996)

26. Wang, B., Zhang, F.: On the precise number of (0, 1)-matrices in $U(R, S)$. Discrete Math. **187**, 211–220 (1998)

27. Woeginger, G.J.: The reconstruction of polyominoes from their orthogonal projections. Inf. Process. Lett. **77**, 225–229 (2001)

# Author Index

Akkeleş, Arif  16
Amelio, Alessia  280
Arslan, Abdullah N.  308

Balázs, Péter  105
Basu, Subhadip  256
Bhatnagar, Shaleen  156
Bhowmick, Partha  40, 93, 212
Bhunre, Piyush K.  212
Biswas, Arindam  184
Biswas, Ranita  93
Bodnár, Péter  105
Brimkov, Boris  30
Brodić, Darko  280
Brunetti, Sara  105

Debled-Rennesson, Isabelle  256
Diaz-del-Rio, Fernando  142

Eppstein, David  117
Escudero, L.M.  229

Gómez-Gálvez, P.  229
Goodrich, Michael T.  117

Han, Yo-Sub  79
Hicks, Illya V.  268

James Immanuel, S.  170
Jarray, Fethi  334, 347
Jimenez, M.J.  229
Jovanovic, Raka  294

Karmakar, Nilanjana  184

Lindblad, Joakim  243
Lukić, Tibor  322

Mahato, Papia  40
Malik, Saleem  156

Mamano, Nil  117
Marčeta, Marina  322
Midya, Abhisek  156
Mikesell, Derek  268
Milivojević, Zoran N.  280
Mir-Mohammad-Sadeghi, Hamid  53
Mondal, Sharmistha  184
Mukhopadhyay, Jayanta  212

Nagar, Atulya K.  170
Nagy, Benedek  16, 53
Nasipuri, Mita  256
Nasser, Hayat  256
Ngo, Phuc  256

Onchis, Darian  142

Palágyi, Kálmán  3
Pani, Alok Kumar  156
Paul, Soumi  256
Perner, Petra  66
Průša, Daniel  79

Real, Pedro  142
Rucco, M.  229

Simian, Dana  294
Sirakov, Nikolay M.  198, 308
Sirakova, Nona Nikolaeva  198
Sladoje, Nataša  243
Šlapal, Josef  132

Thamburaj, Robinson  170
Thomas, D.G.  156, 170
Tlig, Ghassen  334, 347
Tuba, Eva  294
Tuba, Milan  294

Vicente-Munuera, P.  229

Printed in the United States
by Bookmasters

Printed in the United States
By Bookmasters